Optimum Aerodynamic Design
& Parallel Navier-Stokes Computations

ECARP – European Computational Aerodynamics Research Project

Edited by
Jacques Periaux
Gabriel Bugeda
Panagiotis K. Chaviaropoulos
Kyriakos Giannakoglou
Stephane Lanteri
and Bertrand Mantel

Notes on Numerical Fluid Mechanics (NNFM) Volume 61

Volume 61 Optimum Aerodynamic Design & Parallel Navier-Stokes Computations, ECARP-European Computational Aerodynamics Research Project (J. Periaux / G. Bugeda / P. Chaviaropoulos / K. Giannokoglou / S. Lanteri / B. Mantel, Eds.)

Volume 60 New Results in Numerical and Experimental Fluid Mechanics. Contributions to the 10th AG STAB/DGLR Symposium Braunschweig, Germany 1996 (H. Körner / R. Hilbig, Eds.)

Volume 59 Modeling and Computations in Environmental Sciences. Proceedins of the First GAMM-Seminar at ICA Stuttgart, October 12–13, 1995 (R. Helmig / W. Jäger / W. Kinzelbach / P. Knabner / G. Wittum, Eds.)

Volume 58 ECARP – European Computational Aerodynamics Research Project: Validation of CFD Codes and Assessment of Turbulence Models (W. Haase / E. Chaput / E. Elsholz / M. A. Leschziner / U. R. Müller, Eds.)

Volume 57 Euler and Navier-Stokes Solvers Using Multi-Dimensional Upwind Schemes and Multigrid Acceleration. Results of the BRITE/EURAM Projects AERO-CT89-0003 and AER2-CT92-00040, 1989–1995 (H. Deconinck / B. Koren, Eds.)

Volume 56 EUROSHOCK-Drag Reduktion by Passive Shock Control. Results of the Project EUROSHOCK, AER2-CT92-0049 Supported by the European Union, 1993–1995 (E. Stanewsky / J. Délery / J. Fulker / W. Geißler, Eds.)

Volume 55 EUROPT – A European Initiative on Optimum Design Methods in Aerodynamics. Proceedings of the Brite/Euram Project Workshop „Optimum Design in Aerodynamics", Barcelona, 1992 (J. Periaux / G. Bugeda / P. K. Chaviaropoulos / T. Labrujere / B. Stoufflet, Eds.)

Volume 54 Boundary Elements: Implementation and Analysis of Advanced Algorithms. Proceedings of the Twelfth GAMM-Seminar, Kiel, January 19–21, 1996 (W. Hackbusch / G. Wittum, Eds.)

Volume 53 Computation of Three-Dimensional Complex Flows. Proceedings of the IMACS-COST Conference on Computational Fluid Dynamics, Lausanne, September 13–15, 1995 (M. Deville / S. Gavrilakis / I. L. Ryhming, Eds.)

Volume 52 Flow Simulation with High-Performance Computers II. DFG Priority Research Programme Results 1993–1995 (E. H. Hirschel, Ed.)

Volumes 1 to 51 are out of print.
The addresses of the Editors are listed at the end of the book.

Optimum Aerodynamic Design & Parallel Navier-Stokes Computations

ECARP – European Computational Aerodynamics Research Project

Edited by
Jacques Periaux
Gabriel Bugeda
Panagiotis K. Chaviaropoulos
Kyriakos Giannakoglou
Stephane Lanteri
and Bertrand Mantel

vieweg

Vieweg is a subsidiary company of the Bertelsmann Professional Information.

http://www.vieweg.de

Produced by Geronimo GmbH, Rosenheim
Printed on acid-free paper

ISBN 978-3-322-90195-8 ISBN 978-3-322-90193-4 (eBook)
DOI 10.1007/978-3-322-90193-4

Preface

This volume entitled "European Computational Aerodynamics Research Project (ECARP)" contains the contributions of partners presented in two workshops focused on the following areas: Task 3 on Optimum Design and Task 4.2 on Navier Stokes Flow algorithms on Massively Parallel Processors.

ECARP has been supported by the European Union (EU) through the Industrial and Materials Technology Programme, Area 3 Aeronautics, with the Third Research Framework Programme (1990-1994).

Part A of this volume is focused on computational constrained optimization as a follow up of the EU research project "Optimum Design in Aerodynamics", (AERO-89-0026) dealing with more viscous flow based real applications. It provides the reader with a set of optimization tools and referenced data useful in modern aerodynamic design.

Task 3 of the project entitled "Optimum Design" brought together 13 European partners from the academic and industrial aeronautic oriented community showing state of the art expertise in traditional automated optimization software on current computer technology to improve the capability to optimize aircraft shapes.

The book contains a comprehensive set of computerized data issued from an optimum and inverse problems list of 8 test cases extending the currently pursued technology from an inviscid flow base to a viscous one. Critical problems with relevant geometries listed as 2-D nozzle viscous drag minimization (TE1), airfoil reproduction with viscous effects (TE2), inviscid drag minimization (TE3), multi-point 2-D airfoil design (TE4), wing-pylon-nacelle optimization (TE5), transonic single-wing optimization with inviscid/viscous flows (TE6), axisymmetric afterbody viscous drag minimization (TE7) and laminar riblets (TE8) were defined by partners.

Participants sent prior to the workshop their computerized results according to prescribed access rules and unified storage formats to the INRIA database which were treated by the VIGIE graphic visualisation tool during the workshop.

The performance and accuracy of these techniques depending on the quality of flow analysis solvers have been evaluated during the workshop. The database allowed an interactive comparison of results and provided a reliable basis for the real use of those methods in current airframe design engineering work.

Parallel to the development of the project, multi disciplinary/multi criteria strategies motivated by concurrent engineering targeted products of increasing complexity were considered for the first time with the evident interest of global optimization for arbitrary functions of independent integers, discrete or continuous variables.Promising derivative free computations using a novel shared information technology called Genetic Algorithms were presented by NLR and Dassault Aviation during the project and undoubtedly paved the way to future hybridization of deterministic and stochastic methods in modern optimization.

Part B is focused on a database workshop aimed to determine the impact of parallel architectures on the usability of laminar Navier Stokes solvers for compressible viscous flows. Eight participants from the academic and industrial community carried out computations of at least one of the following test cases:

TC1: 2-D laminar flow around NACA0012 airfoil (turbulent TC3 and TC4).

TC2: 3-D laminar flow around the ONERA M6 wing (turbulent TC5).

The main focus of the workshop was solving the compressible Navier Stokes equations at the lowest cost for a given accuracy.The requested output formats for a contribution consisted in filling the following types of outputs: platform identity, cost evaluation, brief description of the methodology used, approximation, solution method, type of mesh, quality of the 2-D output (resp. 3-D), pressure coefficient, skin friction coefficient on the airfoil (resp. at different span sections on the wing).

Classification of contributions was done in terms of various kinds of meshes, methods and approximation, considering structured or unstructured grids, explicit or implicit methods and upwind or centered schemes.

We are under considerable obligation to D. Knörzer, European Commission DGXII for his constant interest in high-tech methodologies of crucial interest to European airframe manufacturers.

Special thanks go to the Coordinator of ECARP, D. Hills and his collaborator, I. Risk, British Aerospace for doing both a very professional managing job all along the two year contractual period with CFD partners and their continuous efforts making well organized technical meetings attractive and interactive between partners of the three main areas of the project.

We express our gratitude to the INRIA Sophia Antipolis staff who contributed significantly to the success of the database workshops by giving access to contributors to the INRIA database and VIGIE graphic software allowing real time presentations.

We also thank partners for providing data which complied to the access rules and storage formats of the ECARP Database .

The editors acknowledge in particular partners from Aerospatiale, Alenia, INRIA, NTUA, NLR, Dassault Aviation, Daimler Benz Aerospace Airbus, for their precious help in the definition, session chair and synthesis of test cases .

The editors are grateful to E.H. Hirschel, Vieweg general editor for accepting to publish ECARP results and his valuable advises. We are also deeply indebted to patience of Vieweg Verlag staff in the long range preparation of the manuscript published in this series.

Special thanks are also due to A. Dervieux, J-A. Desideri, W. Haase and R. Fournier for their stimulating and fruitful discussions in CFD and Data Processing during the whole development of the workshop.

Jacques Periaux
Gabriel Bugeda
Panagiotis K. Chaviaropoulos
Kyriakos Giannakoglou
Stephane Lanteri
Bertrand Mantel
July 1997

Contents **Page**

Contents (continued) **Page**

Contents (continued) **Page**

Part A: Numerical Optimization in Aerodynamic Design

I Introduction

The optimal shape design problem has been always in the hand of engineers before some famous mathematician like Hadamard /1/ at the beginning of the century introduced the first variational principle of a Partial Differential Equation with respect to the shape of an obstacle. With the development of computers Calculus of Variations gave a significant impulse to numerical optimisation techniques based on control theory introduced by J.L. Lions /2/ and E.J. Haug and J. Cea /3/ and presenting the design problem as a control problem.

After many engineers conducted the optimisation of their design for many years by hand and with analysis tools iteratively without many theoretical results on the existence of solution computerized automated design appeared in Aerospace Engineering with the development of specialized optimization packages /4/, libraries and books /5/Pironneau and /6/ Haslinger and Neittaanmaki as references in the field. However progress toward automated design methods has been decreased because of the severe computing costs involved in the optimization process resulting from the important number of flow evaluations required to get converged solution.

Since the designer has generally a quite good idea of the targeted pressure distribution of interest in Aerodynamics, a first class of airfoil design problems named inverse problems appeared in the late 40's introduced by Lighthill /7/ for an incompressible flow with the major drawback that such a feasible airfoil corresponding to the prescribed surface pressure distribution may not necessarily exist. Meanwhile CFD methods did not take much time with the development of computers to mature within the aerospace industry opening the door to the direct optimization approach - for instance, minimization of the drag of an aerodynamic shape operating under different flow conditions - for single or multi -point design problem as described more recently by A. Jameson /8/ and T. Labrujere /9/.

Major improvements in CFD methods and optimization techniques were beneficial to 3-D aircraft shapes optimization, reducing significantly the time cycle for aircraft design through fully automated automated shape optimisation on current computer technology.However in order to continue to be successfully employed on modern civil transport and business jets on more complex configurations operation under real flow conditions it was necessary to investigate new criteria and more robust constrained optimizers.

It was the main goal of the European project EUROPT in which academic and industrial partners developed and assessed mathematical and engineering tools /10/ for solving inverse and optimisation problems with simple inviscid flow solvers.

The goal of this volume is to describe further development and assessment of new methodologies, algorithms for 3-D automated shape design optimisation of real life problems - namely nozzles, wings, wing/pylon/nacelle configuration, afterbody and riblets - operating in potential or Euler flows with viscous corrections.The associated software have been implemented by 13 partners of the

ECARP project and the validation of the design procedures in terms of accuracy and efficiency presented in a database workshop. The solution results of contributors on 8 test cases dealing with reconstruction, inverse and (multi) single point optimization problems are analyzed and discussed using the graphic visualizations tools of the database. The available data illustrate the capabilities of the design software with parameterized non linear geometries in a industrial environment. Part A of the volume contains 5 sections organized as follows: Section 2 entitled definition of selected test cases for the optimum design workshop describes the type of design problem and the selected parameters of each test case . In this section is also included an overview of the ECARP database describing access rules, storage formats and the VIGIE graphic visualization tools.

The different results obtained by ECARP test case contributors with their optimization methodologies are presented in a partner by partner basis in Section 3 as "Contribution to the resolution of the Database Workshop test cases". Each contribution provides computerized data of the test case with a description of the method and numerical results with analysis and references. Section 5 contains a synthesis of the test cases with comparisons of flow analysis solvers, optimization techniques and parameterization techniques used by contributors. Main characteristics and difficulties of each test case are revised. Comparative information on contributions is provided test case by test case with superposed data including pressure distribution, optimized shapes and convergence curves extracted from the database. A conclusion containing the most significant outcomes of the workshop for aerospace industry closes this volume in Section 6. Further remarks on new directions of research on global search for non convex optimization using promising evolutions algorithms/11/ for aircraft manufacturers are mentioned.

Continuous dissemination of the results and their improvements is envisaged to the European academic and industrial community by means of the existing Database via networks.It is one significant contribution accomplished by ECARP in the Optimum Design Area.

References

- /1/ *J. Hadamard* : Leon sur le calcul des variations. Gauthier Villars ,1910.
- /2/ *J.L. Lions*,Optimal Control of systems governed by PDEs, Springer Verlag, New York, 1971.
- /3/ *E.J. Haug, J. Cea* :optimisation of distributed parameter structures vol 1 and 2, Sitjhoff and Noordhoff,1981.
- /4/ *Vanderplaats*, G.N: ADS- A Fortran Program for Automated Design Synthesis. NASA CR 172460, 1984
- /5/ *O. Pironneau*, Optimum shape design for elliptic systems, Springer-Verlag, 1984

- /6/ *J. Haslinger, P. Neittaanmaki*, Finite element approximations for optimal shape design, J. Wiley, 1989.
- /7/ *M. J. Lighthill*, A new method of two-dimensional aerodynamics design, R&M1111, Aeronautical Research Council, 1945.
- /8/ *A. Jameson*, Optimal Shape Design for Aerodynamics, in Optimum Design methods for Aerodynamics, AGARD Report 803, J. Periaux, Special Course Director at the VKI, Rhode Saint-Gense, Belgium, 25-29 april 1994
- /9/ *T. Labrujere*, Residual-Correction Type and Related Computational Methods for Aerodynamic Design: Multi-Point Airfoil Design, in Optimum Design methods for Aerodynamics, AGARD Report 803, J. Periaux, Special Course Director at the VKI, Rhode Saint-Gense, Belgium, 25-29 april 1994
- /10/ EUROPT- A European Initiative on Optimum Design Methods in Aerodynamics, in Notes on Numerical Fluid Mechanics, vol 55, J. Periaux, G. Bugeda, P.K. Chaviaropoulos, T. Labrujere and B. Stoufflet, eds, 1997
- /11/ *D.E. Goldberg*,Genetic Algorithms in Search, Optimization and Machine Learning, Addison Wesley, 1989

II Definition of the problems for the analysis

1 Definition of test cases

1.1 Introduction

The present document contains the definition of the proposed test cases for the "Optimum Design in Aerodynamics" Workshop which was organized within the EU research project ECARP. The goal of this workshop was to provide the ECARP partners a common test-bed for the validation of their design tools and to promote comparisons in terms of accuracy and efficiency of shape design computations using different methodologies.

The ECARP Optimum Design workshop forms a continuation and maturation of the "Optimum Design in Aerodynamics" workshop which was organized within the initial phase of the EU research project and was started in June 1990. In the initial phase seventeen test cases were defined (called T1,...,T17) from which eight were selected as mandatory, nine were optional and three were kept for the qualification phase. Previous experience showed that the number of test-cases should be decreased in order to increase the number of participants and favor deeper comparison of results. In that respect the "ECARP Optimum Design" workshop includes eight, only, test cases all of them considered as mandatory. The term "mandatory" is used in the sense that each partner should perform his computations using data from the above test-cases list instead of defining his own data and test problems. The selection of the new test cases, called TE1,...,TE8, has been done on a basis of maximum possible participation according to the contractual obligations of the partners.

In selecting the new test-cases emphasis was given to the following topics:

- i) Viscous flows (TE1 ,TE2, TE7, TE8)
- ii) Transonic flows (TE3, TE4, TE6)
- iii) Multi-point design (TE4)
- iv) Complicated 3-D flows (TE5, TE6)

The above test cases involve inverse and optimization problems. In an inverse problem the geometry is recovered from the corresponding flow field solution (usually given as a "target pressure" distribution). The optimization problems which are under consideration here are dealing with drag minimization (viscous drag due to the development of the boundary layer or inviscid drag due to the shock waves in the transonic flow regime).

Although the "ECARP Optimum Design" cases are more application oriented and less academic than those of phase I (see for example TE1, TE4, TE5, TE7, TE8), care was taken to avoid complicated flow phenomena such as separated flows, shock boundary layer interaction or viscous dominated 3D-flows. The reasoning for this approach is that the validation of the design tools should rely on flow problems which have a "well defined" direct solution.

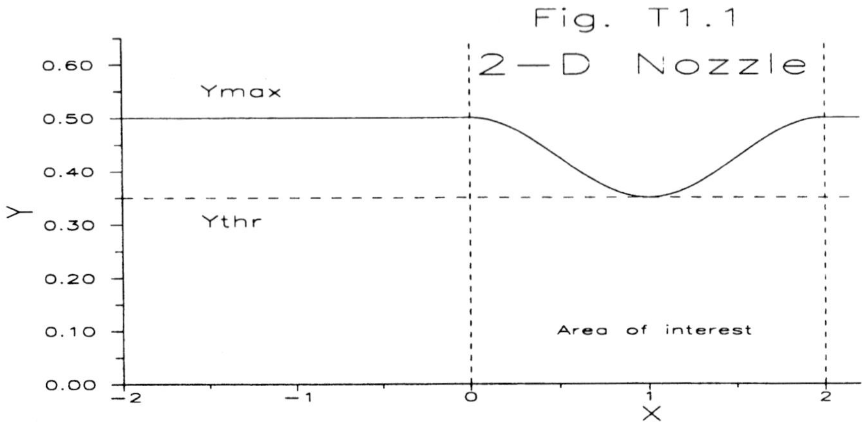

Figure 1 Analysis domain for the TE1 test case

The description of the test cases and the corresponding output formats are given in the following sections.

1.2 Problem TE1. 2-D nozzle viscous drag minimization

1.2.1 Summary

The case is a viscous variant of the test cases T1 and T2 defined in the workshop of Brite Euram project 1082 "Optimum design in aerodynamics". The optimization of a simple symmetric 2-D nozzle geometry under subsonic flow conditions is considered. The inlet and outlet cross sections and the inlet to throat area ratio of the nozzle are fixed acting as geometrical constraints in the optimization process. The cost function is defined as a global measure of the viscous drag along the solid walls in the area of interest $0 \leq x \leq 2$. Compared to T1 and T2 the throat area is now increased in order to avoid flow separation in the decelerating part of the nozzle. Computations will be performed using viscous-inviscid interaction techniques (a weak interaction scheme is suggested) where the inviscid part may be a potential or an Euler code.

1.2.2 Definition of design problem

The nozzle is symmetric with respect to the $y = 0$ line (see fig. 1). The area of interest lies between $x = 0$ and $x = 2$. The half-hight of the inlet and outlet sections is $y_{max} = 0.5$. The half-hight of the throat is $y_{thr} = 0.35$ at $x = 1$. The Mach number is 0.2 and the Reynolds number based on half length (l=1) is $2 * 10^6$. The computational domain is extended for positive and negative x $(-2 \leq x \leq 4)$ for inflow and outflow uniformity reasons. The geometry of the extensions is fixed $(y = y_{max})$.

We are looking for the nozzle shape which respects the geometrical constraints and minimizes the viscous drag on the solid walls in the area of interest $0 \le x \le 2$. The boundary layer development is considered in the area of interest, only, starting with turbulent initial conditions at $x = 0$, in order to avoid incompatibilities due to the laminar to turbulent transition criterion. The initial boundary layer conditions are

$$\begin{aligned} \frac{\delta}{l} &= 4.97 * 10^{-3} \\ \frac{\delta_1}{l} &= 8.64 * 10^{-4} \\ H = \frac{\delta_1}{\Theta} &= 1.431 \end{aligned} \tag{1}$$

where H the form factor, δ_1, Θ and δ the displacement, the momentum and the boundary layer thicknesses based on half-length l.

For an indicative drag computation (C_D) the use of the Squire and Young [Squire] formula as proposed for wake flows is recommended.

$$C_D = \left(\frac{\Theta}{l}\right)\left(\frac{V}{V_\infty}\right)^{.5*(H+5)} \tag{2}$$

where the boundary layer properties are computed at $x = 2$ and V is the local inviscid velocity.

First attempts have shown that the boundary layer optimization using the Squire and Young formula may lead to non-acceptable (wavy) geometries when a large number of control variables is used. For this reason we recommend the use of 3-6 control variables.

1.2.3 Initial geometry

In order to define a common starting point for the design process, the following initial shape is given, which respects the geometrical constraints

$$\begin{aligned} 0 \le x \le 2; \qquad y &= a0 + a1 * sin(\pi * (x - 1.5)) \\ a0 &= (y_{max} + y_{thr})/2 \\ a1 &= (y_{max} - y_{thr})/2 \,. \end{aligned} \tag{3}$$

1.2.4 Results and output formats

Geometry:

- resulting nozzle shape (half-nozzle, $y \ge 0$, $0 \le x \le 2$) y versus x.

Pressure distribution:

- pressure distribution along the solid wall in the area of interest
- $(-C_p)$ versus x , $-C_p = 2 * \frac{(P_\infty - P)}{(\rho_\infty * V_\infty^2)}$.

Boundary layer:

- H : form factor in the area of interest.
- H versus x.

Convergence:

- Value of objective function normalized with its starting value (C_D/C_{Do}) versus number of objective function evaluations (number of direct flow solutions).

1.2.5 *Bibliography*

- [Squire] *H.B. Squire, A.D. Young,* The Calculation of the Profile Drag of Aerofoils. ARCR&M, No 1838, 1938.

1.3 Problem TE2. Airfoil reproduction case with viscous effects

1.3.1 *Summary*

Test case TE2 replaces T4 of the previous workshop. It concerns the reproduction of an airfoil under viscous flow conditions. The sensitive (to its suction side coordinates) Korn airfoil is now replaced by NACA0012 (target profile) and the inflow Mach number is decreased to $M_\infty = 0.3$ in order to assure the subsonic character of the flow field and avoid any shock boundary layer interaction effects. The incidence angle is fixed at 3°. Although the case is rather academic it consists a nice exercise for the validation of the available design tools. Computations will be performed using viscous-inviscid interaction techniques, where the inviscid part may be a potential or an Euler code.

1.3.2 *Definition of design problem*

The NACA0012 profile, given in Table 1, will be reproduced from the "target pressure" distribution which corresponds to the above profile at $M_\infty = 0.3$ incidence $\alpha = 3^o$ and $Re = 3 * 10^6$ (based on chord c). The contributors will use their own analysis method in order to produce the "target pressure distribution".

1.3.3 *Initial geometry*

In order to define a common starting point for the design process, the NACA63215 airfoil given in table T2, is specified as initial guess of the geometry. Figure 2 shows a superposition of the NACA0012 and the NACA63215 profiles.

1.3.4 *Results and output formats*

Geometry:

- target and final profile coordinates (dashed-continuous); y/c versus x/c.

Pressure distribution:

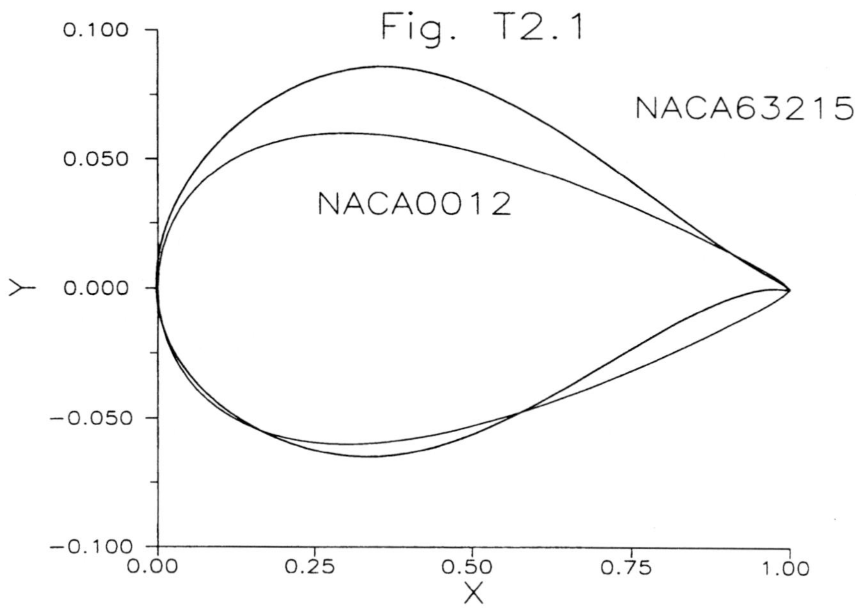

Figure 2 Superposition of the NACA0012 and the NACA63215 profiles

- final pressure coefficient distribution
- $(-C_p)$ versus x , $-C_p = 2 * \frac{(P_\infty - P)}{(\rho_\infty * V_\infty^2)}$.

Convergence history:

- Value of objective function normalized with its starting value versus number of objective function evaluations (number of direct flow solutions)
- Value of $L_2(y)$ versus number of objective function evaluations (number of direct flow solutions).

$L_2(y)$ is defined by

$$L_2(y) = \frac{1}{N}\sqrt{\sum_{i=1}^{N}\left[\left(\frac{y}{c}\right)_i^k - \left(\frac{y}{c}\right)_i^{k-1}\right]^2} \qquad (4)$$

where the superscripts k and $k-1$ refer to two consecutive geometry updates ($k = 0$ is the initial airfoil NACA63215).

Table 1 Definition of NACA0012 profile

X	Y	X	Y	X	Y
1.000000	0.000000	0.248538	-0.059375	0.271550	0.059829
0.988272	-0.002528	0.225764	-0.058627	0.294778	0.060011
0.975358	-0.004636	0.203251	-0.057559	0.318202	0.059943
0.961222	-0.006607	0.181022	-0.056141	0.341802	0.059642
0.945794	-0.008644	0.159105	-0.054336	0.365558	0.059124
0.928964	-0.010801	0.137530	-0.052098	0.389451	0.058406
0.910606	-0.013127	0.116332	-0.049367	0.413464	0.057502
0.890581	-0.015628	0.096424	-0.046213	0.437577	0.056424
0.868736	-0.018298	0.078564	-0.042770	0.461772	0.055186
0.845814	-0.021021	0.062578	-0.039046	0.486032	0.053800
0.822677	-0.023689	0.048321	-0.035032	0.510339	0.052276
0.799343	-0.026304	0.035680	-0.030691	0.534676	0.050625
0.775830	-0.028868	0.024591	-0.025952	0.559024	0.048856
0.752156	-0.031378	0.015073	-0.020683	0.583367	0.046977
0.728337	-0.033827	0.007320	-0.014676	0.607687	0.044997
0.704392	-0.036211	0.001933	-0.007685	0.631967	0.042925
0.680337	-0.038525	0.000000	0.000000	0.656189	0.040765
0.656189	-0.040765	0.001933	0.007685	0.680337	0.038525
0.631967	-0.042925	0.007320	0.014676	0.704392	0.036211
0.607687	-0.044997	0.015073	0.020683	0.728337	0.033827
0.583367	-0.046977	0.024591	0.025952	0.752156	0.031378
0.559024	-0.048856	0.035680	0.030691	0.775830	0.028868
0.534676	-0.050625	0.048321	0.035032	0.799343	0.026304
0.510339	-0.052276	0.062578	0.039046	0.822677	0.023689
0.486032	-0.053800	0.078564	0.042770	0.845814	0.021021
0.461772	-0.055186	0.096424	0.046213	0.868736	0.018298
0.437577	-0.056424	0.116332	0.049367	0.890581	0.015628
0.413464	-0.057502	0.137530	0.052098	0.910606	0.013127
0.389451	-0.058406	0.159105	0.054336	0.928964	0.010801
0.365558	-0.059124	0.181022	0.056141	0.945794	0.008644
0.341802	-0.059642	0.203251	0.057559	0.961222	0.006607
0.318202	-0.059942	0.225764	0.058627	0.975358	0.004636
0.294778	-0.060011	0.248538	0.059375	0.988272	0.002528
0.271550	-0.059829			1.000000	0.000000

1.4 Problem TE3. Transonic airfoil, inviscid drag minimization case

1.4.1 Summary

Test case TE3 is a revised version of T6, originally defined for the workshop of Brite/Euram project 1082 "Optimum design in aerodynamics" (see also Ref.[1]).

Table 2 Definition of NACA63215 profile

X	Y	X	Y	X	Y
1.0000000	0.0000000	0.2603451	-0.0630404	0.2801250	0.0834265
0.9919814	-0.0000313	0.2364209	-0.0616403	0.3037396	0.0846778
0.9819519	0.0001336	0.2122400	-0.0598124	0.3270785	0.0854812
0.9694127	0.0001729	0.1878117	-0.0575238	0.3501364	0.0858395
0.9552994	-0.0001735	0.1631472	-0.0547386	0.3729060	0.0857598
0.9409577	-0.0008494	0.1382586	-0.0514174	0.3953812	0.0852721
0.9263909	-0.0018236	0.1150544	-0.0477824	0.4175575	0.0844173
0.9115973	-0.0030378	0.0951086	-0.0441225	0.4394335	0.0832340
0.8965776	-0.0044346	0.0779834	-0.0404797	0.4610084	0.0817573
0.8813334	-0.0059768	0.0632986	-0.0368962	0.4822829	0.0800185
0.8658642	-0.0076591	0.0507260	-0.0333990	0.5032582	0.0780457
0.8501684	-0.0094857	0.0399834	-0.0299989	0.5239358	0.0758660
0.8342440	-0.0114536	0.0308339	-0.0267005	0.5443185	0.0735068
0.8180881	-0.0135512	0.0230653	-0.0234929	0.5644095	0.0709939
0.8016981	-0.0157620	0.0165334	-0.0203233	0.5842126	0.0683479
0.7850717	-0.0180704	0.0111222	-0.0171382	0.6037308	0.0655862
0.7682065	-0.0204653	0.0067409	-0.0139300	0.6229677	0.0627251
0.7510999	-0.0229379	0.0032696	-0.0106768	0.6419270	0.0597813
0.7337490	-0.0254773	0.0006188	-0.0074608	0.6606128	0.0567708
0.7161500	-0.0280695	-0.0013743	-0.0044177	0.6790293	0.0537093
0.6982994	-0.0306983	-0.0025596	-0.0004652	0.6971810	0.0506119
0.6801935	-0.0333473	-0.0022266	0.0054482	0.7150721	0.0474932
0.6618291	-0.0360017	-0.0009671	0.0089686	0.7327074	0.0443670
0.6432026	-0.0386471	0.0011331	0.0126414	0.7500912	0.0412465
0.6243111	-0.0412689	0.0041070	0.0163868	0.7672282	0.0381449
0.6051513	-0.0438531	0.0080537	0.0201648	0.7841229	0.0350763
0.5857205	-0.0463849	0.0130512	0.0240153	0.8007803	0.0320548
0.5660156	-0.0488484	0.0192337	0.0279173	0.8172050	0.0290921
0.5460337	-0.0512262	0.0267020	0.0319210	0.8334006	0.0261959
0.5257718	-0.0534985	0.0355645	0.0361163	0.8493705	0.0233716
0.5052272	-0.0556425	0.0460458	0.0405183	0.8651178	0.0206255
0.4843972	-0.0576336	0.0584073	0.0450987	0.8806462	0.0179680
0.4632801	-0.0594461	0.0729410	0.0498403	0.8959598	0.0154104
0.4418743	-0.0610533	0.0899763	0.0547251	0.9110625	0.0129629
0.4201792	-0.0624270	0.1099073	0.0597126	0.9259576	0.0106337
0.3981946	-0.0635370	0.1331907	0.0647465	0.9406469	0.0084301
0.3759207	-0.0643491	0.1582307	0.0694095	0.9551351	0.0063590
0.3533598	-0.0648242	0.1830758	0.0733786	0.9694176	0.0044302
0.3305168	-0.0649328	0.2077040	0.0767178	0.9820902	0.0028317
0.3073972	-0.0646705	0.2320967	0.0794811	0.9920973	0.0013408
0.2840055	-0.0640427	0.2562412	0.0817059	1.0000000	0.0000000

TE3 addresses the problem of inviscid drag minimization in transonic flows. The far field flow conditions are $M_\infty = 0.73$ and the angle of attack $\alpha = 2°$. Starting the RAE 2822 airfoil it is asked to optimize the profile shape in terms of drag minimization while maintaining the lift of the original profile. The flow will be modeled by the full potential or the Euler equations. Viscous effects are not included for simplicity reasons.

1.4.2 Definition of the design problem

We propose two alternative ways to formulate the minimization problem

a) Improvement of existing profile via an inverse problem: Firstly, the problem is formulated as a perturbation of an inverse problem. The target pressure distribution P_{tar} is taken to be the actual pressure distribution of the initial profile predicted by the numerical solution of the flow equations. The inclusion of this target distribution will force the method to generate a profile with lift coefficient close that of the initial profile. Drag coefficient C_D is added to the cost function so that the expression of the cost function I is

$$\begin{aligned} I_{tar} &= \int (C_p - C_{p_{tar}})^2 d\frac{s}{c} \\ I &= I_{tar} + \beta * C_D \end{aligned} \tag{5}$$

where I_{tar} is the cost function corresponding to the pure inverse problem. The addition of a drag penally now causes the method to reshape the profile to reduce its drag where beta is a parameter. It is asked to evaluate the sensitivity of the minimization procedure and of the solution to the magnitude of this parameter.

b) Improvement of existing profile via a minimization: Secondly, we consider the previous problem as a pure minimization problem expressed as:

Minimize $I = C_D$ under the constraint of a given lift C_L.

1.4.3 Initial geometry

The initial profile for the minimization exercise is the RAE 2822 airfoil, specified in Table 3.

1.4.4 Results and output formats

Geometry:

- initial and final profile coordinates (dashed-continuous); y/c versus x/c .

Pressure distribution:

- initial and final pressure coefficient distributions
- $(-C_p)$ versus x , $-C_p = 2 * \frac{(P_\infty - P)}{(\rho_\infty * V_\infty^2)}$.

Convergence:

Table 3 Definition of RAE2822 profile

X	Y	X	Y	X	Y
1.0000000	0.0000000	0.2221680	0.0537662	0.2580510	-0.0559978
0.9959220	0.0010185	0.2042840	0.0520800	0.2764110	-0.0570496
0.9908080	0.0020974	0.1865050	0.0502664	0.2949820	-0.0579169
0.9844620	0.0033714	0.1688060	0.0483126	0.3137900	-0.0585898
0.9766020	0.0049170	0.1511660	0.0461898	0.3328590	-0.0590213
0.9668700	0.0068107	0.1335710	0.0438014	0.3522150	-0.0591898
0.9548150	0.0091071	0.1160000	0.0411334	0.3718800	-0.0590510
0.9398860	0.0118994	0.0984315	0.0381756	0.3918750	-0.0585691
0.9213920	0.0152705	0.0808600	0.0348389	0.4122210	-0.0577111
0.8984720	0.0192785	0.0646139	0.0313443	0.4329390	-0.0564815
0.8700710	0.0240370	0.0507680	0.0279842	0.4540500	-0.0549000
0.8389200	0.0289940	0.0389946	0.0247281	0.4755760	-0.0529597
0.8085740	0.0335884	0.0290141	0.0215556	0.4975460	-0.0507070
0.7790000	0.0377808	0.0206190	0.0183730	0.5199830	-0.0481567
0.7501810	0.0416085	0.0136518	0.0151011	0.5429130	-0.0453082
0.7220950	0.0451056	0.0080702	0.0115957	0.5663690	-0.0422229
0.6947090	0.0482047	0.0037569	0.0079290	0.5903760	-0.0388995
0.6680270	0.0511649	0.0009308	0.0040219	0.6149889	-0.0355421
0.6419900	0.0536774	0.0000000	0.0000000	0.6401920	-0.0318437
0.6165880	0.0558723	0.0009342	-0.0040305	0.6660470	-0.0280477
0.5917950	0.0577384	0.0036593	-0.0079937	0.6925870	-0.0242002
0.5675890	0.0592968	0.0080125	-0.0116162	0.7198450	-0.0203256
0.5439460	0.0605419	0.0136091	-0.0150900	0.7478510	-0.0164679
0.5208410	0.0614837	0.0205598	-0.0183872	0.7766430	-0.0127027
0.4982530	0.0621638	0.0289036	-0.0216825	0.8062630	-0.0091534
0.4761580	0.0625996	0.0388395	-0.0249594	0.8367350	-0.0058432
0.4545310	0.0628399	0.0505752	-0.0282977	0.8680990	-0.0029292
0.4333470	0.0628760	0.0644001	-0.0316657	0.8967780	-0.0007898
0.4125810	0.0627469	0.0806307	-0.0351212	0.9199830	0.0004634
0.3922060	0.0624733	0.0981861	-0.0384027	0.9387510	0.0010867
0.3721970	0.0620483	0.1157410	-0.0412911	0.9539250	0.0012587
0.3525290	0.0614741	0.1333050	-0.0438607	0.9661870	0.0012062
0.3331730	0.0607654	0.1508900	-0.0461675	0.9760930	0.0010080
0.3141050	0.0599320	0.1685180	-0.0482350	0.9840960	0.0007453
0.2952980	0.0589826	0.1862060	-0.0501200	0.9905600	0.0004715
0.2767240	0.0579220	0.2039790	-0.0518376	0.9957810	0.0001990
0.2583600	0.0567215	0.2218610	-0.0533717	1.0000000	0.0000000
0.2401830	0.0553349	0.2398770	-0.0547586		

- Value of objective function normalized with its starting value versus number of objective function evaluations (number of direct flow solutions).

1.4.5 Bibliography

- [1] *A. Jameson*, Automatic Design of Transonic Airfoil to Reduce the Shock-Induced Pressure Drag 31st Israel Annual Conference Aviation and Aeronautics, February 21-22, 1990.

1.5 Problem TE4. Multi-point 2-D airfoil design

1.5.1 Summary

The present test case concerns the problem of designing an airfoil with specific properties at different operating conditions. The test case is a redefinition of test case T8 as defined for the workshop in EUROPT I. As a starting point for the test case definition two largely different airfoils are selected, one typical high lift airfoil operating at low speed and one typical low drag airfoil operating at high speed. The present two-point airfoil design test case aims at the design of an airfoil, which combines the favorable performances of both high lift and low drag airfoils. Analogous design problems will be encountered at e.g. the design of a transport aircraft outer wing section or at the design of a helicopter rotor blade section.

1.5.2 Problem definition

Objective function

A two-point design problem is defined as the minimization of the objective function:

$$F(\alpha_1, \alpha_2, x(s), y(s)) = \sum_{i=1}^{2} \left[W_i \int_0^1 \left(C_p^i(s) - C_{p_{tar}}^i(s)\right)^2 ds \right] \tag{6}$$

where s is the fractional arc-length measured along the airfoil contour to be defined and where the angle of attack (α) is defined with respect to the x-axis of the coordinate system in which the contour points are specified. The angles of attack (α_1), (α_2) and the coordinate functions x(s) and y(s) are the design variables. The weight factors (W_i) are given constants. (C_p^i) represents the actual pressure distribution at design condition i (defined by angle of attack (α_i), Mach number (Ma_i) and Reynolds number (Re_i)). ($C_{p_{tar}}^i$) represents the target pressure distribution at design condition i.

Target pressure distributions

The target pressure distributions are obtained by performing two analysis computations:

For design condition 1 : perform an analysis computation for ($\alpha_1 = 10.8$ at ($Ma_1 = 0.2$) and ($Re_1 = 5 * 10^6$) for the airfoil specified by means of Table 4. The resulting pressure distribution along the airfoil contour serves as target pressure distribution for the first design condition.

For design condition 2 : perform an analysis computation for (α_2 = 1.0 at (Ma_2 = 0.77) and ($Re_2 = 1 * 10^7$) for the airfoil specified by means of Table 5. The resulting pressure distribution along the airfoil contour serves as target pressure distribution for the second design condition.

Weight factors

The weight factors (W_1) and (W_2) should be chosen as follows :

1. (W_1 = 1) and (W_2 = 0), which makes the design problem a single-point reconstruction problem for subsonic flow,
2. (W_1 = 0.5) and (W_2 = 0.5), which makes the design problem a two-point design design problem aiming at a combination of desired subsonic and transonic properties,
3. (W_1 = 0) and (W_2 = 1), which makes the design problem a single-point reconstruction problem for transonic flow.

Initial geometry

As a common starting point for the design process the NACA4412 airfoil as given on Table 6. is specified as initial geometry.

1.5.3 Results and output formats

In order to enable direct comparison of results for each computation the following data should be presented : 1. The resulting airfoil shape, airfoil chord along the x-axis : y/c versus x/c

2. The pressure distributions along the contour of the resulting airfoil for both design conditions : C_p versus x/c

3. Convergence history : a) Value of objective function normalized with its starting value versus number of objective function evaluations (number of direct flow solutions) b) Value of $L_2(y)$ versus number of objective function evaluations (number of direct flow solutions), where $L_2(y)$ is defined by

$$L_2(y) = \frac{1}{N}\sqrt{\sum_{i=1}^{N}\left[\left(\frac{y}{c}\right)_i^k - \left(\frac{y}{c}\right)_i^{k-1}\right]^2}. \qquad (7)$$

The superscripts refer to two consecutive geometry updates (k=0 is the initial airfoil NACA4412) and y is the coordinate of a point from the airfoil geometry sets as defined above.

1.5.4 Bibliography

- [1] *van Egmond, J.A., Labrujère, Th.E. and van der Vooren, J.*, Test cases for a workshop within the Brite Euram project BE1082 "Optimum design in aerodynamics" NLR CR 90329 L.

Table 4 Definition of the high lift profile

X	Y	X	Y	X	Y
1.0000000	0.0000000	0.2309588	-0.0657025	0.2895326	0.0881946
0.9988834	-0.0000677	0.2034101	-0.0642503	0.3202961	0.0839482
0.9955388	-0.0002688	0.1771861	-0.0624757	0.3518622	0.0791939
0.9899809	-0.0005957	0.1524038	-0.0605055	0.3840899	0.0742062
0.9822346	-0.0010325	0.1291739	-0.0584882	0.4168353	0.0691538
0.9723346	-0.0015489	0.1076003	-0.0565314	0.4499522	0.0640855
0.9603249	-0.0021046	0.0877792	-0.0546339	0.4832925	0.0589940
0.9462594	-0.0027575	0.0697992	-0.0525527	0.5167075	0.0538866
0.9302008	-0.0036389	0.0537406	-0.0501394	0.5500478	0.0488010
0.9122208	-0.0048263	0.0396751	-0.0473565	0.5831647	0.0437819
0.8923997	-0.0063468	0.0276655	-0.0440813	0.6159101	0.0388975
0.8708261	-0.0082393	0.0177654	-0.0398350	0.6481378	0.0341803
0.8475962	-0.0104996	0.0100191	-0.0324439	0.6797039	0.0296926
0.8228139	-0.0131595	0.0044613	-0.0209754	0.7104674	0.0254750
0.7965899	-0.0161949	0.0011166	-0.0095479	0.7402909	0.0215593
0.7690412	-0.0196288	0.0000000	0.0000000	0.7690412	0.0179885
0.7402909	-0.0234260	0.0011166	0.0090043	0.7965899	0.0147531
0.7104674	-0.0275538	0.0044613	0.0183788	0.8228139	0.0118776
0.6797039	-0.0319436	0.0100191	0.0272777	0.8475962	0.0093432
0.6481378	-0.0365148	0.0177654	0.0358753	0.8708261	0.0071653
0.6159101	-0.0411570	0.0276655	0.0444941	0.8923997	0.0053496
0.5831647	-0.0457497	0.0396751	0.0531655	0.9122208	0.0038990
0.5500479	-0.0501580	0.0537406	0.0616488	0.9302008	0.0027936
0.5167075	-0.0542392	0.0697992	0.0696638	0.9462594	0.0019975
0.4832925	-0.0578731	0.0877792	0.0770109	0.9603249	0.0014960
0.4499522	-0.0609781	0.1076003	0.0834877	0.9723346	0.0011036
0.4168353	-0.0634999	0.1291739	0.0888298	0.9822346	0.0007192
0.3840899	-0.0653877	0.1524038	0.0927475	0.9899809	0.0004004
0.3518622	-0.0666411	0.1771861	0.0949745	0.9955388	0.0001743
0.3202961	-0.0672643	0.2034101	0.0954416	0.9988834	0.0000428
0.2895326	-0.0672819	0.2309588	0.0942597	1.0000000	0.0000000
0.2597091	-0.0667402	0.2597091	0.0917073		

Table 5 Definition of the low drag profile

X	Y	X	Y	X	Y
1.0000000	0.0000000	0.2309588	-0.0517751	0.2895326	0.0621137
0.9988834	-0.0000259	0.2034101	-0.0497342	0.3202961	0.0633637
0.9955388	-0.0001031	0.1771861	-0.0472836	0.3518622	0.0643886
0.9899809	-0.0002307	0.1524038	-0.0444447	0.3840899	0.0650782
0.9822346	-0.0004068	0.1291739	-0.0412616	0.4168353	0.0653836
0.9723346	-0.0006385	0.1076003	-0.0378254	0.4499522	0.0652344
0.9603249	-0.0009137	0.0877792	-0.0343007	0.4832925	0.0645609
0.9462594	-0.0012381	0.0697992	-0.0308336	0.5167075	0.0633419
0.9302008	-0.0016081	0.0537406	-0.0275538	0.5500478	0.0616386
0.9122208	-0.0020186	0.0396751	-0.0243983	0.5831647	0.0595515
0.8923997	-0.0024352	0.0276655	-0.0209139	0.6159101	0.0571880
0.8708261	-0.0031422	0.0177654	-0.0170160	0.6481378	0.0546379
0.8475962	-0.0042213	0.0100191	-0.0128884	0.6797039	0.0518838
0.8228139	-0.0057361	0.0044613	-0.0089655	0.7104674	0.0489281
0.7965899	-0.0077285	0.0011166	-0.0061722	0.7402909	0.0457464
0.7690412	-0.0101906	0.0000000	0.0000000	0.7690412	0.0423713
0.7402909	-0.0131140	0.0011166	0.0089232	0.7965899	0.0388351
0.7104674	-0.0164601	0.0044613	0.0187306	0.8228139	0.0351994
0.6797039	-0.0201800	0.0100191	0.0251819	0.8475962	0.0315170
0.6481378	-0.0241938	0.0177654	0.0306903	0.8708261	0.0278474
0.6159101	-0.0284326	0.0276655	0.0351921	0.8923997	0.0242358
0.5831647	-0.0327959	0.0396751	0.0385315	0.9122208	0.0207372
0.5500479	-0.0371338	0.0537406	0.0410028	0.9302008	0.0172055
0.5167075	-0.0413065	0.0697992	0.0434826	0.9462594	0.0137356
0.4832925	-0.0451533	0.0877792	0.0461645	0.9603249	0.0104835
0.4499522	-0.0485247	0.1076003	0.0489310	0.9723346	0.0075282
0.4168353	-0.0512904	0.1291739	0.0515776	0.9822346	0.0049478
0.3840899	-0.0533304	0.1524038	0.0539270	0.9899809	0.0028419
0.3518622	-0.0545449	0.1771861	0.0559413	0.9955388	0.0012874
0.3202961	-0.0548945	0.2034101	0.0576633	0.9988834	0.0003535
0.2895326	-0.0544609	0.2309588	0.0592147	1.0000000	0.0000000
0.2597091	-0.0533836	0.2597091	0.0607076		

Table 6 Definition of the NACA4412 profile

X	Y	X	Y	X	Y
1.0000000	0.0000000	0.2309588	-0.0280498	0.2895326	0.0951233
0.9988834	-0.0000180	0.2034101	-0.0294187	0.3202961	0.0964762
0.9955388	-0.0000745	0.1771861	-0.0304225	0.3518622	0.0970059
0.9899809	-0.0001767	0.1524038	-0.0310518	0.3840899	0.0967447
0.9822346	-0.0003351	0.1291739	-0.0313348	0.4168353	0.0957474
0.9723346	-0.0005604	0.1076003	-0.0311917	0.4499522	0.0940824
0.9603249	-0.0008607	0.0877792	-0.0305724	0.4832925	0.0918301
0.9462594	-0.0012374	0.0697992	-0.0295082	0.5167075	0.0890826
0.9302008	-0.0016805	0.0537406	-0.0279206	0.5500478	0.0858959
0.9122208	-0.0021650	0.0396751	-0.0256581	0.5831647	0.0822873
0.8923997	-0.0026463	0.0276655	-0.0228082	0.6159101	0.0782804
0.8708261	-0.0030937	0.0177654	-0.0194548	0.6481378	0.0739161
0.8475962	-0.0035250	0.0100191	-0.0151557	0.6797039	0.0692510
0.8228139	-0.0039795	0.0044613	-0.0102591	0.7104674	0.0643468
0.7965899	-0.0045167	0.0011166	-0.0052078	0.7402909	0.0592768
0.7690412	-0.0051934	0.0000000	0.0000000	0.7690412	0.0541271
0.7402909	-0.0060125	0.0011166	0.0058120	0.7965899	0.0489808
0.7104674	-0.0069608	0.0044613	0.0122024	0.8228139	0.0438981
0.6797039	-0.0080199	0.0100191	0.0190414	0.8475962	0.0388430
0.6481378	-0.0091800	0.0177654	0.0260443	0.8708261	0.0337771
0.6159101	-0.0104376	0.0276655	0.0328087	0.8923997	0.0287004
0.5831647	-0.0117871	0.0396751	0.0395788	0.9122208	0.0236563
0.5500479	-0.0132036	0.0537406	0.0463612	0.9302008	0.0188194
0.5167075	-0.0146426	0.0697992	0.0530709	0.9462594	0.0143910
0.4832925	-0.0160592	0.0877792	0.0595219	0.9603249	0.0105008
0.4499522	-0.0174546	0.1076003	0.0657006	0.9723346	0.0072180
0.4168353	-0.0188751	0.1291739	0.0716125	0.9822346	0.0045664
0.3840899	-0.0203642	0.1524038	0.0770983	0.9899809	0.0025403
0.3518622	-0.0219173	0.1771861	0.0820187	0.9955388	0.0011188
0.3202961	-0.0234852	0.2034101	0.0863255	0.9988834	0.0002781
0.2895326	-0.0250217	0.2309588	0.0899943	1.0000000	0.0000000
0.2597091	-0.0265293	0.2597091	0.0929493		

1.6 Problem TE5. Wing-pylon-nacelle optimization case

1.6.1 Summary

The case addresses the optimization problem of a complicated 3-D configuration which includes a wing, a fuselage and a pylon. The name of the configuration to optimize is A528E. It includes the A528 wing the A528 nacelle and a geometrical pylon generated from the initial cambered A528 pylon. The objective of the optimization is the modification of the wing geometry in the presence of the pylon and the nacelle, aiming to obtain the target pressures of the clean wing. The computations will be performed for Mach=0.5 and 2° angle of attack, using panel methods for the inviscid part and, optionally, boundary layer codes for viscous corrections.

1.6.2 Definition of the design problem

The definition of TE5 is the following:
"minimize the perturbation (pressure deviations with respect to the wing-body configuration) induced on the wing by the engine installation, by modifying the wing geometry".

The fuselage has not been included into the A528E configuration, in order to limit the size of the influence matrix of the panel method called during the optimization process. As TE5 aims to reduce the interference between a wing and a power plant, the fuselage would have increased uselessly the cost of the computation.

The A528E configuration is described by a panel mesh. The smooth wing used for computing the target in terms of Cp-distribution is also described by a panel mesh. The panel meshes are available to the partners in a tar formatted file on a floppy disk, along with a technical paper describing the topology of the meshes and the storage format of the data. A supplementary file describing the pylon with the help of iso-z cuts is also available.

The computed Cp-distribution of the wing-body configuration will serve as target distribution. In order to be consistent with the aerodynamic model used within the optimization process, each participant must compute the target Cp with his own panel method.

Within the optimization process the pylon geometry will be kept fixed and only the wing geometry will be modified. No geometrical constraints will be taken into account with the exception of the wing, pylon and nacelle trailing edge thickness which is considered to be equal to zero.

1.7 Problem TE6. Transonic single-wing optimization case with inviscid flow

1.7.1 Summary

The case concerns the optimization of a single wing configuration under transonic flow conditions. Due to the transonic character of the flow, a field method should be used for the simulation. Computations will be performed using a full potential code or an Euler code. A boundary layer code may, optionally, be coupled with the inviscid solver to take into account the viscous effects.

1.7.2 Definition of design problem

The ONERA M6 wing will be "optimized" at $M_\infty = 0.84$ and $\alpha = 3.06°$. in order to reduce inviscid drag with constraints on the lift coefficient and on the maximum thickness of span-wise wing shape cuts. The lift coefficient of the optimized wing has to be greater or equal to the ONERA M6 wing lift coefficient computed by contributors own analysis method. The maximum thickness must be the one of the original wing. The platform is kept fixed.

1.7.3 Initial geometry

The starting geometry for the design process is the ONERA M6 wing shape.

1.7.4 Results and output formats

Geometry: Initial and final profile coordinates (dashed-continuous) at a fixed number of span-wise locations located at 0, 20, 40, 60, 80, 100 percents of the span. Coordinates are normalized with the local platform c(y), y being the span-wise direction.
Global coefficients: Lift, drag and pitch moment coefficients. For the computation of these coefficients, use: Mean aerodynamic chord = 0.800681 c(0) Reference coordinates for moments computation = (0.588293,0.,0.) c(0) c(0) is the root section platform chord. It is assumed that the platform leading edge of the root section is located at the axis origin.
Pressure distribution: Initial and final pressure coefficient distribution versus x normalized with the local platform chord, $(-C_p)$ versus x/c(y), at the previously defined span-wise locations.
Convergence history: Value of the objective function versus number of objective function evaluations (number of direct flow solutions).

1.7.5 Description of the initial geometry

The initial geometry is an untwisted wing obtained by the reproduction of a unique (scaled) symmetrical 2-D shape span-wise. The wing platform has the form of a trapezium with a 30° leading edge sweep angle. If we assume that the root section platform chord c(0) is equal to one, the platform can be defined as follows:

- Root section leading edge location: (0., 0., 0.)
- Root section trailing edge location: (1., 0., 0.)
- Tip section leading edge location: (0.8568, 1.484, 0.)
- Tip section trailing edge location: (1.416794, 1.484, 0.)

where y represents the span-wise direction.

Table 7 Root section of the Onera M6 wing

X	Y	X	Y	X	Y
1.0000000	0.0000000	0.1963300	-0.0436900	0.1963300	0.0436900
0.9938000	-0.0007700	0.1773600	-0.0425500	0.2150700	0.0446800
0.9858900	-0.0017500	0.1581300	-0.0412500	0.2336200	0.0455400
0.9758100	-0.0029900	0.1386000	-0.0397500	0.2520100	0.0462900
0.9629700	-0.0045800	0.1187600	-0.0380000	0.2702700	0.0469300
0.9466100	-0.0066100	0.1000400	-0.0361100	0.2884400	0.0474700
0.9257500	-0.0091800	0.0837000	-0.0342400	0.3065600	0.0479100
0.8991700	-0.0123900	0.0694500	-0.0324100	0.3246400	0.0482500
0.8696600	-0.0158000	0.0570100	-0.0306400	0.3427400	0.0484900
0.8410000	-0.0189400	0.0461600	-0.0289500	0.3608900	0.0486200
0.8131800	-0.0218300	0.0367100	-0.0272800	0.3791200	0.0486500
0.7861500	-0.0245000	0.0285300	-0.0254200	0.3974600	0.0485600
0.7598900	-0.0269900	0.0215400	-0.0232000	0.4159500	0.0483600
0.7343600	-0.0293300	0.0156900	-0.0205900	0.4346200	0.0480300
0.7095300	-0.0315300	0.0109200	-0.0176500	0.4535000	0.0475800
0.6853600	-0.0336000	0.0071500	-0.0145300	0.4726400	0.0469900
0.6618300	-0.0355400	0.0042900	-0.0113400	0.4920700	0.0462700
0.6388900	-0.0373600	0.0022300	-0.0082400	0.5118100	0.0454200
0.6165100	-0.0390400	0.0008800	-0.0052500	0.5319100	0.0444200
0.5946600	-0.0405900	0.0002200	-0.0024900	0.5523900	0.0432900
0.5732900	-0.0420100	0.0000000	0.0000000	0.5732900	0.0420100
0.5523900	-0.0432900	0.0002200	0.0024900	0.5946600	0.0405900
0.5319100	-0.0444200	0.0008800	0.0052500	0.6165100	0.0390400
0.5118100	-0.0454200	0.0022300	0.0082400	0.6388900	0.0373600
0.4920700	-0.0462700	0.0042900	0.0113400	0.6618300	0.0355400
0.4726400	-0.0469900	0.0071500	0.0145300	0.6853600	0.0336000
0.4535000	-0.0475800	0.0109200	0.0176500	0.7095300	0.0315300
0.4346200	-0.0480300	0.0156900	0.0205900	0.7343600	0.0293300
0.4159500	-0.0483600	0.0215400	0.0232000	0.7598900	0.0269900
0.3974600	-0.0485600	0.0285300	0.0254200	0.7861500	0.0245000
0.3791200	-0.0486500	0.0367100	0.0272800	0.8131800	0.0218300
0.3608900	-0.0486200	0.0461600	0.0289500	0.8410000	0.0189400
0.3427400	-0.0484900	0.0570100	0.0306400	0.8696600	0.0158000
0.3246400	-0.0482500	0.0694500	0.0324100	0.8991700	0.0123900
0.3065600	-0.0479100	0.0837000	0.0342400	0.9257500	0.0091800
0.2884400	-0.0474700	0.1000400	0.0361100	0.9466100	0.0066100
0.2702700	-0.0469300	0.1187600	0.0380000	0.9629700	0.0045800
0.2520100	-0.0462900	0.1386000	0.0397500	0.9758100	0.0029900
0.2336200	-0.0455400	0.1581300	0.0412500	0.9858900	0.0017500
0.2150700	-0.0446800	0.1773600	0.0425500	0.9938000	0.0007700

Table 8 Tip section of the Onera M6 wing

X	Y	X	Y	X	Y
1.4167939	0.0000000	0.9667440	-0.0244660	0.9667440	0.0244660
1.4133220	-0.0004310	0.9561210	-0.0238280	0.9772380	0.0250210
1.4088920	-0.0009800	0.9453520	-0.0231000	0.9876260	0.0255020
1.4032480	-0.0016740	0.9344150	-0.0222600	0.9979240	0.0259220
1.3960570	-0.0025650	0.9233050	-0.0212800	1.0081500	0.0262810
1.3868960	-0.0037020	0.9128220	-0.0202210	1.0183250	0.0265830
1.3752140	-0.0051410	0.9036710	-0.0191740	1.0284719	0.0268290
1.3603300	-0.0069380	0.8956920	-0.0181490	1.0385960	0.0270200
1.3438040	-0.0088480	0.8887250	-0.0171580	1.0487320	0.0271540
1.3277550	-0.0106060	0.8826490	-0.0162120	1.0588959	0.0272270
1.3121760	-0.0122250	0.8773570	-0.0152770	1.0691050	0.0272440
1.2970390	-0.0137200	0.8727770	-0.0142350	1.0793750	0.0271930
1.2823340	-0.0151140	0.8688620	-0.0129920	1.0897300	0.0270810
1.2680370	-0.0164250	0.8655860	-0.0115300	1.1001850	0.0268970
1.2541330	-0.0176570	0.8629150	-0.0098840	1.1107570	0.0266450
1.2405970	-0.0188160	0.8608040	-0.0081370	1.1214761	0.0263140
1.2274210	-0.0199020	0.8592020	-0.0063500	1.1323560	0.0259110
1.2145751	-0.0209210	0.8580490	-0.0046140	1.1434110	0.0254350
1.2020420	-0.0218620	0.8572930	-0.0029400	1.1546659	0.0248750
1.1898060	-0.0227300	0.8569230	-0.0013940	1.1661350	0.0242420
1.1778390	-0.0235250	0.8568000	0.0000000	1.1778390	0.0235250
1.1661350	-0.0242420	0.8569230	0.0013940	1.1898060	0.0227300
1.1546659	-0.0248750	0.8572930	0.0029400	1.2020420	0.0218620
1.1434110	-0.0254350	0.8580490	0.0046140	1.2145751	0.0209210
1.1323560	-0.0259110	0.8592020	0.0063500	1.2274210	0.0199020
1.1214761	-0.0263140	0.8608040	0.0081370	1.2405970	0.0188160
1.1107570	-0.0266450	0.8629150	0.0098840	1.2541330	0.0176570
1.1001850	-0.0268970	0.8655860	0.0115300	1.2680370	0.0164250
1.0897300	-0.0270810	0.8688620	0.0129920	1.2823340	0.0151140
1.0793750	-0.0271930	0.8727770	0.0142350	1.2970390	0.0137200
1.0691050	-0.0272440	0.8773570	0.0152770	1.3121760	0.0122250
1.0588959	-0.0272270	0.8826490	0.0162120	1.3277550	0.0106060
1.0487320	-0.0271540	0.8887250	0.0171580	1.3438040	0.0088480
1.0385960	-0.0270200	0.8956920	0.0181490	1.3603300	0.0069380
1.0284719	-0.0268290	0.9036710	0.0191740	1.3752140	0.0051410
1.0183250	-0.0265830	0.9128220	0.0202210	1.3868960	0.0037020
1.0081500	-0.0262810	0.9233050	0.0212800	1.3960570	0.0025650
0.9979240	-0.0259220	0.9344150	0.0222600	1.4032480	0.0016740
0.9876260	-0.0255020	0.9453520	0.0231000	1.4088920	0.0009800
0.9772380	-0.0250210	0.9561210	0.0238280	1.4133220	0.0004310

The root and tip wing section shapes coordinates are described in Tables 7 and 8 respectively. The shapes of intermediate wing sections are obtained by linear interpolation.

1.8 Problem TE7. Axisymmetric afterbody viscous drag minimization

1.8.1 Summary

The optimization of an axisymmetric afterbody under transonic flow conditions is considered. The cost function is defined as a global measure of the viscous drag along the solid walls while the thrust is kept to a desired level. This is accomplished through the specification of the exit Mach numbers of both the primary and secondary streams of the engine and the corresponding cross-sections. Computations will be performed using a weak viscous-inviscid interaction technique.

1.8.2 Definition of design problem

A typical axisymmetric bypass afterbody configuration is shown in fig. 3. Free stream (flight conditions) are specified to be $M_\infty = 0.8$ and altitude h=10.000 ft. The compression ratio $(p_{ts}/p_{t\infty})$ of the secondary stream is 1.5 while for the primary one $(p_{tp}/p_{t\infty})$ is set to 5.0. The stagnation temperature ratios are related to the stagnation pressure ratios through the isentropic relations. The Mach number at the primary and secondary throat sections is fixed to 0.9. To achieve a desirable thrust level, the inlet to throat cross sections ratio and Mach numbers are fixed as geometrical and flow constraints within the overall design procedure. In order to maintain the basic geometrical characteristics of the engine the locations of the starting points of sections 1 to 5 (which are sought to be optimized) as well as their arc lengths are kept fixed. The shape optimization of sections 1 to 5 is carried out with the objective of minimum viscous drag assuming fully turbulent flow conditions. The Reynolds number based on the nacelle diameter (4 m) is $4 * 10^6$. The drag coefficient of each individual section will be estimated with the Squire and Young Formula (see TE1).

1.8.3 Initial geometry

The initial geometry is shown in fig. 3 and in Tables 9 to 13.

1.8.4 Results and output formats

Geometry

- resulting afterbody shape: z versus R.

Mach number distribution

- The "inviscid" Mach number distribution along the solid walls of sections 1 to 5: M versus z.

Boundary layer

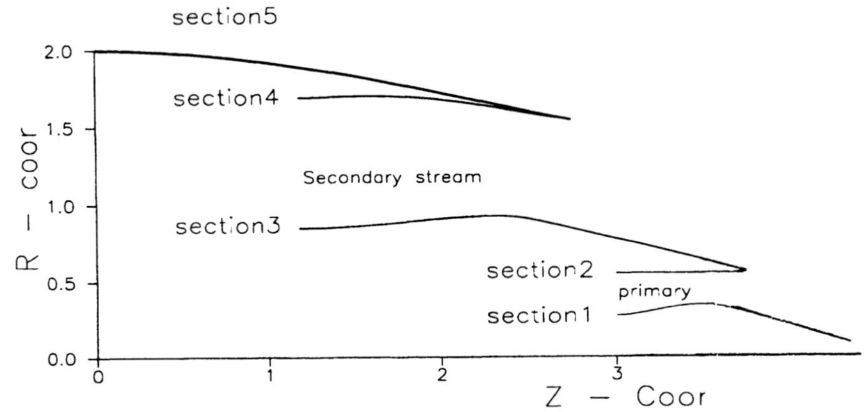

Figure 3 Afterbody configuration

- - H: form factor along the solid walls: H versus z.

Convergence

- Values of objective function normalized with its starting value (C_d/C_{do}) versus number of objective function evaluations.

1.9 Problem TE8. Laminar riblets

1.9.1 The geometry

To test the efficiency of riblets we shall consider the flow in a channel with approximate rectangular cross section. It is limited by two vertical parallel plane walls, two horizontal walls with riblets (the floor and the ceiling) and a fiction inflow and outflow plane section. The riblets are small and either smooth or shaped like a saw tooth. The Reynolds number is based on the horizontal velocity in the center of the channel, U on the inflow cross section (Re=UD/n ... (n=m/r)). Experimental and numerical data are available for the saw tooth riblets with the following values: Re=4200,L=pD,d=l=0.1135D The number of riblets, n, is large, the computation should not depend upon n, but if a value is required, we shall take nd=0.289pD

1.9.2 Boundary conditions

- At inflow boundary: $u = (0, 0, U)$, which is a constant velocity field entering the channel.
- At outflow boundary: The user decides. For example: $n\theta_\eta u - pn = 0$.
- On the vertical walls: $u.n = 0$ and $\theta_\eta uYn = 0$.

- On the riblets planes $u = 0$.

1.9.3 Test Case 8.1: Direct simulation

The flow is laminar without turbulence model and stationary. Periodicity and symmetry allows to reduce the computational domain to one riblet or even a half riblet, limited by two planes of symmetry further cut in half by the horizontal plane of symmetry so that only one riblet wall is considered. The user is free to choose his computational domain so long as his approximations are compatible with the problem. The Reynolds number Re=4200 corresponds to a riblet height of $y^+ = 20$ in wall units. The mesh should be tuned accordingly.

Output

- 1. wall shear stress and pressure on the riblet plane.
- 2. Velocity level plot for each 3 components at three cross section of the computational domain corresponding to 2=L/4, L/2, 3L/4.

1.9.4 Test Case 8.2: Optimization

The saw tooth riblet does not reduce the drag when the flow is laminar but it does when the flow is turbulent. However simulations require more than 200.000 mesh points and so optimization is beyond the reach of present numerical methods. Thus it is proposed to see whether there exists a laminar riblet. For this one must:

Minimize the drag at the riblet wall for a given cross section area. This is also beyond our possibility in 3D so we could optimize the drag per unit length at the cross section z=3L/4 for a given z-component of the velocity, u3, and then reset the riblet shape over the whole domain to the shape just found and iterate the process. This corresponds to 0. Take initial riblet shape to be a flat plate.

Loop

- 1. Solve 3D Navier-Stokes; compute $\theta_3 u_3$ and $u_3\theta_3\nu$ where $u = (u_1, u_2, u_3)$, $V = (u_1, u_2)$.
- 2. Solve in 2D at z=3L/4: min $\int_s |\theta_\eta u_3| ds$:
  $-\nu\Delta V + V.\Delta V + \Delta p = -u_3\theta u_3, ...\Delta.V = \theta_3 u_3$ for a given constant cross section area, where S is the riblet surface.
- 3. Update the entire riblet surface to the new shape found in step 2 and iterate.

Table 9 Coordinates of section 1

X	Y	X	Y	X	Y	X	Y
3.0000	0.2700	3.5101	0.3433	4.0110	0.2271	4.5109	0.1066
3.0138	0.2700	3.5230	0.3423	4.0235	0.2241	4.5234	0.1036
3.0268	0.2702	3.5358	0.3407	4.0360	0.2211	4.5359	0.1006
3.0397	0.2707	3.5485	0.3385	4.0485	0.2181	4.5484	0.0975
3.0526	0.2715	3.5611	0.3360	4.0610	0.2151	4.5609	0.0945
3.0655	0.2727	3.5737	0.3332	4.0735	0.2121	4.5734	0.0915
3.0783	0.2741	3.5862	0.3301	4.0860	0.2091	4.5859	0.0885
3.0911	0.2757	3.5986	0.3270	4.0985	0.2060	4.5984	0.0855
3.1039	0.2775	3.6111	0.3238	4.1110	0.2030	4.6109	0.0825
3.1167	0.2795	3.6236	0.3207	4.1235	0.2000	4.6234	0.0795
3.1294	0.2817	3.6360	0.3176	4.1360	0.1970	4.6359	0.0764
3.1420	0.2841	3.6485	0.3146	4.1485	0.1940	4.6484	0.0734
3.1547	0.2866	3.6610	0.3115	4.1610	0.1910	4.6609	0.0704
3.1672	0.2893	3.6735	0.3085	4.1735	0.1880	4.6734	0.0674
3.1798	0.2921	3.6860	0.3055	4.1860	0.1850	4.6859	0.0644
3.1923	0.2950	3.6985	0.3025	4.1985	0.1819	4.6984	0.0614
3.2049	0.2979	3.7110	0.2995	4.2110	0.1789	4.7109	0.0584
3.2174	0.3007	3.7235	0.2965	4.2235	0.1759	4.7234	0.0553
3.2300	0.3036	3.7360	0.2935	4.2360	0.1729	4.7359	0.0523
3.2425	0.3064	3.7485	0.2904	4.2485	0.1699	4.7484	0.0493
3.2551	0.3092	3.7610	0.2874	4.2610	0.1669	4.7609	0.0463
3.2676	0.3121	3.7735	0.2844	4.2735	0.1639	4.7734	0.0433
3.2802	0.3149	3.7860	0.2814	4.2860	0.1608	4.7859	0.0403
3.2927	0.3177	3.7985	0.2784	4.2985	0.1578	4.7984	0.0373
3.3053	0.3204	3.8110	0.2754	4.3110	0.1548	4.8109	0.0342
3.3179	0.3231	3.8235	0.2724	4.3235	0.1518	4.8234	0.0312
3.3305	0.3256	3.8360	0.2693	4.3360	0.1488	4.8359	0.0282
3.3432	0.3281	3.8485	0.2663	4.3485	0.1458	4.8484	0.0252
3.3559	0.3304	3.8610	0.2633	4.3610	0.1428	4.8609	0.0222
3.3686	0.3326	3.8735	0.2603	4.3735	0.1397	4.8734	0.0192
3.3813	0.3346	3.8860	0.2573	4.3860	0.1367	4.8859	0.0162
3.3941	0.3365	3.8985	0.2543	4.3984	0.1337	4.8984	0.0132
3.4069	0.3383	3.9110	0.2513	4.4110	0.1307	4.9109	0.0101
3.4197	0.3398	3.9235	0.2482	4.4234	0.1277	4.9234	0.0071
3.4326	0.3411	3.9360	0.2452	4.4359	0.1247	4.9359	0.0041
3.4455	0.3421	3.9485	0.2422	4.4485	0.1217	4.9484	0.0011
3.4584	0.3429	3.9610	0.2392	4.4609	0.1186	4.9612	0.0000
3.4713	0.3435	3.9735	0.2362	4.4735	0.1156	4.9741	0.0000
3.4842	0.3438	3.9860	0.2332	4.4859	0.1126	4.9871	0.0000
3.4972	0.3438	3.9985	0.2302	4.4984	0.1096	5.0000	0.0000

Table 10 Coordinates of section 2

X	Y	X	Y	X	Y
3.0000000	0.5500000	3.2813900	0.5500000	3.5646701	0.5500000
3.0086200	0.5500000	3.2918799	0.5500000	3.5751400	0.5500000
3.0191300	0.5500000	3.3023601	0.5500000	3.5855801	0.5500000
3.0296199	0.5500000	3.3128500	0.5500000	3.5960100	0.5500000
3.0401199	0.5500000	3.3233399	0.5500000	3.6064501	0.5500000
3.0506101	0.5500000	3.3338301	0.5500000	3.6168699	0.5500000
3.0611100	0.5500000	3.3443201	0.5500000	3.6272700	0.5500000
3.0716100	0.5500000	3.3548100	0.5500000	3.6376500	0.5500000
3.0820999	0.5500000	3.3653200	0.5500000	3.6480100	0.5500000
3.0925901	0.5500000	3.3758099	0.5500000	3.6583600	0.5500000
3.1031001	0.5500000	3.3863101	0.5500000	3.6686699	0.5500000
3.1135900	0.5500000	3.3968000	0.5500000	3.6789701	0.5500000
3.1240799	0.5500000	3.4073000	0.5500000	3.6892400	0.5500000
3.1345699	0.5500000	3.4177899	0.5500000	3.6995101	0.5500000
3.1450601	0.5500000	3.4282899	0.5500000	3.7097800	0.5500000
3.1555500	0.5500000	3.4388001	0.5500000	3.7200501	0.5500000
3.1660399	0.5500000	3.4492900	0.5500000	3.7303200	0.5500000
3.1765299	0.5500000	3.4597900	0.5500000	3.7405901	0.5500000
3.1870000	0.5500000	3.4702799	0.5500000	3.7508600	0.5500000
3.1974900	0.5500000	3.4807799	0.5500000	3.7611201	0.5500000
3.2079799	0.5500000	3.4912801	0.5500000	3.7713900	0.5500000
3.2184601	0.5500000	3.5017700	0.5500000	3.7816601	0.5500000
3.2289500	0.5500000	3.5122600	0.5500000	3.7919300	0.5500000
3.2394400	0.5500000	3.5227499	0.5500000	3.8022001	0.5500000
3.2499199	0.5500000	3.5332401	0.5500000	3.8124700	0.5500000
3.2604101	0.5500000	3.5437300	0.5500000	3.8227401	0.5500000
3.2709000	0.5500000	3.5542099	0.5500000	3.8299999	0.5500000

Table 11 Coordinates of section 3

X	Y	X	Y	X	Y
1.1799999	0.8500000	2.0900900	0.9148300	2.9813600	0.7805360
1.2105401	0.8500850	2.1167600	0.9167300	3.0071700	0.7735500
1.2372800	0.8503160	2.1434400	0.9184830	3.0329800	0.7665640
1.2640100	0.8506720	2.1701300	0.9200960	3.0587800	0.7595780
1.2907400	0.8511820	2.1968300	0.9215520	3.0845900	0.7525920
1.3174700	0.8518920	2.2235301	0.9227940	3.1104000	0.7456070
1.3441900	0.8528160	2.2502501	0.9237320	3.1362100	0.7386210
1.3709000	0.8539290	2.2769799	0.9243090	3.1620100	0.7316350
1.3976099	0.8551890	2.3037200	0.9245260	3.1878200	0.7246490
1.4243100	0.8565620	2.3304501	0.9243390	3.2136300	0.7176630
1.4510100	0.8580350	2.3571801	0.9236180	3.2394400	0.7106770
1.4777000	0.8596130	2.3838799	0.9222210	3.2652400	0.7036910
1.5043800	0.8613070	2.4105201	0.9200520	3.2910500	0.6967050
1.5310500	0.8631230	2.4370899	0.9170660	3.3168600	0.6897190
1.5577199	0.8650560	2.4635601	0.9132720	3.3426700	0.6827330
1.5843800	0.8670960	2.4899001	0.9087200	3.3684700	0.6757470
1.6110300	0.8692280	2.5161200	0.9034930	3.3942800	0.6687610
1.6376700	0.8714390	2.5422201	0.8977030	3.4200900	0.6617750
1.6643100	0.8737190	2.5682199	0.8914730	3.4459000	0.6547890
1.6909400	0.8760650	2.5941401	0.8849280	3.4717000	0.6478030
1.7175699	0.8784790	2.6200099	0.8781760	3.4975100	0.6408170
1.7441900	0.8809760	2.6458499	0.8712990	3.5233200	0.6338310
1.7708000	0.8835680	2.6716700	0.8643550	3.5491199	0.6268450
1.7974000	0.8862650	2.6974800	0.8573810	3.5749300	0.6198590
1.8239900	0.8890560	2.7232900	0.8503960	3.6007400	0.6128740
1.8505700	0.8919130	2.7491000	0.8434100	3.6265500	0.6058880
1.8771600	0.8947920	2.7749000	0.8364240	3.6523499	0.5989020
1.9037400	0.8976470	2.8007100	0.8294380	3.6781600	0.5919160
1.9303300	0.9004390	2.8265200	0.8224520	3.7039700	0.5849300
1.9569300	0.9031430	2.8523200	0.8154660	3.7297800	0.5779440
1.9835401	0.9057410	2.8781300	0.8084800	3.7555799	0.5709580
2.0101600	0.9082220	2.9039400	0.8014940	3.7813900	0.5639720
2.0367899	0.9105710	2.9297500	0.7945080	3.8072000	0.5569860
2.0634401	0.9127770	2.9555500	0.7875220	3.8299999	0.5500000

Table 12 Coordinates of section 4

X	Y	X	Y	X	Y
1.1799999	1.6911401	1.7184600	1.6993200	2.2500200	1.6383400
1.1995400	1.6912200	1.7341800	1.6986200	2.2655301	1.6354700
1.2152700	1.6914400	1.7498900	1.6978800	2.2810400	1.6325800
1.2309901	1.6918000	1.7656100	1.6970600	2.2965500	1.6296500
1.2467200	1.6922899	1.7813200	1.6961900	2.3120501	1.6266900
1.2624400	1.6928700	1.7970200	1.6952699	2.3275399	1.6237200
1.2781600	1.6935300	1.8127300	1.6942900	2.3430300	1.6207300
1.2938700	1.6942300	1.8284301	1.6932400	2.3585300	1.6177300
1.3095900	1.6949500	1.8441300	1.6921400	2.3740201	1.6147300
1.3253100	1.6956700	1.8598200	1.6909699	2.3895099	1.6117100
1.3410200	1.6963700	1.8755100	1.6897200	2.4050000	1.6087101
1.3567400	1.6970299	1.8911999	1.6883800	2.4204900	1.6057100
1.3724600	1.6976200	1.9068700	1.6869600	2.4359801	1.6027000
1.3881900	1.6981601	1.9225399	1.6854300	2.4514699	1.5997000
1.4039100	1.6986400	1.9382000	1.6838300	2.4669600	1.5967000
1.4196399	1.6990700	1.9538600	1.6821100	2.4824600	1.5936800
1.4353600	1.6994700	1.9694999	1.6803100	2.4979501	1.5906800
1.4510900	1.6998399	1.9851400	1.6784300	2.5134399	1.5876800
1.4668200	1.7002000	2.0007701	1.6764801	2.5289299	1.5846699
1.4825400	1.7005100	2.0163901	1.6744800	2.5444200	1.5816700
1.4982700	1.7008200	2.0320101	1.6724200	2.5599101	1.5786700
1.5140001	1.7010900	2.0476301	1.6703300	2.5754001	1.5756600
1.5297300	1.7013100	2.0632401	1.6682000	2.5908899	1.5726500
1.5454600	1.7014800	2.0788400	1.6660300	2.6063900	1.5696501
1.5611900	1.7016000	2.0944400	1.6638300	2.6218801	1.5666500
1.5769200	1.7016799	2.1100399	1.6615601	2.6373701	1.5636400
1.5926501	1.7017000	2.1256299	1.6592500	2.6528599	1.5606400
1.6083800	1.7016799	2.1412101	1.6568800	2.6683500	1.5576400
1.6241100	1.7016000	2.1567800	1.6544300	2.6838400	1.5546300
1.6398400	1.7014500	2.1723499	1.6519099	2.6993301	1.5516200
1.6555700	1.7012200	2.1879001	1.6493200	2.7148199	1.5486200
1.6712900	1.7008801	2.2034400	1.6466600	2.7303200	1.5456100
1.6870199	1.7004499	2.2189801	1.6439400	2.7500000	1.5420001
1.7027400	1.6999201	2.2344999	1.6411700		

Table 13 Coordinates of section 5

X	Y	X	Y	X	Y
0.0000000	2.0000000	0.5144720	1.9754900	1.6490300	1.8000500
0.0115980	1.9998699	0.5429860	1.9730800	1.6772200	1.7939600
0.0231959	1.9997000	0.5714940	1.9705499	1.7054000	1.7877899
0.0347945	1.9995199	0.5999960	1.9679300	1.7335800	1.7816100
0.0463917	1.9993100	0.6284930	1.9651901	1.7617500	1.7753600
0.0579891	1.9990700	0.6569820	1.9623400	1.7899200	1.7690800
0.0695865	1.9988400	0.6854650	1.9593700	1.8180799	1.7627600
0.0811837	1.9985600	0.7139410	1.9562900	1.8462400	1.7563900
0.0927804	1.9982899	0.7424100	1.9530900	1.8743900	1.7499700
0.1043770	1.9979800	0.7708710	1.9498100	1.9025400	1.7435300
0.1159730	1.9976600	0.7993250	1.9463900	1.9306800	1.7370400
0.1275690	1.9973100	0.8277710	1.9428600	1.9588100	1.7304900
0.1391650	1.9969600	0.8562080	1.9392200	1.9869400	1.7238899
0.1507600	1.9965900	0.8846380	1.9354900	2.0150599	1.7172800
0.1623550	1.9962000	0.9130600	1.9316300	2.0431800	1.7106301
0.1739500	1.9958000	0.9414720	1.9276700	2.0713000	1.7039800
0.1855440	1.9953600	0.9698760	1.9236200	2.0994201	1.6973500
0.1971380	1.9949200	0.9982710	1.9194400	2.1275401	1.6907099
0.2087310	1.9944500	1.0266600	1.9151800	2.1556599	1.6840700
0.2203240	1.9939700	1.0550400	1.9108400	2.1837800	1.6774100
0.2319170	1.9934601	1.0834100	1.9064200	2.2118900	1.6707500
0.2435090	1.9929399	1.1117800	1.9019001	2.2400100	1.6641001
0.2551000	1.9924001	1.1401300	1.8973100	2.2681301	1.6574200
0.2666910	1.9918400	1.1684800	1.8926200	2.2962401	1.6507500
0.2782820	1.9912601	1.1968200	1.8878400	2.3243501	1.6440600
0.2898710	1.9906600	1.2251500	1.8829600	2.3524699	1.6373800
0.3014610	1.9900500	1.2534699	1.8780000	2.3805799	1.6306700
0.3130490	1.9894000	1.2817800	1.8729399	2.4086900	1.6239700
0.3246370	1.9887500	1.3100801	1.8678000	2.4368000	1.6172500
0.3362250	1.9880800	1.3383800	1.8625799	2.4649100	1.6105300
0.3478120	1.9873800	1.3666600	1.8572900	2.4930201	1.6038001
0.3593980	1.9866700	1.3949400	1.8519200	2.5211201	1.5970600
0.3709830	1.9859400	1.4232100	1.8464500	2.5492301	1.5902900
0.3825680	1.9852000	1.4514700	1.8409200	2.5773301	1.5835299
0.3941530	1.9844400	1.4797200	1.8353100	2.6054299	1.5767400
0.4057360	1.9836600	1.5079600	1.8296100	2.6335199	1.5699400
0.4173190	1.9828600	1.5361900	1.8238300	2.6616199	1.5631000
0.4289010	1.9820499	1.5644100	1.8179600	2.6896999	1.5562201
0.4574300	1.9799700	1.5926200	1.8120500	2.7177801	1.5492800
0.4859540	1.9777900	1.6208301	1.8060800	2.7500000	1.5420001

2 ECARP data base: access rules and storage formats

2.1 Introduction

This document gives a presentation of the **ECARP** data base. It is divided into three parts. First, the data base hierarchy is described making a clear distinction between the public part of the data base and the incoming part where the contributions will be dropped. The second part details the two data storage formats that are required for a subsequent treatment by the `VIGIE` graphic visualisation tool during the workshop. This document ends with the description of the typical contribution scheme.

Each potential contributor has its own way of storing data depending on its numerical scheme and its computer facilities, so unified storage formats have to be defined. This unified procedure copes with (we hope) all possible kind of data storage : multi-block mixed structured/unstructured meshes. To make the data base a success, it is necessary that each contributor makes the effort to write an interface code which translates its output formats into the required formats.

2.2 Data base overview

The **ECARP** data base can be viewed as a set of informations related to the problems studied in the **BRITE EURAM ECARP** project; more precisely, the data base will be used to store solutions (in a sense that will be define in the next sections) to the test cases selected for the two workshops of interest to the **ECARP** project : Optimum Design and MPP (Massively Parallel Processing) for Navier-Stokes flows.

Two kinds of information are contained in the **ECARP** data base (see Fig. 1) :

- public informations : this guide, two documents for the technical description of the test cases (one for each workshop, including the description of the relevant quantities that will define a solution) and input data (geometry definitions, meshes). Note that in Fig. 1 the complete structure of the **public/data** directory has not been detailed; in particular there exist additional sub-directories that contain the data related to the selected Optimum Design and MPP for Navier-Stokes flows test cases;

- private informations : report, binary and ASCII files relatives to the solutions.

Concerning the Optimum Design workshop, we recall that the potential contributions include contributions to the **ECARP** project and, in some cases, contributions to the **ECARP TOP-UP** project. Table 1 summarizes the splitting of contributions; I1,V1 (resp. I2, V2) correspond to contributions to the Inviscid and Viscous test cases of the **ECARP** (resp. **ECARP TOP-UP**) project.

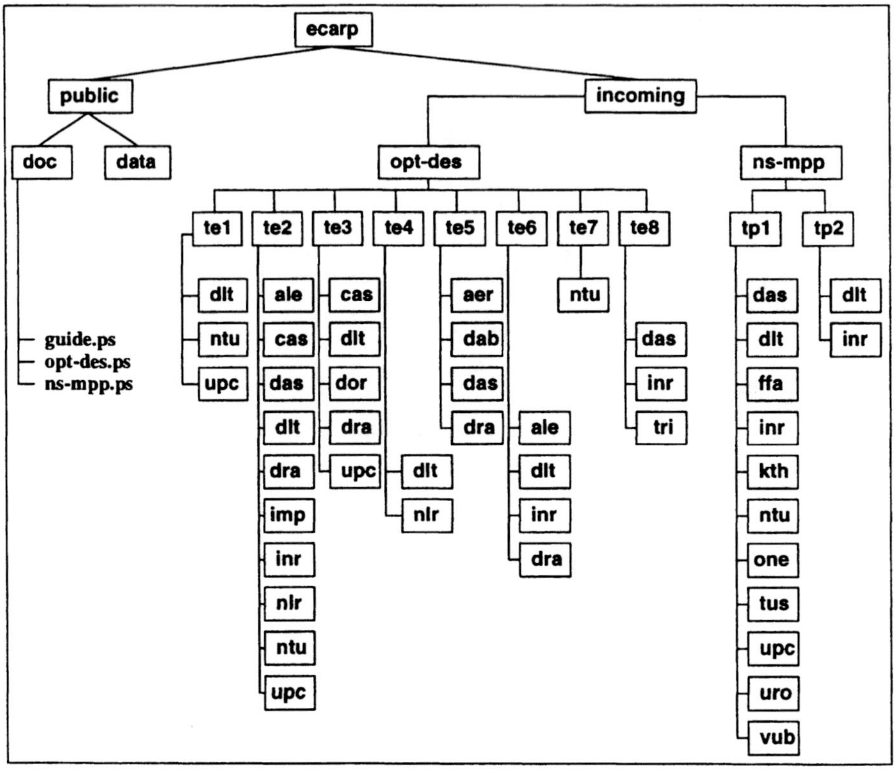

Figure 1 **ECARP** data base structure

Table 1 Contributions to the **ECARP** and **ECARP TOP-UP** projects

	te1	**te2**	**te3**	**te4**	**te5**	**te6**	**te7**	**te8**
aer					I1			
ale		V2				I1V2		
cas		V2	I2V2					
dab					I1V2			
das		V2			I1V2			
dlt	V2	V1	I1V2	I1V2		I1		V1
dor			I1V2					
dra		V2	I2		I2	I2V2		
imp		V1						
inr		V2(2)				I1		V1
nlr		V2		I1V2				
ntu	V2	V1					V1	
tri								V2
upc	V1	V1	I1					

2.3 Solution storage formats

2.3.1 Binary or not?

In order to reduce the amount of data stored in the data base and to be transferred over networks, the following standard has been chosen:

standard IEEE 32 bits binary format

This standard corresponds to the one used on SUN and SILICON GRAPHICS workstations. Fortran translation routines are available on most other computers. Contributors should check on their computer systems the appropriate routines. Two examples are given here:

- on CONVEX computers, native binary −− > IEEE binary, use "rcvtir", IEEE −− > native, use "ircvtir". Check the manual;
- on CRAY computers, there exist at least to ways of doing this :
  1. native binary −− > standard IEEE, use "cray2ieg", IEEE−− > native, use "ieg2cray". Notice that Cray simple precision is 64 bits, so that real numbers lowers than 10^{-32} or bigger than 10^{32} must be filtered to prevent errors. Check the manual (man ieg2cray);
  2. Use of the "assign" Cray function. This solution is more simple than the previous one and does not need any modification of the program that creates the files to be stored. We give an example : assume that the executable *a.out* creates (or reads) a binary sequential file named fort.10 by the statement

     ```
     write(10) (x(i),i=1,N) (or read(10) (x(i)i=1,N))
     ```

     then one can automatically write/read the file fort.10 in IEEE format in a batch job with the following instructions :

     ```
     assign -Ff77 -Nieee u:10
     a.out
     ```

2.3.2 General write format

Many different numerical methods are in use for the simulations under consideration : finite volume, finite difference, finite element, using structured or unstructured meshes, cell centered/point vertex, one or several element types. It has been attempted to propose a format which accounts for all possible methods.

> **Important note** : in all what follows, it is assumed that the stored flow variables are located at the nodal points. Contributors that use cell centered schemes must store the coordinates of the cell nodes.

The following binary storage format is required for subsequent isolines and isosurfaces visualisation:

```
c
c -----------------------------------------------------------------
c NOTE : in this example, there are only two variables (named q
c        and qfr hereafter); q contains the set of variable in
c        the whole field, it has nnu unknowns for each grid point.
c        The variable qfr contains other variable on a specific
c        surface S of the mesh (inflow, outflow, wall, $\cdots$).
c        It contains nnufr variables to be stored at the grid
c        points of the surface S.
c        The unit number(s) below (unit1 and unit2) are arbitrary
c        and to be selected by the participant.
c -----------------------------------------------------------------
c
c     For structured meshes
c
      REAL coor1(ndimax,idimax,jdimmax,kdimax)
      REAL q1(idimax,jdimax,kdimax,nnumax)
      REAL q1fr(idifrmax,jdifrmax,kdifrmax,nnufrmax)
c
c     For unstructured meshes
c
      REAL coor2(ndimax,nnodesmax)
      REAL q2(nnodesmax,nnumax)
      REAL q2fr(6,nnufrmax,nfacmax)
c
      REAL curv(npointmax,ncurvmax)
      INTEGER nnperlement(6), ietyp(nelemax), iconnec(6*nelemax)
      INTEGER iconnecp(nelemax), ibegfr(npatchmax),
              iendfr(npatchmax)
      INTEGER jbegfr(npatchmax), jendfr(npatchmax),
      INTEGER kbegfr(npatchmax), kendfr(npatchmax)
      CHARACTER*30 name(100)
      INTEGER npoint(npointmax)
c
c -----------------------------------------------------------------
c ndimax                 : space dimension (1,2 or 3)
c idimax,jdimax,kdimax : maximum of indices in the i,j,k
c                          direction, structured meshes
c idifrmax,jdifrmax,
c kdifrmax               : maximum of indices in the ifr,jfr,kfr
c                          direction, structured mesh for S
c nnodesmax              : maximum number of nodes, unstructured
c                          meshes
c nnodesfrmax            : maximum number of nodes on S,
c                          unstructured meshes
```

```
c nelemax              : maximum number of element, ustructured
c                        meshes
c nnumax,
c nnufrmax             : maximum number of variables
c nfacmax              : maximum number of elements on S
c npatchmax            : maximum number of patches (structured
c                        meshes)
c coor1,coor2          : x,y,z coordinates
c q1,q2,q1fr,q2fr      : array of variables
c nnperlement          : number of nodes per element
c ietyp                : define for a given element its type
c iconnec              : connectivity table
c iconnecp(nelemax)    : connectivity table pointer
c iconnecpfac(nfacmax) : connectivity table pointer for the
c                        frontier
c nufac(6,nfacmax)     : connectivity table for the frontier
c logfac(nfacmax)      : logic type of the frontier elements
c name                 : name of the variables in file unit2
c curv                 : values of variables in file unit2
c npoint               : number of value on the curves curv(*,*)
c ibegfr,...kendfr     : extreme indices of surfaces where surface
c                        values are asked for.
c
c note : the dimension 6*nelemax for iconnec has to be related to
c the size of nnperlement.
c -----------------------------------------------------------------
c
      OPEN (unit1, FILE='filename', FORM='UNFORMATTED')
c
c     Read the number of blocks.
c     Note that blocks should be either structured
c     or unstructured, but one database file can contain both
c     structured and unstructured blocks.
c
      READ(unit1) nblock
c
      DO 103 nb=1,nblock
c
c        Read the type of block, and number of unknowns stored
c        for this block
c
c        itype = 0: Unstructured mesh
c                1: Structured mesh
c        nnu   = number of unknowns stored
c        nnufr = number of unknown stored on the frontier
```

```
c
        READ(unit1) itype, nnu, nnufr, npatch
c
        IF (itype .EQ. 1) THEN
c
c          Case of a structured mesh
c
c          idim,jdim,kdim : indexes for the nodes
c          ndim           : dimension of the problem
c          ibegfr,jbegfr,kbegfr : first indexes of the nodes of
c                                 the frontier
c          iendfr,jendfr,kendfr  : last indexes of the nodes of
c                                  the frontier
c          npatch : number of frontier per block

c
           READ(unit1) idim, jdim, kdim, ndim
           READ(unit1) ((((coor1(l,i,j,k), l=1,ndim),
     &                                 i=1,idim), j=1,jdim),
     &                                            k=1,kdim)
           READ(unit1) ((((q(i,j,k,nu), i=1,idim),
     &                               j=1,jdim), k=1,kdim),
     &                                          nu=1,nnu)
           DO 2 np=1,npatch
              READ(unit1) ibegfr(np), iendfr(np), jbegfr(np),
     &                    jendfr(np), kbegfr(np), kendfr(np)
              READ(unit1) ((((q1fr(i,j,k,nu),
     6                          i=ibegfr(np),iendfr(np)
     &                          j=jbegfr(np),jendfr(np)),
     &                          k=kbegfr(np),kendfr(np)),
     &                          nu=1,nnufr)
 2         CONTINUE
c
        ELSE
c
c          Case of an unstructured mesh
c
c          nelemtypes : number of element types
c          nelem      : number of elements
c          nnodes     : number of nodes
c          nfac       : number of facets on the frontier
c          nnodesfr   : number of nodes on the frontier where
c                       data is stored they correspond to type
c                       of frontiers that are given by the next
c                       paragraph, problem per problem
```

```
c
         READ(unit1) nnodes, nelem, nelemtypes, nfac, nnodesfr,
                     ndim
         READ(unit1) ((coor(l,i),l=1,ndim), i=1,nnodes)
         READ(unit1) ((q(i,nu),i=1,nnodes), nu=1,nnu)
c
c        The following element names implicitly define the
c        element characteristics :
c
c        HE8    8-noded (linear) 3D hex-cube cell
c        TE4    4-noded (linear) 3D tetrahedral cell
c        T3     3-noded triangular 2D cell
c        Q4     4-noded quadrilateral 2D cell
c        P5     5-noded pyramid element
c        P6     6-noded prismatic element
c
c        itypel     : element type number in table
c        (order is  1=HE8, 2=TE4, 3=T3, 4=Q4, 5=P5, 6=P6)
c        Note : the ordering of the nodes are defined
c        implicitely, so that the faces of the element itypel
c        are known at priory (see the Fig. \label{oreint}
c        below)
c
         nnperelement(1) = 8
         nnperelement(2) = 4
         nnperelement(3) = 3
         nnperelement(4) = 4
         nnperelement(5) = 5
         nnperelement(6) = 6
c
c        Read the connectivity table
c
         iptr = 0
c
         DO 203 iel=1,nelem
c
            READ(unit1) itypel
c
            ietype(iel)   = itypel
            iconnecp(iel) = iptr + 1
c
            READ(unit1) (iconnec(iptr+i),
                         i=1,nnperelement(itypel)
c
            iptr = iptr + nnperelement(itypel)
```

```
c
203         CONTINUE
c
c           Logic of facet
c
c           logfac = 1 : entrance facet
c           logfac =-1 : symmetry facet
c           logfac = 2 : outflow facet
c           logfac = 3 : freestream facet
c           logfac = 4 : wall nodes (no slip condition)
c           logfac = 5 : wall nodes (velocity=0)
c
            iptr = 0
c
            DO 204 ifac=1,nfac
c
               READ(unit1) itypel
c
               ietypfac(ifac)    = itypel
               iconnecpfac(ifac) = iptr + 1
c
               READ(unit1) logfac(ifac)
               READ(unit1) ((nufac(l,fac),
                         l=1,nnperelement(itypel))

204         CONTINUE
c
c           Values on frontiers
c
            DO 205 ifac=1,nfac
c
              itypel = ietypfac(ifac)
c
              READ(unit1) ((q2fr(l,nn,ifac),
     &                    l=1,nnperelement(itypel)),nn=1,nnufr)
c
205         CONTINUE

         ENDIF
c
103   CONTINUE
c
      CLOSE(unit1)
c
```

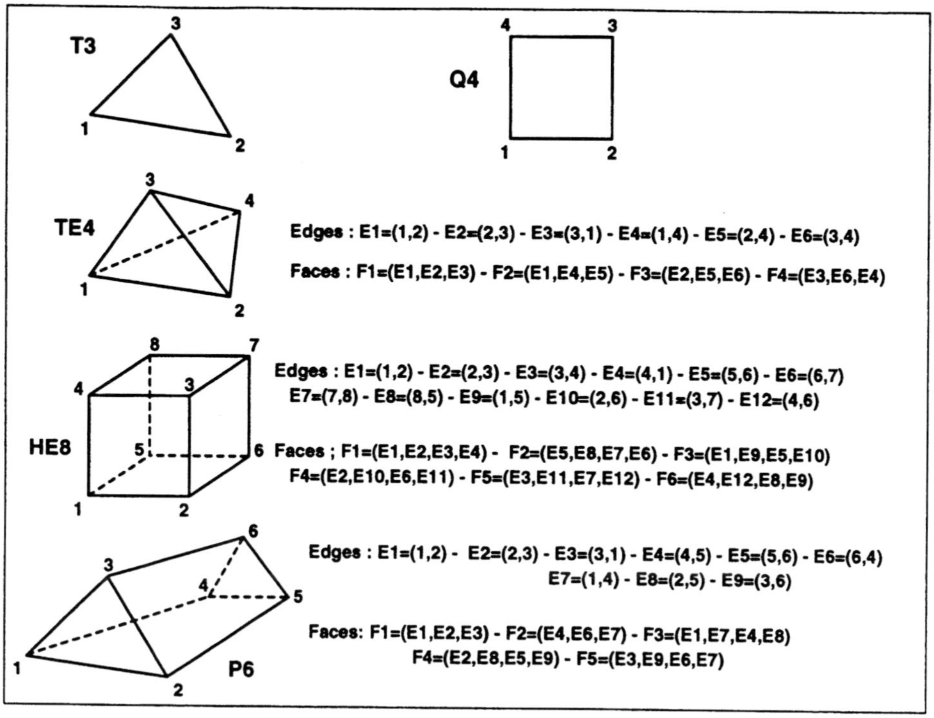

Figure 2 The elements : nodes, egdes and faces

2.3.3 One-dimensional curves write format

An ascii storage format will be used for one-dimensional curves associated to a function `s=f(t)`. We shall make the distinction between the three following situations :

S1 : `t` is a coordinate and `s` is associated to another coordinate (i.e. geometry shape `y=f(x)`) or to some physical coefficient (pressure coefficient, friction coefficient, $\cdots$);

S2 : `t` is a work unit (non-linear iteration, optimisation iteration, $\cdots$) and `s` denotes some residual;

S3 : `t` is a number of processing units and `s` denotes some function related to parallel efficiency (i.e. speed-up, scalability, $\cdots$).

The following is a simple example of the storage format that will be used. The first data block defines the range of values of the argument `t` (for instance, X-coordinates). This data block immediately follows a line containing the keyword **points** and the number of stored values of `t`. We then have several data blocks, each beginning with a line that contains the keyword **value** and the identifier of the function `s`. Each of these blocks gives the set of `s` values asoociated to the `t` values given in the first block.

```
# Every line beginning with a '#' is a comment line
points 4
-12.0
-6.0
 0.0
 1.0
value alpha
 12.0
 3.0
 11.0
-4.0
value beta
 5.0
-2.0
 3.0
12.0
```

Contributors should take care to the following points :

- the values of `t` must be given in increasing order;
- contributions should consist of different files, one for each of the three situations **S1** to **S3** depicted above. We encourage the use of comment lines in order to describe the content of the files;
- predefined identifiers must be used to be associated to the keyword **value**:

S1 :

y-shape :	`y=f(x)`, geometry shape for two-dimensional geometries.
y-shape-*loc* :	`y=f(x)`, geometry shape for three-dimensional geometries *loc* indicates the spanwise location (0, 20, 40, 60, 80, 100).
h-fact :	H=`f(x)`, boundary layer form factor.
cp-coef :	C_p=`f(x)`, pressure coefficient for two-dimensional geometries.
cp-coef-*loc* :	C_p=`f(x)`, pressure coefficient for three-dimensional geometries *loc* indicates the spanwise location (0, 20, 40, 60, 80, 100).
cf-coef :	C_f=`f(x)`, skin-friction coefficient for two-dimensional geometries.

S2 :

enrg-res :	convergence history for the energy equation versus the number of non-linear iterations.
obj-fct :	value of objective function versus number of objective function evaluations.
L2-fct :	value of $L_2(y)$ versus number of objective function evaluations.
drag-fct :	value of drag versus number of objective function evaluations.
lift-fct :	value of lift versus number of objective function evaluations.
pitch-fct :	value of pitch versus number of objective function evaluations.

S3 :

cpu-proc :	execution time versus the number of processing units.
com-proc :	communication time versus the number of processing units.
sp-proc :	speed-up versus the number of processing units.

2.4 A typical contribution scheme

2.4.1 Convention on file names

General format : `gensol.data`

1D curves format : we shall make here the distinction between each of the three situations **S1** to **S3** :

S1 : `1D-coor.data`

S2 : `1D-unit.data`

S3 : `1D-proc.data`

In addition, if 1D curves have to be given at several sections/stations of a test geometry then this should be done using several files i.e. 1D-coor*loc*.data where *loc* indicates the section/station location (0, 20, 40, 60, 80, 100).

Reports : for any technical report accompanying a contribution we propose to name the corresponding files as :

Ascii : `inst-tstc.ascii`

LaTeX : `inst-tstc.ltx`

PostScript : `inst-tstc.ps`

where `inst` is a three-letter prefix identifying the contributing institution (see Table 2) and `tstc` is one of **te1** to **te8** and **tp1** to **tp2**.

Table 2 Abbreviated names for the contributing institutions

aer	Aerospatiale
ale	Alenia
cas	CASA
dab or das	Daimler Benz Aerospace Airbus
dlt	Dassault
dor	Dornier
dra	DRA
ffa	FFA
imp	Imperial College
inr	INRIA
kth	KTH
nlr	NLR
ntu	NTUA-LTT
one	ONERA
tri	Tritech
tus	TU Stuttgart
uro	University of Roma
upc	UPC of Barcelona
vub	VUB of Brussels

2.4.2 Storing a contribution

Access to the data-base will be done by **ftp** on the serveur **thetis.inria.fr** (SUN4/SS10 workstation). In order to do so please get in touch with Stephane Lanteri (E-mail : lanteri@sophia.inria.fr, Tel : 33-93-65-77-34) in order to obtain the login and passwd informations. Note that the default directory is **incoming** and that you will have to change explicitely to the other sub-directories as well as to the **public** directory (refer to Fig. 1 in doing so).

Once you are connected on the serveur they you can retrieve informations from the **public** directory while your contribution should be put under the **incoming** directory. In the latter case please note that :

- it is essential that you point into the correct sub-directory,
- the transfer should take place using the binary tansfer mode,
- as soon as your file has been deposited in the targeted sub-directory it will be automatically renamed so that you won't be able to retrieve it (nevertheless, you will be able to verify the correctness of the transfer by checking the size of the transferred files),
- you have to inform your coordinating manager that you have contributed so that your contribution can be checked and validated.

3 ECARP data base: visualization tools

In numerous domains of science and computational engineering (e.g. computational fluid dynamics, electromagnetics, optimization, etc) workshops are organized around well-specified test problems proposed to experimentation and/or modelling and/or computation with the twofold purpose of assessing the knowledge of the physical solutions to these problems, as well as the state-of-the-art in the numerical techniques in current use in the specialized areas of concern (e.g. Navier-Stokes solvers for direct simulation of turbulence).

Associated with the organization of such workshops, it appeared natural to develop a simple data base computer environment to serve several purposes. One, to facilitate the dissemination of information concerning the proposed test problems, test-case precise technical definitions, output requirements and available technical complements (computer package for models and grids in particular) can be accessed from specialized servers. In this way the environment may facilitate, and thus encourage a larger participation. Second, contributions (mostly computations) are digitalized according to specific formats permitting the corresponding solutions to be visualized interactively during the workshop and compared, thus enhancing the efficiency of the workshop development. Although this requirement necessitates a larger initial effort from contributors, as a result, the comparisons and the syntheses of the workshop are highly facilitated and made immediate and quantitative. Third, the conpendium of these solutions constitute an evolutive archive of the state-of-the-art in a specialized area of advanced engineering or science. Fourth, when dealing with areas of technology that are not extensively taught in universities (e.g. hypersonics), the database may serve as an educational tool offering young researchers or engineers an access to reference solutions to generic problems representative of critical issues.

Typically, a contribution to the database is made of a small abstract providing a basic description of the employed method of investigation. For a computational solution this implies some details on the approximation method (Finite-Difference, -Volume, -Element scheme or other, central/upwind, structured/unstructured mesh, single/multi-block data, etc...) and solution method (implicit, multigrid, preconditioners, etc). The grid is defined by the list of node coordinates, and the basic properties (density, momentum components and energy) at nodal points are stored. A rather general graphics software ("VIGIE") currently under continuing development at Sophia Antipolis, permits to visualize iso-value contours (or domains), plots in adjustable cross-sections, local enlargements (zooms), etc. One main originality of this software is to permit with only one interactive operation to activate some application (rotation, zoom, cuts, etc) in several windows simultaneously, each window visualizing one independent contribution. An example of this is provided by Figures 1 and 2 which are related to 4 independent calculations of a shock-schock interaction problem presented at [1]. In particular, the first computation was made with an unstructured grid. Details of the flow are visualized by a zoom of the same domain area in Figure 2. As a result the effect of the grid structure and density can be comparatively appreciated. Also, individual one-dimensional plots can be superimposed easily in another window. It is also possible for 3D large applications to run the software in parallel mode (using the PVM environment). Several machines are then operated, each one being associated with one contribution calculates its own window. Another machine, the master, collects the information for the visualization in parallel; in this mode the different plotting areas are filled in parallel. Clearly, these functionalities have been designed to facilitate cross-comparison of contributions including some of experimental nature also, and interactive preparation of test-case synthesis.

This environment has been utilized in the framework of the ECARP Workshops on Parallel Computing for Navier-Stokes Flows and Optimum Design in Aerodynamics. For this purpose, the VIGIE software has been adapted and used to also visualize speed-up plots and geometrical ("optimum") shapes, in addition to flowfields, thus demonstrating its versatility.

References

[1] US/Europe conference on high speed flow fields and first us-europe high-speed flow field database workshop. University of Houston, Houston, Texas, USA, November 6-9, 1995.

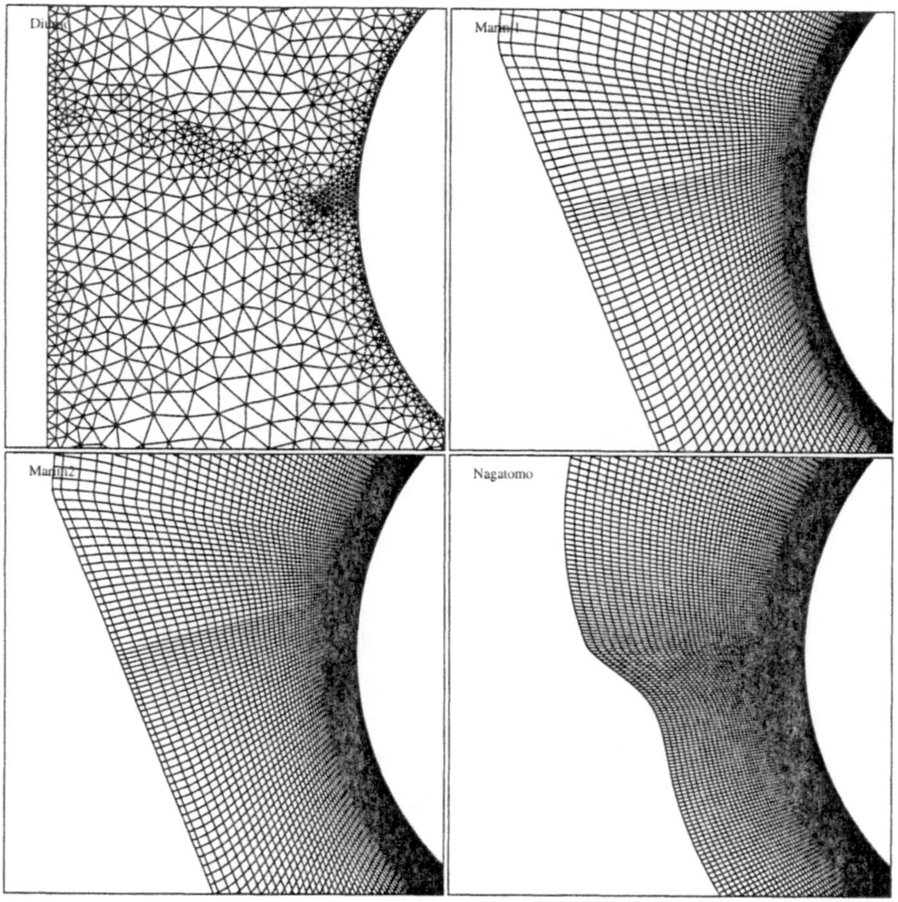

Figure 1. Shock-shock Interaction Problem – Zoom of meshes employed by different contributors (Diurno, Marini (×2) and Nagatomo)

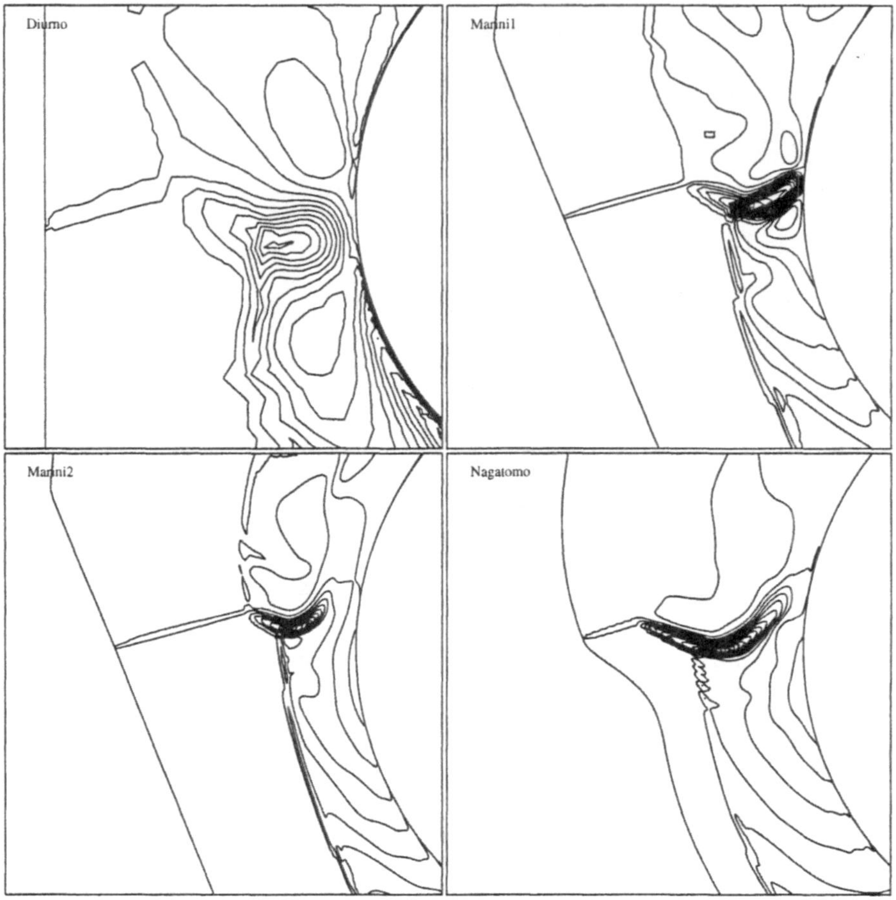

Figure 2. Shock-shock Interaction Problem – x momentum isolines (solutions by Diurno, Marini (×2) and Nagatomo)

III Contributions to the Resolution of the Data Workshop Test Cases

The following sections contain the different contributions from the ECARP consortium to the resolution of the workshop test cases defined in section II. Each contribution contains a theoretical description of the optimization technique used by each partner and the results obtained by using this techniques for the resolution of some test cases.

The contributions are ordered according with the following list of partners:

- 1. Aerospatiale
- 2. Alenia
- 3. CASA
- 4. Dasa - M
- 5. Dasa - Airbus
- 6. Dassault Aviation
- 7. Dornier
- 8. INRIA Sophia Antipolis
- 9. INRIA Rocquencourt
- 10. NLR
- 11. NTUA
- 12. UPC
- 13. DRA

The above numeration of the partners will also be used in the section corresponding to the synthesis of the workshop results.

1 On the optimisation of a wing-pylon-nacelle configuration

Thierry Fol, Aerospatiale Aircraft Business

Abstract

Optimum design tools have been used to reduce the difference between the pressure distribution on the clean wing and the pressure distribution on the engined wing of the AS28E configuration. Aerospatiale have developped a shape parametrization code and have coupled a panel method with an industrial minimization package to achieve the optimization task. Aerospatiale has defined the configuration used by three partners in order to test their optimum design tools. Even if the optima found by the partners are not strictly identical, they have proposed similar shape modifications.

1.1 Optimum Design at Aerospatiale within ECARP

The optimization of 2D aerodynamic shapes is today available. Some industrial softwares are able to parametrize an airfoil and to find out the combination of parameters that minimize the drag for a given lift and a given minimal thickness. The flow solver of such optimization package involves an Euler code coupled with a boundary layer code.

The optimization of 3D aerodynamic shapes within an industrial environnement has still to be developped. The main difficulties concern the parametrization of complex 3D configuration and the cost of each computation of each flow solution.

Aerospatiale Aircraft Business has decided to develop a shape optimization loop that involved a 3D panel method solver coupled with a general minimization package used as black box. In order to reduce the computational cost of the flow variables required by the optimizer, each geometrical degree of freedom is simulated by an injection distribution that avoids remeshing and recomputing the matrix of influence at each iteration [1].

In the framework of the ECARP/Optimum design project, AEROSPATIALE has been in charge of a developpement task concerning the treatment of geometrical constraints, an evaluation task concerning the evaluation of its ability to optimize complex 3D shapes and a definition task of the TE5 test case.

1.2 Technical Progress achieved during ECARP

1.2.1 Treatment of the Geometrical Constraints

An aerodynamics designer takes two main kinds of constraints into account when he modifies a shape: aerodynamic constraints such as a minimum lift coefficient, and geometrical constraints such as a minimum relative thickness. In order to make the optimizer OPT10 able to compute the geometrical constraints for the most general configurations, three supplementary subroutines have been implemented and two new common blocks have also been added to the program. Then some execution tests have been performed on the forward end of a fuselage for minimizing the maximum Mach number without reducing the volume of the cabin.

The evaluation of geometrical constraints requires to know at each iteration the current shape of the body to be optimized. This shape is not directly available during the optimization process because it is represented by a transpiration law. These laws are the inputs of the optimizer OPT10 [1]. Three new modules have been implemented in order to:

- compute the functions based on geometrical constraints.
- organize the new geometrical data flow required by the constraints computation.
- organize the new data flow describing the location of the geometrical constraints.

1.2.2 Parametrization of The Optimized Shapes

A tool that generates automatically a set of degrees of freedom must be developped. The research space based on this set influences strongly the effectivnessof the optimization. A group of aerodynamic designers has been interviewed in order to establish the guidelines of a tool that generates automatically 3D degrees of freedom within the CAD environment they used. This tool has been developped. In the case of the parametrization of a wing the main steps for generating the degrees of freedom are:

- selection of the parts of the wing to modify
- selection of the wing sections that pilot the deformations (master lines)
- selection of the elementary transformation of each master line: twist, thickness, camber, local curvature .

A code that allows to explore quickly the optimization problem has also been developped in order to analyze the sensibility of each degree of freedom and to find an interesting starting point for the optimization run. This code is based on the same technique as the one used for evaluating a cost function in the optimum design code OPT10. Each degree of freedom is simulated by a distribution of injections [1].

1.3 TE5 - Workshop Definition

1.3.1 Configuration to be optimized

The name of the configuration to be optimized is AS28E. It includes the AS28 wing, the AS28 nacelle and a symmetrical pylon generated from the initial cambered AS28 pylon.

The fuselage has not been included into the AS28E configuration, in order to limit the size of the influence matrix of the panel method called during the optimization process. As TE5 test case aims to reduce the interference between a wing and a power-plant, the fuselage would have increased uselessly the cost of the computation.

The TE5 test case as been proposed by Aerospatiale which is responsible for the engine integration in the AIRBUS worksharing. For this reason the minimization of the influence of the power on the wing has been selected as industrial test case for the ECARP Optimum Design workshop.

1.3.2 Objective function

The computed Cp-distribution of the smooth wing will serve as target distribution. In order to be consistent with the aerodynamic model used within the optimization process, the target Cp- distribution should be computed with the help of a panel method by all partners. The flow conditions will be Mach = 0.5, angle of attack $\alpha = 2°$.

1.3.3 Degrees of freedom

The pylon geometry will be kept fixed. The wing geometry only will be modified.

1.3.4 Constraints

No geometrical constraint will be taken into account. The evolution of the lift coefficient of the optimized configuration is implicitly controlled by the chosen objective function. Its value can not be very different than the one of the smooth wing .

1.3.5 Who contributed to TE5 ?

As TE5 is an industrial test case, three industrial partners have finally been involed in this work :

- Deutsche Airbus (H. Schwarten)
- DASA (L. Fornasier)

- Aerospatiale (T. Fol)

1.3.6 Data Exchange

The geometries of the engined wing and the clean wing required by each partners in order to perform the optimization were represented by a 3194 skin panels mesh stored on a floppy disk as an ascii file, at the begining of the workshop, because Aerospatiale was not connected to Internet. As soon it has been possible DASA, Deutsche Airbus and Aerospatiale have exchanged data with the help of electronic connection such as the IBM TSO net or e-mail.

1.3.7 Problems Encounted by the Partners with the definition of TE5

Only one release of the initial TE5 data set has been necessary. The wing/pylon intersection line of the initial TE5 engined wing intersected the leading edge line of the wing. This led to a locally bad mesh. A small modification executed by Aerospatiale of the pylon without modification of the relative position of the wing and the nacelle allowed to avoid this trouble.

The panel mesh generator of Aerospatiale produces a panel distribution on the pylon and on the wing that didn't match because the panel method used at Aerospatiale treated that type of configuration. This meshing technique produced small gaps near the wing/pylon intersection and near the pylon/nacelle intersection. As Deutsche Airbus and DASA couldn't compute accuratly that type of mesh, they introduced an additional panel stripe in the mesh near the intersections lines.

1.4 TE5 Partners Methods

1.4.1 Methods used for solving TE5

The classical ingredients of an optimization tool are a flow solver, a minimization strategy and a surface module. Even if each partners used a panel method, three different optimization techniques have been tested during the workshop.

1.4.2 Parametrization of the Shape to be Optimized

	Aerospatiale	DASA	Deutsche Airbus
Number of dof	28	32x10=320	7x15=105
Type of dof	Designers well-known transformations applied to five selected wing sections : Twist Thickness Local curvature Local camber Each degree of freedom is converted into transpiration	Modifications of 10 selected sections : Parameters of B-Spline fitting controlling vertical displacement of pane corner points (32 dof per section 4th order polynomial) + smooth curves along spanwise only Trailing edge fixed	Modification of 7 selected sections: Twist(trailing edge fixed) Ordinates of control points of B-Splines that define wing. Linear interpolation in spanwise direction

1.4.3 Flow Solver : Three different panel methods

	Aerospatiale	DASA	Deutsche Airbus
Name	FP3D: inhouse code	HISSS: inhouse code	VSAERO
panel mesh	Gap accepted	Gap not accepted	Gap not accepted

1.4.4 Minimization Strategy

Aerospatiale	DASA	Deutsche Airbus
Industrial minimization package used as black box Transpiration technique avoids remeshing and recomputing of the influence matrix	Augmented Lagrangian Multigradient algorithm	Levenberg-Marquardt method Unification of steepest descent and inverse Hessian Method

1.5 TE5 - AEROSPATIALE results

The difference between the flow field on the AS28E engined wing and the one on the AS28E clean wing is characterized by a local lost of lift on the outboard wing, an overspeed located at the inboard wing section aside the pylon and an influence of the pressure recovery of the pylon on the lower surface of the engined wing.

The mesh used for the optimization runs contains 2130 skin panels instead of the 3194 panels used for the definition of the geometry of the TE5 test case . This avoids to write the influence matrix on a disk, speeding up the optimization process.

A local minimum has been found after two optimization runs. The first one involved ten degrees of freedom modifying the twist and the thickness of the engined wing. The second one involved local camber and curvature modifications. The computing performances are reported on fig. 1.b. The total computing cost on a C92 Cray is 580 sec for 277 evaluations of the cost function, whereas only one direct analysis of the flow requires 150 sec with the same mesh on the same machine. This proves the effectiveness of the implementation of the transpiration law technique.

In terms of Cp distribution the optimized shape is close to the target but not identical to the target (cf. fig. 2 and 3). The local lift has been recovered, the overspeeds aside the pylon have been limited and the influence of the pressure recovery of the pylon has decreased. However few degrees of freedom have been used (28) for solving this quite inverse problem, and their evolution have been limited in order to find out a realistic wing without crazy spanwise gradient of thickness . This appoach does not allow for example to control very accuratly the pressure distribution on lower surface of the inboard wing between x/c=0.15 and x/c=0.35. It is only possible to improve the shape without matching the target pressure distribution. Each partner achieved his work on TE5 in time, an the results have been stored as a set of 1-D curves on the INRIA data-base. The results are analysed in term of geometry and Cp distribution .

1.6 TE5 - Analysis of Workshop results

Each partner achieved his work on TE5 in time, an the results have been stored as a set of 1-D curves on the INRIA data-base. The results are analysed in term of geometry and Cp distribution .

1.6.1 Cp deviation

The target pressure distribution has been performed by each partners. Without this procedure the results obtained by Deutsche Airbus would have been very different from the results of the other partners. The suction peak computed by Deutsche Airbus is much greater than the one computed by DASA or Aerospatiale.

The proposed test case presents wing-engine interferences which amplitude is small, because the diameter of the engine is not very important and because the nacelle is relavely far away from the wing. However the classical phenomenas due to the engine are visible : lost of list on the outboard wing because of a diminution of the suction peak, overspeed on the lower surface of the inboard wing, influence of the pylon pressure recovery on the lower surface.

The three methods have been able to reduce the Cp deviation and to converge to Cp distribution close to the target. Each method has been able to produce an optimized Cp distribution that fitted well the target Cp distribution at the upper surface. At the lower surface of the engined wing the results are also good between x/c =0.50 and x/c = 0.90 where the influence of the pylon pressure recovery is important.

However each mehod seemed to meet some difficulties to reduce the Cp deviation at the inboard lower surface of the wing between x/c=.05 and x/c=0.5 especially close to the pylon. The sign of the Cp deviation changes rapidly along spanwise and chordwise : it could explain why this region is sensitive to the limitation of the three methods :

-Aerospatiale : few degrees of freedom

-DASA : smoothing constraints

-Deutsche Airbus : linear depending of the sections

1.6.2 Optimized geometries

The wing sections optimized by each partners are not identical. The difference of parametrization is one of the main reasons of those differences. However each partners has found the same direction of modification espacially in the outboard wing region. In the inboard region the differences between the optimized shapes are much important. This is certainly due to the difficulties encounted by the optimization code to reduce the Cp deviation in this region.

Reference

- [1] ECARP/TEC/3.1.2.1.A/93 report (25 August 1993)

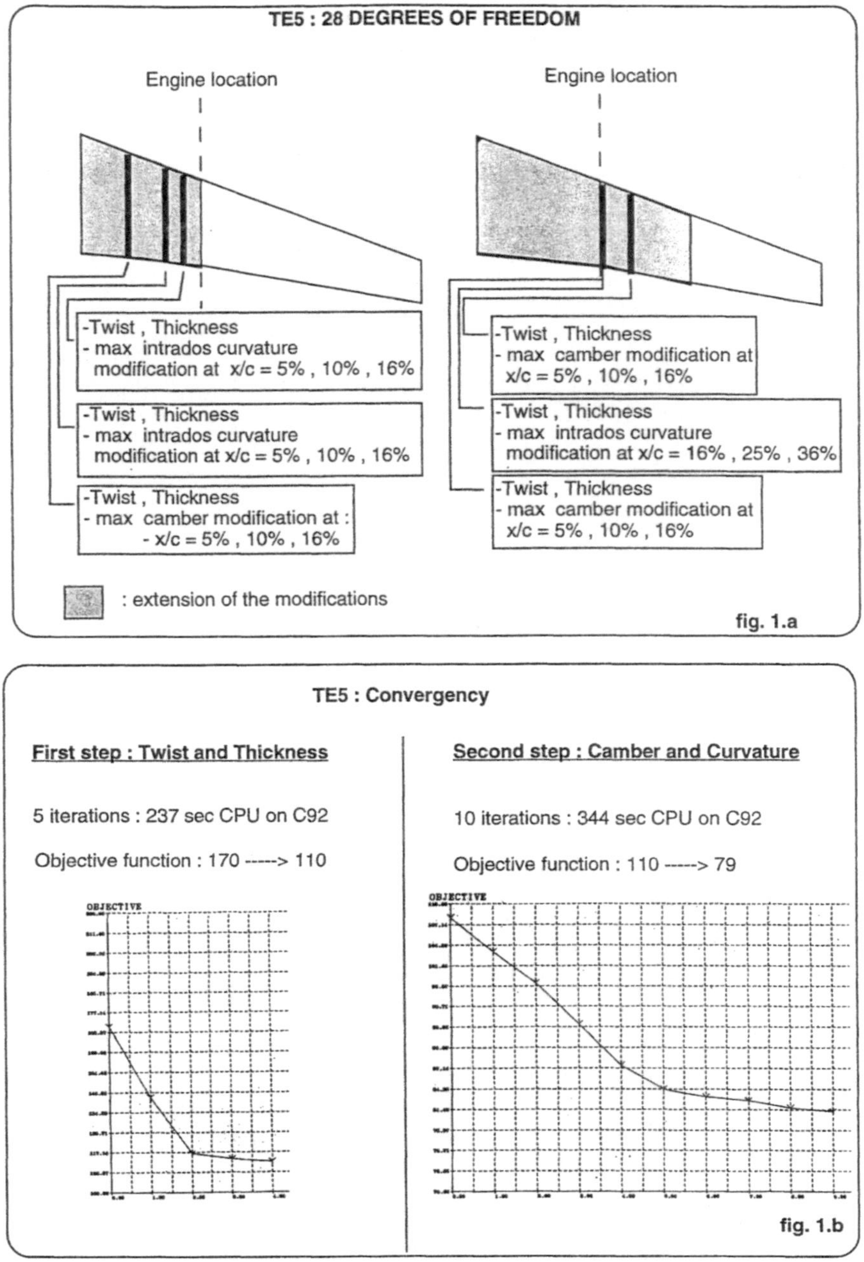

Figure 1 Degrees of freedom and history of the optimization process

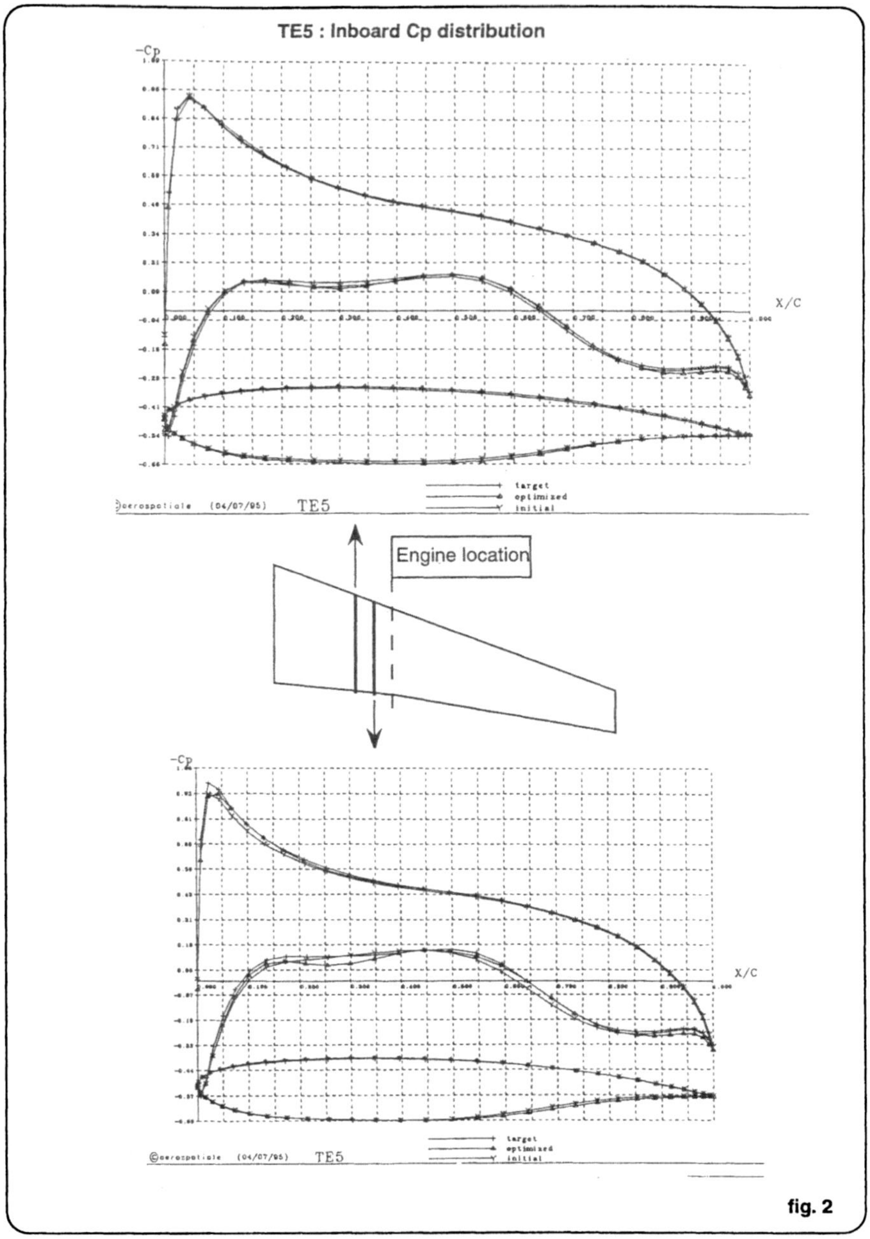

Figure 2 Inboard wing : initial, optimized and target Cp distribution

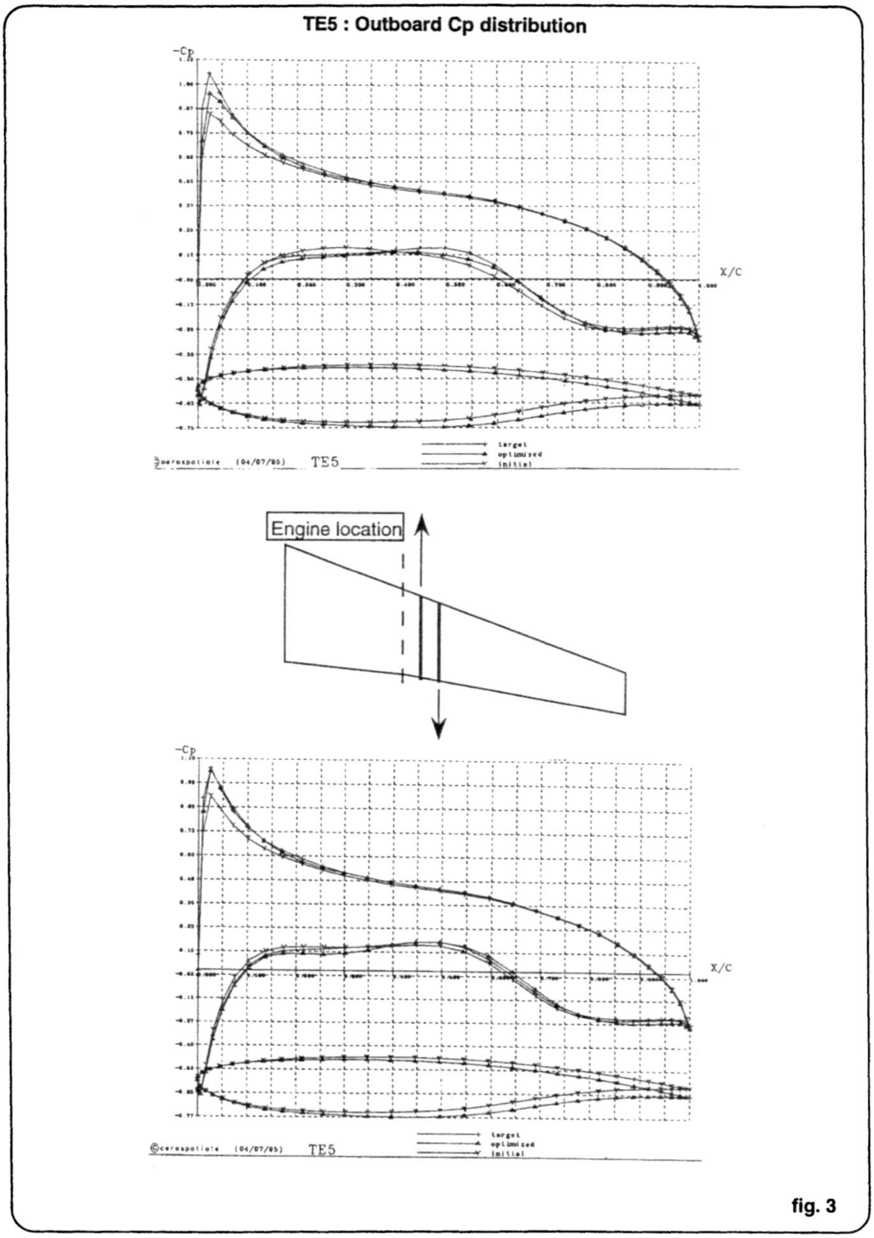

Figure 3 Outboard wing : initial, optimized and target Cp distribution

2 Optimization of Transonic Airfoils and Wings

V. Selmin, Alenia Aeronautica, Torino

2.1 Optimization of Transonic Airfoils

Method description.

The method that has been developed solves the problem of minimizing a non-linear objective function subject to non-linear constraint function, by selecting the optimum set of values for the independent design variables, upon which the objective and constraint functions depend. Finite difference methods are used to compute the gradients of the objective and constraint functions with respect to each of the design variables. From these gradients, a line search direction is defined, and the appropriate combination of changes in the design variables is performed in a series of steps. The cycle of gradient determination followed by line search is repeated to reduce the objective function, while still satisfying the constraint functions. The objective and constraint functions can be based upon any function for which a value can be extracted from the geometric model or the CFD solution.

The optimization procedure is composed by an optimization module, an aerodynamic module and an interface module that also handles the geometric modification of the geometry.

1. *Optimization module.* The method of Zoutendijk which is based on the feasible direction algorithm has been selected to solve constrained problems. In the optimization code, the gradients are calculated by a first order forward finite difference unless a variable is at is upper bound. In this case, a first order backward finite difference step is used.

2. *Aerodynamic module.* The optimization routine is coupled with an aerodynamic code which gives the values of the aerodynamic coefficients in order to calculate their derivatives with respect to the design variables. Based on these derivatives, the optimization routine chooses the most suitable modification direction. Since a large number of aerodynamic coefficients evaluations is expected, it is important to use a low-computational-cost aerodynamic method whose formulation is able to treat transonic flows. A code that solves the full potential equation has been selected.

3. *Geometry parametrization.* The representation of the airfoil shape is done in terms of a spline approximation. In this work, the so-called Bézier curve segment approximation has been adopted. Each of the lower and upper sides of the airfoil is approximated using the spline representation. The design variables are the ordinates of the spline control nodes at fixed x-locations.

We refer to section III.6.2 for a detailed description of the adopted methodology.

Numerical results and analysis.

The test case addresses the problem of inviscid drag minimization in transonic flows. The far field conditions are $M_\infty = 0.73$ and the angle of attack $\alpha = 2°$. Starting from the RAE2822 airfoil, it is asked to optimize the shape in terms of drag minimization maintaining the lift of the original profile.

A Bézier curve segment approximation using 14 control points for both upper and lower airfoil surfaces was adopted to approximate the airfoil shape, which results in taking 24 design variables.

The results of the optimization problem are illustrated in figures 1 to 3. About 150 objective function evaluations are needed to reach the optimum. The C_p distribution on the leeward side of the optimized airfoil is characterized by an increase of the suction peack and a reduction of the shock jump; the windward side remains nearly unchanged. The maximum thickness location moves towards the trailing edge.

Table 1 summarizes the performances of the two airfoils.

Table 1. Comparison of the performances of the two airfoils.

	C_l	C_d	C_m	thick.
RAE2822 airfoil	0.8670	0.0031	-0.1342	0.121
Optim. airfoil	0.8730	0.0009	-0.1250	0.115

The new airfoil satisfies the optimization problem. It allows to maintain the value of the lift coefficient with the additional advantage of reducing drag. The values of the pitch moment and of the maximum thickness are slightly reduced.

The computation has been performed using a 161 × 33 O-topology grid, which is rebuilt each time the airfoil shape is modified. The CPU time is about 30 minutes on an INDIGO 912D workstation.

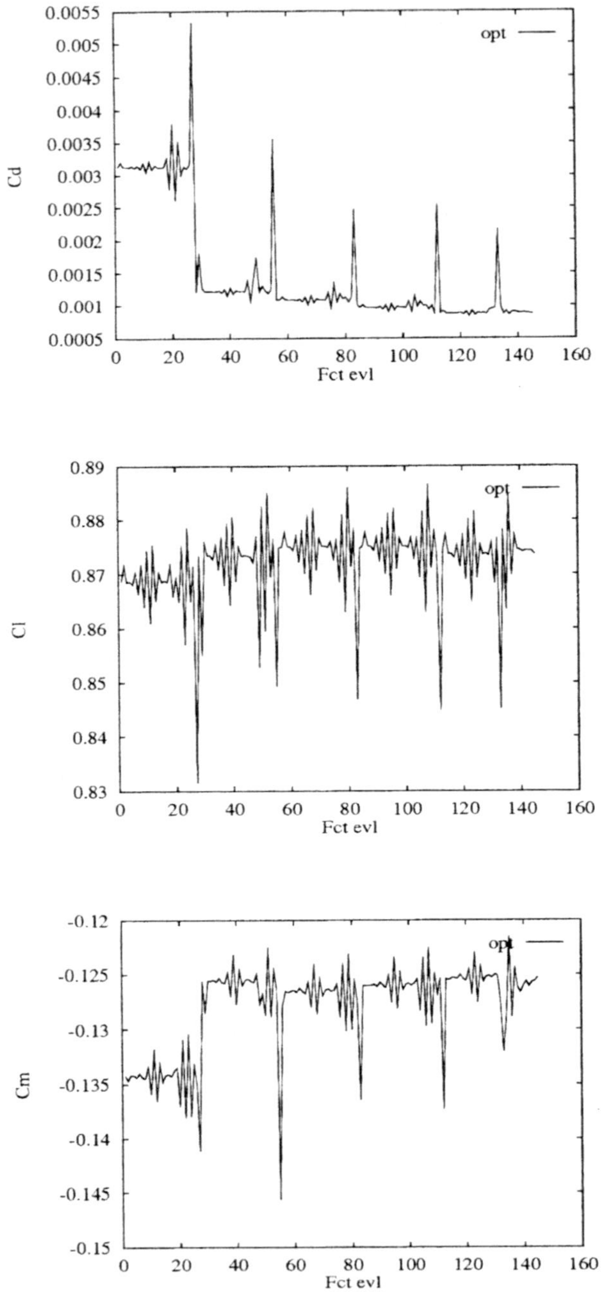

Figure 1: Integral coefficients history.

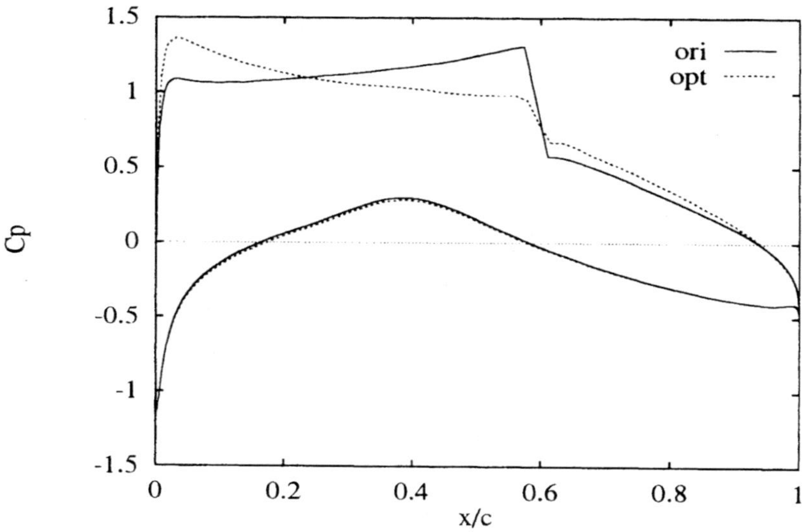

Figure 2: C_p distributions.

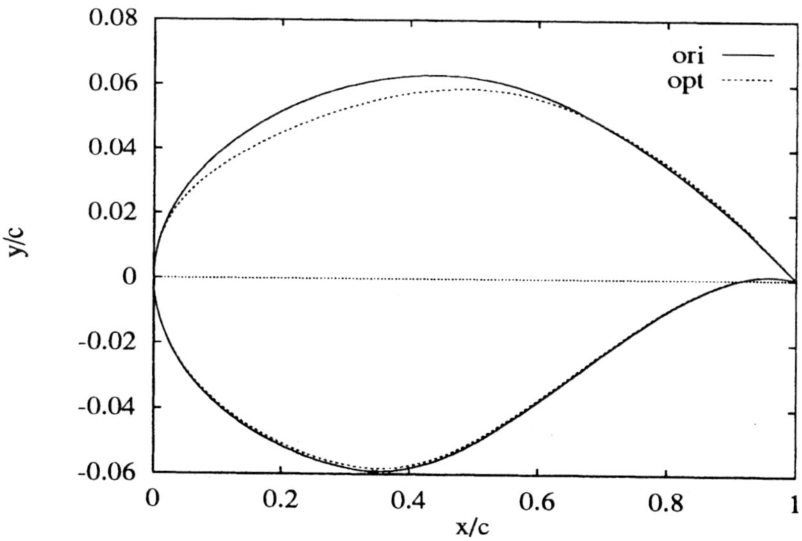

Figure 3: Comparison between shapes.

2.2 Optimization of Transonic Wings

Method description.

Introduction. The method that has been developed solves the problem of minimizing a non-linear objective function subject to non-linear constraint function, by selecting the optimum set of values for the independent design variables, upon which the objective and constraint functions depend. Finite difference methods are used to compute the gradients of the objective and constraint functions with respect to each of the design variables. From these gradients, a line search direction is defined, and the appropriate combination of changes in the design variables is performed in a series of steps. The cycle of gradient determination followed by line search is repeated to reduce the objective function, while still satisfying the constraint functions. The objective and constraint functions can be based upon any function for which a value can be extracted from the geometric model or the CFD solution. Examples include the thickness to chord ratio of the wing section, lift coefficient and drag coefficient.

The optimization procedure is composed by an optimization module, an aerodynamic module and an interface module that also handles the geometric modifications of the wing. The aerodynamic analysis module usually is a well assessed CFD computer code whose reliability has a strong impact on the final result. On the other hand, the code must have a low computational cost in order to maintain the overall optimization cost at a reasonable level. However, highly sophisticated CFD computer codes can be used to verify the aerodynamic characteristics of the optimized wing.

To solve the optimization problem, the values assigned to the design variables $\mathbf{X}$ must be found so as to minimize the objective function $F(\mathbf{X})$ while maintaining that the possible constaint functions $G_j(\mathbf{X})$ are ≤ 0. The designer must derive these functions and choose the design variables that govern the transformation of the wing geometry.

Objective functions and constraints. The general aerodynamic design requirements need to be translated into mathematical formulas for constraints and objective functions. A simple expression for the objective function includes the linear combination of aerodynamic coefficients (C_d, C_m, C_l, C_p) at one or more operating conditions of the wing (n). For instance,

$$F(\mathbf{X}) = \sum_{i=1}^{n} \{ \; K_{1,i}\, C_l(\alpha_i, M_{\infty i}, \mathbf{X}) + K_{2,i}\, C_d(\alpha_i, M_{\infty i}, \mathbf{X}) + K_{3,i}\, C_m(\alpha_i, M_{\infty i}, \mathbf{X}) + K_{4,i}\; \Phi\,(C_p(\alpha_i, M_{\infty i}, \mathbf{X})) \; \} \qquad (1)$$

where the index i defines the operating condition in which the coefficient is used. C_l represents the lift coefficient, C_d is the drag coefficient, C_m is the momentum coefficient and Φ is a function of the pressure coefficient C_p.

Constraints on the wing geometry and on its aerodynamic characteristics are used. The aerodynamic constraints are used to control the values of the design coefficients (typically C_l or C_m), while geometric constraints are used both to satisfy structural requirements and to indirectly control aerodynamic characteristics in off-design conditions. The following expression is used for the constraints:

$$G = K(\frac{V}{\hat{V}} - 1) \tag{2}$$

where V is an aerodynamic coefficient or a geometric parameter and $\hat{V}$ is a limit imposed on it. K represents a multiplication factor which is used to give a relative weight to each equation of the constraint set; in general, it is set to be equal to unity.

Minimization algorithm. Some optimization routines coming from the commercial package ADS [1] are usually used for minimization purpose. For constrained problems, a method based on the feasible direction algorithm [2] has been selected.

At each iteration q of the minimization process, the design vector $\mathbf{X}$ is updated according to the formula $\mathbf{X}^q = \mathbf{X}^{q-1} + \omega \mathbf{S}^q$, where $\mathbf{S}^q$ is a unit vector representing a search direction in a space having as many dimensions as the design variables, and where ω defines the displacement in the direction $\mathbf{S}^q$. The search direction is built in such a way that it will reduce the objective function without violating the constraint for some finite move, and thus can be defined as a "constrained steepest descent" direction.

Any vector $\mathbf{S}$ which reduces the objective function is called a usable direction. Clearly, such a direction will make an angle greater than 90° with the gradient vector of the objective function. This suggests that the dot product of $\nabla F(\mathbf{X})$ and $\mathbf{S}$ should be negative. The limiting case is when the dot product is zero, in which case the vector $\mathbf{S}$ is tangent to the plane of constant objective function. Mathematically, the usability requirement becomes

$$\nabla F(\mathbf{X}) \cdot \mathbf{S} \leq 0 . \tag{3}$$

A direction is called feasible if, for some small move in that direction, the active constraint will not be violated. That is, the dot product of $\nabla G_j(\mathbf{X})$ with $\mathbf{S}$ must be nonpositive. Thus

$$\nabla G_j(\mathbf{X}) \cdot \mathbf{S} \leq 0 . \tag{4}$$

The search direction $\mathbf{S}$ is then determinated by considering the following problem:

Minimize:

$$\nabla F(\mathbf{X}) \cdot \mathbf{S} \tag{5}$$

Subject to:

$$\nabla G_j(\mathbf{X}) \cdot \mathbf{S} \leq 0 \qquad (6)$$

$$\mathbf{S} \cdot \mathbf{S} \leq 1 \qquad (7)$$

The last inequality has been introduced to ensure that the search vector remains bounded. Solving this problem gives a search direction which is tangent to the critical constraint boundaries, unless the objective can be reduced more rapidly by moving away from one or more constraints.

In the optimization code, the gradients are calculated by a first order forward finite difference unless a variable is at its upper bound. In this case, a first order backward finite difference step is used.

Wing surface definition. The wing is defined by its shape at some spanwise cuts. Linear interpolation between two sections is adopted, i.e. section points at the same normalised x-station are joined with straight lines. The definition of the wing surface requires at least the knowledge of the root and tip sections shapes. The wing surface will depend of a set of design variables $\mathbf{X}$, which are split into a global and a local part, i.e. $\mathbf{X} = (\mathbf{X}^{glob}, \mathbf{X}^{loc})$.

The global part of the design variables specifies the location of the wing section with regard to a global coordinates system. It consists in five parameters $\mathbf{X}^{glob} = (x_g, y_g, z_g, \theta, c)$. The point of coordinates $\mathbf{r}_g \equiv (x_g, y_g, z_g)$ locates a point of the chord around which the wing section can be rotated. It is defined as follows:

$$\mathbf{r}_g = \mathbf{r}_{le} + \tau(\mathbf{r}_{te} - \mathbf{r}_{le}) \qquad (8)$$

where $\mathbf{r}_{le}$ and $\mathbf{r}_{te}$ represent the location of the leading and the trailing edges, respectively, and τ is a user specified parameter ($\tau \in [0, 1]$). The angle θ sets the chord angle (angle between the chord and the x-axis of the reference frame) and c scales the chord. The global parameters identify the planform, the twist and the anhedral/dihedral angle of the wing. Some of these parameters can be active or can be kept fixed.

The local part of the design variables defines the shape of the wing section. Each section is defined with respect to a local frame, the section coordinates system, centred at the section leading edge and having the x-axis coincident with the section chord. Normalised coordinates are introduced by scaling each length with the chord length. The representation of the wing section shape is done in terms of a spline approximation. Let $\mathbf{r}(t)$ be the position vector of the curve in the local frame, then we set

$$\mathbf{r}(t) = \sum_{i=0}^{n} B_{i,n}(t)\mathbf{r}_i^B \quad ; \quad 0 \leq t \leq 1 \qquad (9)$$

where $\mathbf{r}_i^B \equiv (x_i^B, \bar{y}_s, z_i^B)$ is the position vector of the ith vertex of the defining curve polygon and $B_{i,n}(t)$ is a general basis function. Bézier curve segment and B-splines approximations [4] have been implemented. The coordinate $\bar{y}_s$ is frozen and locates the wing section plane.

Two different strategies can be used for the representation of the wing section shape. In the first, the airfoil is split into camber line and thickness distribution, and each of them are approximated using the above spline representation. We recall that the thickness distribution of an airfoil is symmetric and, consequently, can be approximated by one spline. This approach allows to kept constant the camber line or the thickness distribution. In the second, expansion (9) is done separately for the lower and upper sides of the airfoil. It is more general than the previous one, but we have less control on the geometry.

The number n and the abscissae $\bar{x}_i^B$ of the control polygon vertices are given in input. The deviations of the actual vertices ordinates from the initial ones $(\bar{z}_i^B)$ represent design variables:

$$\mathbf{r}_i^B = (\bar{x}_i^B, \bar{y}_s, \bar{z}_i^B + X_i \bar{z}_i^B). \tag{10}$$

As the x-locations of the control vertices are specified, we have first to invert the equation $x = x(t)$ in order to compute the corresponding t-parameter values. This has to be done only once, because the x-locations are kept fixed during calculations. The ordinates $\bar{z}_i^B$ can then be obtained by imposing that the curve passes through $n+1$ points of the initial shape.

In the optimization procedure, we prefer to use the Bézier curve segment approximation because changes in the position of a control polygon vertex often maintain a smooth curve, which is not always the case using cubic B-splines. Moreover, the use of cubic B-splines often leads to a wavely representation of wing section profiles and requires a smoothing procedure.

Flow solver. The optimization routine is interfaced with an aerodynamic code which provides the values of the aerodynamic coefficients in order to compute their derivatives with respect to the design variables. Based on those derivatives, the optimization routine chooses the most suitable modification direction. The more reliable the aerodynamic code is, the more credible the design. But, since a large number of aerodynamic coefficients evaluations is expected, it is also important to use a low-computational-cost aerodynamic method whose formulation is able to treat transonic flows. A code that solves the full potential equation has been selected. Weak viscous effects can be taken into account by using a boundary layer correction.

The full potential equation is frequently used for solving transonic flow problems. In developing the full potential equation, the existence of the velocity potential requires that the flow be irrotational. Furthermore, Crocco's equation requires that no entropy production occurs. Thus no entropy changes are allowed across shocks. At first glance, this appears to be a poor assumption. However, experience has shown that the full potential and Euler solutions do not differ significantly if the component of the Mach number normal to the shock is close to one. The entropy production across a weak shock is dependent on the normal

Mach number, M_n, and may be written as

$$\frac{\Delta s}{R} \approx \frac{2\gamma}{\gamma+1}(M_n^2 - 1)^3. \tag{11}$$

This shows that the assumption of no entropy change across a shock is reasonable so long as the normal component of the Mach number is sufficiently close to one. It is important to note that the restriction is on the normal component of the local Mach number and not on the freestream Mach number.

The code solves the full potential equation in quasi-linear form,

$$\begin{aligned}(a^2 - u^2)\Phi_{xx} + (a^2 - v^2)\Phi_{yy} &+ (a^2 - w^2)\Phi_{zz} \\ &- 2uv\Phi_{xy} - 2uw\Phi_{xz} - 2vw\Phi_{yz} = 0 \quad (12)\end{aligned}$$

in the space about a semi-wing mounted on a wall, using a non-conservative finite difference discretization coupled with a relaxation method [3]. In Eq. (12),

$$u = \frac{\partial\Phi}{\partial x} \quad , \quad v = \frac{\partial\Phi}{\partial y} \quad , \quad w = \frac{\partial\Phi}{\partial z}$$

are the cartesian components of the velocity vector and Φ is the velocity potential. The speed of sound a may be obtained from the energy equation

$$\frac{a^2}{\gamma-1} + \frac{u^2+v^2+w^2}{2} = \text{constant}. \tag{13}$$

Although complex 3-D laminar/turbulent layer computations on wing alone configurations can be carried out at Alenia, a simplified approach has been selected in order to maintain the computing within acceptable limits. The approach is based on the evaluation of the laminar and turbulent boundary layer characteristics along wing strips parallel to the symmetry plane by using the 2-D integral method described below. The limits of the approach are:

- the wing planform must not be excessively swept in order to maintain the crossflow velocities small,
- the accuracy of the boundary layer computation is reduced across oblique shocks, but it is maintained if the shock is "normal" to the wing strip,
- the shock is not explicitly taken into account and no continuation down stream boundary layer separations is provided.

An improved approach is needed to solve the abovementioned limitations.

Laminar solutions are obtained by using Twaites method which has been extended for compressible flows on adiabatic walls by Rott and Crabtree. In the method, the momentum thickness is computed by integrating the following equation

$$Re_\theta^2 = 0.45\,\nu_0 \left(\frac{T_0}{T_e}\right)^3 u_e^{-6} \int_0^x \left(\frac{T_0}{T_e}\right)^{1.5} u_e^5\,dx \tag{14}$$

where ν_0 is the cinematic viscosity evaluated at the stagnation point and where the ratio T_0/T_e between the stagnation and edge temperature is computed using the isentropic formula

$$\frac{T_0}{T_e} = 1 + \frac{\gamma - 1}{2} M^2. \tag{15}$$

The shape parameter H, the displacement thickness θ and the skin friction C_f are obtained from closure relations [5].

For turbulent flows, the Lag-Entrainment method of Green extended to adiabatic compressible flows by Green, Weeks and Brooman [5] has been selected to predict the boundary layer characteristics of 2-D attached flows from the knowledge of the surface velocity and pressure distributions. This integral method is efficient and fast. It does not involve any coordinate transformation and, for what concerns the lag equation, it follows the interpretation of the Morkovins's hypothesis[1] due to Bradshaw and Ferriss [6].

Various transition criteria based on the analysis of experimental data sets can be found in the literature. The criterion of Abu-Ghannam and Shaw [7] has been selected. It is based on the determination of a critical value for $Re_\theta \equiv U_e\theta/\nu_e$ which depends of the level of turbulence in the free stream Tu and on the gradient parameter λ_θ according to

$$Re_\theta^{crit} = 163 + \exp\left(F(\lambda_\theta) - \frac{F(\lambda_\theta)}{6.91} Tu\right) \quad , \quad \lambda_\theta = Re \frac{du_e}{dx} \theta^2 \left(\frac{T_e}{T_0}\right)^{1/2}. \tag{16}$$

The critical values for Re_θ are close to 400 for compressible flows. The turbulent flow is assumed to start at the transition point and no transitional flow is computed.

References

[1] *Vanderplaats, G.N.* (1987): ADS - A Fortran Program for Automated Design Synthesis. User Manual.

[2] *Zoutendijk, M.* (1960): Methods of Feasible Directions. Elsevier Publishing Co, Amsterdam.

[3] *Jameson, A. and Caughey, D.A.* (1977): Numerical Calculation of the Transonic Flow past a swept wing. ERDA Rep. COO-3077-140 Courant Institute New York University.

[4] *Farin, G.* (1989): Curves and Surfaces for Computer Aided Geometric Design. Academic Press.

[5] *Green, J.E., Weeks, D.J. and Brooman, J.W.F.* (1973): Improved entrainment method for calculating boundary layers and wakes in compressible flow. A.R.C. R & M 3791.

[1] The turbulent structure is not affected by compressibility.

[6] *Bradshaw, P. and Ferris, D.H.* (1971): Calculation of boundary layer development using the turbulent energy equation; compressible flow on adiabatic walls. Journal Fluid Mech., **46**, pp 83-110.

[7] *Abu-Ghannam, B.J. and Shaw, R.* (1980): Natural transition of boundary layers - The effects of turbulence, pressure gradient and flow history. Journal of Mech. Eng. Sci., **22**, pp 213-228.

Numerical results and analysis.

The test case addresses the problem of inviscid drag minimization for 3-D transonic flows. The ONERA M6 wing will be "optimized" at $M_\infty = 0.84$ and $\alpha = 3.06°$ in order to reduce drag with constraints on the lift coefficient, pitch moment and on the maximum thickness of spanwise wing shape cuts. The lift coefficient of the optimized wing has to be greater or equal to the ONERA M6 wing lift coefficient provided by the analysis method. The pitch moment and the maximum thickness to local chord ratio of the original wing has to be maintained. The planform is kept fixed.

A Bézier curve segment approximation using 10 control points for each upper and lower surfaces was adopted to approximate a wing section shape, which results in taking 16 design variables per control wing section. Two shape modification strategies have been followed. The first (OPT1) modifies only the root wing section which is scaled spanwise. In the second (OPT2), both root and tip wing sections are modified.

The results of the optimization problem are illustrated in figures 1 to 8. About 130 objective function evaluations are needed to reach the optimum for the first case and about 360 for the second. The C_p distribution on the leeward side of both optimized wings is characterized by a reduction of the overall C_p level and of the shocks jumps. The level of the suction peacks decreases which should result in an improvement of the "viscous" behaviour of the optimized wings. The original upper surface lambda shock pattern reduces to one weak shock. The pressure on the windward side is increased in order to reach the target lift coefficient. The maximum thickness location moves towards the trailing edge.

Table 1 summarizes the performances of the three wings.

Table 1. Comparison of the performances of the wings.

	C_l	C_d	C_m	thick.
ONERA M6	0.2828	0.0124	0.0058	0.0968
OPT1	0.2832	0.0094	0.0057	0.0968
OPT2	0.2824	0.0095	0.0057	0.0968

Both optimized wings satisfy the optimization constraints. They allow to maintain the value of the lift coefficient, pitch moment and maximum thickness to local chord ratio with the additional advantage of reducing drag. The two optimized wings have similar characteristics. The shape of the wing OPT2 is nevertheless closer to those of the original wing.

The computations have been performed using a $161 \times 29 \times 33$ C-H topology grid, which is rebuilt each time the wing shape is modified. The CPU time on an INDIGO 912D workstation is about 19 hours and 54 hours for wings OPT1 and OPT2, respectively.

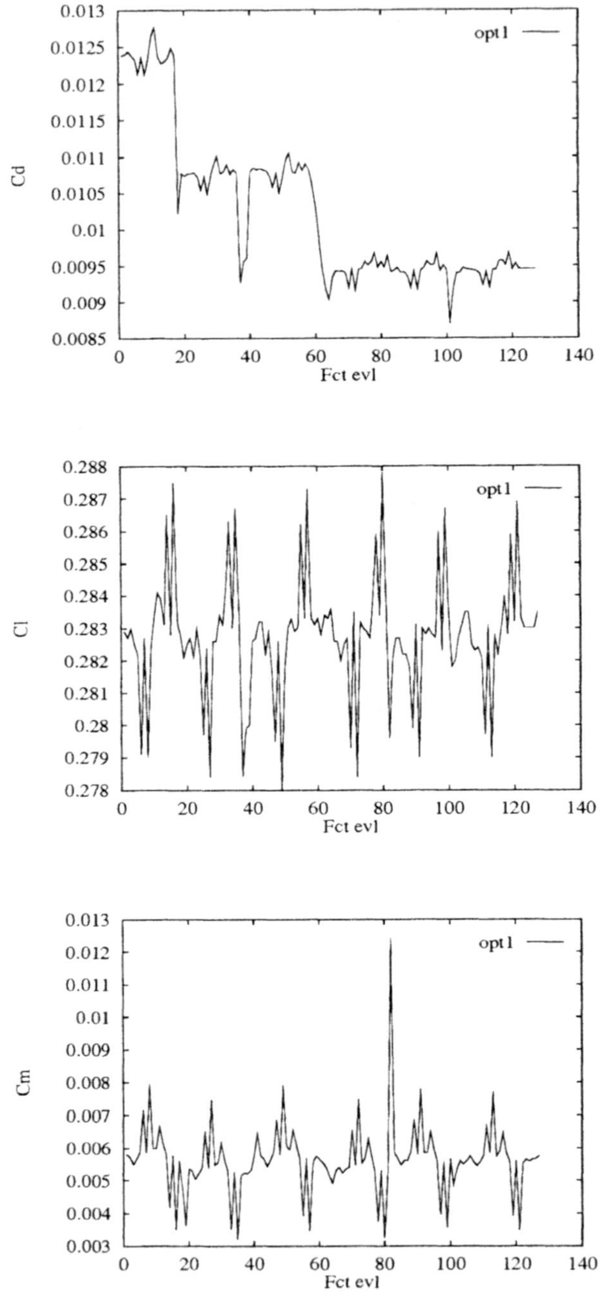

Figure 1: Integral coefficients history for root section optimization.

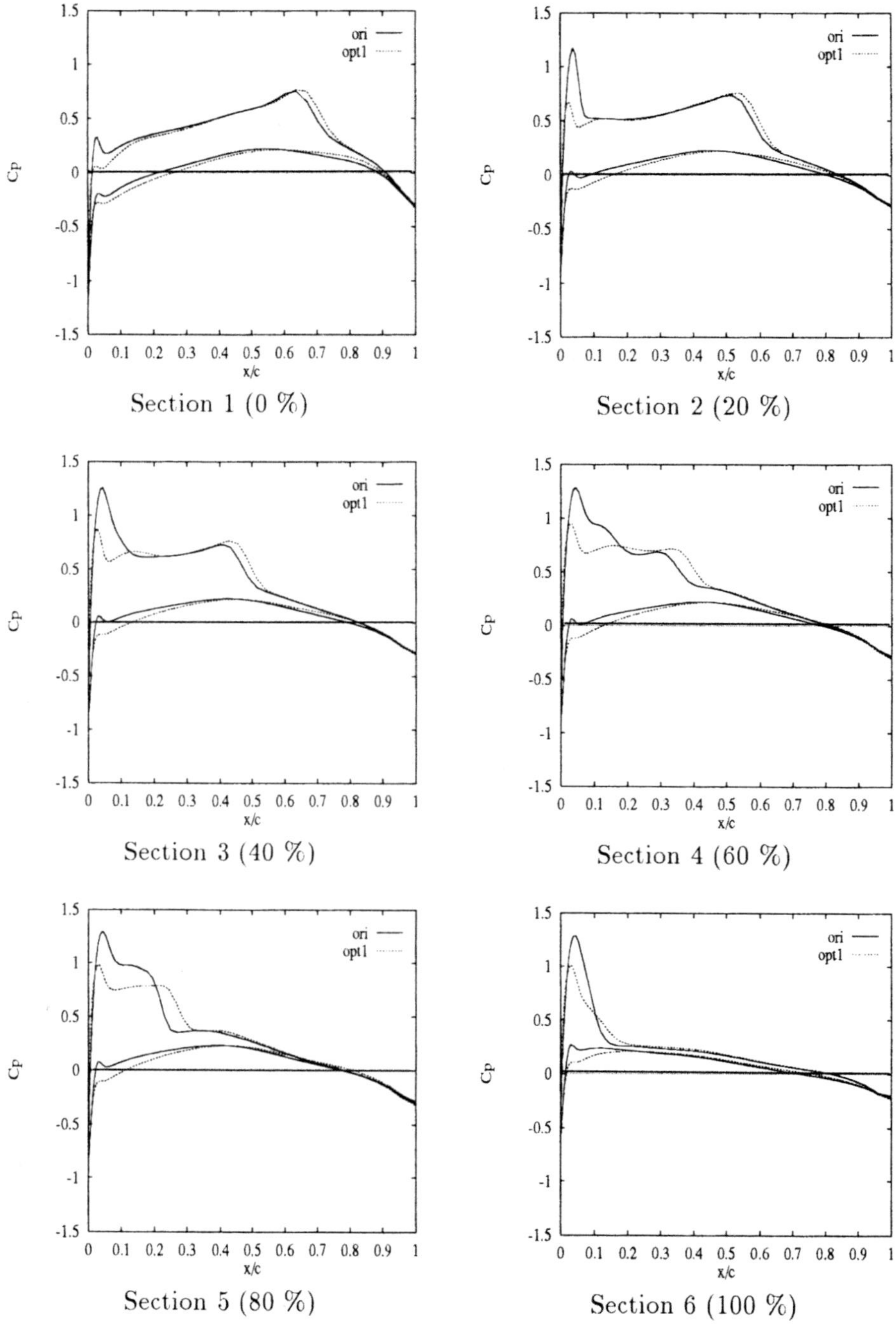

Figure 2: Spanwise C_p distribution for root section optimization.

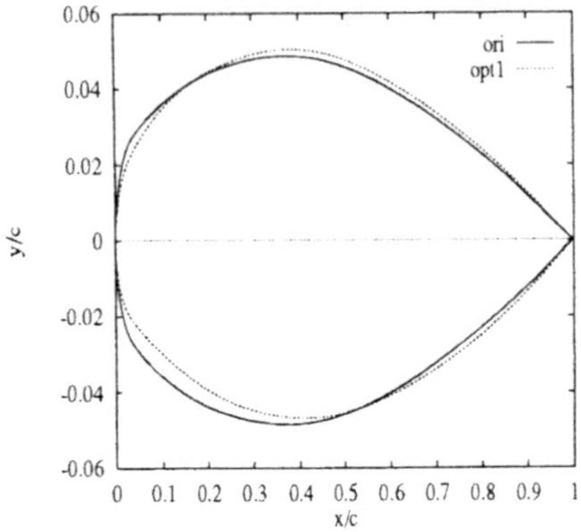

Section 1 (0 %)

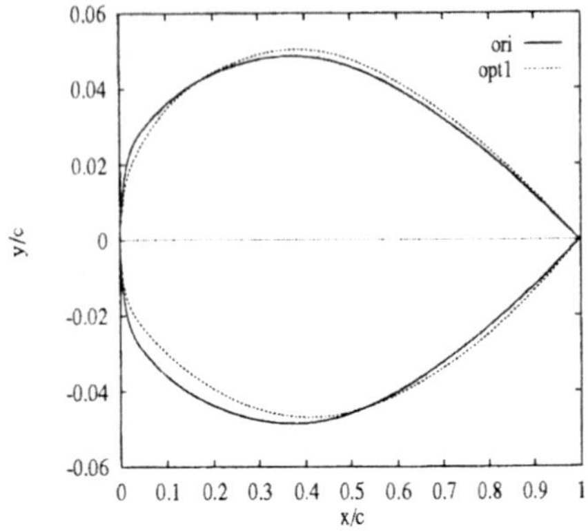

Section 2 (20 %)

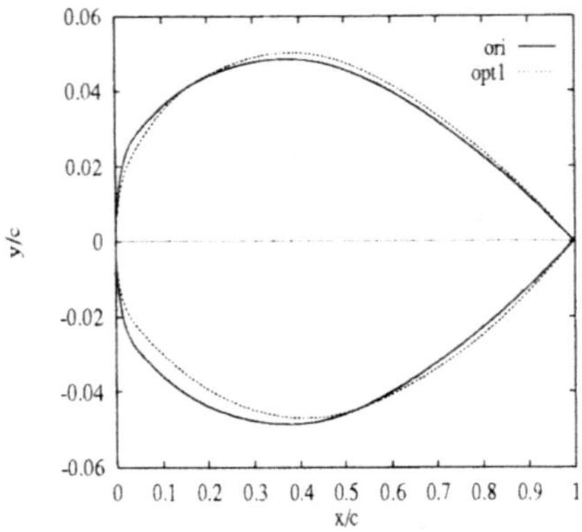

Section 3 (40 %)

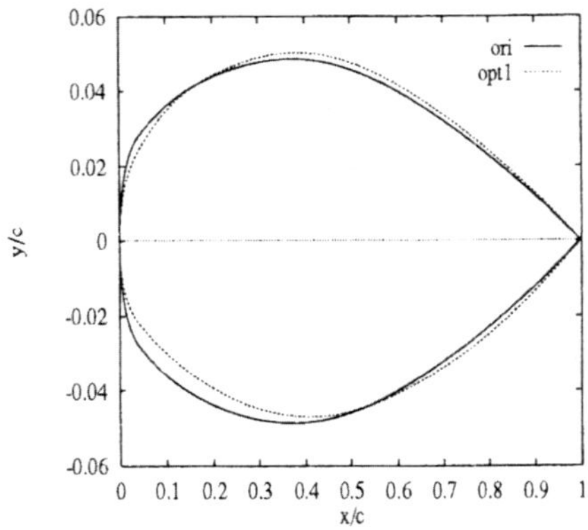

Section 4 (60 %)

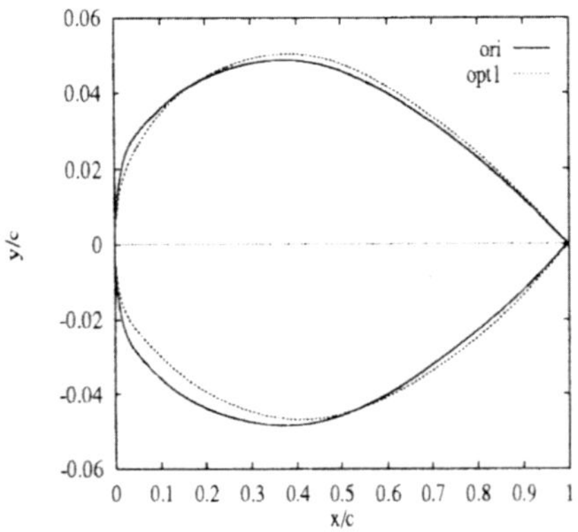

Section 5 (80 %)

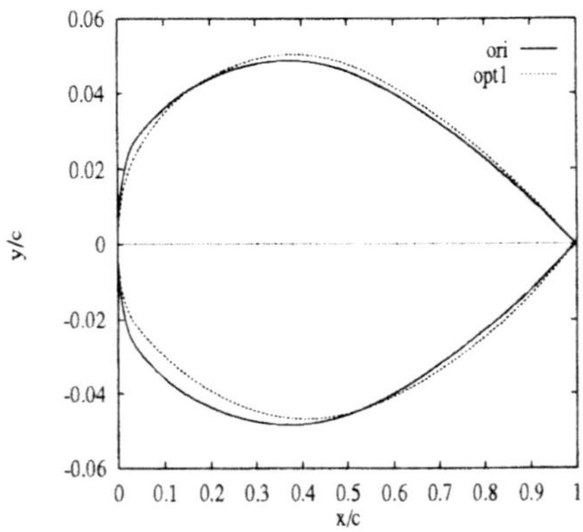

Section 6 (100 %)

Figure 3: Wing sections shape for root section optimization.

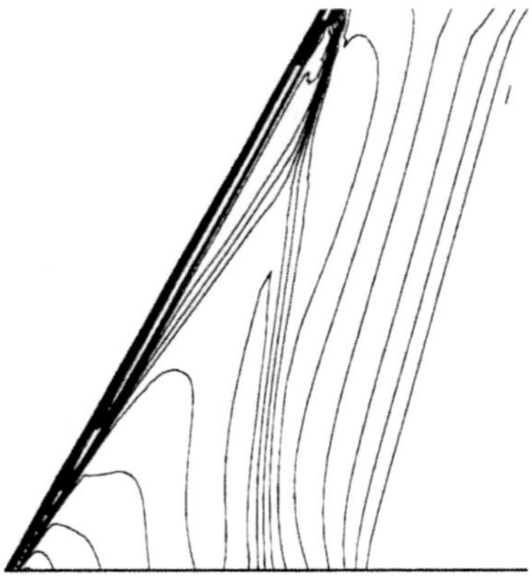

Original wing: upper surface

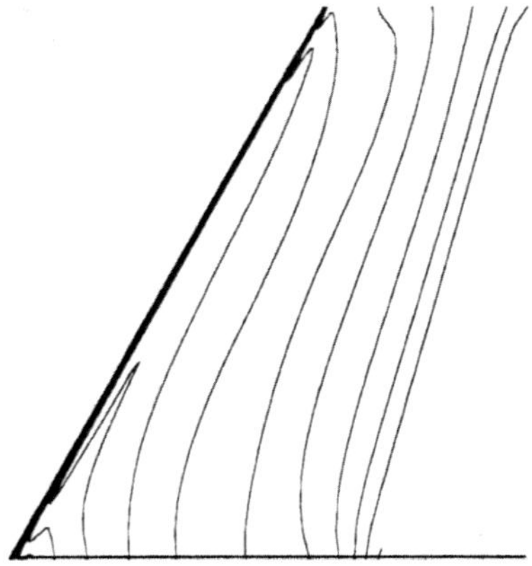

Original wing: lower surface

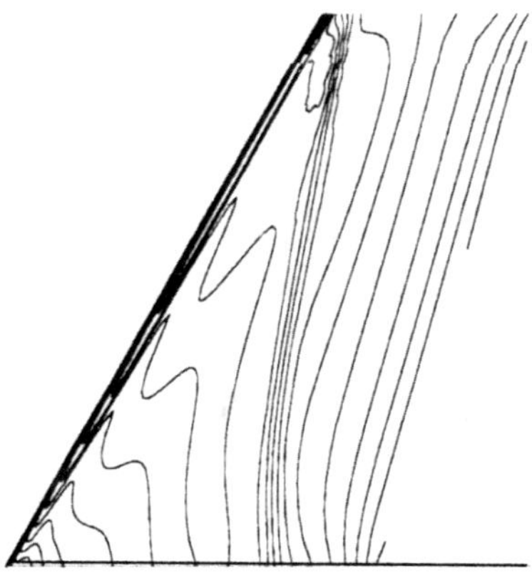

Optimized wing: upper surface

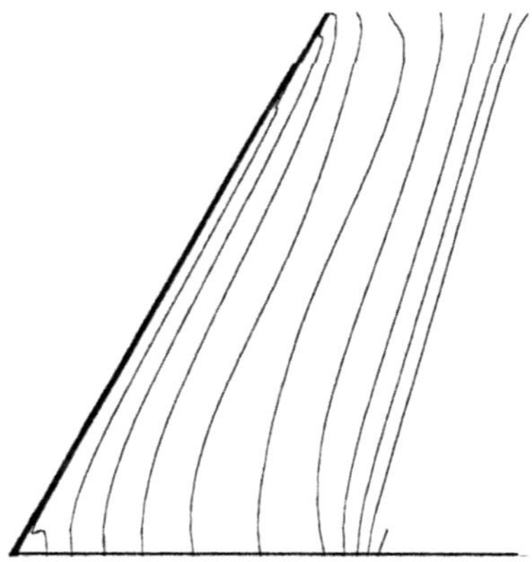

Optimized wing: lower surface

Figure 4: Iso-C_p contours for root section optimization.

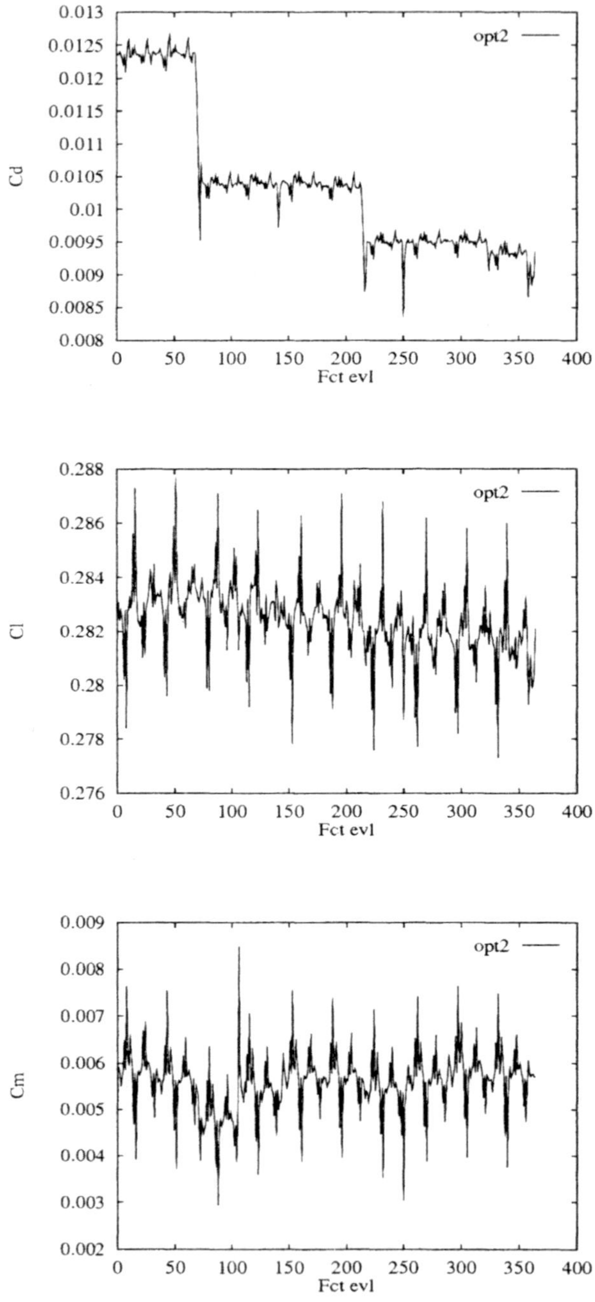

Figure 5: Integral coefficients history for root and tip sections optimization.

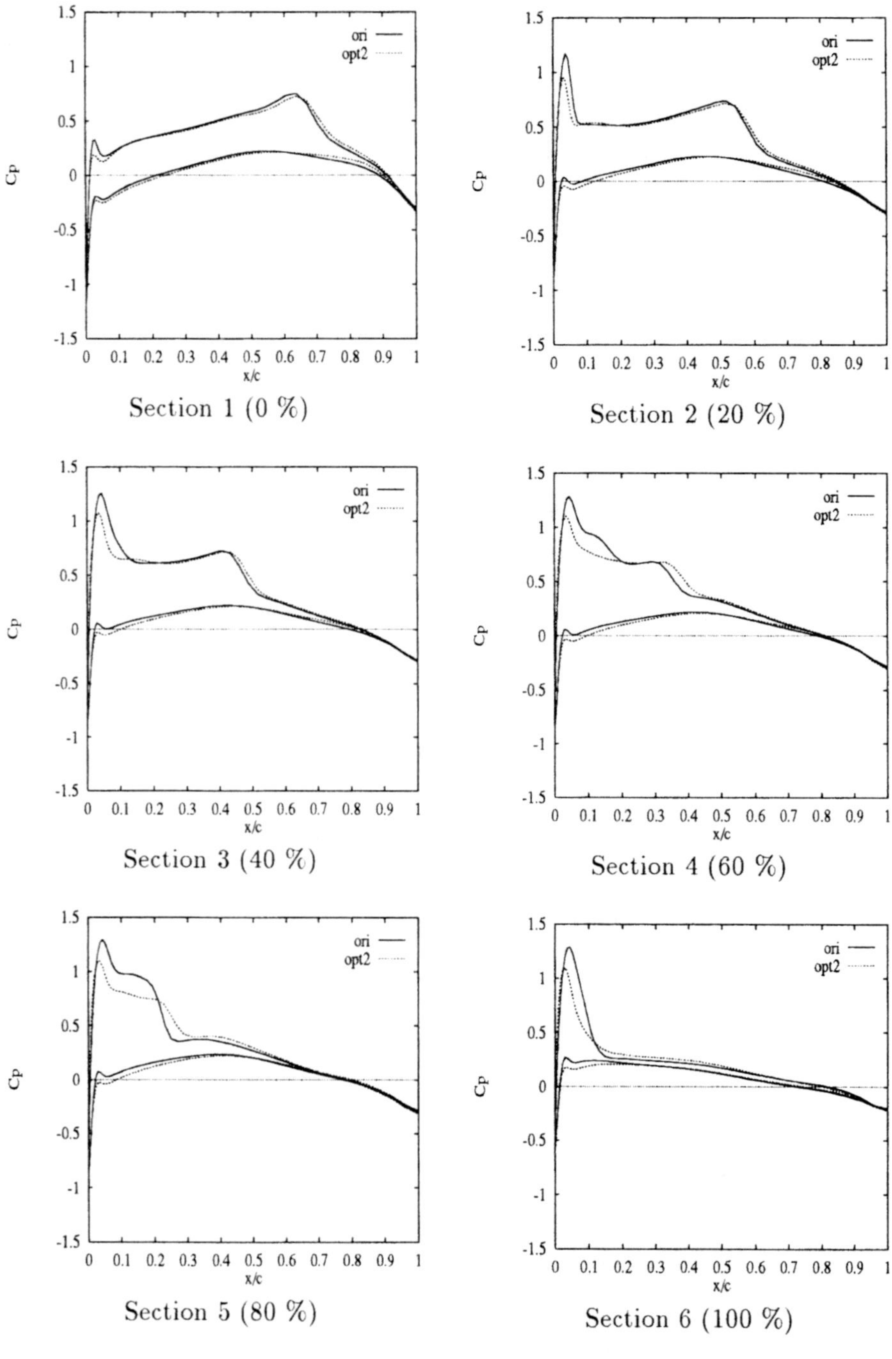

Figure 6: Spanwise C_p distribution for root and tip sections optimization.

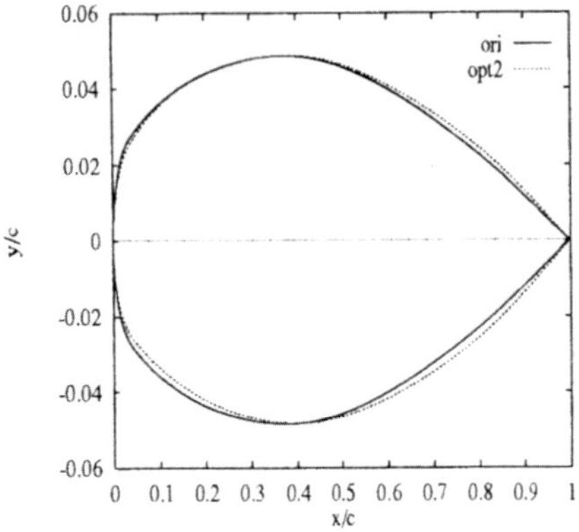

Section 1 (0 %)

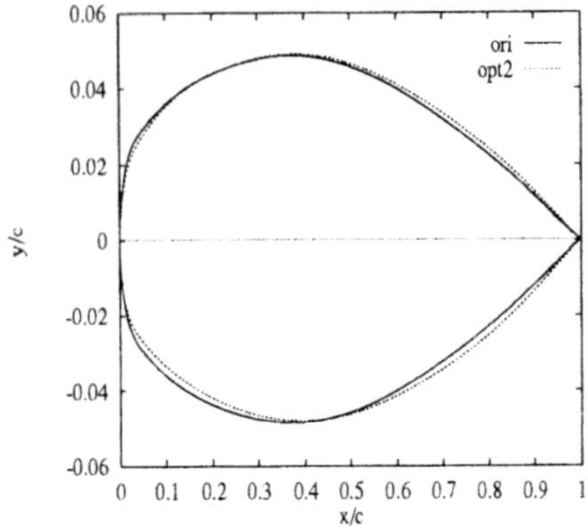

Section 2 (20 %)

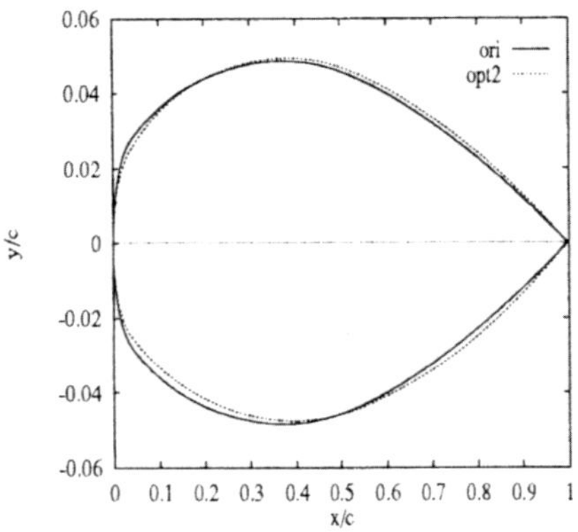

Section 3 (40 %)

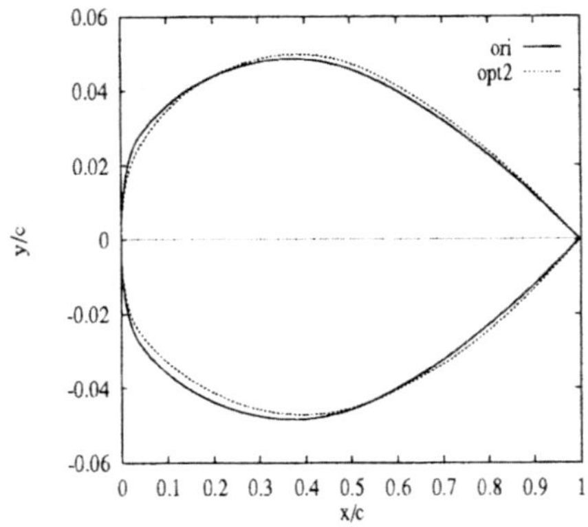

Section 4 (60 %)

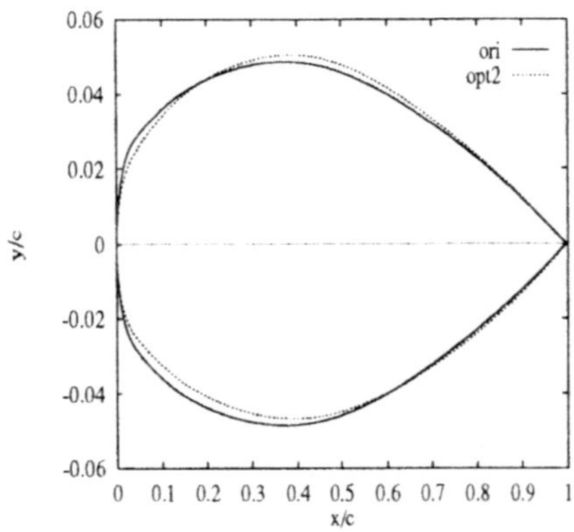

Section 5 (80 %)

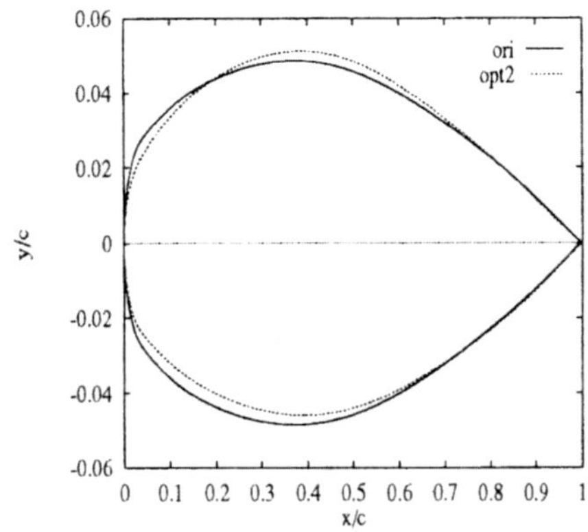

Section 6 (100 %)

Figure 7: Wing sections shape for root and tip sections optimization.

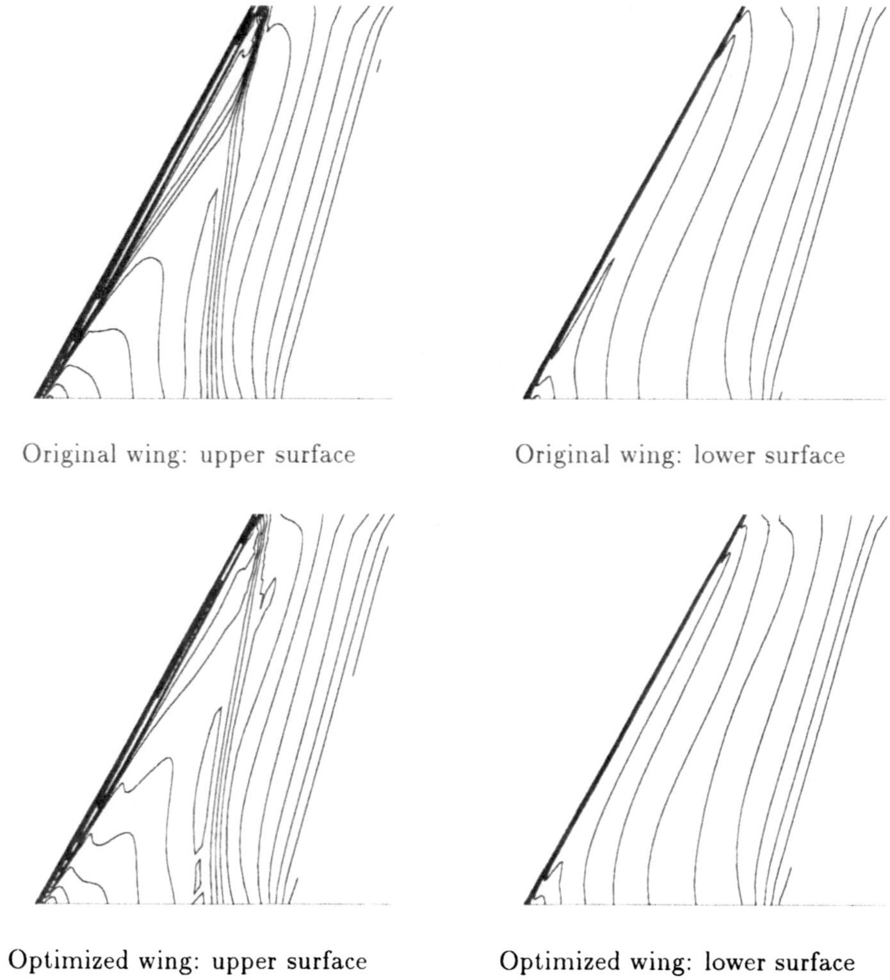

Figure 8: Iso-C_p contours for root and tip sections optimization.

3 Airfoil optimization using Viscous/Inviscid Coupling Code

J. M. Alonso, J. M. de la Viuda and A. Abbas, CASA

3.1 Introduction

This report briefly describes the work performed by CASA within the inverse design and optimization task of the ECARP project. The objective of the present work has been to optimize airfoil geometry by minimizing a predefined objective function. An Euler code coupled with boundary layer is used together with an optimization routine in an iterative procedure to reach an objective design goal. Results for ECARP Workshop test cases TE2 and TE3 are presented and discussed.

3.2 Methodology

A optimization procedure for aerodynamic configurations is generaly consisted of an aerodynamic module, optimization module and an interface module. The first module is a CFD computer code for the aerodynamic analysis of the considered geometry. The optimization module is employed to minimize an objective function F(x), possibly subjected to certain constraints. The interaction module takes care of the geometrical modifications and derivatives calculations. Example of such optimization procedure can be seen in Vanderplaats (1979) and Hicks (1981).

3.2.1 Aerodynamic Analysis Code

Aerodynamic coefficients and flow variables are calculated using a viscous inviscid interaction (VII) code. This method solves the steady Euler equations in conservative formulation. The viscous effects are included through a coupled integral boundary layer which can be subjected to separation.

The discrete Euler equations are obtained by means of a finite volume technique applied on a structured mesh in which one of the grid lines in the flow direction corresponds to the flow streamlines. This approach is similar to the one adopted by Giles,et al (1985). In this method the continuity equation reduces to a statement that the mass flux in each streamline tube is constant and the energy equation is simply the conservation of enthalpy along the tube. This means that the flow perturbation is propagated through pressure field and geometry of the streamline tubes.

The viscous effects are simulated using a two-equations integral formulation, based on dissipation closure, for turbulent and laminar flow, (Le Balleur (1981)). The boundary layer equations are strongly coupled to the inviscid Euler equation using the dissplacement thickness concept.

The discrete Euler equations together with discrete boundary layer equations are solved by means of a Newton solution procedure using a direct Gaussian block-elimination method.

3.2.2 Geometrical modifications

The key point of the present methodology is to define an easy way to model changes in the geometry of aerodynamic shapes as a function of a set of parameters.

The geometry of a perturbed airfoil has been defined as a function of the normalized arc length as follows :

$$\vec{f}(s) = \vec{f}_0(s) + \vec{n}(s)\left[\sum_{i=1}^{i=N-1} A_i Tn(k_i, s) + BR(\rho, s)\right]$$

where :

- $\vec{f}_0(s)$ is the initial geometry airfoil.
- $\vec{n}(s)$ represents the unit normal vector to the initial surface. airfoil.
- $Tn(k_i, s)$ is the normalized Tchebicheff polynomial of k_i-th order.
- A_i is the parameter associated with $Tn(k_i, s)$.
- $R(\rho, s)$ is a normalized function to bump the LE. (ρ represents the curvature radius at the leading edge).
- B is the parameter associated with the bump.

The k-th normalized Tchebicheff polynomial $Tn(k, s)$ is defined as follows :

$$\begin{aligned} X &= 1 - 2 * S \\ \theta &= \arccos(X) \\ Tn(k,s) &= \begin{cases} (X \quad - \cos(k * \theta))/k & \text{if k odd} \\ (1 \quad - \cos(k * \theta))/k & \text{if k even .} \end{cases} \end{aligned}$$

The function $R(\rho, s)$ has been defined as:

$$R(\rho, s) = s^2(1 - s)\rho e^{-2\rho s}.$$

This definition leads to a vector $\{P\}$ of N components (modification parameters) which can control any geometrical change. Since both the Tchebicheff and the bump function are continuous with continuous derivatives in the whole range, the modified geometry of the airfoil will have no abrupt discontinuities neither in function nor in its derivatives.

Linearizing the geometrical perturbation around the initial geometry, one can get the flow solution over the initial shape and also over the N modified shapes obtained giving infinitesimal values to every modification parameter and treating them independiently one with respect to the others. Therefore, the solution and its linearized sensibilities with respect to the whole set of N geometrical modification spatial dimensions can easly be obtained.

This combination of parametrized geometrical changes and the derivatives of flow solution with respect to the parametric spatial coordinates is the basic tool used in inverse design and optimization. For both problems an iterative procedure has been developed.

3.2.3 Inverse Design

The starting data for Inverse Design are :

- A target function (i.e. $Cp_t(s)$).
- An initial geometry $\vec{f}_0(s)$.
- Initial flow solution $Cp_a(s)$.
- The derivatives of the flow solution with respect to every geometrical modification spatial dimension $\frac{\partial Cp_a(s)}{\partial \delta_i}$.

The objective is to find a geometrical shape which reproduces as best as possible the target function.

One posible methodology is to solve the problem of minimizing the functional of cuadratic error $U[Cp]$:

$$U[Cp] \quad = \quad \oint (Cp_t(s) - Cp(s))^2 ds \,.$$

If we linearize the Cp(s) distribution as :

$$Cp(s) = Cp_a(s) + \sum_i^N \frac{\partial Cp_a(s)}{\partial \delta_i} \cdot P_i \,.$$

We can establish the following linear system of equations in which the unknowns are the elements of the gometrical modification control vector $\{P\}$ (amplitudes related to every modification DOF), that minimizes the functional U.

$$\begin{cases} \oint \left(Cp_t - Cp_a - \sum_i^N \frac{\partial Cp_a}{\partial \delta_i} \cdot P_i \right) \cdot \frac{\partial Cp_a}{\partial \delta_j} \cdot ds = 0 \\ j = 1..N \end{cases}$$

$$\Rightarrow$$

$$\{P\} = [A]^{-1} \cdot \{B\}$$

where

$$[A] = \begin{bmatrix} \oint \left(\frac{\partial Cp_a}{\partial \delta_1}\right)^2 ds & \oint \frac{\partial Cp_a}{\partial \delta_1}\frac{\partial Cp_a}{\partial \delta_2} ds & \dots & \oint \frac{\partial Cp_a}{\partial \delta_1}\frac{\partial Cp_a}{\partial \delta_N} ds \\ \oint \frac{\partial Cp_a}{\partial \delta_2}\frac{\partial Cp_a}{\partial \delta_1} ds & \oint \left(\frac{\partial Cp_a}{\partial \delta_2}\right)^2 ds & \dots & \oint \frac{\partial Cp_a}{\partial \delta_2}\frac{\partial Cp_a}{\partial \delta_N} ds \\ \dots & \dots & \dots & \dots \\ \oint \frac{\partial Cp_a}{\partial \delta_N}\frac{\partial Cp_a}{\partial \delta_1} ds & \dots & \dots & \oint \left(\frac{\partial Cp_a}{\partial \delta_N}\right)^2 ds \end{bmatrix}$$

$$\{B\} = \begin{Bmatrix} \oint (Cp_t - Cp_a)\frac{\partial Cp_a}{\partial \delta_1} ds \\ \oint (Cp_t - Cp_a)\frac{\partial Cp_a}{\partial \delta_2} ds \\ \dots \\ \oint (Cp_t - Cp_a)\frac{\partial Cp_a}{\partial \delta_N} ds \end{Bmatrix}.$$

Once a vector $\{P\}$ is found, the shape is updated, and consecutive minimization steps are completed until the requiered quadratic error limit is reached.

3.2.4 Optimization

The starting data for an Optimization process are :

- An initial geometry $\vec{f_0}(s)$.
- Initial flow solution $Cp_a(s)$, and global aerodynamic coefficients.
- The derivatives of flow solution with respect to every modification spatial coordinate $\frac{\partial Cp_a(s)}{\partial \delta_i}$, and the sensibilities of the global aerodynamic coefficientes to infinitesimal displacements along these coordinates.

The objective of the optimization procedure is to find a geometrical shape which minimizes (or maximizes) a merit function (let it be $F(P_1, ..., P_N)$) of the global aerodynamic coefficients, having to take into account that the shape and others aerodynamic coefficients could be constrainted by some geometrical or aerodynamic limits (K constraints ; $Q_j(P_1, ..., P_N) = 0$).

Some usual merit functions to be minimized are Cd, Cd/C_L, $Cd/(C_L * C_M)$, while C_L to be maximized. A typical aerodynamic constraint is a especified C_L.

We have used the well known Stepest Descendent Methodology to find the direction in the parametric space, that modifies the geometry minimizing the functional (F), with the constraints (Q).

For taking into account the constraints we consider the function:

$$U = F + \sum_j^K \lambda_j Q_j .$$

Where the elements $\{\lambda_j\}$ are the K Langrange multipliers.

The Stepest Descendent Methodology chooses the search direction in the parametric space to be opposite to the local gradient.

$$P_i = -\epsilon \frac{\partial U}{\partial \delta_i} \,.$$

Since the magnitude of the elements of the vector $\vec{\nabla U}$ in the parameteric space are often very different, it is strongly recommended (if posible) to scale the parametric coordinates in order to make all the elements of the vector of similar magnitude.
If the scaling factors are assumed to be :

$$\tilde{\delta}_i = a_i \cdot \delta_i \,.$$

Then :

$$\left\{ \frac{\partial F}{\partial \tilde{\delta}_i} \right\} = \left\{ \frac{1}{a_i} \frac{\partial F}{\partial \delta_i} \right\} .$$

Let the vector $\left\{ \tilde{P} \right\}$ be the geometrical modification parameters in the new scaled parametric coordinates.

We have the following equation system to calculate the vector $\left\{ \tilde{P} \right\}$:

$$\left\{ \begin{array}{lcll} \frac{1}{\epsilon}\tilde{P}_i + \sum_j^K \lambda_j \frac{\partial Q_j}{\partial \tilde{\delta}_i} & = & -\frac{\partial F}{\partial \tilde{\delta}_i} & (i = 1,..,N) \\ \sum_i^N \tilde{P}_i \frac{\partial Q_j}{\partial \tilde{\delta}_i} & = & 0 & (j = 1,..,K) \,. \end{array} \right.$$

In matricial notation:

$$\left\{ \begin{array}{c} \left[\begin{array}{c|c} 1/\epsilon & {B_{ij}}^T \\ \hline B_{ji} & 0 \end{array} \right] \cdot \left\{ \begin{array}{c} \tilde{P}_i \\ \hline \lambda_j \end{array} \right\} = \left\{ \begin{array}{c} -\frac{\partial F_i}{\partial \tilde{\delta}_i} \\ \hline 0 \end{array} \right\} \\ \text{where} \\ [B_{ji}] = \left[\frac{\partial Q_j}{\partial \tilde{\delta}_i} \right] . \end{array} \right.$$

The general solution to this system is :

$$\left\{ \begin{array}{lcl} \{\lambda_j\} & = & \left[B_{ji} {B_{ij}}^T \right]^{-1} \cdot \left\{ [B_{ji}] \left\{ -\frac{\partial F}{\partial \tilde{\delta}_i} \right\} \right\} \\ \left\{ \tilde{P}_i \right\} & = & \epsilon \cdot \left\{ -\frac{\partial F}{\partial \tilde{\delta}_i} - [{B_{ij}}^T] \cdot \{\lambda_j\} \right\} . \end{array} \right.$$

The selection of the modification amplitude in the search direction is made with some trial and error loop : A low amplitude would produce very slow convergence in the minimization proccess whilst a high amplitude would even not produce any convergence at all, since in such a case second order terms would weight enough to make the linearizations senseless.

Once a vector $\{P\}$ is found, the geometry shape is updated, a new flow calculation is performed, a new constraint Jacobian is calculated and consecutive minimization steps are made until the desired goal is reached.

3.3 Test Cases

The above described procedures for optimization and inverse design have been applied to the two test cases TE2 and TE3 which were defined for the optimization workshop. Results of these test cases are given in the next section.

3.3.1 Test Case TE2 (Inverse Design)

For this test case, starting from NACA63215 airfoil geometry, an airfoil geometry which reproduces, as close as possible, the pressure distribution of a target airfoil (NACA0012) has to be found.
The design conditions are M=0.3, $\alpha = 3(^o)$, Re=3E6.
The requiered results are:

- Target and final airfoil coordinates.
- Final coefficient pressure distribution : (-Cp = f(x/c)).
- Cp cuadratic error integration normalized with its starting value, versus the number of flow solution calculations.
- Value of the function L2 versus the number of flow solution calculations, with this function defined as :

$$L2(m) = \frac{1}{N}\sqrt{\sum_i^N \left[\frac{y^m(x^t{}_i)}{c} - \frac{y^t{}_i}{c}\right]^2}$$

  where :

  - m is the number of iteration.
  - N is the number of points defining the target airfoil (coordinates (x_i^t, y_i^t)).
  - $y^m(x^t{}_i)$ represents the Y-coordinate of a point of the airfoil reached in the m iteration, at the same X-coordinate as that of the target i-point.

- Lift, drag and pitching moment coefficients versus the number of flow solution calculations.

As can be seen in the figures (1,2) the IsoMach lines for the final and NACA0012 airfoils are almost identical.
Figures (3,4) show that after as few as two cycles the shape and Cp distribution are indistinguish.

Figure 5 shows that the objective function (cuadratic error of Cp distribution) reduces dramatically in just one steep. This has an adverse effect: the right hand terms that drive the geometry modification procces vanishes while still remain a small residual in the L2 function (fugure 6.) and Cd function (figure 8). However we believe that after 6 or 7 cycles the problem is reasonably solved.

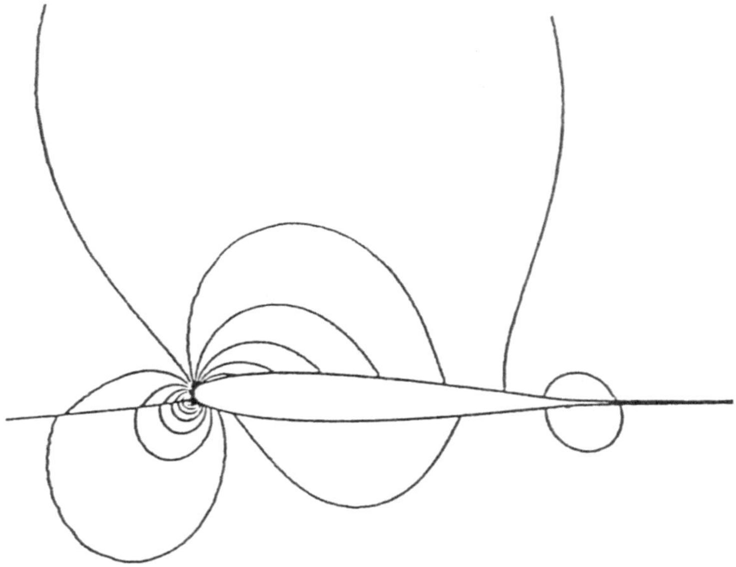

Figure 1. Isomach lines target airfoil (NACA0012).

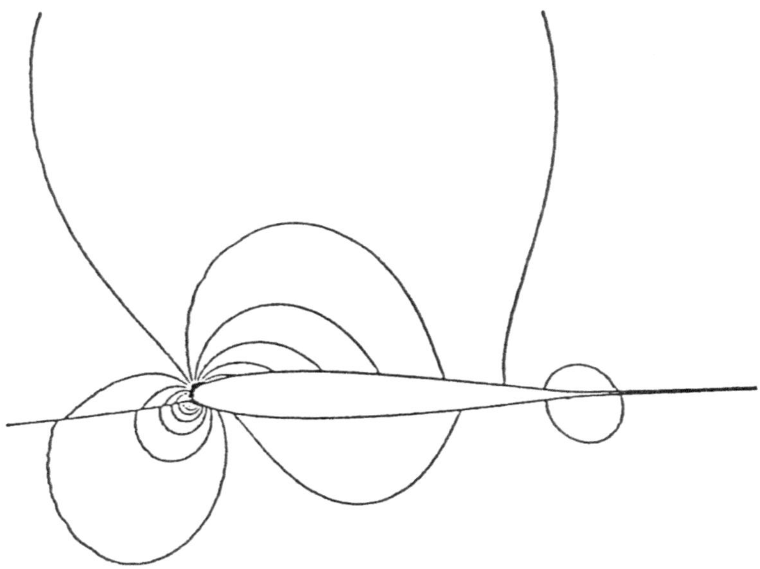

Figure 2. Isomach lines final airfoil.

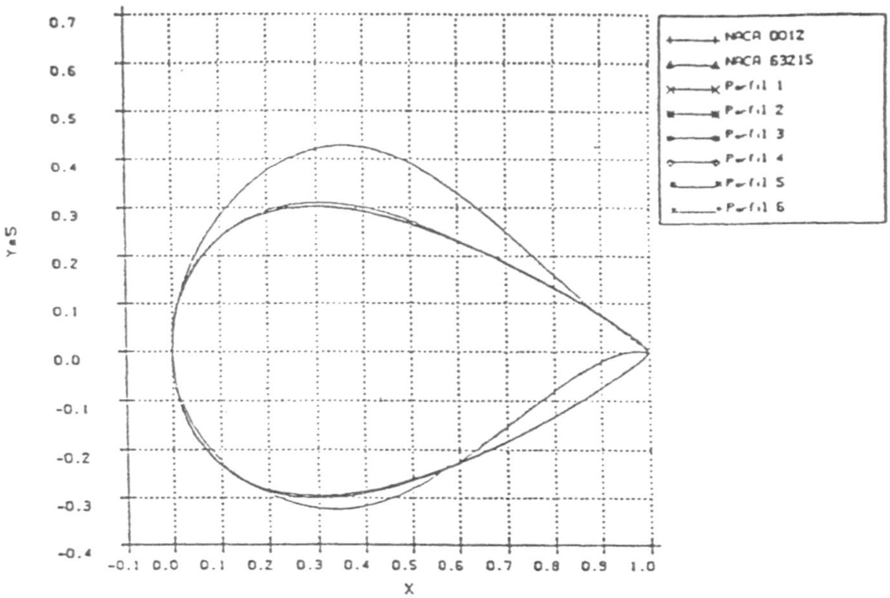

Figure 3. Airfoil shape evolution.

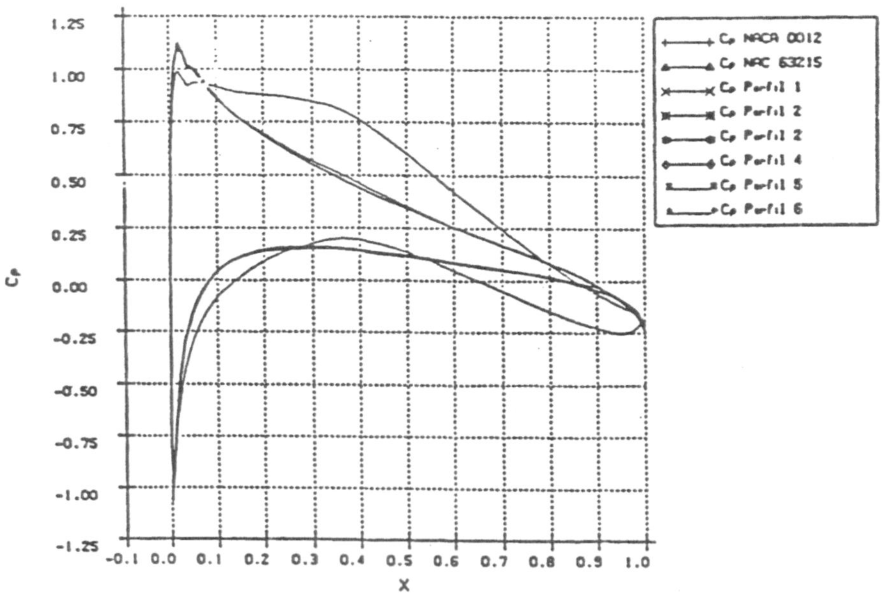

Figure 4. Cp evolution.

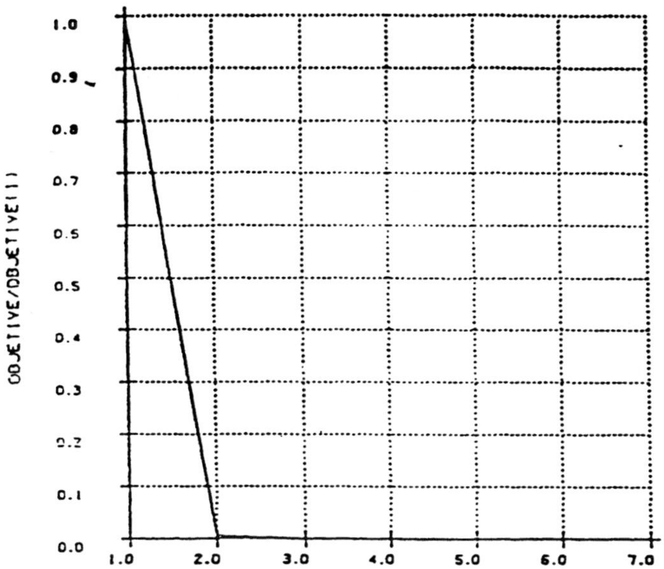

Figure 5. Cp cuadratic error integration evolution.

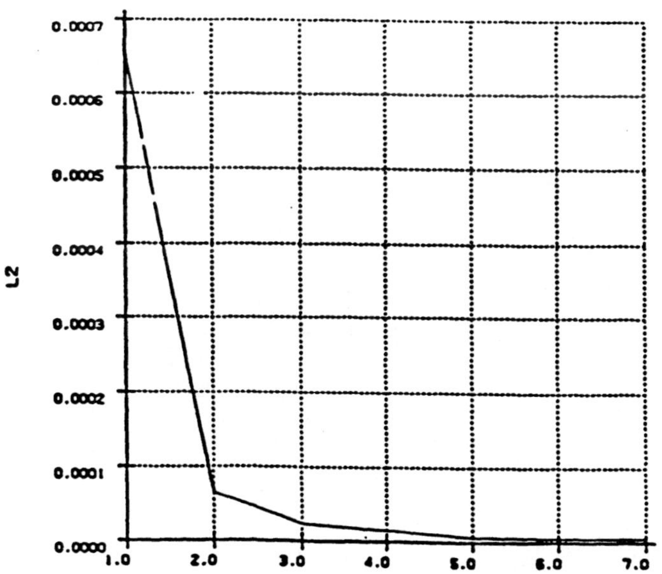

Figure 6. L2 function evolution.

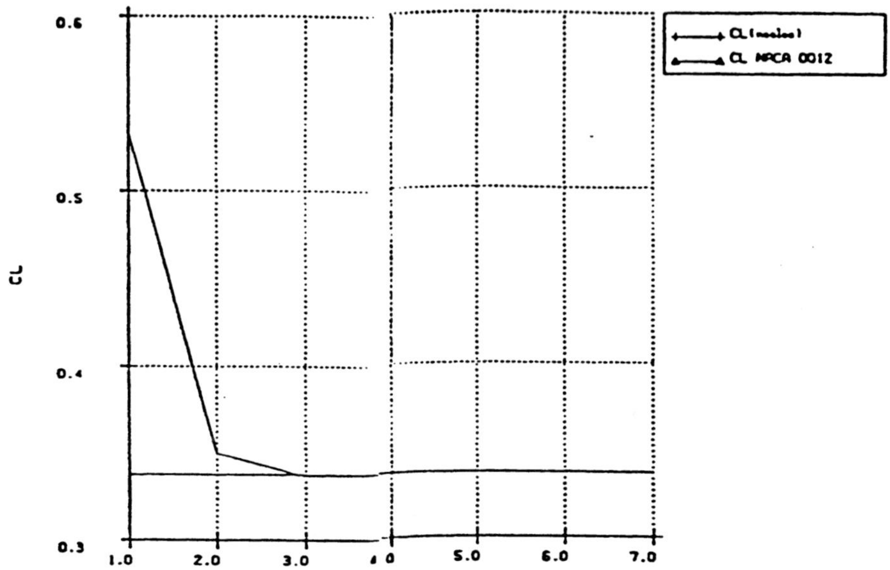

Figure 7. C_L evolution.

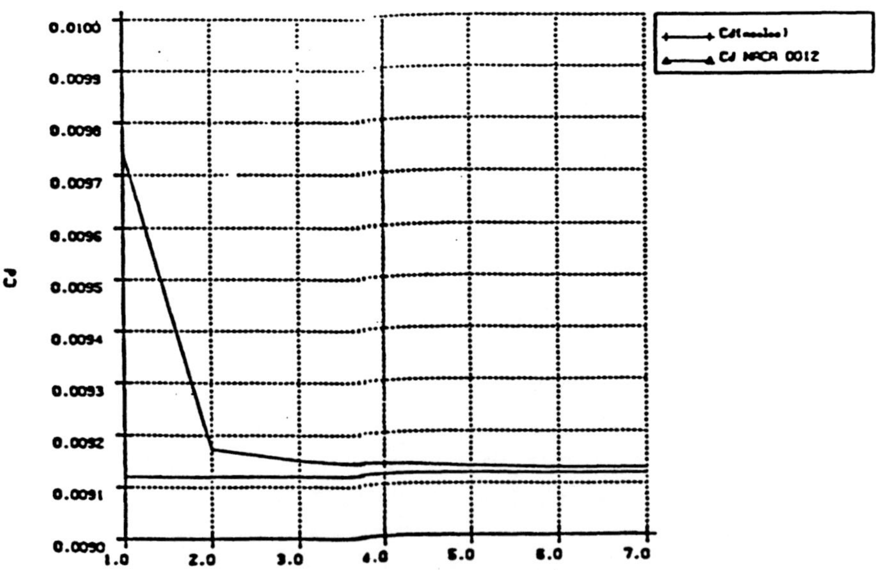

Figure 8. C_d evolution.

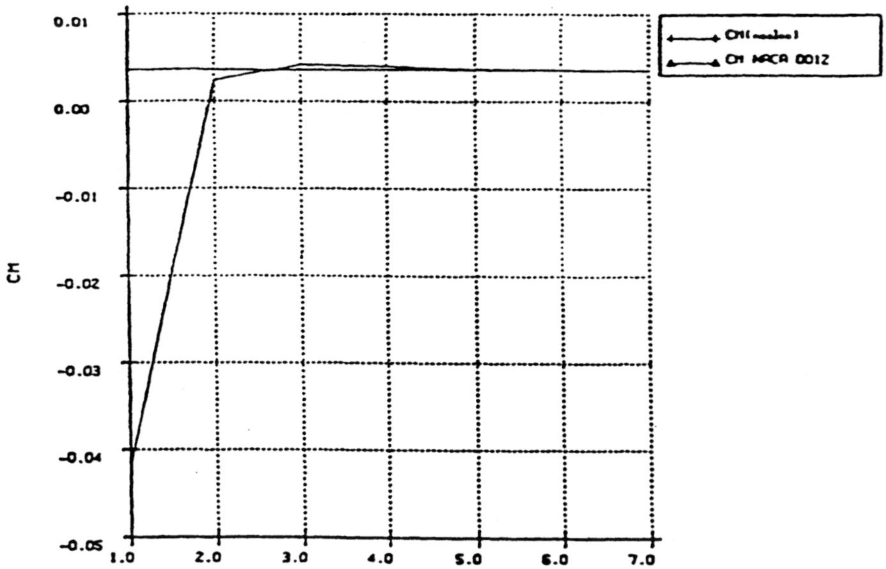

Figure 9. C_M evolution.

3.3.2 Test Case TE3 (Optimun Design)

For this test case starting from the airfoil RAE2822 at M=0.73 and $\alpha = 2(^o)$, a new airfoil geometry has to be found, which has a minimum wave drag while the lift is mantained to its original value ($C_L = 0.8871$). Viscous effects are not included.

The requiered results are:

- Initial and final airfoil coordinates.
- Initial and final coefficient pressure distribution : (-Cp = f(x/c)).
- Lift, drag and pitching moment coefficients versus the number of flow solution calculations.
- For the final solution and over the complete grid Mach number distribution. (Isomach lines).
- Also for the final solution and over the complete grid Cp distribution. (IsoCp lines).

It was decided to consider this problem as a pure optimization application instead of other possibilities proposed.

After 5 optimization cycles a remeshing was made to increase the grid density (from a grid with 150x30 nodes to 200x40).

Even though after 49 optimization cycles were completed zero drag had been reached, since the original and the airfoil Cp distribution presented a small oscilation around (0.7c) on its suction surface, a cycle of inverse design requiring a pressure distribution without such oscillation was made. A second remeshing to increase the grid density and numerical resolution was found necessary (to 300x50 nodes).

Ten more cycles were performed on the finest grid, driving eventualy the drag numerically to zero while the lift to the fixed original value.

The effects of remeshings are clearly shown in the coefficient evolutions (figures 12,13,14).

Although, the numerically obtained drag for this optimized airfoil is zero, the IsoMach picture (figure 17.) shows that there is still a weak shock wave. We believe that with highly dens meshes, and consecuently more computational time, the shock wave could eventualy vanish.

At any cycle we have achieved some convergence but we think it is not good enough to drive the design proccess fairly.

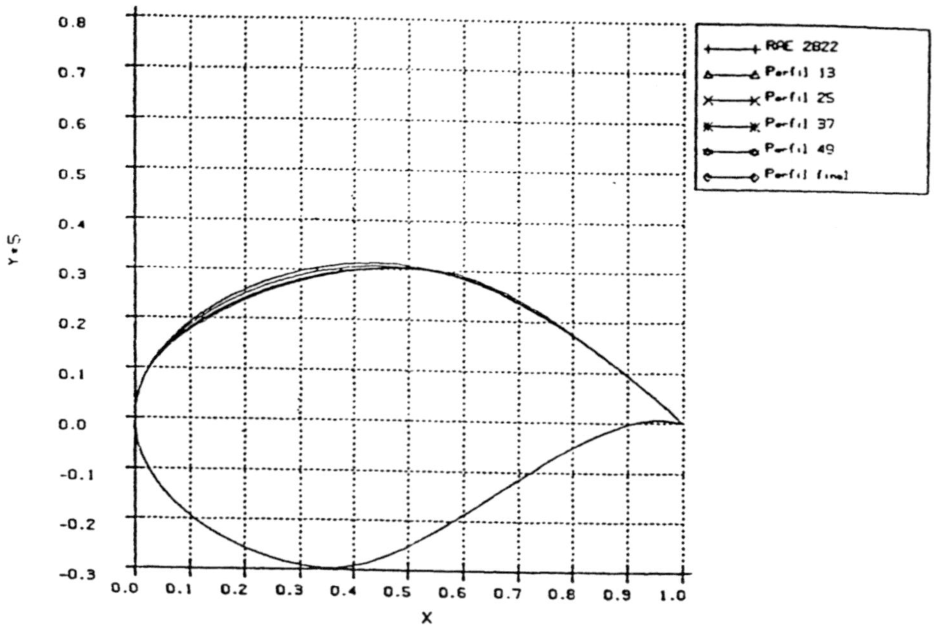

Figure 10. Airfoil shape evolution. (1/6 cycles).

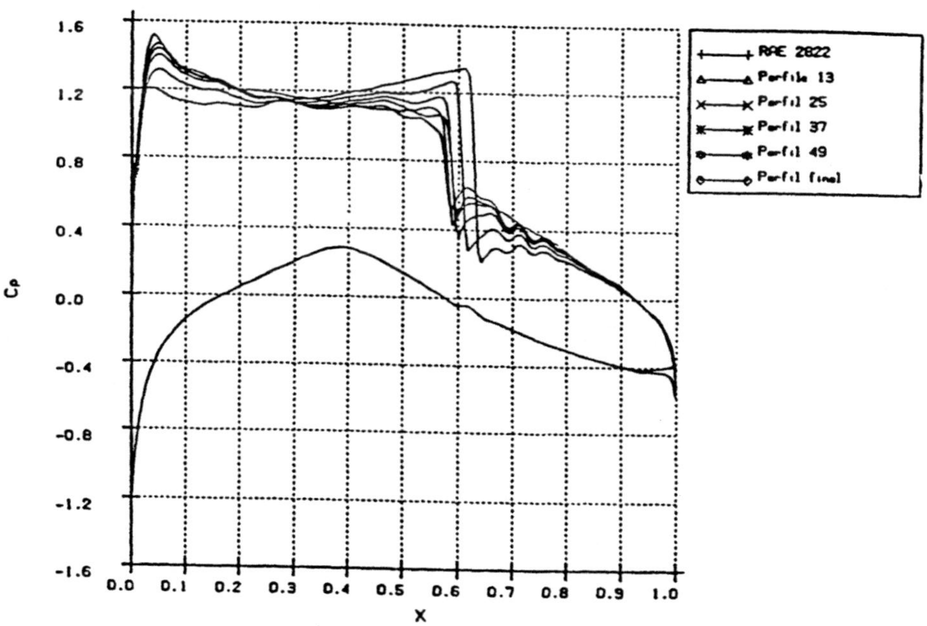

Figure 11. Cp distribution evolution. (1/6 cycles).

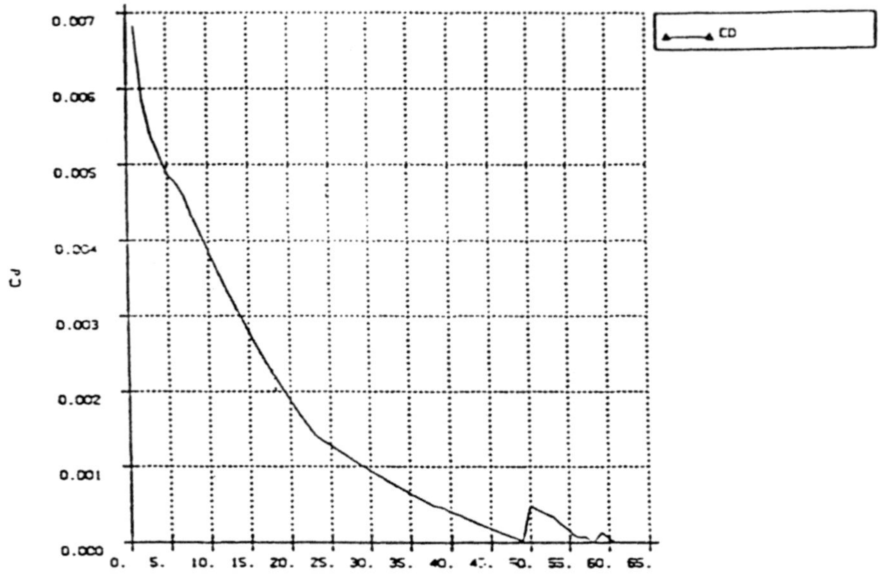

Figure 12. C_d **evolution.**

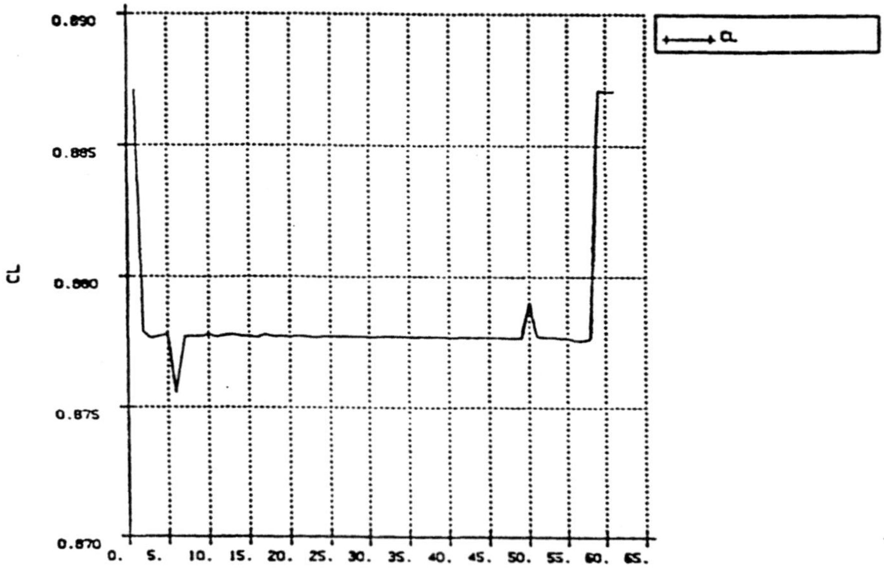

Figure 13. C_L **evolution.**

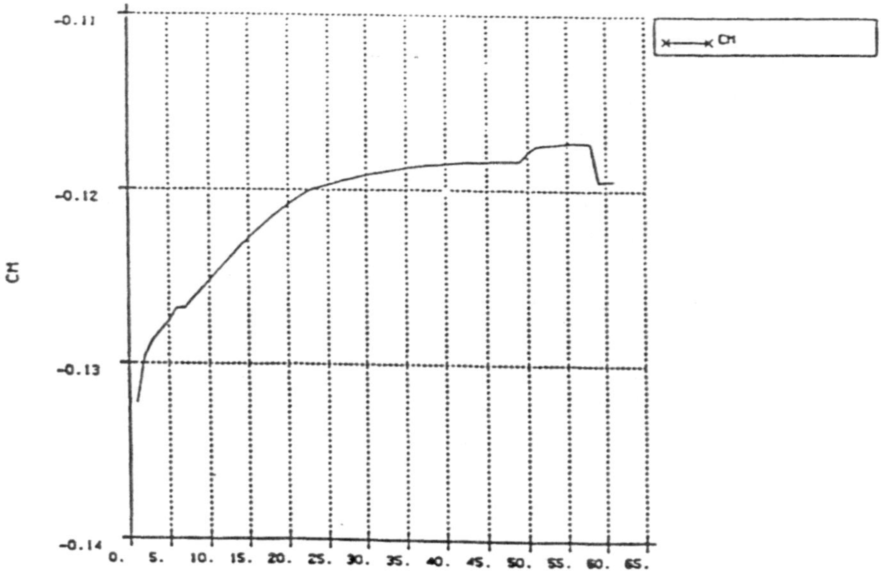

Figure 14. C_M evolution.

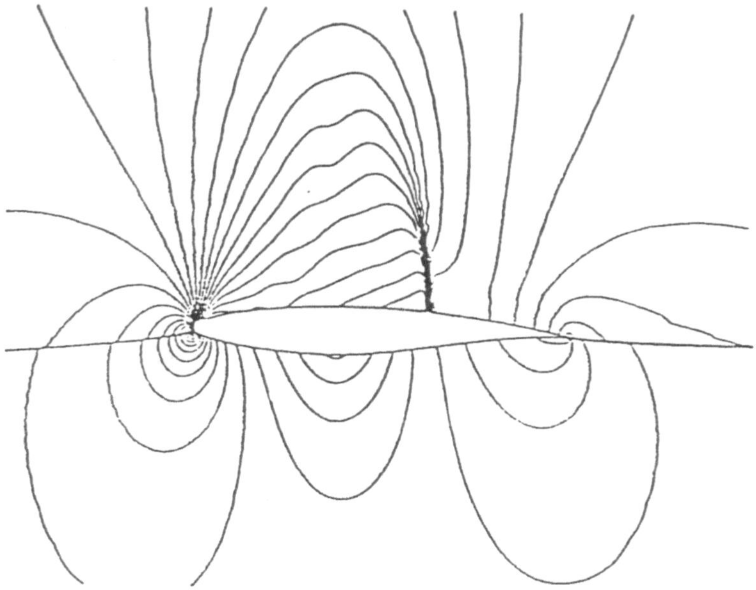

Figure 15. IsoMach lines initial airfoil (RAE2822).

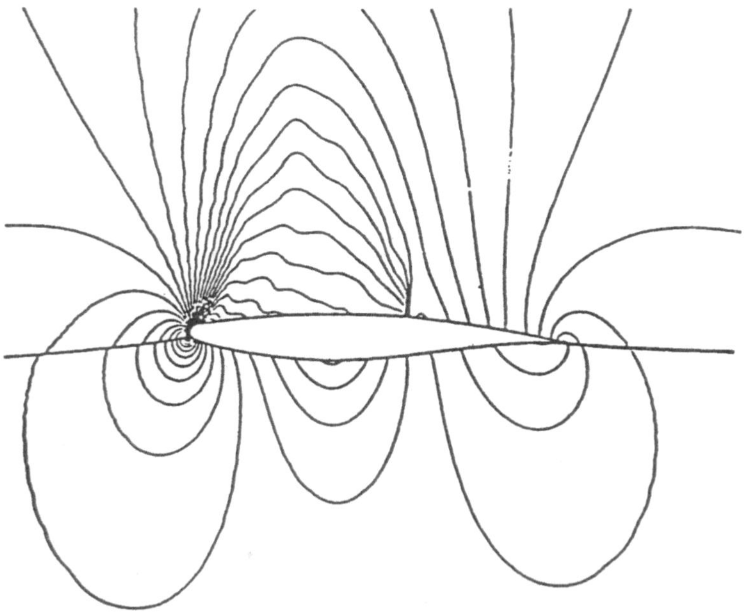

Figure 16. IsoMach lines final airfoil.

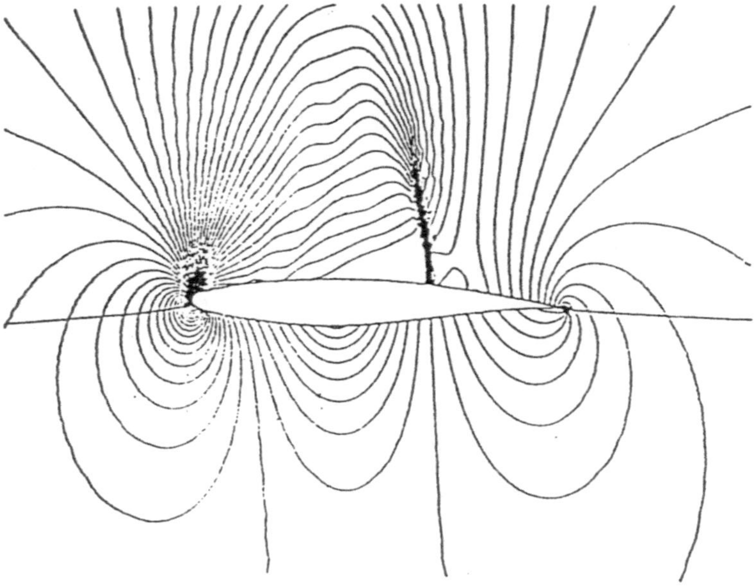

Figure 17. IsoCp lines initial airfoil (RAE2822).

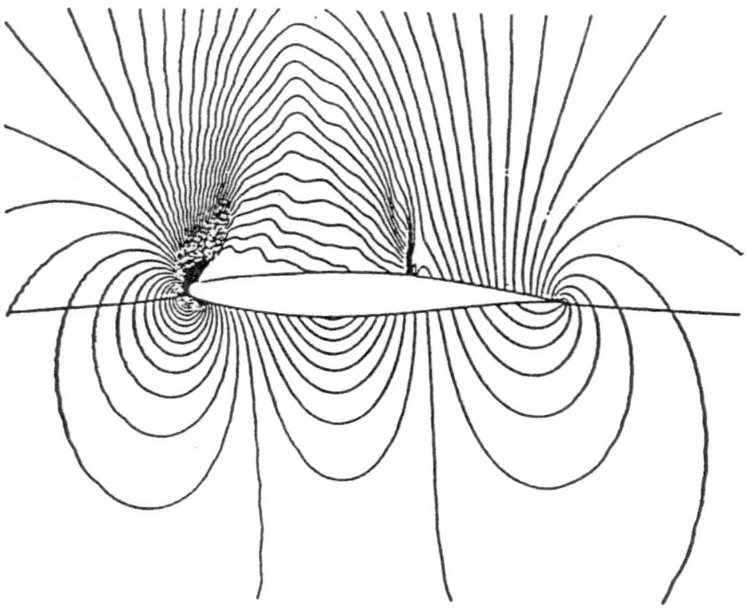

Figure 18. IsoCp lines final airfoil.

Bibliography

Hicks, R.M. (1981) Transonic Wing Design Using Potential Flow Codes - Successes and Failures. SAE paper 810565.

Giles, M.B., Drela, M., Thompkins, W.T. (1985) Newton Solution of Direct and Inverse Transonic Euler Equations. AIAA, 85-1530.

Vanderplaats, G.N. (1979) An efficient Algorithn for Numerical Airfoil Optimization. AIAA, 79-0079.

Le Balleur, J.C. (1981) Strong Matching Method for Computing Transonic Viscous Flows Including Wakes and separations on Lifting Airfoils. La Recherche Airospatiale 1981-83.

4 Optimization of the Wing–Pylon–Nacelle Testcase TE5 by HISSS–D, a Panel Method–Based Design Tool

Luciano Fornasier
Daimler–Benz Aerospace AG, München, Germany

4.1 Introduction

The aerodynamic design process is aimed at defining the overall aircraft shape that maximises the aerodynamic performance for the sizing mission(s). With the global design parameters as general configuration arrangement and wing planform and volume being determined by parametric studies driven by aerodynamic, propulsion and structural considerations, the task of the detailed aerodynamic design consists in optimizing the aircraft external surface for performance at the design condition(s) subject to given constraints and requirements. With the advent of modern computational methods the designer has powerful tools for driving the aerodynamic process by direct analysis of different geometries. However, a systematic investigation of geometrical modifications is still a very time–consuming process for practical purposes. Instead, 'ad hoc' numerical methods can be used for generating the sought–after geometry starting from an initial configuration. In the 'classical' inverse formulation, a specified pressure distribution is used to derive – either in an analytical way or by use of an iterative approach– the geometry suitable to yield – for the given flow conditions – the target surface pressures. Apart from the fact that this kind of methodology is 'de facto' limited to the two–dimensional case, inverse methods suffer inherently from lack of geometrical control. Since the final shape depends directly on the 'target' pressure only, much care must be exercised in order to prevent the generation of undesired geometrical characteristics. In the optimization approach, 'admissable' geometrical variations are computed for minimisation of proper aerodynamic objective functions – e.g. drag, least square deviation from a target pressure distribution with consideration of both aerodynamic and geometrical constraints. This class of methods allows the user to retain a good control on the geometry, in principle 'trading' the best allowable geometrical modification from the initial configuration against the minimum deviation from the desired target. In principle, any flow solver may be embedded in the optimization process, where it is used for calculating the value of the objective function during the iterations and the derivatives of the functional with respect to the design variables when a gradient–based optimization algorithm is used. Hence, the computation for an optimization task depends on the unitary cost of one flow analysis, the number of design variables used and the amount of analysis calculations required.

A design method for arbitrary aircraft configurations at subsonic and supersonic speeds has been developed by this author. This method – called HISSS–D – couples the fast–convergence characteristics of inverse methods with the constraint–capabilities of the optimization methods. Due to the very flexible geometrical modelling capabilities of the aerodynamic analysis code, HISSS–D is well suited for the solution of complex 3–dimensional problems. Moreover, the subsonic and supersonic capability of the analysis code can be exploited for multi–point optimization studies.

Within the frame of the Brite–Euram Project 2003 – ECARP – Optimum Design Area – some basic studies on the influence of the choice of design variables on convergence characteristics of the optimisation process used in HISSS–D have been carried out. In the second period of the project, the method has been applied to the testcase TE5 of the Workshop, a wing–pylon–nacelle integration exercise defined by the ECARP Project Partner Aerospatiale.

The following of the paper is organized as follows. First, the HISSS–D method is briefly presented. Then the results from the application of HISSS–D to the wing–pylon–nacelle optimization testcase are discussed.

4.2 The HISSS–D Design Tool

HISSS–D is a three–dimensional optimization procedure built on top of an in–house developed panel method for subsonic and supersonic flow regimes – called HISSS – which is used to provide the analysis of the initial geometry configuration and the gradients of the target functional in the optimisation loop. The two components of HISSS–D are presented briefly in the following.

4.2.1 The Analysis Module

HISSS – Higher–order Subsonic/Supersonic Singularity Method (Fornasier, 1987) – is a panel method developed at Dasa–M – formerly MBB–Military Aircraft Division – following the formulation pioneered by the Boeing PANAIR code (Ehlers et al, 1979) for overcoming the numerical stability problems experienced by low–order panel codes when calculating supersonic flows (Fornasier, 1984). Both codes use linear distributions of sources and quadratic distributions of doublets distributed on planar subelements which build up a quadrilateral panel. While PANAIR subdivides each panel into eight triangular subpanels, four triangular subpanels and one quadrilateral subpanel lying in the panel average plane are used in HISSS, thus reducing the computational effort in assembling the influence coefficients by approximately one third. Set of panels carrying the same kind of singularities – i.e. sources, doublets or a combination of sources and doublets – are assembled together to form a network and the whole configuration is modelled by a set of networks distributed along its external surface. Boundary conditions are imposed at each panel control point for determining the strength of the singularity distributions. Networks carrying both sources and doublets requires specification of two boundary conditions. A Neumann–type boundary condition is normally used for specifying the value of the normal component of local velocity – which must be set to zero on solid walls – while a boundary condition of Direchlet–type is used to prescribe zero perturbation potential inside closed volumes. It can be shown that by this combination of boundary conditions the local strength of the sources and of the doublets are minimized, thus generally allowing to reduce the discretization errors. In supersonic flows, this formulation allows to avoid the propagation of disturbances inside the configuration, a numerical problem experienced using surface singularity distributions carrying sources or doublets. Although most of these features are strictly necessary only when computing supersonic flows, they contribute to make the codes robust with respect to spacing variations and local mesh distortions, with overall positive impact on modelling flexibility and accuracy of results in applications to complex geometries.

4.2.2 *The Optimization Module*

The above presented panel method HISSS is used to provide the value of the objective function and of its gradients with respect to the design variables defined by the user. As far as the dependency of the aerodynamic characteristics on the shape of the configuration surface is concerned, the optimization is 'linearised', in the sense that the influence coefficients induced by the singularity distributions are not updated during the iterations, i.e. the singularity distributions are 'frozen' at their location of the initial geometry. Hence the variation of the surface pressure due to the perturbation of the surface geometry depends only on the variation of the local normal vector at the control points of the panels. At each iteration step the value of the panel normal depends on the panel edge coordinates, which may be considered the primary design variables. Finally, using the chain differentiation rule and exploiting the inverse matrices of the aerodynamic influence coefficients a direct relationship between the surface pressure and the position of the mesh node points describing the configuration surface is established for the actual flow conditions, i.e. Mach number and angle–of–attack.

Possible choices for the definition of the objective function are the least–square deviation from a target pressure distribution and/or the total inviscid drag coefficient (pressure drag + lift–induced drag). A multigradient technique – the variable metric method of reference (Vanderplaats, 1984) – is used to minimize the objective function subject to a number of aerodynamic and geometrical constraints. Aerodynamic constraints encompass global lift and pitching moment coefficients. Since the analysis code is inviscid, bounds on attainable minimum pressure and streamwise pressure gradients can be specified as additional constraints. During the iterative modification of the initial geometry toward the "optimal" one the technique of the Lagrangian multipliers is used to account explicitly for geometrical constraints, which control total configuration volume, wing spar box area at several span stations, minimum (positive) airfoil thickness and leading edge radii distribution. Weak viscous effects can be taken into account by introduction of surface transpiration for modelling boundary layer thickness.

Finally, a penalty term based on the value of the fourth–order geometrical derivative calculated at the location of the design variables is taken into account in the functional for damping out numerical instabilities possibly leading to surface waviness, a problem related to the linearization of the surface–geometry relationship.

In the first period of the ECARP project, some basic studies have been carried out aiming at assessing the influence of the type of design variables and of different search algorithms on the convergence characteristics of the unconstrained minimisation constraints procedure used in HISSS–D. Results have clearly shown the superiority of the multi–gradient method over simpler steepest descent and conjugate gradient methods with respect to overall convergence and robustness characteristics. As far as the definition of design variables is concerned the possibility of using the control points of Bezier spline, B–spline and non–uniform rational B–spline fittings as alternative to the standard choice of panel corner points has been implemented.

4.3 Optimization of the Wing–Pylon–Nacelle Test Case TE5

During the ECARP Project, the HISSS–D method has been applied to the testcase TE5 of the Workshop, a wing–pylon–nacelle configuration defined by the project partner

Aerospatiale. The optimization task – as agreed with the ECARP partners Aerospatiale and Dasa–Airbus, which have applied their own optimisation tools to the testcase TE5, consists of the redesign of a typical subsonic transport wing in the region close to the (underwing) power plant installation. Target of the optimisation exercise is the minimisation of the interference due to the presence of the pylon and nacelle on the wing pressure distribution for the reference flow condition Mach 0.50 and 2.0 degs angle of attack. The conditions for the optimization exercise have been defined in order to allow a fair comparison of the results obtained by the partners using rather equivalent but yet different optimisation strategies and numerical implementation. Accordingly, the target pressure distribution has to be derived from the analysis at the reference flow conditions of the wing geometry without pylon and nacelle, and the redesign area has been limited to a portion of the wingspan adjacent to the pylon position.

The configuration geometry was provided by Aerospatiale in form of a panel grid for the wing alone and the wing/pylon/nacelle configurations. Beside the conversion from the Aerospatiale format into the proper HISSS–D format, the wing/pylon/nacelle panelling had to be modified in order to repair the edge mismatches between the pylon patch and the panel halo used to connect the pylon to the wing surface, see Figure 1. This was necessary in order to comply with the more stringent panelling rules of HISSS–D in comparisons to the low–order panel methods used by Aerospatiale and possibly Dasa Airbus. For the sake of compatibility with the original data, the modification of the grid was limited to the inner contour of the halo, and consisted in moving the edge points either to the closest wing corner points or to project them onto the closest wing panel edge. For the sake of simplicity, these edge points were not allowed to move during the optimization iterations.

Starting from the modified panelling, two "optimum" geometry solutions have been produced by using two different sets of design variables and associated geometrical constraints:

- Variant 1: vertical displacement of panel corner points in the redesign area used as design variables; fourth–order damping for geometrical smoothness along grid lines both chord– and spanwise.
- Variant 2: vertical displacement of the parameters of B–spline fittings in chord direction used as design variables, together with smoothing constraints in span directions only.

Figure 2 presents the deviation of the surface pressure of the wing–pylon–nacelle from the equivalent one obtained for the wing alone configuration. As can be seen, the interference effects are concentrated at the wing leading edge, where the minimum pressure peak of the wing–pylon–nacelle are decreased with respect to the wing alone solution – with possible negative effects on the transonic characteristics – and at the intersection of the pylon with the lower side of the wing, evidenced by a weak recompression area. Globally, these effects are – for the testcase flow conditions – quite weak and have a strongly local, three dimensional character. The rapid decay of the disturbances away from the pylon footprint justifies the assumption to limit the redesign area to the dashed region shown in Figure 2.

Figure 3 and 4 show the equivalent deviations computed for the optimized geometries, Variants 1 and 2. The comparison with figure 2 shows that the deviation levels have been largely reduced. In particular, the loss of expansion peak at the wing leading edge and the distortions of the isobar pattern over the wing upper surface have been almost completed recovered by both solutions. At the wing lower surface the effects due to the presence of the pylon could only be marginally attenuated, most likely due to the requirement of 'smooth' geometrical modifications. Both Variant 1 and 2 show the same qualitative trend, both in terms of recover of pressure deviations and of geometry modifications, Figure 5 through 9. The isoclines of the vertical displacements are represented in the Figures 5 and 6, while comparison of Figure 8 and 9 with Figure 7 allows to get an impression of the involved geometry modifications by presenting the original and the distorted grids in the redesign area.

The comparison of the analysis results for the original geometry and for the two optimized solutions are presented in form of iso–Mach distributions in the Figures 10, 11 and 12. As can be seen, both Variant 1 and 2 are suitable to produce a slight increase of the maximum local Mach numbers. At the section immediately outboard of the pylon the local maximum Mach numbers increase from 0.72 (original geometry) to 0.76 and 0.756 (Variant 1 and 2).

4.4 Conclusion

The panel–based HISSS–D design tool has been applied to the wing–pylon–nacelle test case TE5. Results from the design exercise – which consisted in the recovery of the wing–alone pressure distribution for the wing in presence of pylon and nacelle – have shown that the method is able to produce 'feasible' geometrical modifications which satisfy at best the design goal. Due to explicit consideration of geometrical and aerodynamic constraints – which were within the scope of the TE5 exercise, it is thought that HISSS–D can be successfully applied to 'real' project applications.

4.5 Bibilography

Fornasier, L. (1987): HISSS – A Higher–Order Panel Method for Subsonic and Supersonic Attached Flow About Arbitrary Configurations. In: Panel Methods and Fluid Mechanics, Notes on Numerical Fluid Mechanics , Vol. 21. Vieweg Verlag.

Ehlers, F.E., et al. (1979): 'A Higher Order Panel Method for Linearized Supersonic Flow. NASA CR–3062.

Fornasier, L. (1984): Treatment of Supersonic Configurations by an Updated Low–Order Panel Method. J. of Aircraft, **21**.

Vanderplaats, G.N. (1984): Numerical Optimization Techniques for Engineering Design. Mc Graw–Hill Company.

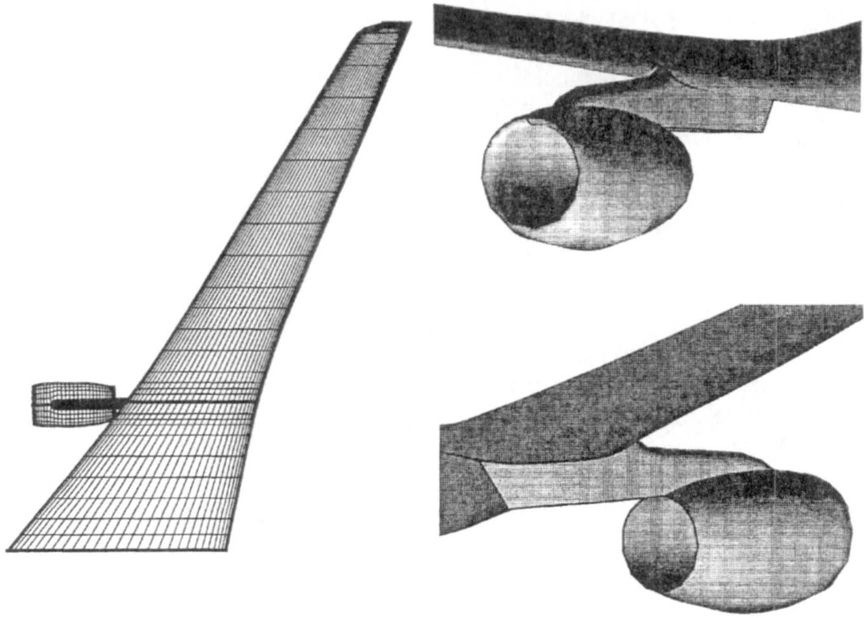

Figure 1 Initial Geometry for test case TE 5 – Panel Grid

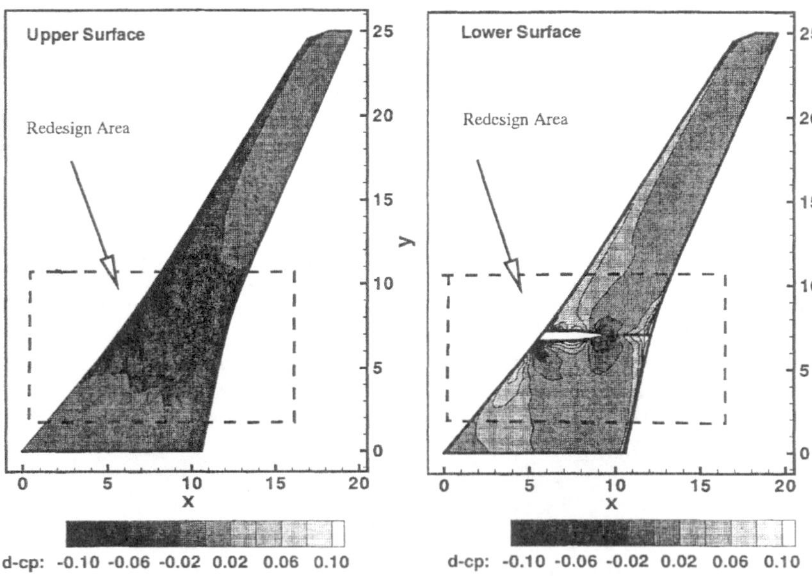

Figure 2 Pressure Deviation for Initial Geometry

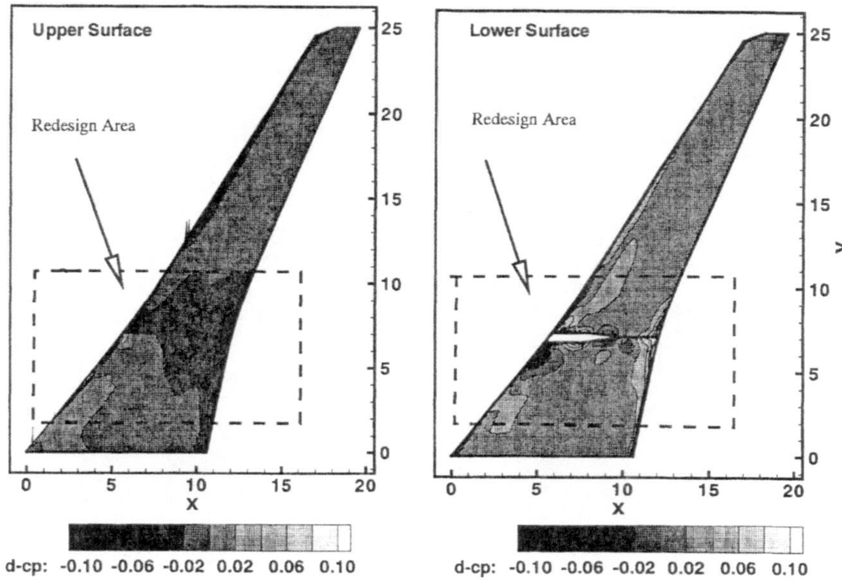

Figure 3 Pressure Deviation for Optimized Geometry – Variant 1

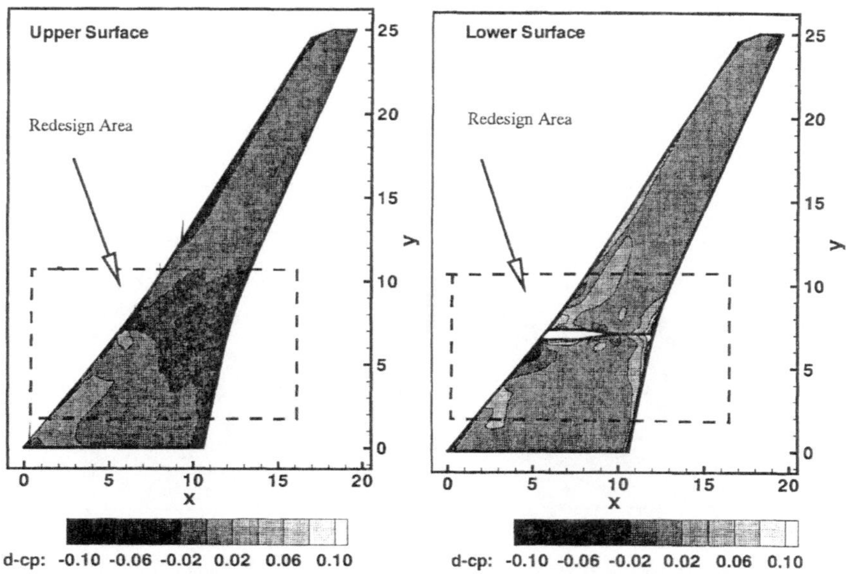

Figure 4 Pressure Deviation for Optimized Geometry – Variant 2

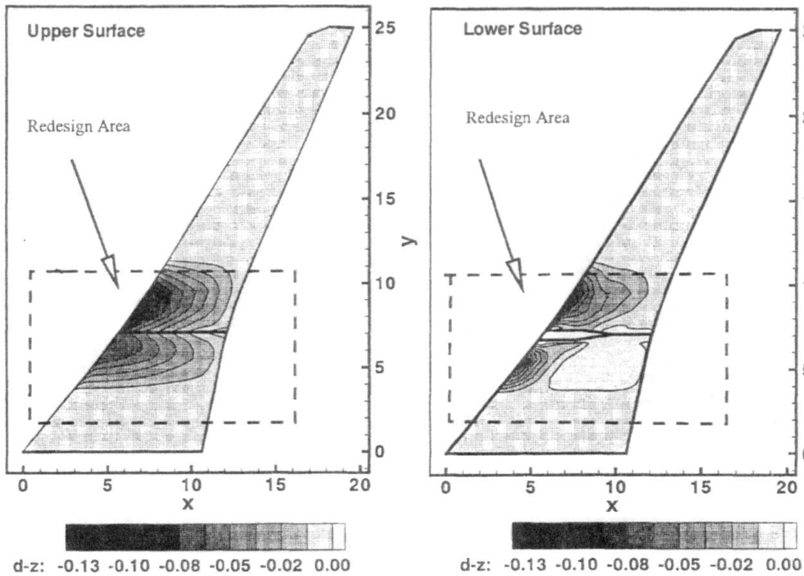

Figure 5 Vertical Displacement of Design Variables for Optimized Geometry – Variant 1

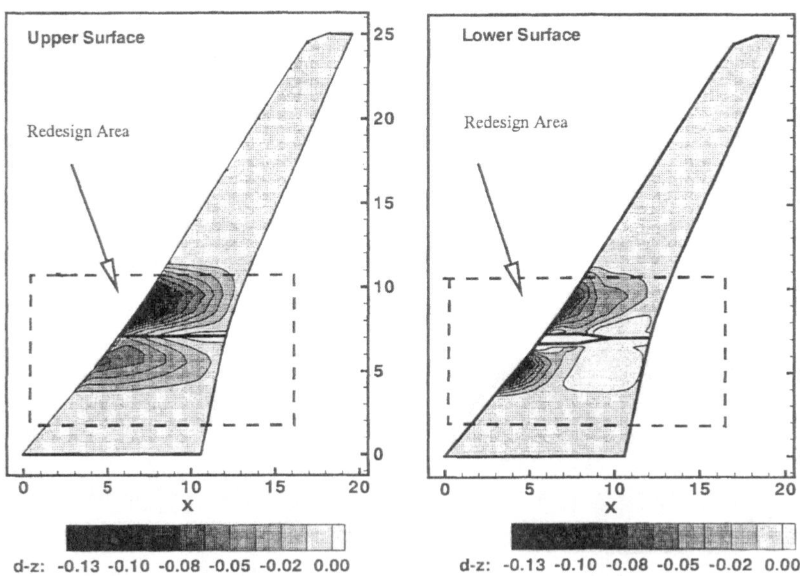

Figure 6 Vertical Displacement of Design Variables for Optimized Geometry – Variant 2

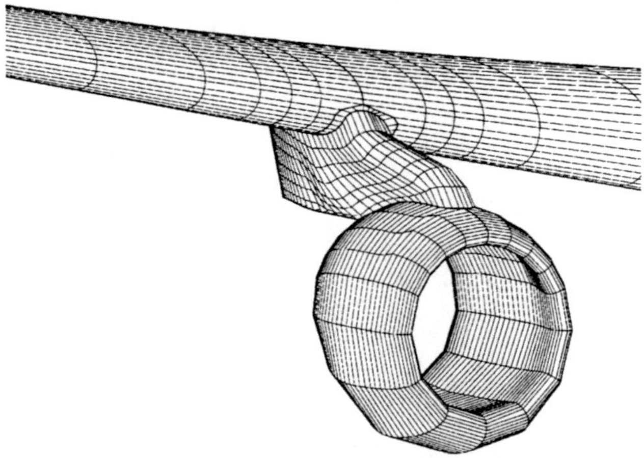

Figure 7 Panel Grid for Initial Geometry for Test Case TE5

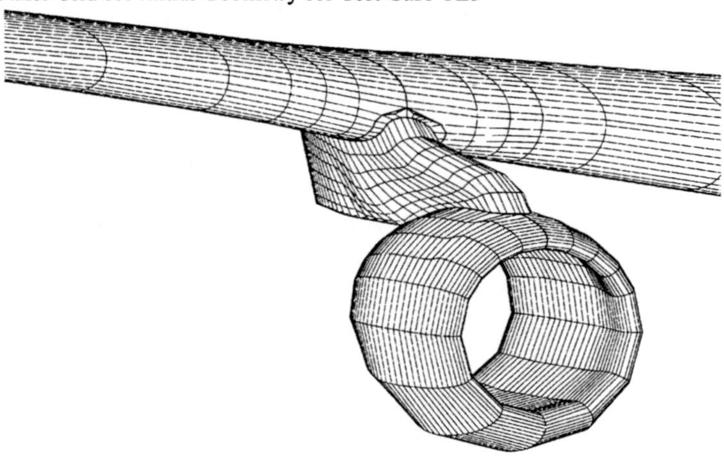

Figure 8 Panel Grid for Optimized geometry – Variant 1

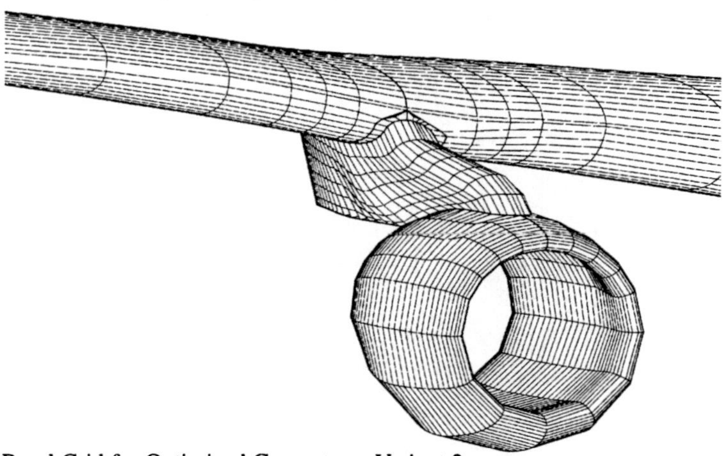

Figure 9 Panel Grid for Optimized Geometry – Variant 2

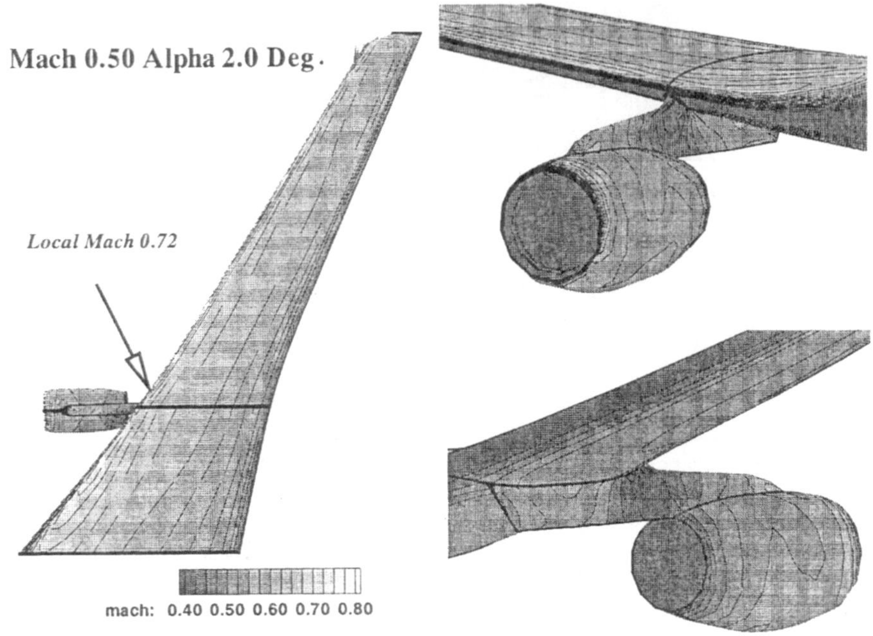

Figure 10 Mach Distribution for Initial Geometry

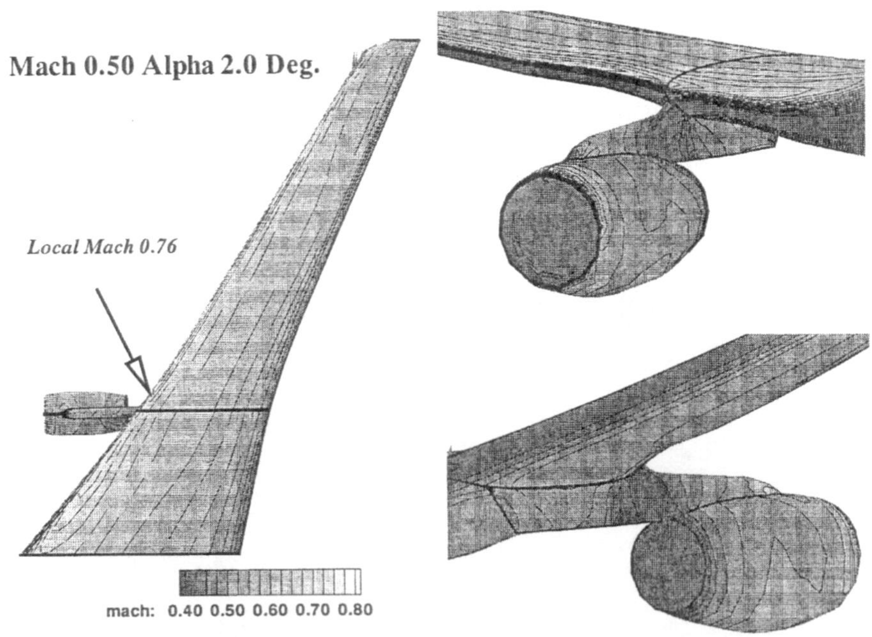

Figure 11 Mach Distribution for Optimized Geometry – Variant 1

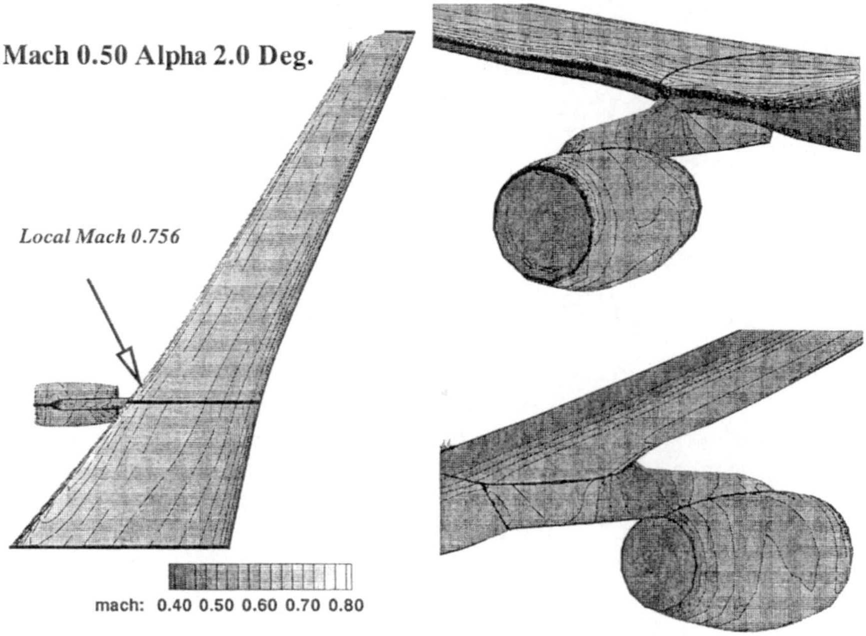

Figure 12 Mach Distribution for Optimized Geometry – Variant 2

5 Design of a Wing-Pylon-Nacelle Configuration

Helmuth Schwarten, Daimler-Benz Aerospace Airbus GmbH, Bremen, Germany

Methodology

Problem Formulation

The design task of producing a wing–pylon–nacelle configuration with specified wing pressures ('target pressures') is formulated as a discrete minimization problem. The wing surface is given in terms of a set of parameters $\vec{a} = (a_1, \ldots, a_n)$ (design parameters, to be explained later), whose definition range supplies all potentially occuring surfaces during design procedure. The flow code provides the corresponding pressures, or, expressed as a functional relationship, transforms parameters to pressures[1].

Thus, the pressure coefficients $\{c_{p_i}\}, i = 1, \ldots, m$ at a set of control points may be regarded as given functions of these parameters

$$c_{p_i} = c_{p_i}(a_1, \ldots, a_n)\,, i = 1, \ldots, m\,.$$

The design task is taken to be solved, if such parameters have been determined, that minimize the sum of weighted, squared pressure deviations

$$E(\vec{a}) = \sum_i \sigma_i \left(c_{p_i}(a_1, \ldots, a_n) - c_{p_i}^{target}\right)^2 \stackrel{!}{=} min.$$

This procedure solves the invers design approximately, but reduces trouble with conflicting or non–physical target pressures. Especially, omitting problem target pressures at stagnation region or the trailing edge region by setting the corresponding weights to zero, has proven to be very useful. Normally, target pressures are prescribed on surface patches under design. But this can be 'decoupled'. For instance, at a wing–body–conjuntion target pressures could be prescribed on inner wing sections, while changing the geometry of the body [1].

A further advantage of this 'invers design by direct optimization' is, that one's favorite flow code could be forced to do invers tasks, because the minimum search requires no special aerodynamic formulation. Unfortunately this has to be paid with long calculation times. Therefore, 3D–designs are restricted to potential flow. In the present implementation (panel method for subsonic flow) configurations up to ca. 3000 panels with about 100 design parameters take 6 to 8 CPU–hours on a HP735 workstation for inviscid runs and approximately double the time for runs with viscous corrections.

[1] The Jacobian of this transformation plays a fundamental role in the minimization process, see section minimization procedure.

The method is a generalization of a 2D–version, that has been proven to work very well for both subsonic and transonik flow [2], [3]. The main topics of the design method should now be presented in some detail. These are (i) the flow model, (ii) the minimization procedure and (iii) the surface representation.

Flow model

In the present case aerodynamic calculations are based on subsonic potential flow. The well known and widely spread VSAERO–code [4], [5] is chosen as 'aerodynamic driver' for the following reasons: treatment of full aircraft configurations, inclusion of nonlinear effects by wake iterations, user friendly input organization as well as fast calculation and reliable results.

The panel model is based on potential formulation for the perturbation velocity. It uses quadrilateral panels with constant source and doublet singularities. Viscosity is treated using a transpiration model with normal velocities derived from integral boundary layer calculations along streamlines.

As this code is still under investigation, it is essential, to have aerodynamic progress automatically available for invers runs. This was another reason, to chose the optimization formulation for inverse tasks.

Within this framework of invers design, the panel code can be regarded as a black box, which delivers pressure coefficients on demand. Any other panel code could be used as well.

Minimization procedure

The minimization of the sum of weighted, squared pressure residuals

$$E(\vec{a}) = \sum_k \sigma_k \left(c_{p_k}(\vec{a}) - c_{p_k}^{\text{target}}\right)^2$$

is done with a strategy due to LEVENBERG and MARQUARDT [6], [7]. It is a unification of two quite different procedures, the inverse Hessian method, which is good in the vicinity of the minimum, and the steepest descent method, which is good far from the minimum.

In the inverse Hessian method it is assumed, that the Taylor–expansion up to second order is a reasonable approximation of E in the vicinity of a current minimum location $\vec{a}_c$. A refinement $\vec{a}$ of this position is calculated in the following manner. The Taylor–expansion up to second order about the current location reads

$$E(\vec{a}_c + \vec{a}) \approx E(\vec{a}_c) + \sum_i \left.\frac{\partial E}{\partial a_i}\right|_{a_c} a_i + \frac{1}{2} \sum_{i,j} \left.\frac{\partial^2 E}{\partial a_i \partial a_j}\right|_{a_c} a_i a_j \,. \tag{1}$$

For abbreviation let

$$\beta_i \equiv -\frac{1}{2} \left.\frac{\partial E}{\partial a_i}\right|_{a_c} = -\sum_k \sigma_k \left(c_{p_k}(\vec{a}_c) - c_{p_k}^{\text{target}}\right) \left.\frac{\partial c_{p_k}(\vec{a})}{\partial a_i}\right|_{a_c} \tag{2}$$

be the negativ gradient of E (apart from a factor 1/2) at the current location $\vec{a}_c$ and further

$$\alpha_{ij} \equiv \frac{1}{2}\left.\frac{\partial^2 E}{\partial a_i\, \partial a_j}\right|_{a_c} \simeq \sum_k \sigma_k \left(c_{p_k}(\vec{a}_c) - c_{p_k}^{\mathrm{target}}\right) \left.\frac{\partial c_{p_k}(\vec{a})}{\partial a_i}\right|_{a_c} \left.\frac{\partial c_{p_k}(\vec{a})}{\partial a_j}\right|_{a_c} . \qquad (3)$$

The matrix α_{ij} is known as Hessian matrix (apart from the factor 1/2). In the last formula the term proportional to the second derivative $\partial^2 c_{p_k}(\vec{a})/\partial a_i \partial a_j$ is neglected. The Taylor-expansion (1) becomes

$$E(\vec{a}_c + \vec{a}) \approx E(\vec{a}_c) - 2\sum_i \beta_i a_i + \sum_{i,j} \alpha_{ij} a_i a_j .$$

From the condition for the minimum we get

$$\frac{\partial E}{\partial a_k} = 0 \quad\Longrightarrow\quad -2\sum_i \beta_i \delta_{ik} + \sum_{i,j} \alpha_{ij}(\delta_{ik} a_j + a_i \delta_{jk}) = 0\,, \qquad (4)$$

where the Kronecker symbol δ_{ik} stems from

$$\frac{\partial a_i}{\partial a_k} = \delta_{ik} = \begin{cases} 1 & i = k \\ 0 & \text{else} \end{cases} .$$

Carriing out the sums in (4), using the Kronecker symbol definition and the symmetry of the Hessian, we are left with

$$\sum_l \alpha_{kl}\, a_l = \beta_k . \qquad (5)$$

The refinement $\vec{a}$, calculated from this equation, updates the current minimum position $\vec{a}_c$ according to $\vec{a}_{new} = \vec{a}_c + \vec{a}$, which is the minimum position exactly, if the function E is just of second order.

For a steepest descent step, a step along the negativ gradient, the formula for the refinement reads

$$a_k = const. \times \beta_k \qquad (6)$$

with a suitably choosen constant, small enough not to exhaust the downhill direction. The last two equations (5) and (6) could be combined, as was pointed out by Marquardt. He showed, that the diagonal elements of the Hessian set the *scale* of the gradient. This is plausible from dimensionality considerations. E is nondimensional so β_k has the dimension of $[1/a_k]$, as may be seen from the definition (2). Therefore, the constant in the steepest descent formula (6) must have the dimension of $[a_k^2]$ i. e. is proportional to $1/\alpha_{kk}$. This can be expressed as a formula by introducing a factor λ (Marquardt-factor) and rewriting the steepest descent formula as

$$a_k = \frac{1}{\lambda \alpha_{kk}} \beta_k \qquad \text{or} \qquad \lambda \alpha_{kk} a_k = \beta_k .$$

Now both equations can be combined by defining a new matrix α' by the following prescription

$$\begin{aligned} \alpha'_{kk} &= \alpha_{kk}(1+\lambda) \\ \alpha'_{kl} &= \alpha_{kl}, \quad k \neq l \end{aligned}$$

and we are left with the single equation

$$\sum_l \alpha'_{kl}\, a_l = \beta_k \, . \tag{7}$$

If λ is large, the matrix α' is forced to be diagonally dominant and a steepest descent step is performed. On the other hand, as λ approches zero an inverse Hessian step is done.

The main idea of the Levenberg–Marquardt–minimization strategy is the adjustment of the factor λ during minimization. The calculation starts with a modest value of λ (e. g. 0.01) and is increased (more 'steepest descent'), if the least square sum worsens: $E(\vec{a}_c + \vec{a}) \geq E(\vec{a}_c)$, whereas λ is decreased (more 'Hessian'), if the least square sum diminishes. This method works very well in practise. Fig. 1 shows a flow chart of the main steps of the minimization strategy.

A crucial point is the determination of the Jacobian $\partial c_{p_k}/\partial a_i$. It means the sensitivity of the kth pressure to a variation of the ith parameter. This matrix is calculated numerically, row by row, by perturbing the parameters one after another. The perturbations must be specified very carefully! They have to be large enough to get significant pressure changes, but small enough, to get a reasonable gradient.

The above presentation of the Levenberg–Marquardt–method is taken from [7]. The computer program printed there has not been tried out however, because we use a somewhat more sophisticated implementation [8]. Therefore, no recommodation of that special version can be given.

Surface representation

The wing surface is built up from a number of spanwise design sections, which are connected by linear interpolation. If $\vec{r}_0(t)$ and $\vec{r}_1(t)$ are representations of wing sections at any two span locations, the surface between them is given by

$$\vec{r}(t,\lambda) = (1-\lambda)\vec{r}_0(t) + \lambda\vec{r}_1(t).$$

Such a surface patch is called a *regular* surface, because it could thought to be obtained by continous movement of a straight line, connecting section points of equal t–parameter.

Each design section is defined relativ to a local frame, the section coordinate system, centered at the sections leading edge and the x-axis coincident with the section chord. Relativ to this coordinate system the profil, separately for the lower and upper side, is given by a B–spline expansion

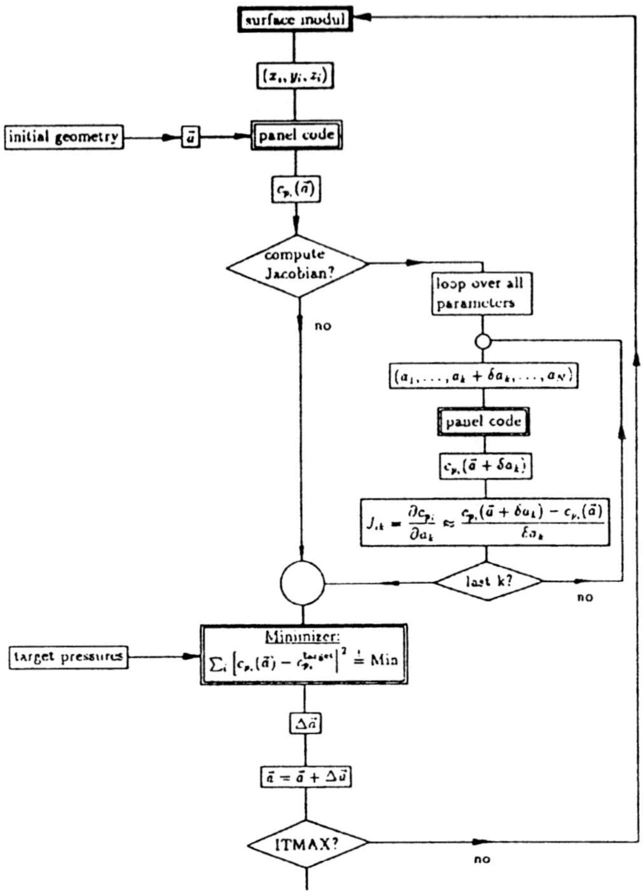

Figure 1 Flow diagram of inverse iteration loop.

$$\vec{r}(t) = \sum_{i=0}^{n} \vec{d_i}\, N_i^k(t)\,,$$

where $N_i^k(t)$ is the ith B-Spline function of order k and $\vec{d_i}$ are the so-called control points. They are nothing more than the expansion coefficients due to the basis functions $N_i^k(t)$ ('B-splines').

For given values of n ($n+1$ control points) and k, the complete curve is composed of $n-k+2$ partial curves. These are polynomials of order k (degree $k-1$), which are connected with C^2-continuity. The highest possible order is $k = n+1$, which gives just one partial curve for the whole domain (*Bézier*-representation).

The control points (more exactly: their ordinates relativ to the local frame, because they are shifted perpendicular to the chord) form one part of the design parameters. The other part is the set of chord angles of the design sections.

The B–spline technique is a convenient tool in computational geometry, where it is mainly used for the interactive design of curves. It's most popular feature is the 'local control', which means, that a variation of a control point influences only a part of the curve while maintaining the C^2–continuity. Some further highlights of B–spline properties are the linear precision, the strong convex hull property and the variation diminishing property.

Their meaning can be found in nearly any textbook on B–splines such as [9], [10] or [11]. They are enumerated here to demonstrate, that these functions are neither any common–or–garden functions nor a speciality of interactive curve designers, but refer to a very sophisticated method of function representation.

From our experience cubic B–splines with six to ten movable control points per profil half are sufficient to generate profils, which fit even unusual target pressure distributions.

Constraints

Geometrical constraints may be maintained by fixing selected control points, thereby generating curves with special geometric properties. The simplest example is the maintenance of a specified trailing edge thickness just by locating control points there and keeping them fixed. This constraint is fulfilled exactly, because the spline curve will pass through the first and last control point by construction.

This is not true for inner control points. Here the curve lies within the convex hull of $k+1$ nearby control points (convex hull property). As curves, being constructed from cubic B–splines, will closely follow the control point, it is possible, to maintain geometrical contraints at inner points *approximately* by fixing suitable control points or restricting their movement to a vertical intervall by a parameter transformation.

Constraining the chord angle is much more simpler. It is just done by appliing the section rotation or not.

Dynamical constraints could be introduced by adding further terms to the minimization functional. For instance, a lift control, weighted by a factor τ, can be done by minimizing the augmented functional

$$E(\vec{a}) = \sum_i \sigma_i \left(c_{p_i}(a_1, \ldots, a_n) - c_{p_i}^{\text{target}}\right)^2 + \tau\left(c_A(a_1, \ldots, a_n) - c_A^{\text{target}}\right)^2.$$

Intersection of surfaces

Designing a wing in presence of a pylon means (among others), that the intersection between wing and pylon has to be recalculated after each minimization step. Generally one has to solve a surface–surface intersection, which could be reduced here to a line–surface intersection problem ('piercing model'). It is defined by piercing the wing with pylon rays, whose direction is taken from the uppermost two sections of the initial configuration, see Fig. 2. Further, there could be seen an additional patch between wing and pylon ('pylon-surround-patch'), which is located on the wing and plays the part of a buffer patch. It provides 'space' for the intersection line as the wing rotates, and serves for appropriate panel matching between wing and pylon.

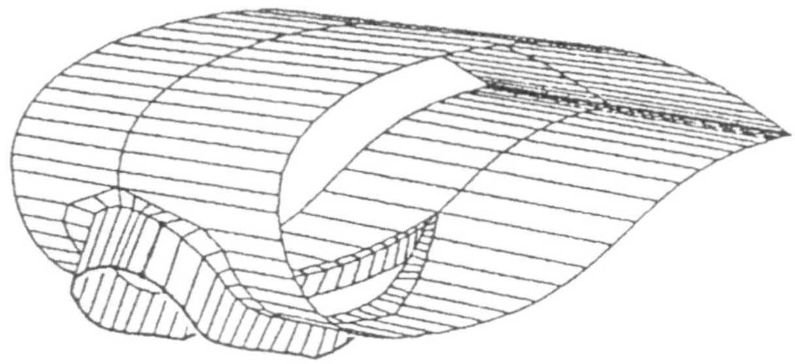

Figure 2 Wing-pylon-intersection by piercing technique and pylon-surround patch.

Analysis of results

Inviscid design

The design task is to meet the clean-wing c_p-distribution with a full supplied wing-pylon-nacelle configuration. To this end seven design sections are defined in the interference region, see Fig. 3 and Fig. 4. As will be recognized from the changes of the design profils, it is allowed to restrict the wing design region in this way (see Fig. 4), because outside the interference is too small, to be resolved by numerical parameter stepping.

Each design section is described by 15 parameters: the chord angle and seven B-spline control ordinates for each the lower and upper half. Alltogether 105 design parameters are available to fit 720 target pressures, distributed on 12 pressure control stations (the middle between two wing sections) with each 60 control points.

The comparison of the target pressures, the initial (iter=0) and the final ones (iter=4) for the inboard-pylon pressure control columns are shown in Fig. 5; the

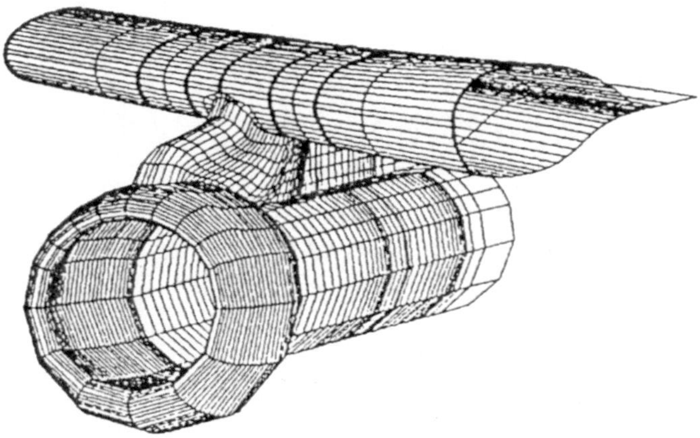

Figure 3 Surface–and wakemodeling in the interference region.

outboard–pylon ones are shown in Fig. 6. Only the lower wing halfes, located at near inbord pylon positions, show some small deviations. At all other locations the c_p–distribution of the designed wing is in accordance with the target one within the thickness of the plotlines.

Fig. 7 displays the resulting additional chord angle distribution. It is not symmetric relativ to the pylon position. It is positiv ('nose up') throughout, reflecting the fact, that the presence of a nacelle reduces the lift compared with the clean wing.

Fig. 8 shows the convergence history of all pressure control columns. The L2–norm of the pressure deviations is displayed in dependence from the 'number of elementary problems', which means a complete run through the flow code. The two diagrams, the left one for inboard–pylon wing sections, the right diagram for outboard–pylon ones, contain a lot of information [2].

First, it could be seen, that four iterations could be done with just one Jacobi-matrix. This is a special feature of our Levenberg–Marquardt–implementation. Second, the initial level of the curves is directly related to the strenght of the interference. Interference is highest in the vicinity of the pylon and significantly higher outboard pylon than inboard. (The reason is, that the pylon is not force free but has a resultant lift directed to to the tip. However, the optimal shape and orientation of the pylon was not a matter of research here.) Last, the pressure stations with a L2–norm of about 10^{-3} are converged (more or less) after one iteration, but not the one with a value greater 10^{-2}, e.g. the first column outboard pylon. This corresponds to the 'linear subsonic aerodynamics'. The

[2]The long horizontal curve segment corresponds to the calculation of the Jacobian.

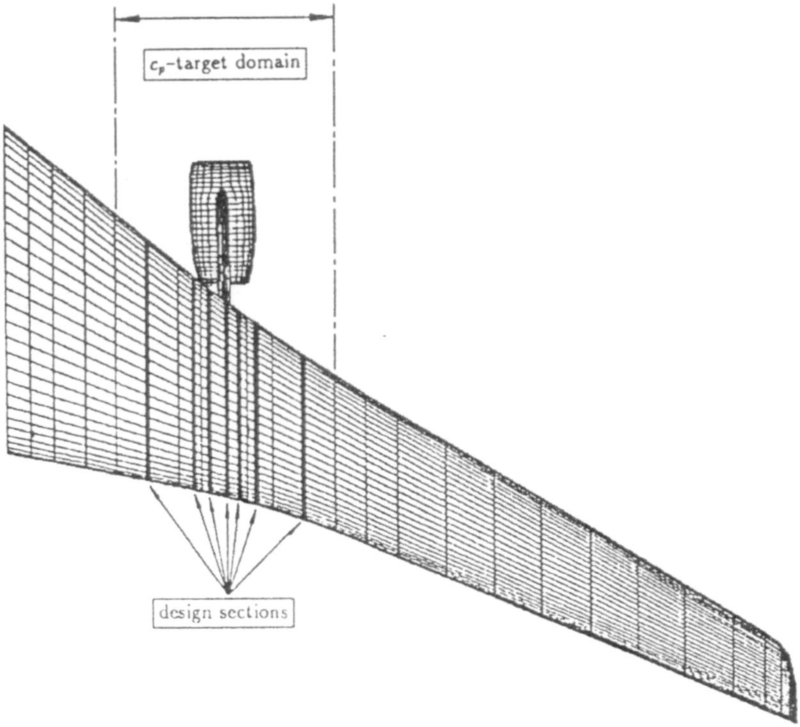

Figure 4 AS28E-configuration prepared for design runs.

pressure coefficients depend 'nearly' linear from the parameters, so design cases with small deviations from the target distribution can be done in one iteration.

Fig. 9 shows the seven design profils in normalized position. Their upper halfs undergo nearly no changes. Here the resultant design rotations of the sections suffice to meet the target pressures. The lower profil halfs are reshaped significantly, however. They show characteristic different behavior relativ to pylon position: the inboard one gets thinner, while the outboard one is thickend. Finally, Fig. 10 shows a perspective view of the initial and final configuration (wakes removed).

Viscous design

As was mentioned above the panel code is capable of viscous corrections, based on boundary layer calculations along streamlines. If any code and calculation time optimizations are ignored, the invers iteration loop — as assessed for inviscid designs — can be left unchanged and it should be possible, to do viscous designs by activating a viscous iteration loop inside the panel code.

This approach demonstrates the feasibility of viscous designs at the expence of computer time. Though using not optimized code it works very well!

The viscous TE5–testcase is done in exactly the same manner as the inviscid one before, but now with viscous clean–wing pressures as target pressures and activated viscous corrections during the invers loop.

Each of Fig. 11 and Fig. 12 show four pressure control cuts, namely inboard–pylon and outboard pylon ones, just as before (compared the corresponding Fig.'s 5 and 6). As was the case in inviscid design, four iterations could be performed with one Jacobian. The achieved agreement with the target pressures is as good as before.

Fig. 13 shows the comparison between the viscous and the inviscid design profils. It is seen, that former thin design profiles are thickend and thick ones are thinned or, with other words, the interference between wing and pylon/nacelle is weakend by viscous effects!

Last, Fig. 14 shows a perspective view of the viscous and inviscid designed configuration.

Bibliography

[1] Schwarten, H., Stuke, H. *Pressure controlled surfaces—a 3D inverse panel method as a design tool.* Computational Methods in Applied Sciences, Ch. Hirsch, J. Périaux, E. Oñate ed., Elsevier, Amsterdam, London, New York, Tokyo, 1992, pp. 261–266.

[2] Schwarten, H. *Ein inverses, subsonisches 2D–Panelverfahren nach der Methode der kleinsten Quadrate zum Entwurf und zur Modifikation von Mehrelementprofilen.* DGLR–Jahrbuch 1987 (1) der Deutschen Gesellschaft für Luft–und Raumfahrt, pp. 163–170.

[3] Greff, E., Forbrich, D., Schwarten, H. *Application of direct inverse analogy method (DIVA) and viscous design optimization techniques.* Proc. 3rd Int. Conf. Inverse Des. Concepts and Optim. in Eng. Sci. ICIDES III, G. S. Dulikravich, ed., 1991

[4] Maskew, B., *Prediction of subsonic aerodynamic characteristics: a case for low order panel methods.* Journal of Aircraft, Vol.19,No2, February 1982, pp. 157-163.

[5] Maskew, B., *Program* VSAERO. *A Computer program for calculating the non–linear aerodynmic characteristics of arbitrary configurations.* NASA–Contract NAS2–11945, December 1984.

[6] Marquardt, D. W. *An algorithm for least–squares estimation of nonlinear parameters.* J. Soc. Indust. Appl. Math., vol. 11, no. 2, (1963), pp. 431–441.

[7] Press, H., Flannery, B., Teukolsky, S., Vetterling, W. *Numerical recipes*, Cambridge University Press, Cambridge 1986.

[8] Kosmol, P. *Methoden zur numerischen Behandlung nichtlinearer Gleichungen und Optimierungsaufgaben.* B. G. Teubner, Stuttgart, 1989.

[9] Yamaguchi, F., *Curves and surfaces in computer aided geometric design.* Springer Verlag, Berlin, Heidelberg, New York, 1988.

[10] Hoschek, J., Lasser, D. *Grundlagen der geometrischen Datenverarbeitung.* B. G. Teubner, Stuttgart, 1989.

[11] Farin, G., *Curves and surfaces for computer aided geometric design.* Academic Press, London, Sydney, Tokio, Toronto, 1990.

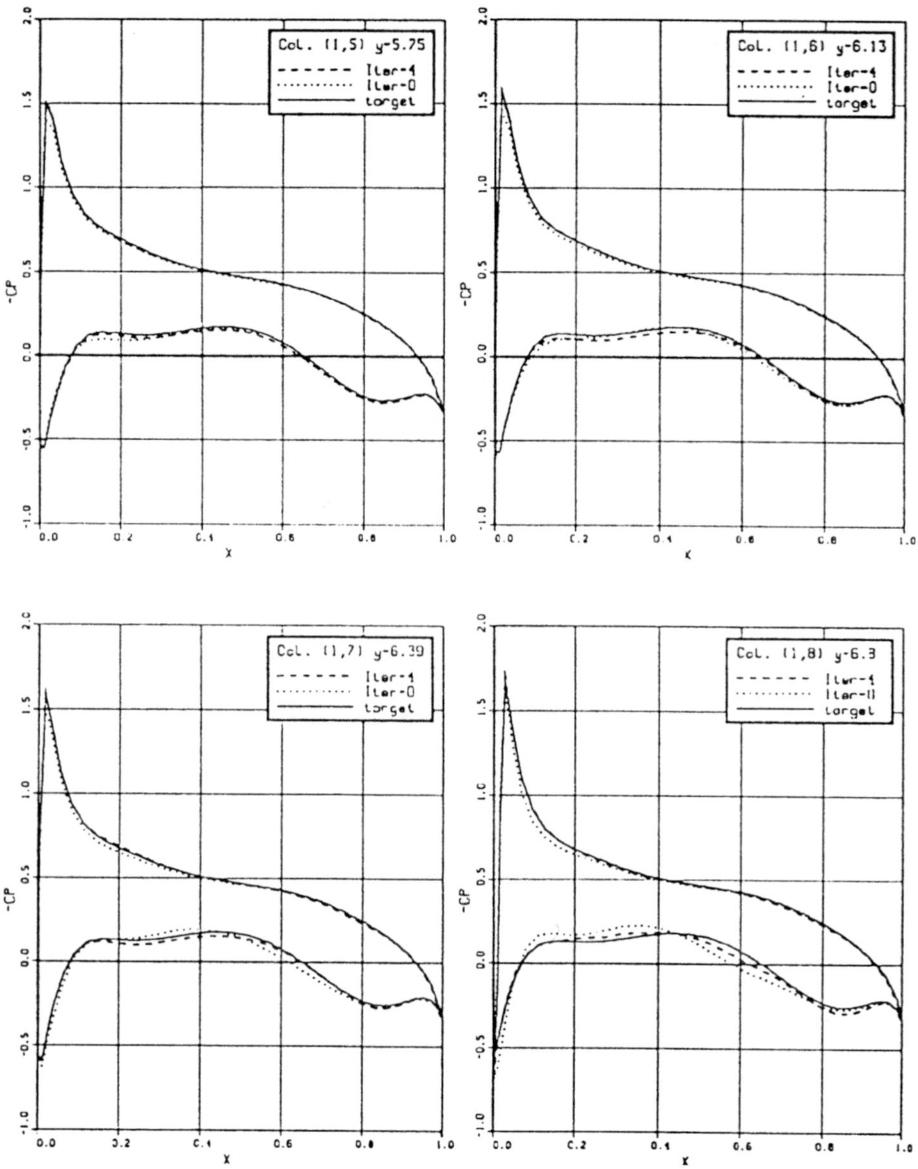

Figure 5 Comparison of initial (Iter=0), final (Iter=4) and target pressures for pressure control cuts located inboard-pylon.

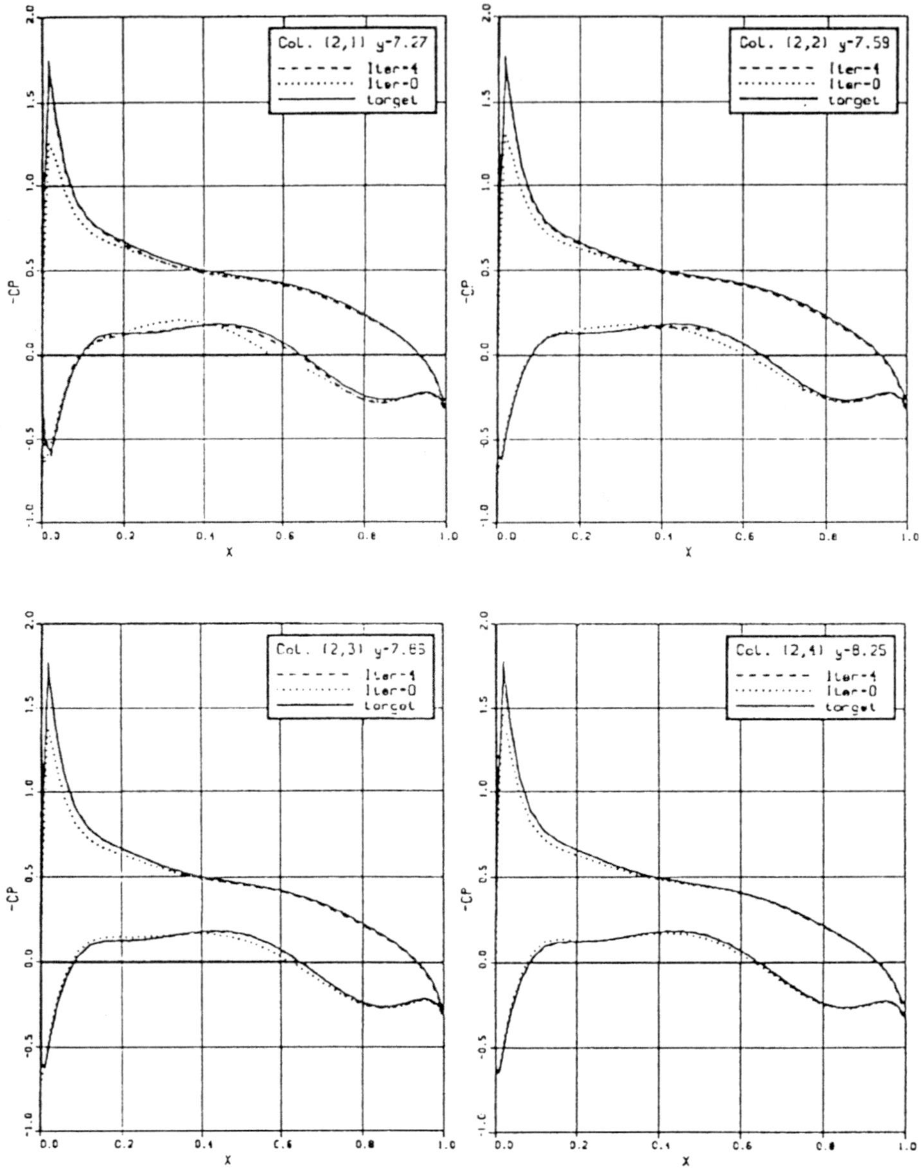

Figure 6 Comparison of initial (Iter=0), final (Iter=4) and target pressures for pressure control cuts located outboard-pylon.

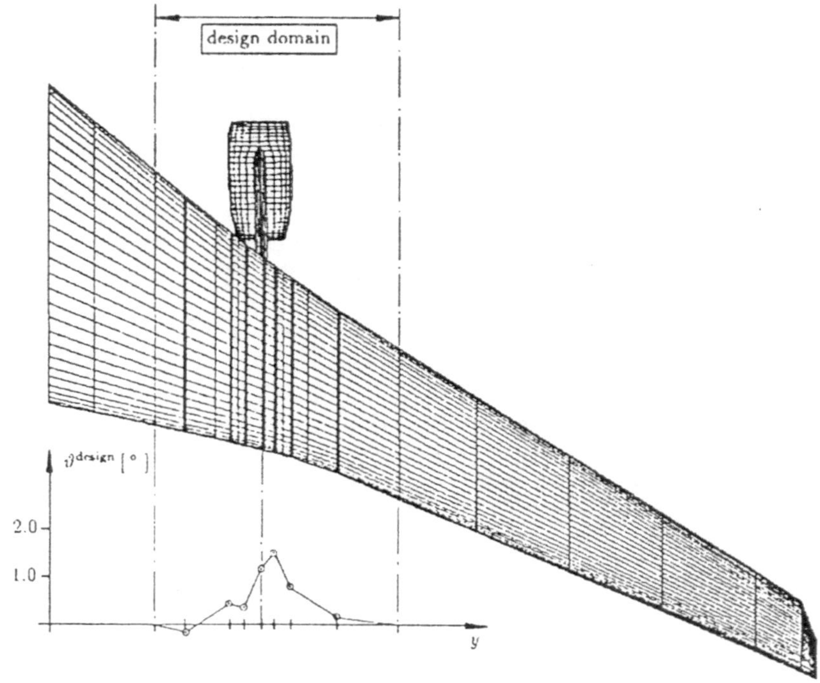

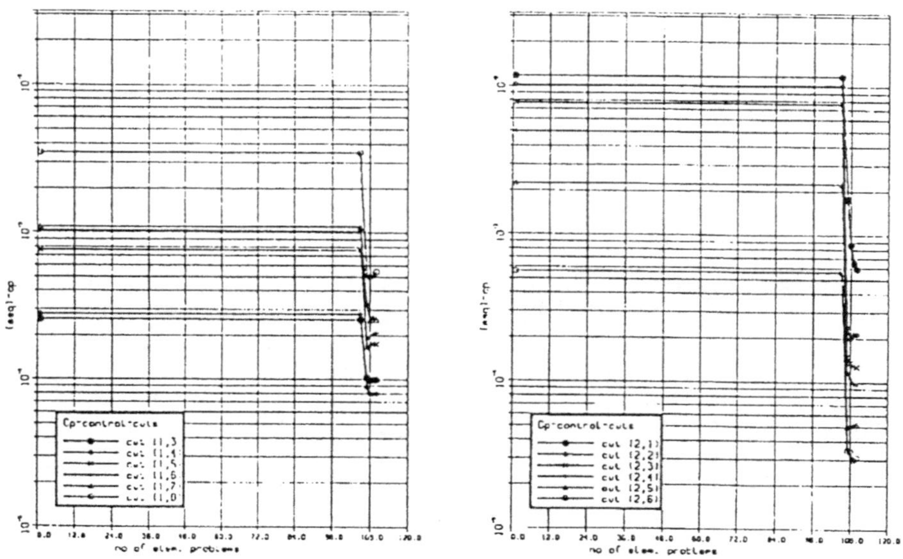

Figure 8 Convergence history of sectional Δc_p-L2-norm at all pressure control cuts. Left: inboard pylon, right: outboard pylon.

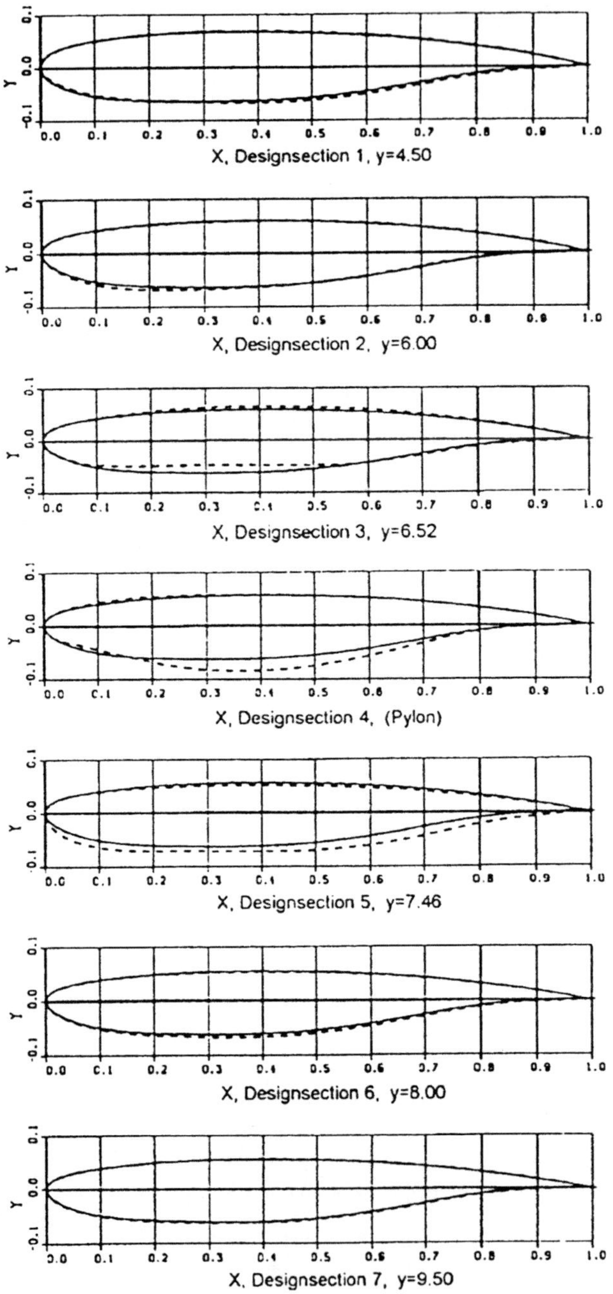

Figure 9 Comparison of initial (solid line) and designed profils (dashed line) in normalized position.

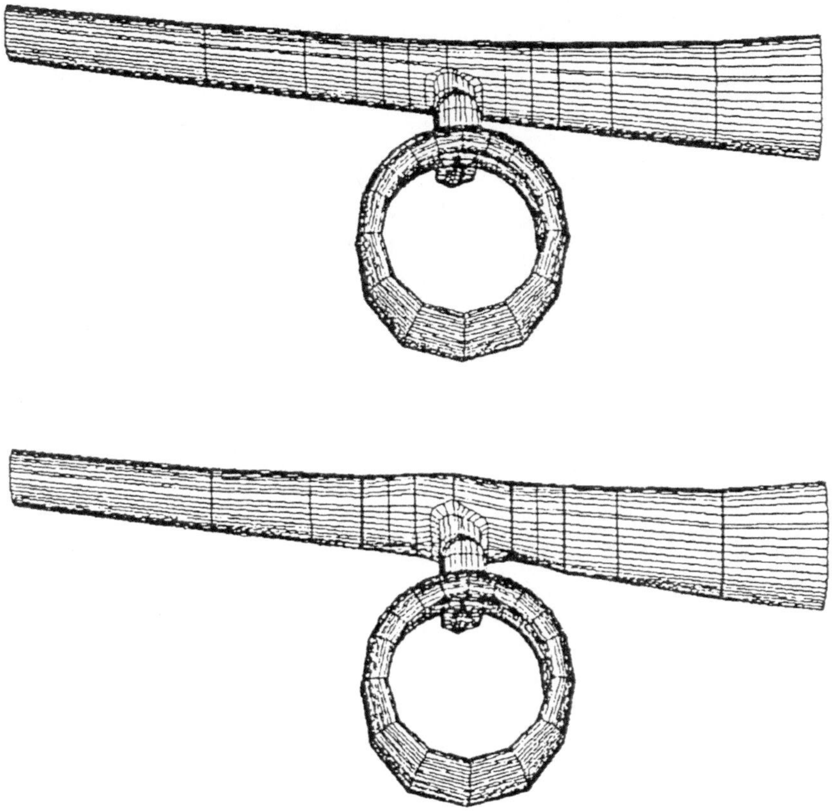

Figure 10 Comparison of initial and designed configuration.

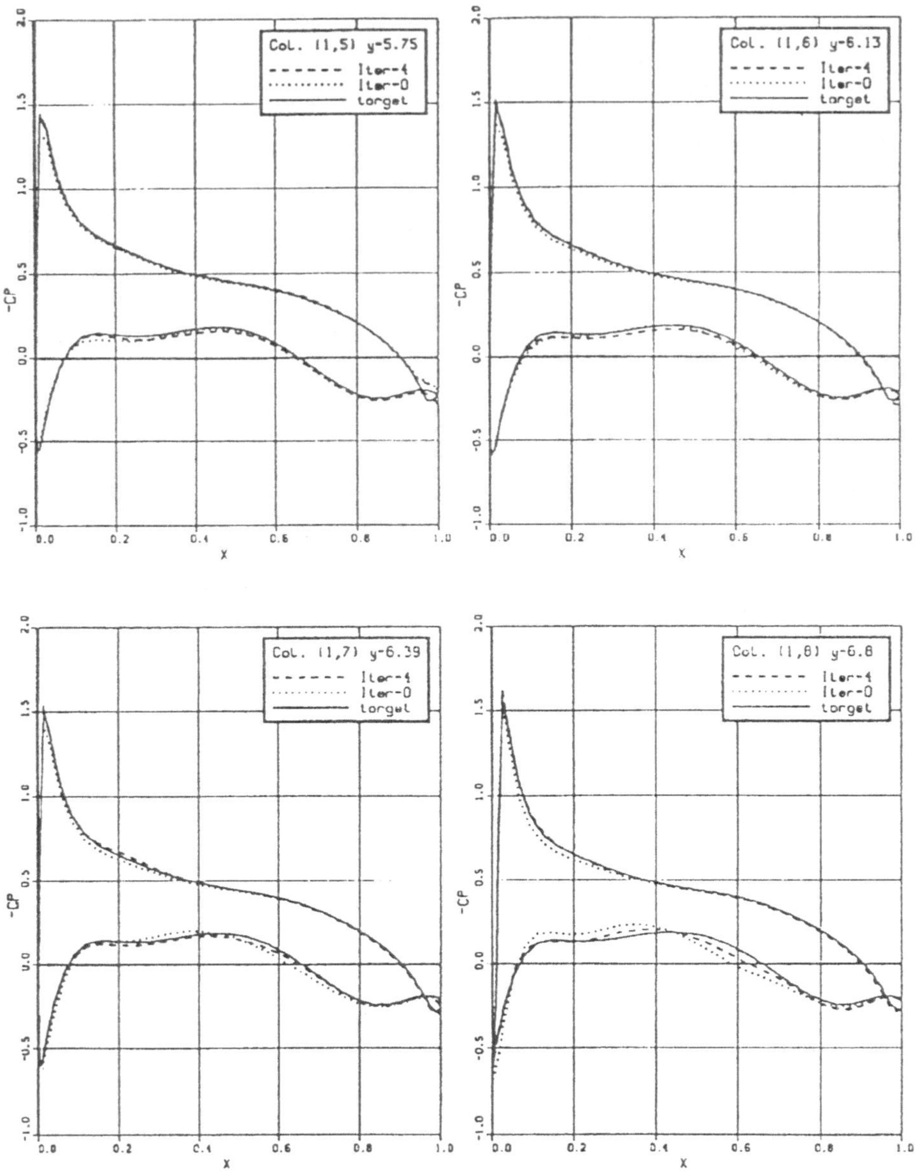

Figure 11 Viscous design. Comparison of initial (Iter=0), final (Iter=4) and target pressures for pressure control cuts located inboard-pylon.

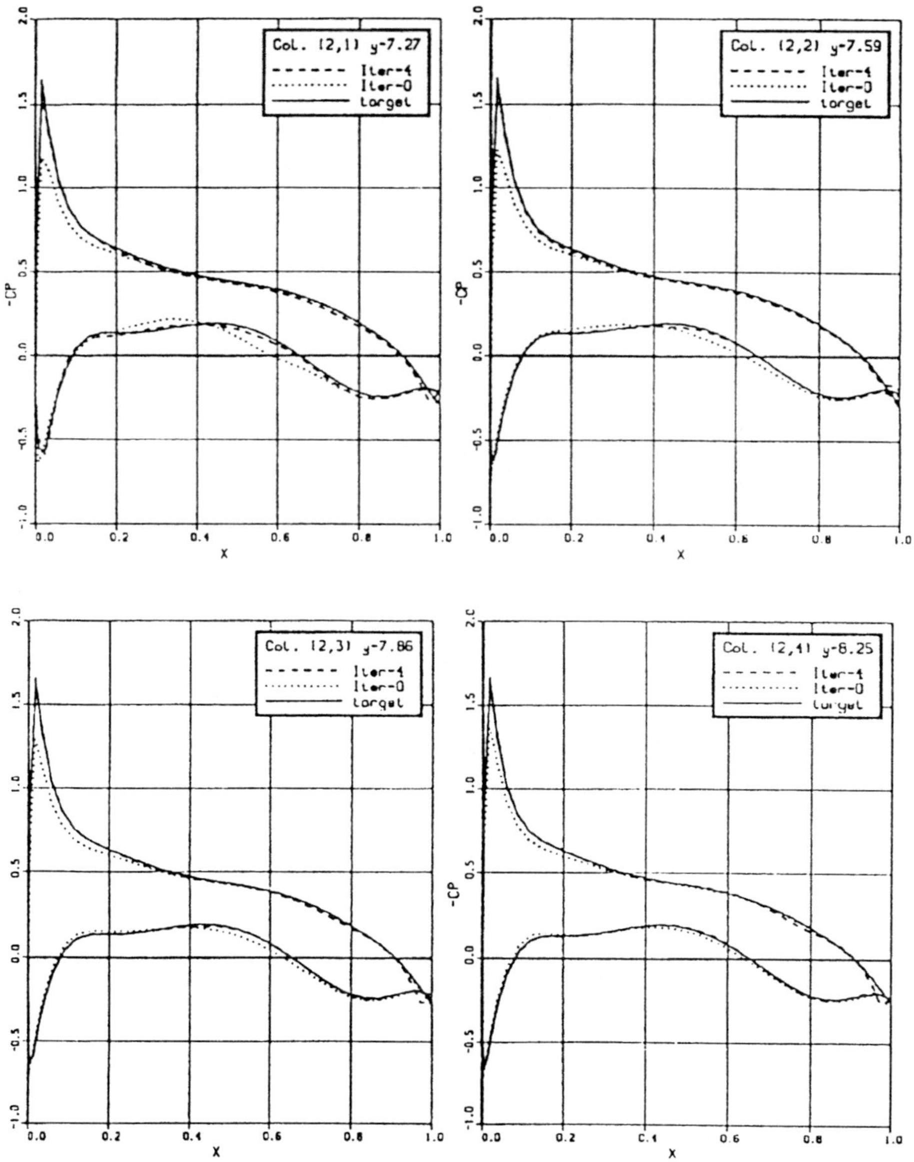

Figure 12 Viscous design. Comparison of initial (Iter=0), final (Iter=4) and target pressures for pressure control cuts located outboard-pylon.

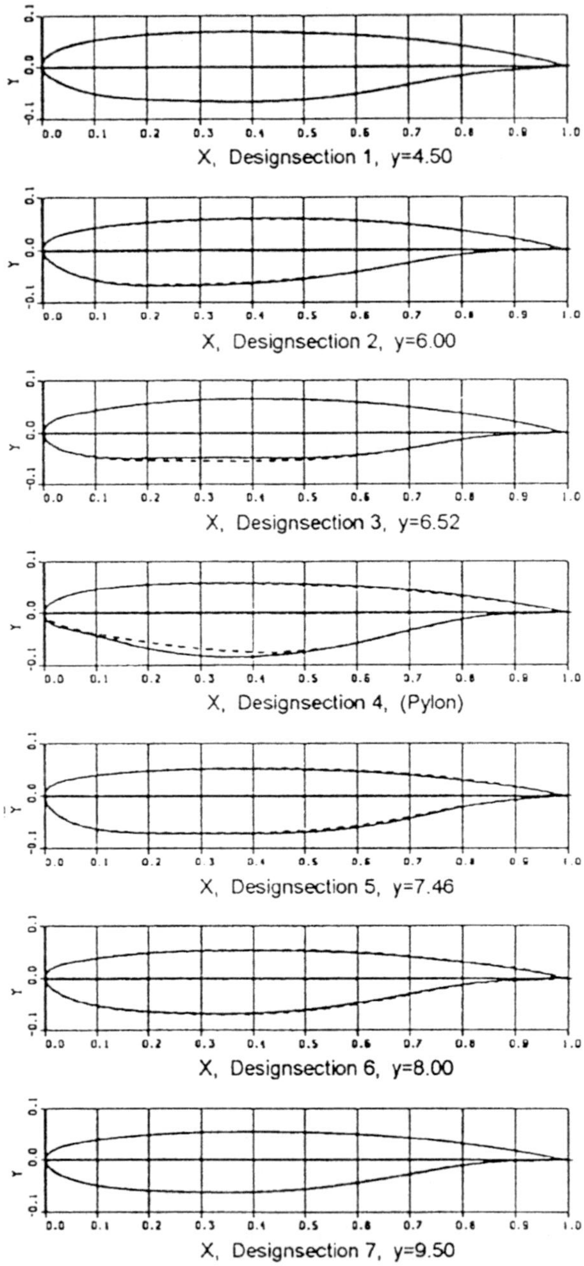

Figure 13 Comparison of inviscid (solid line) and viscous design profils (dashed line) in normalized position.

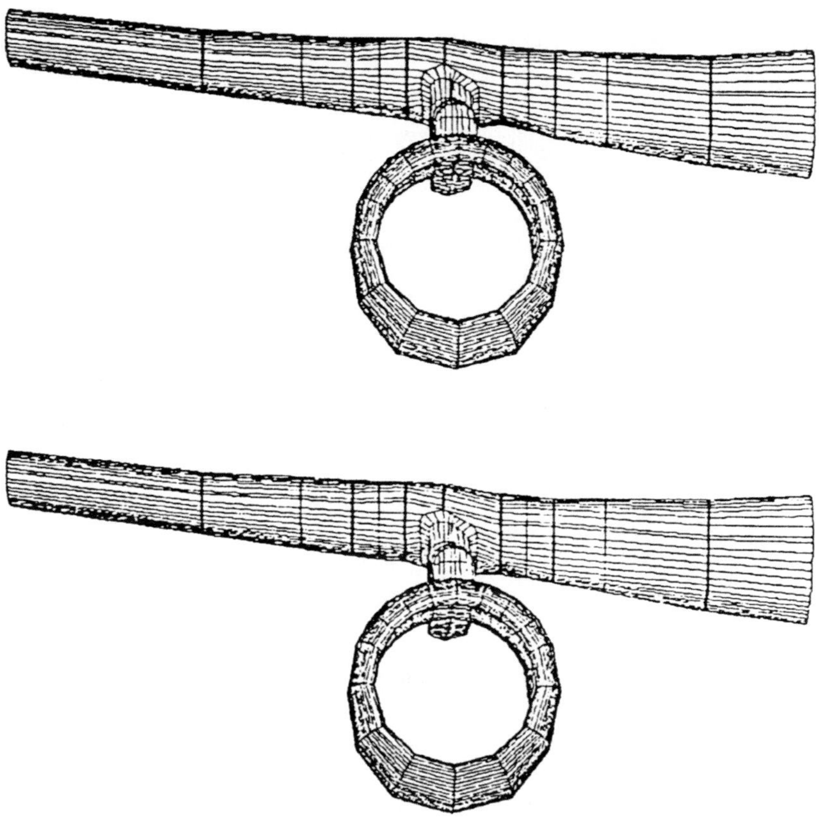

Figure 14 Comparison of inviscid and viscous design configuration.

6 Dassault contribution to the Optimum Design ECARP Project

J. Periaux and B. Stoufflet, Dasault Aviation, St. Cloud Cedex, France

I Introduction

The paper reports on the joint contribution of Dassault Aviation and INRIA to the two-dimensionnal test cases of the Workshop organized within ECARP - Optimum Design task. This includes participation to test-cases TE2, TE3 and TE4.

II Methodology description

We have developped within the project ECARP an optimization procedure capable of treating shape optimization of (multi) - airfoil profiles including the multi-point optimization. Two optimization approaches have been selected and applied in the Workshop. Both techniques rely on the same aerodynamic and the same geometrical parametrization modules. They differ from their optimizer characteristics.

The first approach (α) is based on the control theory applied to the discrete system using an approximate discrete sensitivity analysis. The adjoint equation is constructed directly from the discrete equations and the adjoint operator is nothing else than the transposed matrix of the implicit operator used to accelerate the iterative solution convergence of the state equation. The gradient of the cost function is evaluated without additional flowfield computations. A conjugate gradient method has been selected as descent method with a possible combination with hierarchical optimization. Parametrization is based on piecewise spline definition of the profile.

A second approach (β) avoids any gradient evaluation based on evolutive optimization techniques. A genetic algorithm using real coding of the design variables has been selected. Mutation acts directly on the real representation of the parameters by an adaptive refined gaussian function and cross-over acts on the genes constituted of list of parameters. The selection process is based on an elitist wheel. The associated parametrization relies on a Bezier curve.

The aerodynamic module considered in this Workshop solves the inviscid Euler equations. A weak coupling with a 2D boundary layer computation has been applied for test case TE2. The discretization is based on a Galerkin Finite Volume formulation using triangular meshes. The residual is computed by a centered scheme whereas the implicit operator is based on a first-order upwind discretization. Boundary conditions are treated in the boundary integral evaluation.

As numerical methods considered here are based on unstructured meshes in order to treat general geometries, a key point of the method is the relation between optimization parameters and geometrical characteristics of the mesh. This relation is decomposed in elementary functions; the overall gradient will be the product of these elementary gradients for each of which an appropriate evaluation can be used. One can anticipate to use finite differences restricted to some specific elementary operations to calculate the necessary information. In the present work, this manner has been avoided by using analytical dependance between intervening variables. The strategy chosen here can be easily described. Although an initial mesh is constructed using a mesh generator starting from prescribed boundary points, perturbed meshes arising in the optimization process are determined from variations of the boundary mesh points (of coordinates $(x_i^{(b)}, y_i^{(b)})$) by solving an elasticity equation which will provide a new position of the interior mesh points (of coordinates $(x_i^{(m)}, y_i^{(m)})$). As the topology of the mesh is preserved by this approach, the relation between boundary and interior mesh points is differentiable and the corresponding gradient is easily computed. We have used to build directly the subroutines evaluating the gradient the automatic differentiator ODYSSEE developped by INRIA. The optimization parameters z_i chosen here are explicit functions of the ordinates v_i of the control points (of coordinates (u_i, v_i)) of the cubic splines defining the boundary and the angle of attack α of the profile. Their relations with boundary points is then analytical, thus the gradient is trivially evaluated.

III Contribution to test-case TE2

This test case deals with a reconstruction problem of a NACA0012 profile from a NACA63215 airfoil. A spline discretization of the profiles using 20 control points for the entire profile has been considered.

A first computation to check the accuracy of the gradient of the cost function has been performed when using a first-order accurate scheme for the state equation. In that case, the cost function is divided by 10^5 in about 20 optimization steps. The NACA0012 profile is recovered perfectly.

A second computation based on the solver described in the previous section where the state equation is this time discretized by the centered scheme with a weak coupling with viscous effects is reported. The gradient is no more the exact discretized one but the cost function decreases at each iteration. It is reduced by 10^2 in about 50 optimization steps.

This approximate discrete sensitivity applied to this problem provides a descent direction able to reduce the cost at each step; as it is not an exact gradient, the reduction is altered and is not completely appropriate to this problem.

IV Contribution to test-case TE3

This test case addresses the problem of inviscid drag minimization for transonic flows maintening the lift of the original profile. The far field conditions are $M_\infty = 0.73$ and the angle of attack $\alpha = 2$ for the original computation; the angle of attack has been added to the optimization parameters. The aerodynamic solver is the one described in the first section without viscous effects and without farfield conditions. As the average radius of the boundary of the mesh is less than 8 chords, the computed lift is lower than the one obtained with this correction (checked to be around 0.89. Four computations have been performed to evaluate the sensitivity of the weighting coefficients defining the penalization within the cost function

$$j(\gamma) = \omega_1 \; I_{tar} \; + \; \omega_2 \; C_D \; + \; \omega_3 \; (C_L \; - \; C_L^{target})^2$$

using method (α). A satisfactory combination of weights has been selected to be used with method (β). The results are given in the following table and compare the figures of initial and "optimized" drag and lift.

Method	Cost function
1 (α)	$\omega_1 = 0 \;\; \omega_2 = 1 \;\; \omega_3 = 0$
2 (α)	$\omega_1 = 0 \;\; \omega_2 = 1 \;\; \omega_3 = 10$
2 (β)	$\omega_1 = 0 \;\; \omega_2 = 1 \;\; \omega_3 = 10$

Method	initial drag	final drag	initial lift	final lift	number of flow fields
1	0.0675	0.0055	0.69	0.66	12
2	0.0675	0.00579	0.69	0.673	12
3	0.0675	0.0032	0.69	0.68	120

The optimized profiles are obtained in 4 optimization steps with method (α) which are sufficient to exhibit the reduction of the strength of the shock. The solution remains stable in the following steps. The maximum thickness location moves towards the trailing edge. On the other hand, method (β) provides a better minimum corresponding to a very different profile exhibiting a thicker trailing edge and a significant modification of the windward side. Further investigations including continuing optimization and mesh refinement are under progress.

V Contribution to test-case TE4

This multipoint problem has been treated by the above method (β) and the results presented here should be considered as preliminary.

The number of control points defining the optimization parameters is 35. The behaviour of the cost function with respect to the number of flow evaluation is presented and shows the difficult convergence of the method. An "optimized" profile is nevertheless obtained after 30 optimization steps. This profile is presented and some oscillations visible near the trailing edge indicates that the control is not sufficiently regular.

Shape, surfacic pressure and convergence results of the multi point design of test case TE4 are described on Figs. 3-7.

VI Conclusion

Control theory applied to the discrete system using an approximate discrete sensitivity analysis provides an effective approach to treat shape optimization problems using even complex unstructured mesh simulation tools whereas genetic approach successfully solves these problems without additive troubles and gives interesting alternative answers. Future work will deal with multipoint optimization and constrained optimization based on interior point methods.

Bibliography

[1] *J. Periaux, M. Sefrioui, B. Stoufflet, B. Mantel, E. Laporte*, Robust Genetic Algorithms for optimization problems in Aerodynamic Design, in Genetic Algorithms in Engineering and Computer Science, G. Winter, J. Periaux, M. Galan, P. Cuesta, eds, John Wiley, 1995.

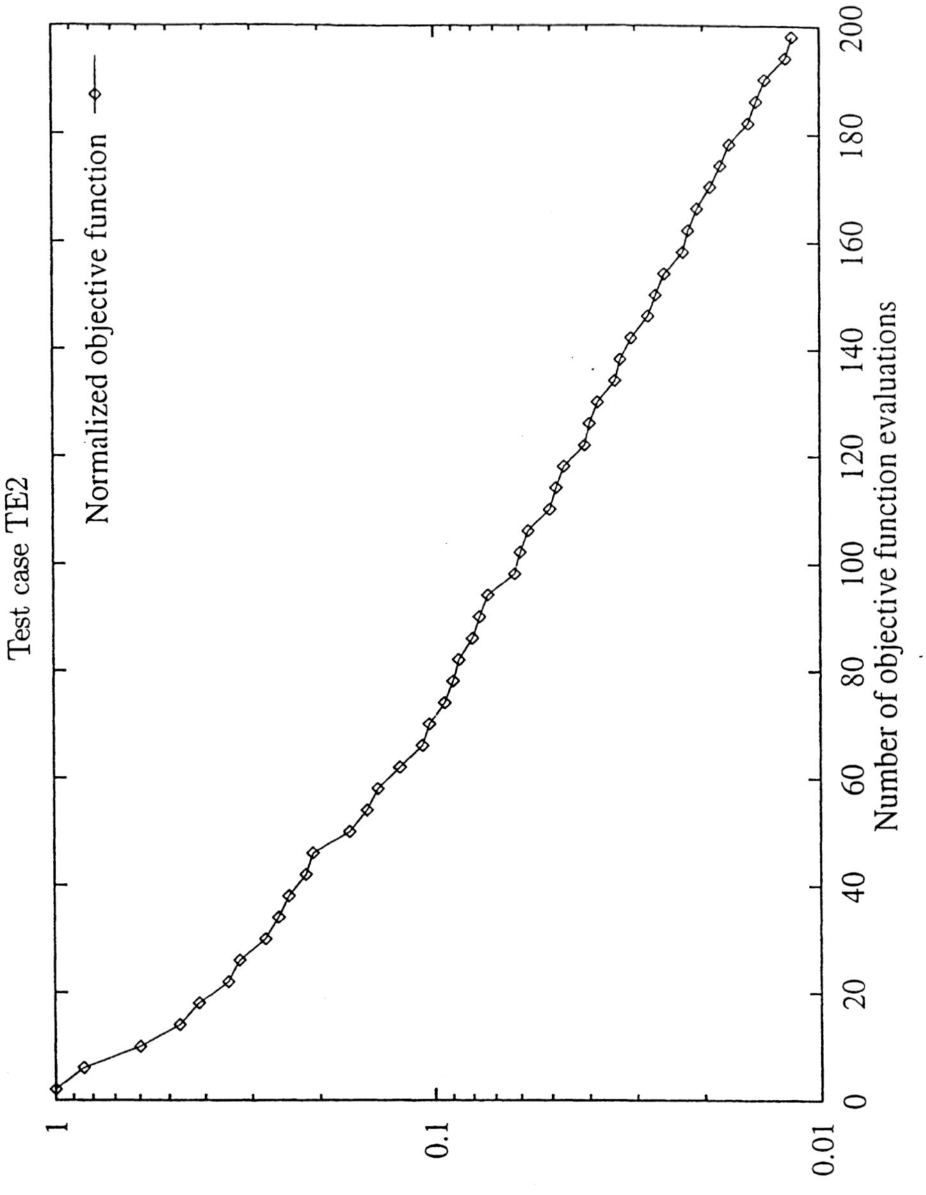

Figure 1 Evolution of the normalized objective function for the TE2 test case

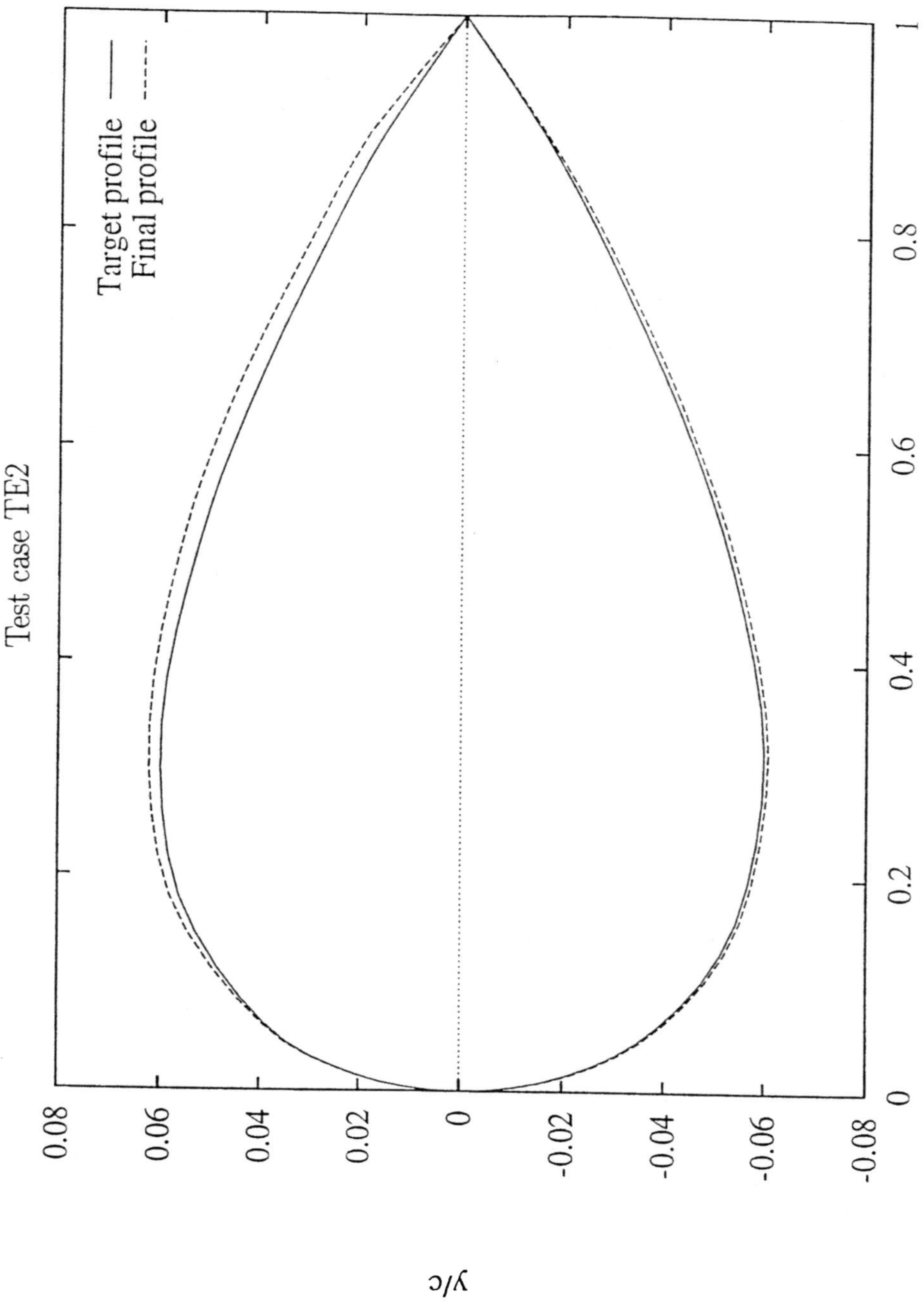

Figure 2 Superposition of the target and the final profiles for the TE2 test case

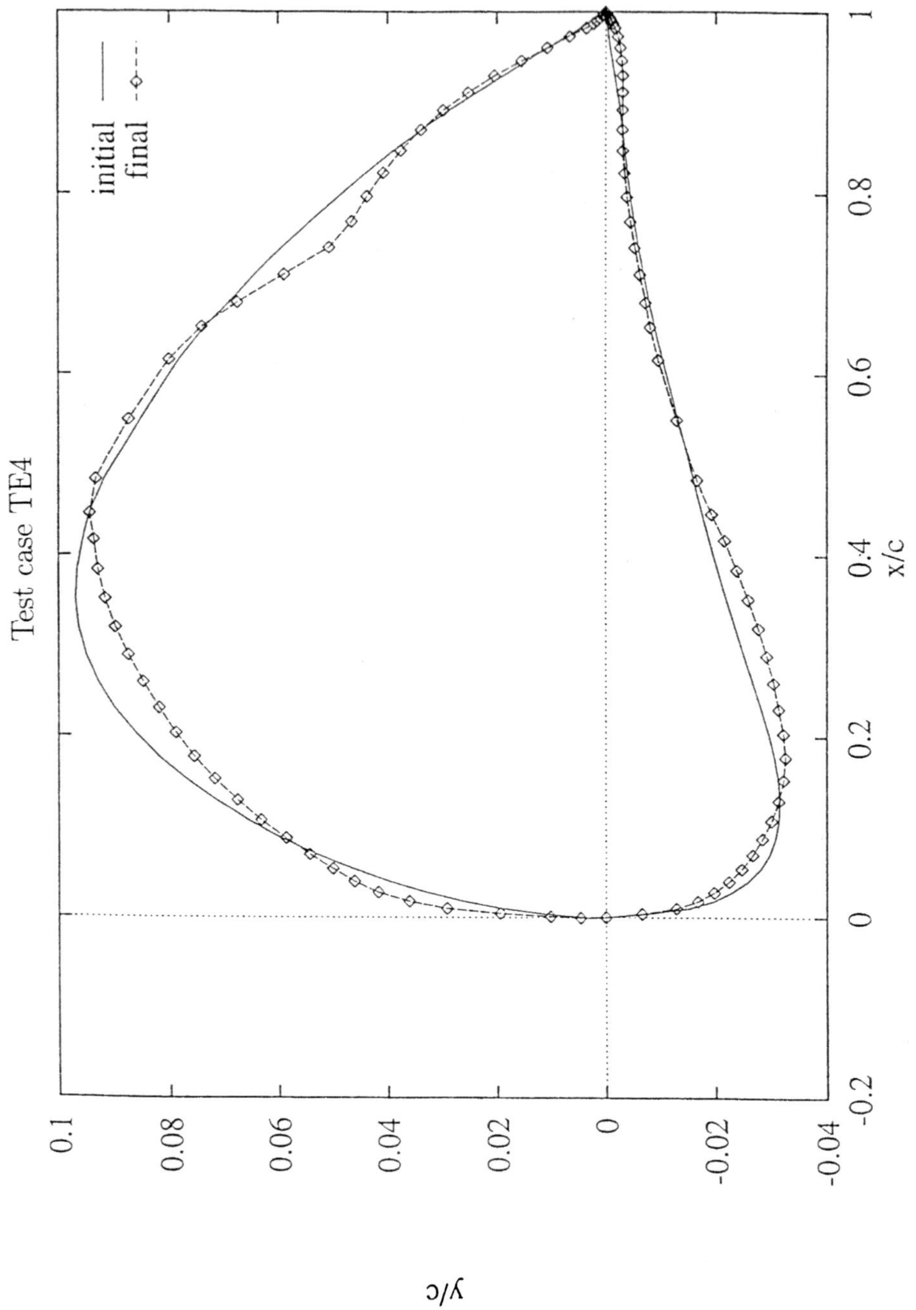

Figure 3 Superposition of the initial and the final profiles for the TE4 test case

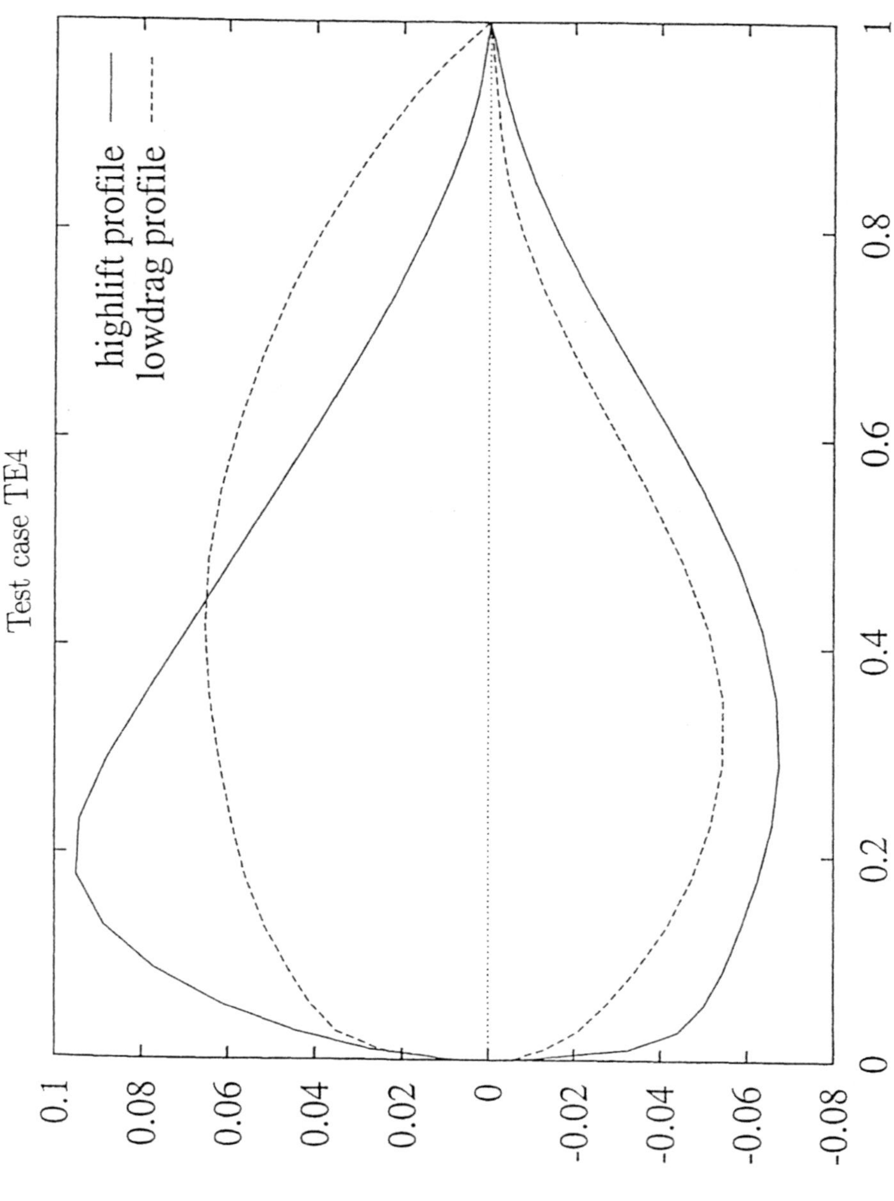

Figure 4 Superposition of the highlift and the lowdrag profiles for the TE4 test case

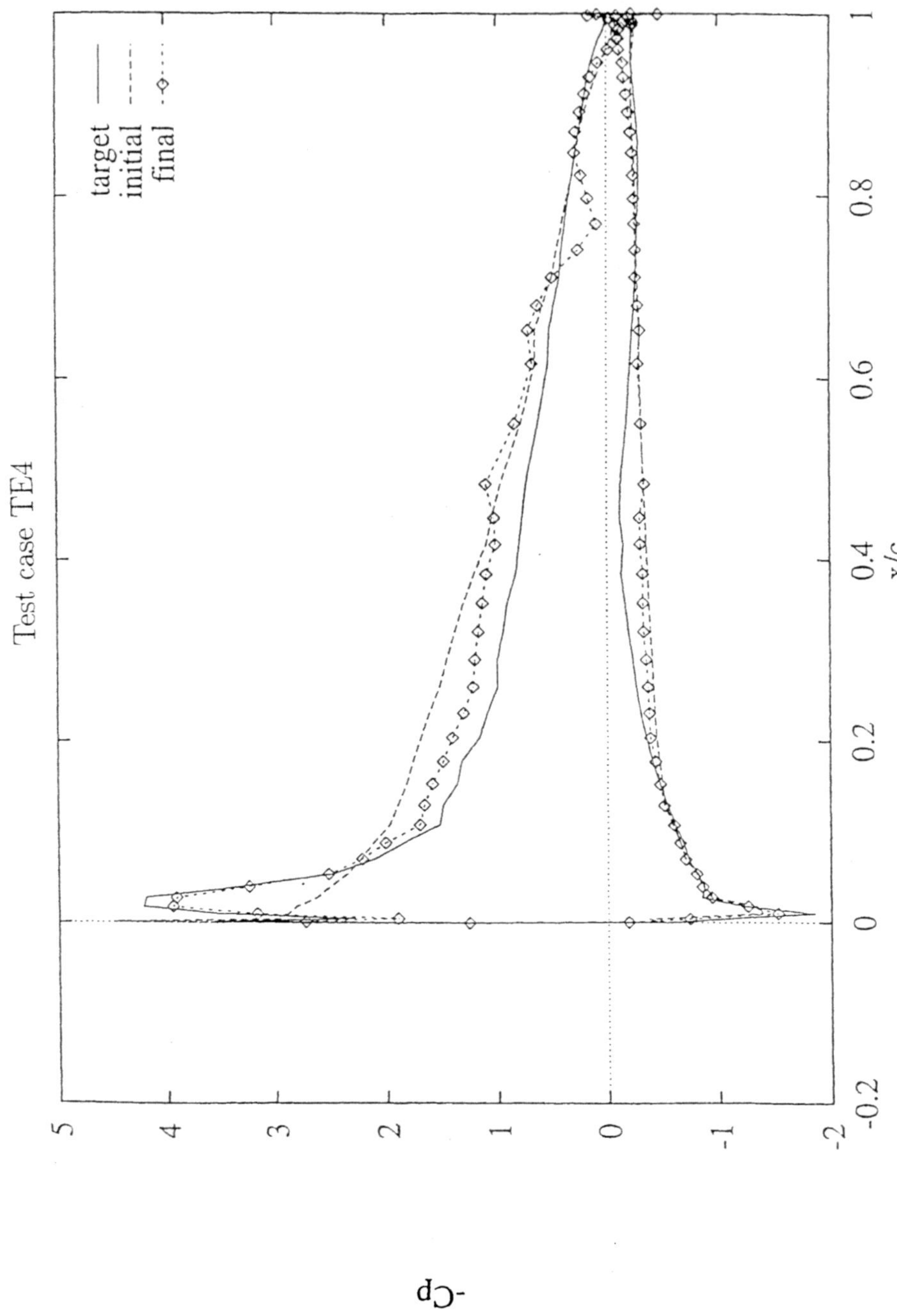

Figure 5 Superposition of the target, the initial and the final Cp distributions for the TE4 test case under subsonic conditions

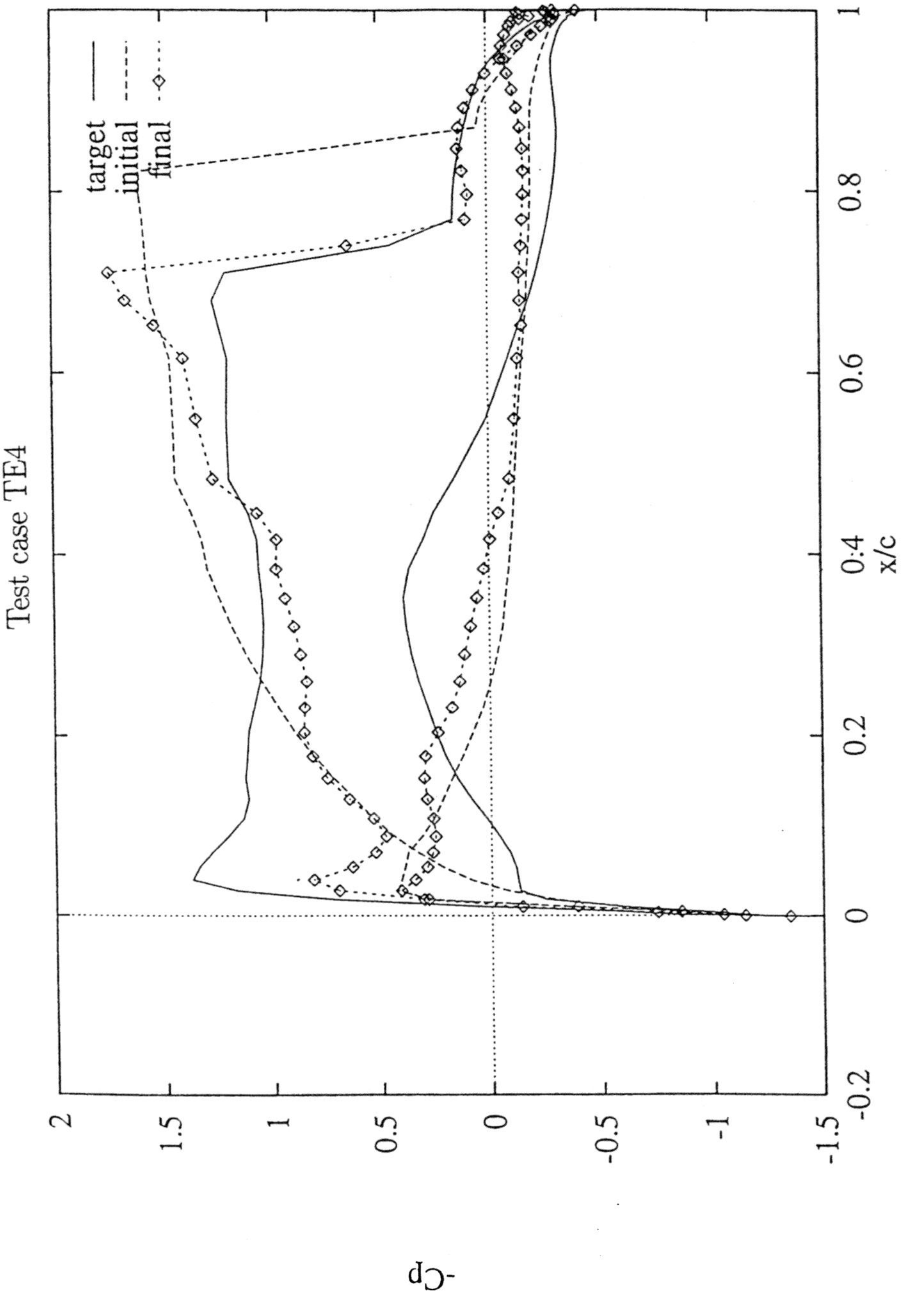

Figure 6 Superposition of the target, the initial and the final Cp distributions for the TE4 test case under transonic conditions

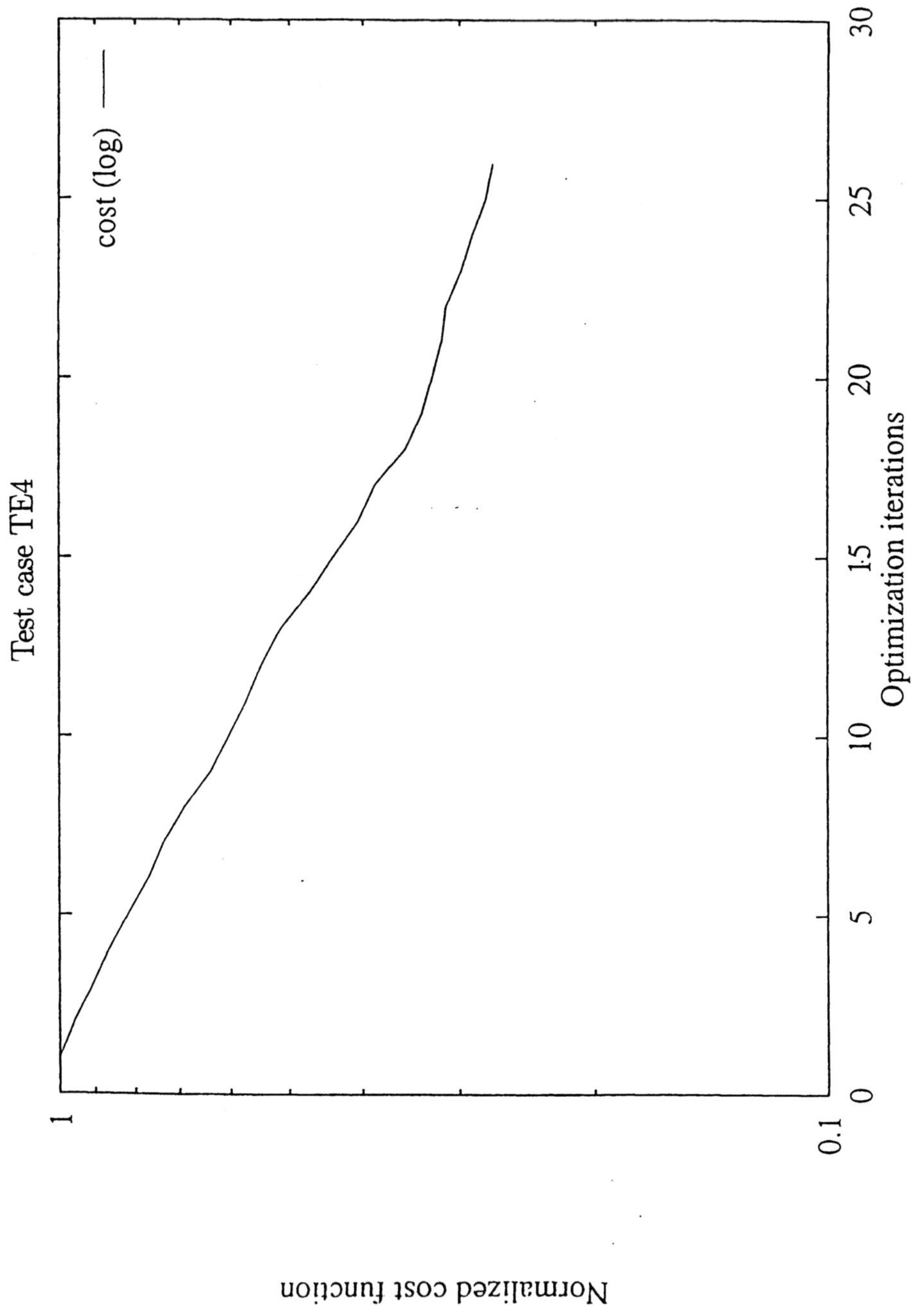

Figure 7 Evolution of the normalized objective function for the TE4 test case

7 Geometry Optimization of Airfoils by Multi-Point Design

K.-W. Bock[1] and W. Haase[2], Dornier Luftfahrt GmbH, Germany

1 Summary

The Dornier method to solve two–dimensional flows by use of a viscous–inviscid approach (VII) is based on the flow prediction code by Drela (1987, 1990) and is coupled with an optimization program, based on the code by Vanderplaats (1979, 1984). Coupling both methods was possible without any changes of the source code. Data exchange between both codes as well as all calculations of objective and constraint functions are performed in one single isolated interface program leading to a very flexible optimization tool. Analysis–computation time was reduced by making direct use of flow and contour data of one optimization step as input data for the VII analysis method.

In the ECARP/Optimum–Design area, the complete – coupled – system has been applied to the shape optimization of airfoils. In order to test the predictive capabilities of the optimization system, tests have been carried out on minimizing drag on airfoils by multi–point design – with and without additional constraints on the pitching moment. The results obtained clearly demonstrated the usefulness and reliability of the method. Furthermore, the present paper describes the results obtained for the mandatory ECARP/Optimum–Design Workshop test cases, TE3 and TE4.

2 The Method

2.1 Introduction

Aerodynamic design is always an optimization process. Some aspects of the aerodynamic behaviour, e.g. lift, may be prescribed quantitatively while at the same time other properties have to reach their extrema, e.g. drag should be at its minimum. Additionally, other properties might be restricted in growth by setting desired constraint conditions.

The method used for optimizing an airfoil shape in order to achieve a defined objective function consists of a flow analysis and an optimization tool. In particular, a viscous–inviscid interactive (VII) approach based on Drela (1987, 1990) has been coupled to an optimization strategy based on Vanderplaats (1979, 1984).

The coupling of both codes can be performed either by an outer– or an inner–loop strategy. In contrast to the inner optimization of inverse methods which might be fast in some applications but is seen as not being flexible enough in multi–point design due to the tremendous effort necessary when new target or constraint functions have to be employed, an outer optimization loop has been selected for the present optimization system. An additional advantage of the outer–loop optimization is by all means its transparency and flexibility, allowing e.g. for easily changing the analysis or optimization code or parts of it, if it proves necessary for improving the optimization systems's predictive capabilities. However, a disadvantage might be an (eventually) increased number of iterations and, thus, computation times.

New addresses: [1]) DLR Göttingen, [2]) Dasa LM, Manching

2.2 The Viscous–Inviscid Approach

Based on Drela (1990), the VII method solves the steady Euler equations in conservative formulation. The Euler method is coupled to a boundary–layer procedure permitting accurate loss prediction in both design and analysis calculations with or without (incipient) separation.

The discretization of the steady Euler equations is based on an intrinsic grid where the coordinate lines correspond to streamlines. The clear advantage of such a streamline mesh is the zero convection across corresponding cell faces. Hence, both the continuity and energy equations may be replaced by conditions of constant stagnation enthalpy and constant mass flux per stream–tube, resulting in only two unknowns per grid point instead of four. Thus the only way signals propagate from one stream–tube to another is through the geometry of the stream–tube and the pressure field, respectively.

The viscous part of the VII method consists of an inverse boundary layer method with displacement thickness or mass defect to be prescribed. The resulting pressure distribution is then compared with the inviscid one and the difference is taken to compute a corrected displacement thickness or mass defect distribution by means of some interaction law formula. This coupling procedure is called semi–inverse.

The non–linear set of equations is solved directly by a Newton method which converges rapidly once the partially converged solution is close to the exact one. The boundary layer equations for all streamwise stations are included in the Newton system and solved simultaneously with the inviscid equations, i.e. fully coupled. An important advantage of the Newton method is its capability to provide sensitivities (gradients), e.g. $dc_L/d\alpha$, $dc_D/d\alpha$, etc., at no additional costs that can be directly used to support the optimization procedure.

2.3 The Optimizer

The optimization code (Vanderplaats, 1984) searches for extrema (hillclimber strategy) of a numerically defined objective function subject to a set of constraint limits. Both objective function and constraint functions are approximated by means of a second order Taylor series with respect to the design variables. This Taylor approximation provides gradients and, therefore, allows for a variation of the design variables in order to reach a minimum in the objective function without violating the pre–defined constraints. In subsequent steps, new analysis results are added to the data basis of the Taylor approximation, consequently becoming more and more accurate in the vicinity of the actual design.

In the past, the optimization code had been applied together with a variety of analysis codes in order to

- minimize forebody wave drag,
- minimize wave drag in supersonic–transport pre–design,
- design missile geometries,
- optimize burning sequences of missiles with respect to mass reduction,
- determe trim conditions for missiles,
- development and optimize empirical relations, and to
- vary certain parameters to achieve best agreement with experiments or exact relations.

3 Discussion of Results

All optimization applications treated in the context of ECARP are concerned with the optimization of airfoil shapes on the basis of minimum drag and/or getting as close as possible to a target pressure distribution. Both single and multi–point designs have been treated and will be discussed in the following.

3.1 Design Parameters

In all cases treated, the design parameters represent the airfoil contour itself. More precisely, the airfoil contours are represented in a parameter form by using a certain number of design points along the lower and upper surface, respectively. In the parameter form, possible airfoil contours read:

$$z(\phi) = \sum_{l=1}^{L} a_l z_l(\phi) + P_{Bezier}(\phi;\phi_m, z_m) + P_{Bezier}(\phi = 0;\phi_m, z_m) \times cos(\phi), \qquad m = 1(1)M \tag{1}$$

with

$$\phi = \pm \arccos(1{-}x) \tag{2}$$

being an adequate transformation of the cartesian contour data. The positive sign indicates the upper surface and the negative sign the lower surface, respectively. The third term in equ. (1) specifies an additional flat plate in cartesian coordinates that can be introduced to eliminate angle–of–attack effects from the second term.

The coefficient a_l denotes the weighting factor of up to L different airfoils $z_l(\phi)$. The parameters ϕ_m and z_m are the Bezier weighting points. Table 1 shows a typical distribution of these points with fixed values of ϕ_m and 8 free values of z_m taken as design parameters.

Table 1 Representation of Airfoil Contour – Bezier Weighting

ϕ_m	-90.	-82.36	-77.03	-70.20	-61.25	-29.30	25.81	60.30	77.18	90.0
	Values fixed from approximation of basic airfoil									
z_m	0.0	z_2	z_3	z_4	z_5	z_6	z_7	z_8	z_9	0.0
	fixed	Values variable, starting value =0.0								fixed

Hence, one airfoil contour – according to L=1 – is represented by 9 design parameters, a and z_2 to z_9; z_1 and z_{10} staying fixed. Initially, all design parameters, z_m, are set to zero and a is set to unity, i.e. the initial airfoil is identical to the basic pre–designed one. The ϕ–coordinates have been selected from a previous investigation, where they were chosen to represent the airfoil – defined by the 10 design points – at its best. In some optimization cases the angle of attack was taken as an additional design parameter, in others the lift was explicitly prescribed and the angle of attack was calculated by the VII analysis method.

3.2 Minimization of Airfoil Drag by 4–Point Design

In order to test accuracy, robustness and reliabilty of the entire optimization system consisting of the analysis code, the optimizer and the coupling procedure, tests have been performed for minimizing airfoil drag. The objective function – minimum drag – had to be fulfilled with and without additonal constraints on the pitching moment. Results on those test cases will be discussed in the following.

Design Without Pitching Moment Constraints

The objective was to minimize the total drag, i.e. the sum of four drag coefficients, according to the four selected design points specified in Table 2. Apart from the constraint to keep the minimum thickness of the airfoil to 0.15 chords, no additional constraints were prescribed.

Table 2 Design Characteristics for Optimization Without Pitching Moment Constraints

Design Point No.	Mach Number	Reynolds number	Lift Coefficient
1	0.4	12.0×10^6	0.3
2	0.4	12.0×10^6	0.6
3	0.7	14.5×10^6	0.1
4	0.7	14.5×10^6	0.4

For the basic airfoil, the viscous–inviscid interaction (analysis) method computed:

$$
\begin{aligned}
c_{D,total} &= C_{D,1} && + C_{D,2} && + C_{D,3} && + C_{D,4} \\
&= 0.003033 && + 0.006536 && + 0.002893 && + 0.003421 && = 0.015883 .
\end{aligned}
$$

The transient behaviour of the optimization process can be taken from Fig. 1. To allow for a smooth start–up procedure of the gradients to be used by the Tayler series, the first optimization steps were prescribed by input. Concerning the target function, it can be recognized that major improvements have been achieved for iteration cycles greater than 47 which is related to the strong nonlinearity of the design parameters. It should pointed out that the coefficients of the Taylor series are complete for design cycles greater than

$$N = 1 + n + n\frac{(n+1)}{2} \tag{3}$$

i.e. for n = 8 design parameters, N equals 45 cycles.

The optimization process was finalized at 125 design cycles, regarding the fact that the VII method failed for the third design point. The reason for that failure was the appearance of separation in the aft part of the airfoil, more exactly: separation on the lower surface in the cusp–like rear–loading area. Nevertheless, the sum of the drag coefficients is reduced by 5.3% or 8.4 drag–counts, i.e. from the original value of $C_{D,total} =$ 0.015883 to $C_{D,total} = 0.015043$ with

$$
\begin{aligned}
c_{D,total} &= C_{D,1} && + C_{D,2} && + C_{D,3} && + C_{D,4} \\
&= 0.002763 && + 0.006106 && + 0.002913 && + 0.003261 && = 0.015043 .
\end{aligned}
$$

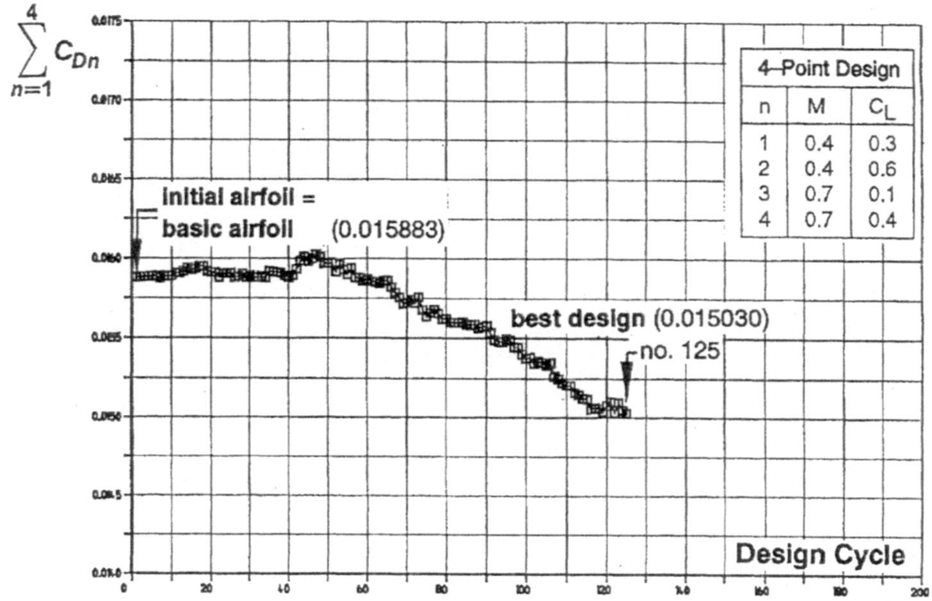

Figure 1 4–Point Design Without Pitching Moment Constraints – Optimization Course of Objective Function

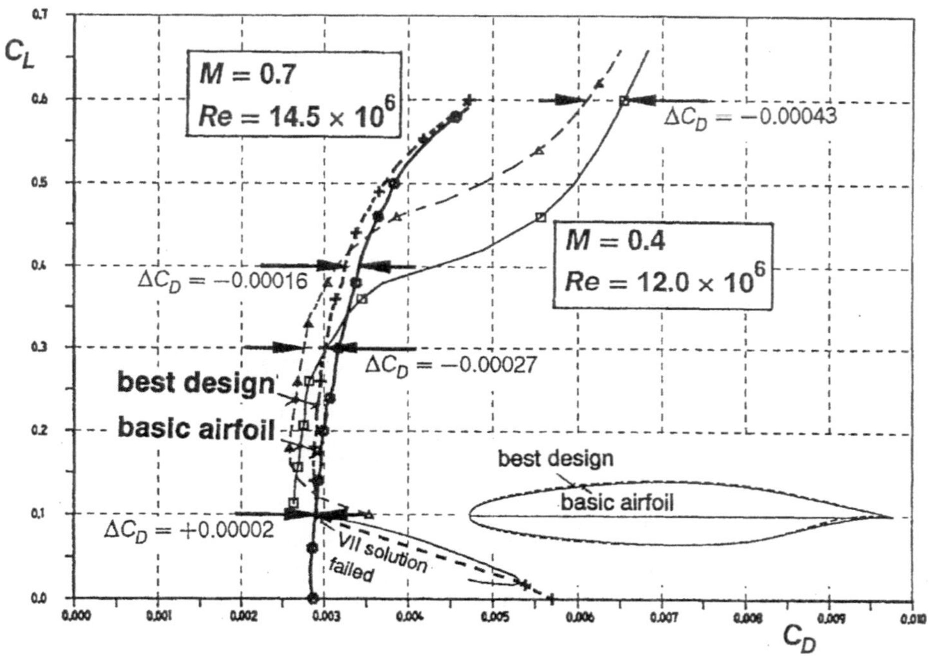

Figure 2 4–Point Design Without Pitching Moment Constraints – Drag Polars for Basic Airfoil and Best Design

Drag polars for both Mach numbers are presented in Fig. 2 together with the optimized airfoil shape represented by the broken lines. The increase in lift is mainly caused by an increase in the rear–loading of the profile, which on the other hand has a negative effect on the pitching moment and – as mentioned above – causes separation in the rear part on the lower surface at negative angles of attack. The decrease in drag is based on the results of three design points (1, 2 and 4) – for the third point a slight increase in drag is predicted.

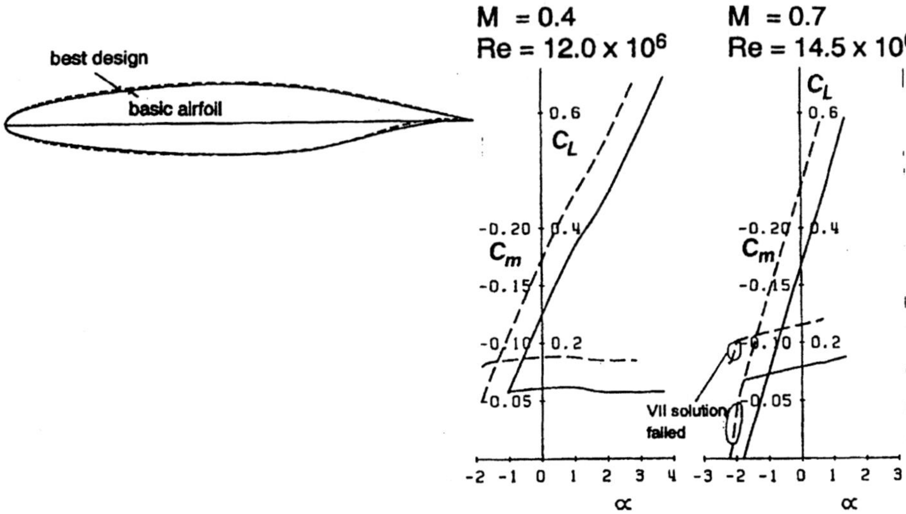

Figure 3 4–Point Design Without Pitching Moment Constraints – Lift and Pitching Moment versus α – Basic Airfoil and Best Design

Additionally, Fig. 3 depicts lift and pitching moment versus the angle of attack for the M=0.4 and M=0.7 test case, respectively. As noted above, the pitching moment is drasticly increased, perhaps unacceptabe with respect to flight meachanics.

Design With Pitching Moment Constraints

In contrast to the optimization procedure presented in the last section, additional constraints are now employed which bound the pitching moments of the four design points described in Table 3.

Table 3 Design Characteristics for Optimization With Pitching Moment Constraints

Design point no.	Mach number	Reynolds number	Lift coefficient	Pitching moment coefficient
1	0.4	12.0×10^6	0.3	≤ -0.06
2	0.4	12.0×10^6	0.6	≤ -0.06
3	0.7	14.5×10^6	0.1	≤ -0.07
4	0.7	14.5×10^6	0.4	≤ -0.08

with the minimum thickness constraint (0.15% of chord) still active.

The values selected for the pitching moment constraints are more or less the values of the basic airfoil, i.e. it should be assumed, that the pitching moment for the optimized airfoil is in the same range as for the basic one. However, an optimization based on constraints which are already close to values for the basic airfoil, i.e. an optimization along those constraints, is – in general – rather time–consuming. In order to decrease the number of the design cycles, the solution from the optimization without pitching moment constraints was taken as the initial solution.

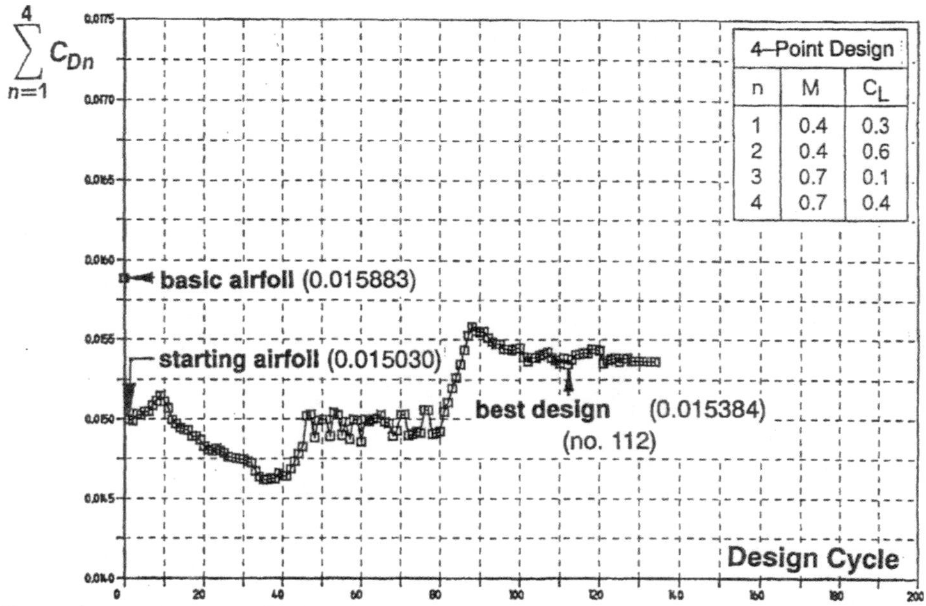

Figure 4 4–Point Design With Pitching Moment Constraints – Optimization Course of Objective Function

Fig. 4 presents the transient sum–of–the–drag behaviour versus the optimization cycles needed to reach the "best design". The optimization procedure was slightly different to the more "straightforward" procedure used for the case without constraints on the pitching moments. In the present case, for speeding up the design process, the pitching moment constraints have been introduced in three subsequent steps. As a result, the target function is reached by three subsequent optimization loops.

Compared to the design loop without the pitching moment constraints, the "best design" including the latter is – as expected – slightly worse. Nevertheless, an improvement of 3.2% or 5.1 drag counts has still been achieved; the sum of the four drag coefficients reads now

$$c_{D,total} = c_{D,1} + c_{D,2} + c_{D,3} + c_{D,4}$$
$$= 0.003323 + 0.006236 + 0.002763 + 0.003051 = 0.015373 .$$

Figs. 5 and 6 depict the resulting polars, the drag polar in Fig. 5, and lift and pitching moment polars in Fig. 6 for both investigated Mach numbers. In contrast to the optimization without pitching moment constraints, an improvement is reached for the high Mach number case, while increase and decrease in drag for the low Mach number case nearly cancel out. Compared to the basic airfoil, the rear–loading in the present case is only slightly increased. Additionally, it can be seen quite easily from Fig. 6 that lift and pitching moments have changed only maginally. In contrast to the optimization without pitching moment constraints, the lower surface is now more smooth in the first half of the airfoil circumventing "rapid" changes in surface curvature.

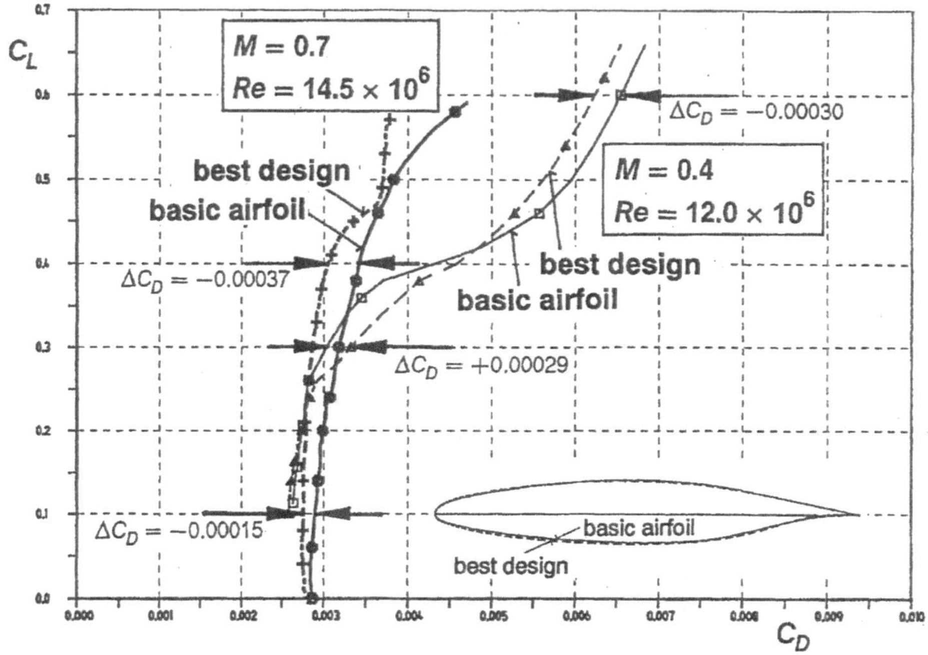

Figure 5 4–Point Design With Pitching Moment Constraints – Drag Polars for Basic Airfoil and Best Design

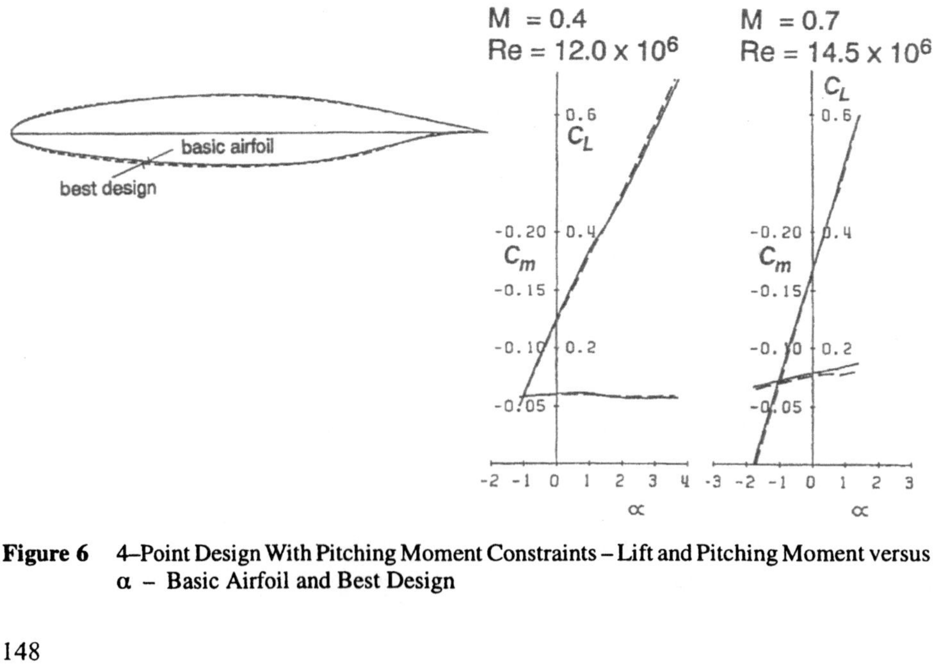

Figure 6 4–Point Design With Pitching Moment Constraints – Lift and Pitching Moment versus α – Basic Airfoil and Best Design

Transonic Airfoil, Inviscid Drag Minimization – Case TE3
The TE3 test case addresses the problem of inviscid drag minimization in transonic flows. Far field conditions chosen are Ma=0.73 and an angle of attack of $\alpha=2^o$. The RAE2822 airfoil has been used as the starting profile. Although this test case was defined as an inviscid one, optimization on the basis of viscous flow has been additionally performed. Three different test cases were run:

- An inviscid single–point optimization based on the above given flow parameters and the constraint that the lift coefficient for $\alpha=2^o$, $c_l=0.869$, had to be kept unchanged.
- A viscous single–point optimization, with the chosen flow conditions Ma=0.73 (as above), a lift coefficient of $c_l=0.654$, a Reynolds number of $Re=9.5\cdot10^6$ and transition fixed at x/c=0.03. The angle of attack was not prescribed. The objective was again to search for a minimum in drag – based on the given lift coefficient.
- A viscous two–point optimization, with the chosen flow conditions Ma=0.73 (as above), two lift coefficients of $c_{l1}=0.6$ and $c_{l2}=0.7$, $Re=9.5\cdot10^6$ and transition fixed at x/c=0.03. The objective is to drive the sum of $c_d(c_{l1})$ and $c_d(c_{l2})$ to a minimum.

As it has been proven to be necessary to introduce an additional constraint for the inviscid flow case, again the minimum thickness of the airfoil, this constraint has been applied to all three subsequent test cases. Particularly, this constraint reads: $t/c \geq (t/c)_{RAE2822}=0.1211$.

Inviscid Single–point Optimization: For the inviscid single–point optimization, the "best design" was reached after 88 iterations resulting – as suggested – in a nearly zero drag. The initial drag coefficient of $c_d=0.006653$ was driven down to 0.000024, i.e. to only 0.36% of the original value. The resulting profile shape allows for a nearly shock–free flow, however, it can be seen, only for the design point. In the near off–design drastic inceases in drag are detected, compare also Fig. 7. It shows very clearly that a single–point optimization might lead to a completely undesirable off–design behaviour.

Viscous Single–point Optimization: The iteration for the viscous single–point optimization was stopped after 143 cycles, achieving a drag reduction of 11.25% based on the initial drag coefficient of $c_d=0.001004$ and an optimized value of 0.000891. The best design resulted in an upward shift of the drag polar, see dashed line in Fig. 7. The profile shape computed, Fig. 8, exhibits a more sharp leading edge region and an increased rear–loading, which might be also – similar to the results presented in section x.3.2 – unacceptable in flight.

Viscous Two–point Optimization: To circumvent the stringent situation of an optimization "around" just one design point, and as the direct consequence to except off–design situations which are contradictory to the achievements reached – as described for the inviscid case – a two–point design has been performed taking into account lift values of 0.6 and 0.7. Again, the angle of attack was not fixed to a certain value. The best design has been achieved after 118 iterations and the drag polar, see dashed–dotted line in Fig. 7, is shifted even more towards increased lift even in the off–design region. However, a major drawback might be seen in the again (slightly) increased rear–loading and in a double–shock system for the c_L=0.6 case as it can be detected from Fig. 9.

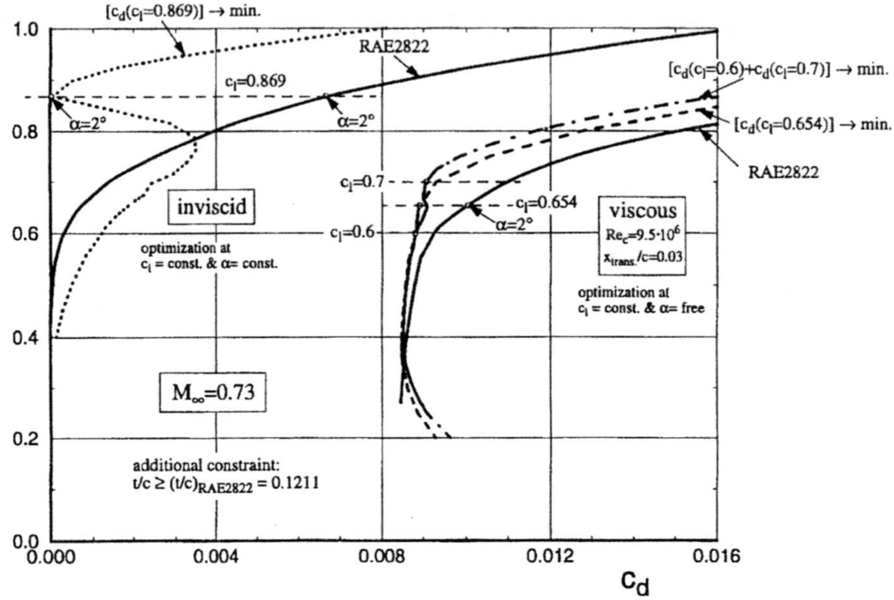

Figure 7 Problem TE3 – Transonic Airfoil Drag Minimization – Comparison of Drag Polars

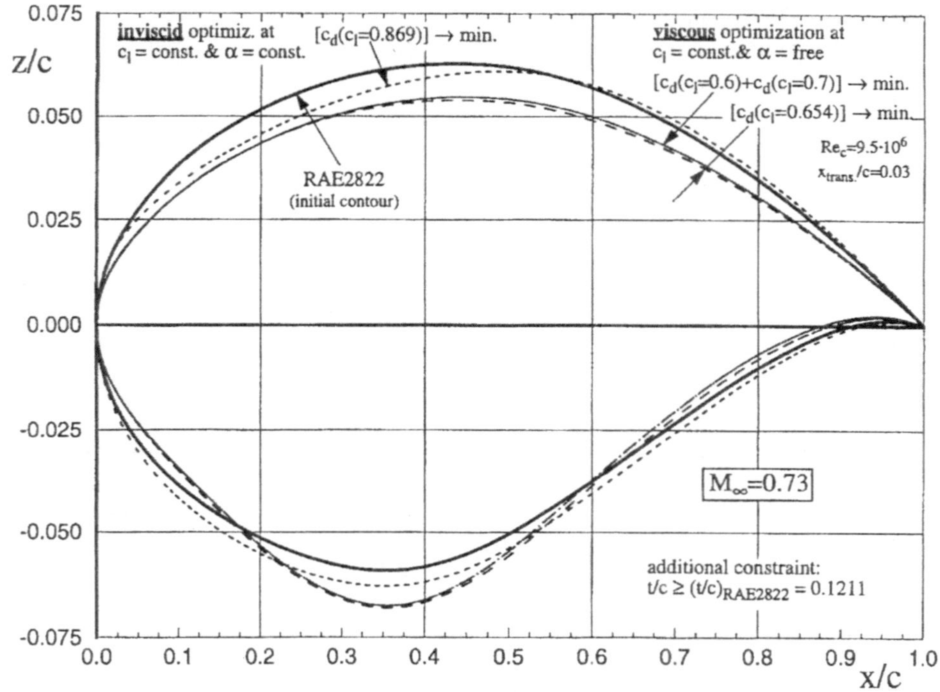

Figure 8 Problem TE3 – Transonic Airfoil Drag Minimization – Comparision of Initial and Optimized Contour

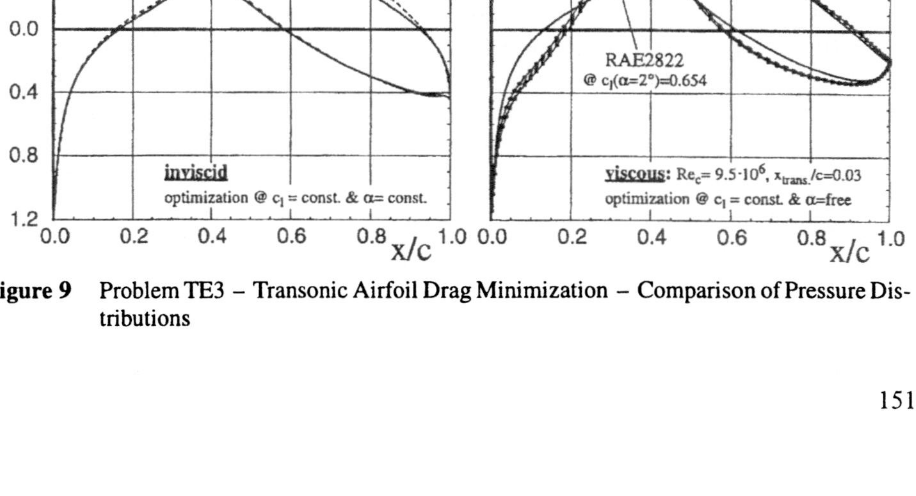

Figure 9 Problem TE3 – Transonic Airfoil Drag Minimization – Comparison of Pressure Distributions

Multi–point 2D Airfoil Design – Case TE4

The actual two–point design problem was defined as the minimization of the objective function

$$F(\alpha_1, \alpha_2, x(s), y(s)) = \sum_{i=1}^{2} \left[W_i \int_0^1 (c_p^i(s) - c_{p,tar}^i(s))^2 ds \right] \tag{4}$$

with the design variables α_1, α_2, x(s) and y(s). The variable s is the arc length measured along the airfoil, W_i denote the weighting factors and $c^i{}_p$ and $c^i{}_{p,tar}$ are the pressure distributions for the actual i'th design condition (at a certain α_i, Mach and Reynolds number) and the target pressure condition at design condition, respectively.

The two different design conditions (i=1,2) were specified as follows:

- i=!: A typical high–lift airfoil is specified (by a given set of geometry data) at subsonic conditions. By analysis computation for α=10.8° incidence, Ma=0.2 and Re=$5 \cdot 10^6$, the calculated pressure distribution serves as the first target pressure for the two–point design exercise.
- i=2: Secondly, the geometry for a typical high–speed airfoil is prescribed. The flow conditions read: α=1.0° incidence, Ma=0.77 and Re=10^7. The calculated pressure distribution is used as the (second) target pressure for the two–point design problem.

Three different calculations have been carried out and are going to be discussed below:

- A first *viscous* optimization taking W_1=1 and W_2=0, aiming at a re–design of the subsonic airfoil.
- A second *inviscid* optimization taking W_1=0.5 and W_2=0.5 for the two–point problem in order to achieve the two target pressures.
- A third optimization taking again W_1=0.5 and W_2=0.5 but now to achieve the two target pressures for *viscous* flow.

Viscous Single–point Optimization – High–lift Target Pressure: The starting airfoil contour for the re–design case has been the NACA4412 airfoil. To support a more rapid convergence of the optimization process, the first 13 iterations (optimization cycles) have been run with a fixed contour, i.e. only incidence variations have been allowed. All subsequent cycles were then run varying the angle of attack plus the airfoil shape. The best design was achieved after 231 iterations.

Fig. 10 depicts the pressure distribution for both lower and upper airfoil surface. Besides the starting design and the design at the N=13 switch, the optimized pressure distribution is compared with the target pressure at α=10.8°. Apart from a slight shift in the location of the stagnation point, causing some minor deviations in the region of the suction peak on the upper surface, the agreement between target and converged pressure is rather good.

However, a second deviation at x/c=0.9 on the upper surface of the best–design airfoil can be more easily detected from Fig. 11 which presents the contours of the starting, best–design and target airfoils. The reason to be blamed for that can be seen in too few design points on the upper surface in that particular area, compare Table 1.

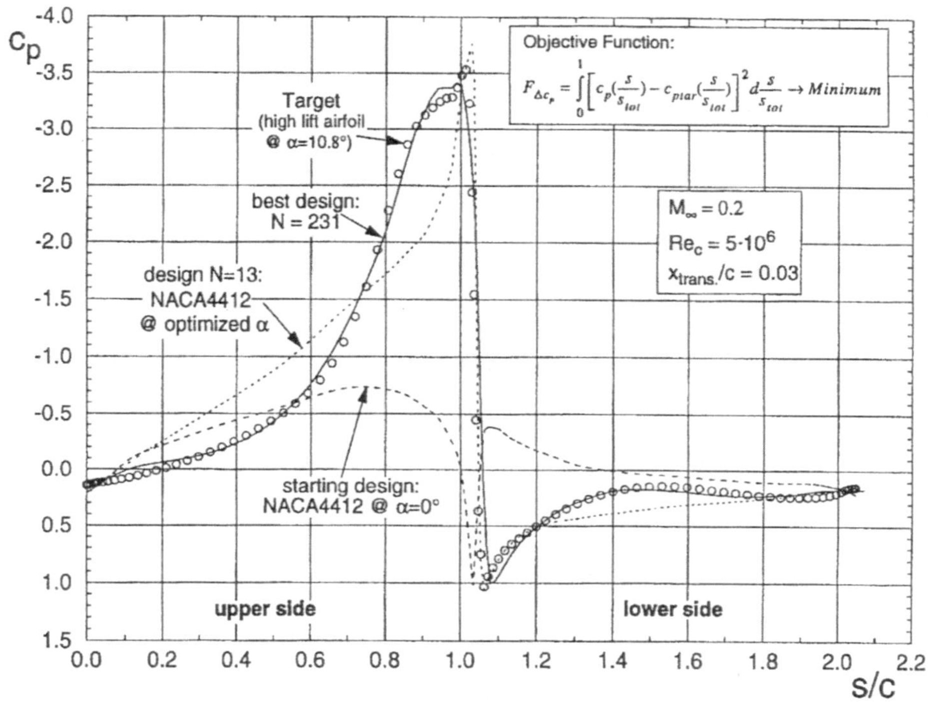

Figure 10 Problem TE4 – Airfoil Optimization to Get High–Lift Target Pressure Distribution – Comparison of Pressure Distributions

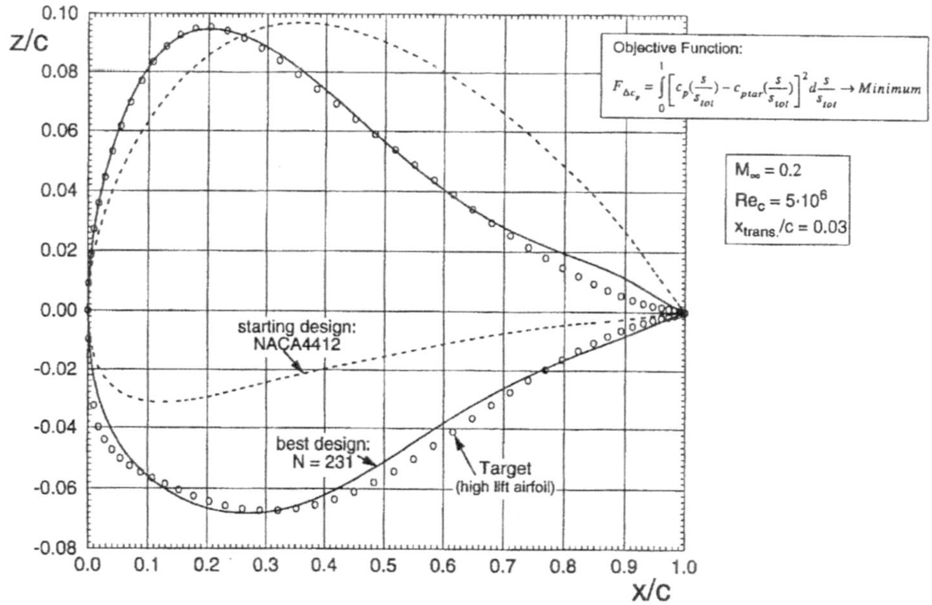

Figure 11 Problem TE4 – Airfoil Optimization to Get High–Lift Target Pressure Distribution – Comparison of Contours

Inviscid Two–point Optimization: The inviscid two–point optimization has been carried out without additional constraints. According to equ. (1) and the fact that now two different airfoil shapes are part of the contour function, 10 design parameters ($a_{target-1}$, $a_{target-2}$, $z_2 - z_9$) have been used. The two angles of attack are design function no. 11 and 12. The starting design contour was initiated by taking the mean values of the two target airfoils, i.e. defining z to be $z=0.5\cdot(z_{target-1-airfoil}+z_{target-2-airfoil})$.

In Fig. 12, target pressure distributions are presented in comparison to those achieved by the optimization procedure. The thick solid and broken lines denote the pressure distributions for the best design after 284 design/optimization cycles for Ma=0.2 and Ma=0.77, respectively. They can be compared with the target pressures, represented by open circles and squares. Very good agreement has been achieved on the lower side of the airfoil – which is shock free for the Ma=0.77 case – and the stagnation point location. On the upper side, the agreement with the target pressures is at least reasonable. The mean square errors, additionally plotted as thin solid and broken lines for the subsonic and transonic case, respectively, correspond to the deviations between the target airfoil and the converged – optimized – airfoil. The main reason for partially missing the target pressure distributions is the fact that the pressure characteristics of the two extremely different airfoils cannot be combined in one single airfoil.

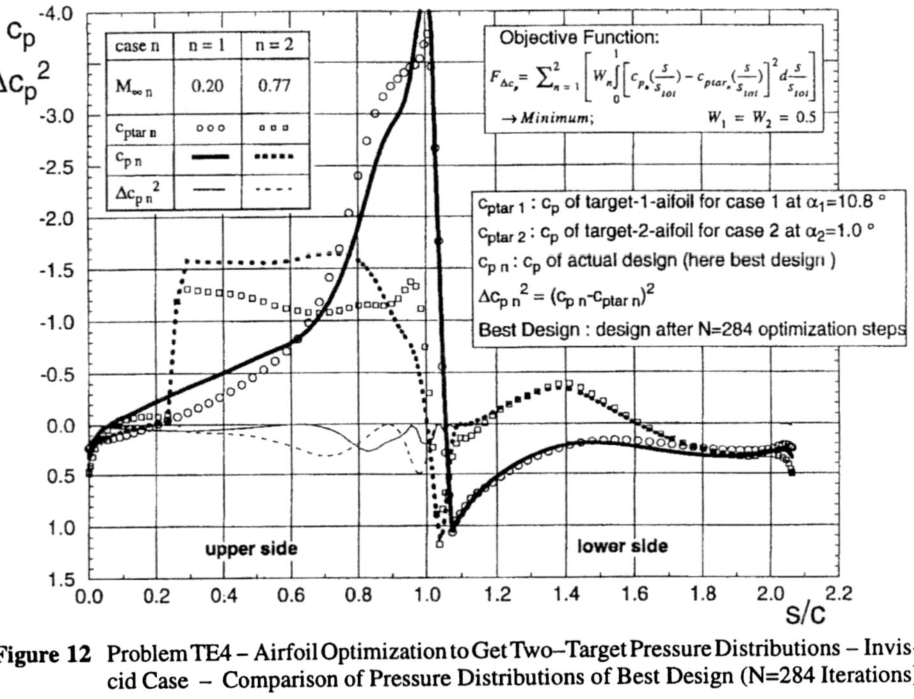

Figure 12 Problem TE4 – Airfoil Optimization to Get Two–Target Pressure Distributions – Inviscid Case – Comparison of Pressure Distributions of Best Design (N=284 Iterations)

The optimized airfoil shape is presented in Fig. 13 together with the starting design and the the target–1 and target–2 airfoils. Due to the fact that weighting factors of 0.5 have been used one would assume that the best–design airfoil shape is somewhat in between the two initial airfoil shapes. Comparing the best–design and the two starting airfoil shapes – achieved by taking the mean z–value of those – one can easily recognize that the starting design and the best design are rather similar. On the other hand, the best–design airfoil shape is getting closer to the high–lift airfoil shape on the upper surface, particularly in the leading edge region, and reaches the low–drag (transonic) airfoil shape in the rear part of the lower surface. Hence, it is promoting both extremely different flow types where ever possible.

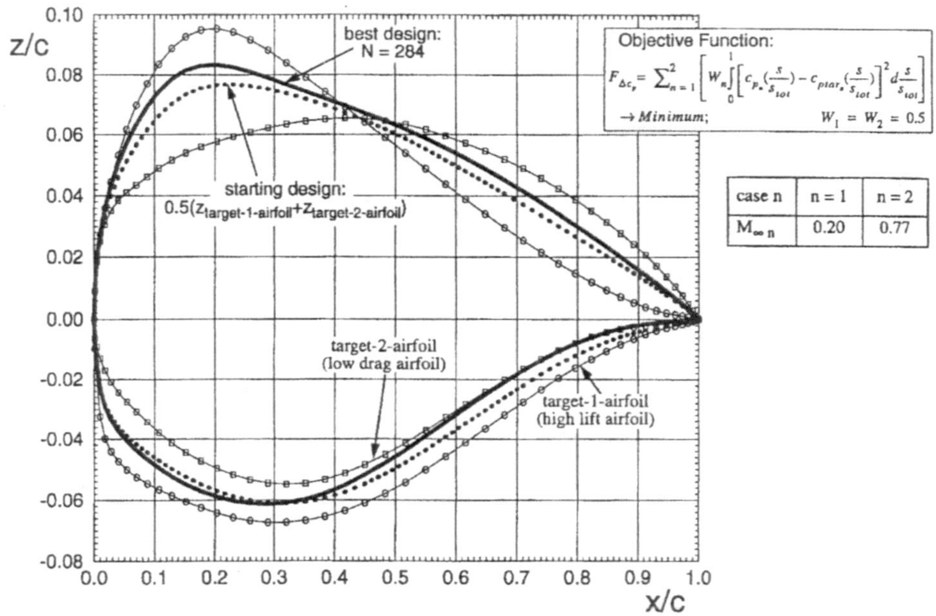

Figure 13 Problem TE4 – Airfoil Optimization to Get Two–Target Pressure Distributions – Inviscid Case – Comparison of Contours

Viscous Two–point Optimization: The most challenging test case, of course, has been the viscous two–point design case. It uses the flow parameters presented above. Transition was fixed at 3% chord on lower and upper surface. The optimized solution had been obtained after 296 iterations using again the NACA4412 airfoil shape as the starting design – resulting in 11 design parameters – 2 angles of attack and 8 contour parameters.

For the objective function, F, Fig. 14 exhibits the transitional behaviour of the optimization process that reaches the best design at N=296 optimization cycles.

Pressure coefficient distributions are presented in Fig. 15 for both targets. The thick solid and broken lines indicate again the optimized pressures for Ma=0.2 and Ma=0.77, respectively, and can easily be compared with the target pressures represented by the open symbols. The corresponding mean square errors are given as thin solid and broken lines. The low Mach number case agrees pretty well with the target pressure, however, the transonic low–drag results are clearly off on the upper airfoil surface. Again the weak agreement is a consequence of the effort to combine contradictory pressure distributions of two extremely different airfoils to just one airfoil.

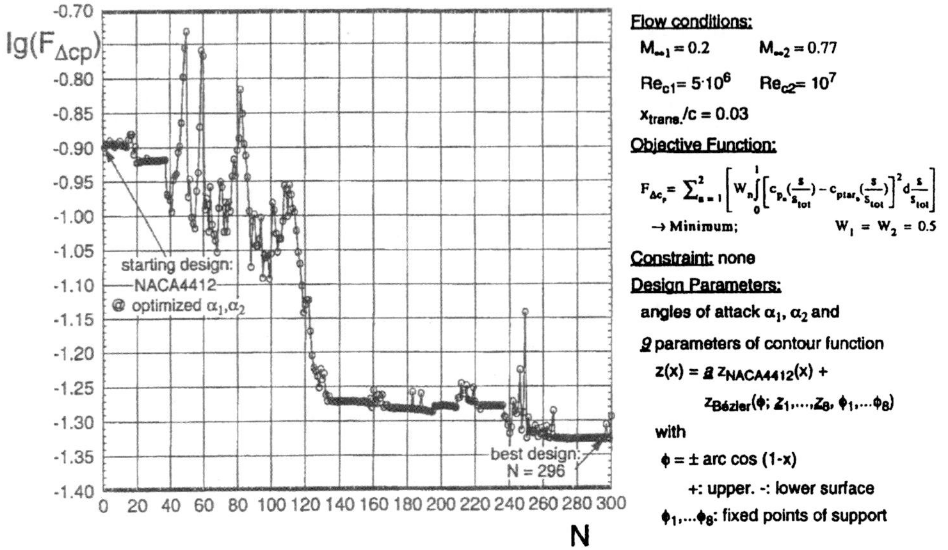

Figure 14 Problem TE4 – Airfoil Optimization to Get Two–Target Pressure Distributions – Viscous Case – Optimization Course of Objective Function

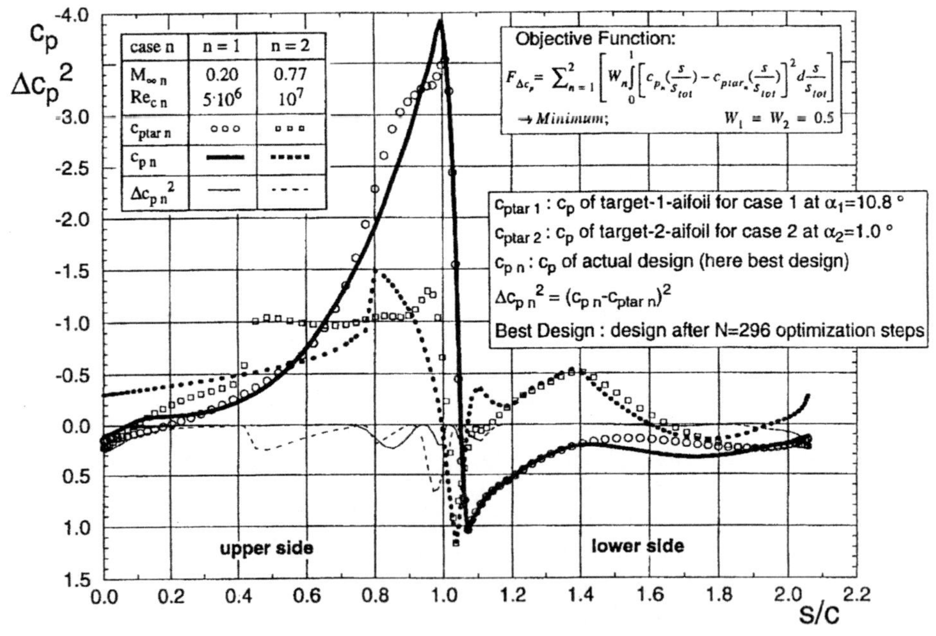

Figure 15 Problem TE4 – Airfoil Optimization to Get Two–Target Pressure Distributions – Viscous Case – Comparison of Pressure Distributions for Best Design (N=296 Iterations)

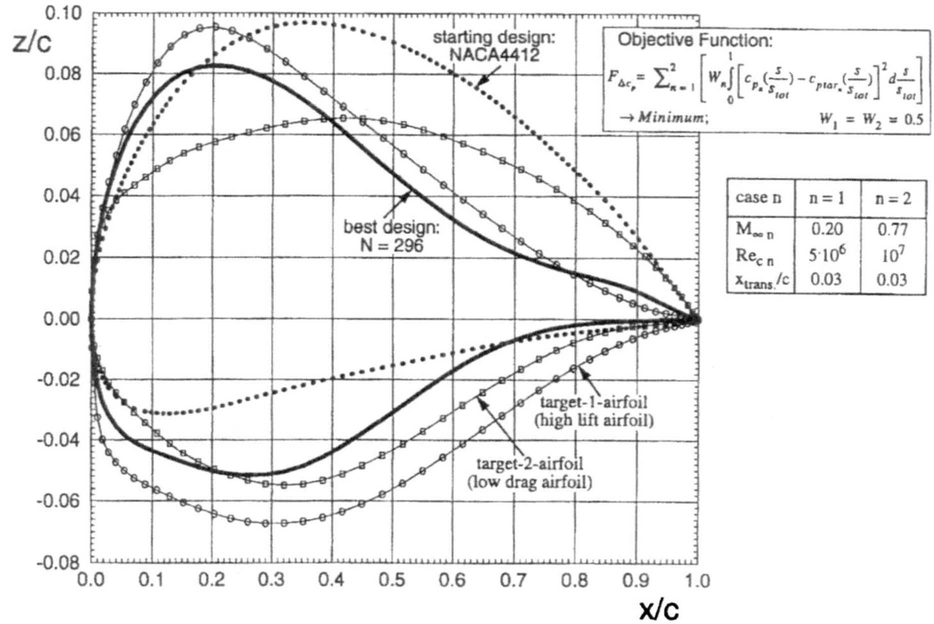

Figure 16 Problem TE4 – Airfoil Optimization to Get Two–Target Pressure Distributions – Viscous Case – Comparison of Contours

Comparing the shapes of the best–design, starting, target–1 and target–2 airfoils in Fig. 16 given above, allows to draw a similar conclusion for the best–design upper–aifoil shape as for the inviscid case presented in the last paragraph, i.e. the optimized contour follows more or less the high–lift (target–1) airfoil contour. The differences in the circulation around the airfoil, indicated by the differences in the trailing edge pressures, for sure influence the position of the shock on the upper surface. In order to produce a low–drag airfoil for the transonic objective and a high–lift airfoil for the subsonic flow case, the optimized airfoil consequently exhibits a strong rear loading. Whether such a strong rear loading is acceptable according to flight mechanic constraints – as noted already – might be questionable. However, no constraint on the pitching moment coefficient had been applied for the TE4 computations.

4 Conclusion

The present contribution describes work performed within the context of the ECARP/ Optimum Design area on the optimization of airfoils by contour variation using a multi–point design method. Optimization had been carried out in order to reduce drag and/or to get as close as possible to a prescribed target pressure distribution. Additional constraints on pitching moment and minimum airfoil thickness have been applied where necessary or adequate.

The design method made use of both a viscous–inviscid interaction (analysis) method and an optimization tool. These methods were connected in an outer–loop manner leading to an accurate and reliable tool for multi–point design purposes. Computation times for the analysis side have been considerably reduced by storing grid and flow data of each optimization step as an initial information for the next one.

The method had been applied to various applications and in particular to airfoil shape optimization as presented in the current contribution. For the optimization of airfoils – in order to reduce the total drag – the method had proven very good predictive capabilities and, in general, the results obtained for the challenging test cases of the ECARP/ Optimum Design area form a sound basis for future applications. Moreover, the results obtained, will definitely justify an extended use in an industrial design context.

5 Bibliography

Drela, M., Giles, M.B. (1987): ISES: A Two–Dimensional Viscous Aerodynamic Design and Analysis Code. AIAA–87–0424.

Drela, M. (1990): A User's Guide to ISES V4.2. MIT Computational Fluid Dynamics Laboratory.

Vanderplaats, G.N. (1979): Approximation Concepts for Numerical Airfoil Optimization. NASA–TP–1370.

Vanderplaats, G.N. (1984): ADS – A Fortran Program for Automated Design Synthesis. NASA CR 172460, October 1984.

8 Riblet Optimisation

Patrick Le Tallec and Olivier Pironneau, INRIA, Le Chesnay Cedex, France

I Introduction

The goal of the present study is the drag reduction of riblets. The problem to solve is to minimise the drag of the riblet with respect to its shape ω, while keeping a constant flow area.

In theory, the flow is governed by the three-dimensional Navier-Stokes equations. Shape optimisation with such a state equation is out of reach. Therefore, we have proposed to compute the flow by using the incompressible three-dimensional Parabolized Navier-Stokes Equations (PNS).

The problem to solve is then

$$Min_{\omega \in S} J(\omega) \qquad (P_m)$$

with function cost

$$J(\omega) = \frac{\nu}{2} \int_{z=0}^{z_{max}} \int_{\omega} |\nabla v|^2 + |\nabla u_3|^2 d\omega dz$$

and state variable $\vec{v} = (u_3, v, p_t)$ solution of the parabolic flow equation (obtained by neglecting all diffusion effects along the axial direction $0x_3$)

$$\begin{aligned}
&\nabla.v + u_{3,3} = 0,\\
&u_3 v_{,3} + v.\nabla v - \nu \Delta v + \nabla p_t = 0,\\
&u_3 u_{3,3} + v.\nabla u_3 - \nu \Delta u_3 + p_{t,3} = 0,\\
&\vec{v}|_{x_3=0} = \vec{v}^0,\\
&\vec{v}|_{\Gamma_1} = 0,\ \vec{v}.n|_{\Gamma \setminus \Gamma_1} = 0,\ \frac{\partial(\vec{v}.\tau)}{\partial n}|_{\Gamma \setminus \Gamma_1} = 0.
\end{aligned}$$

Above, S is the set of admissible controls defining the two-dimensional shape ω of the riblet. In abstract form, we will rewrite the state equation as the variational equation

$$a(\vec{v}, \vec{\varphi}) = 0, \forall \vec{\varphi}.$$

We solve the above constrained optimisation problem by a quasi-Newton Herskovits algorithm. This requires an analytic calculation of the gradient of J that we obtain by introducing and computing an adjoint state.

The purpose of this report is to describe the numerical solution of this parabolized state flow equations, to describe the calculation procedure for the adjoint state, to introduce the full optimisation algorithm which has been used and to present validation results.

II Numerical Solution of the Flow Equation

1 Formulation

Our goal is to compute the **steady laminar incompressible** flow of a fluid inside a channel delimited by two planes of riblets. More precisely, to test the efficiency of riblets we shall consider the flow in a semi-infinite channel with approximate rectangular cross section ω as shown in Figure (1) with two horizontal walls with riblets (the floor and the ceiling). The riblets are small and shaped like a saw tooth.

We model this flow by using the three-dimensional Navier-Stokes equations while neglecting all diffusion effects along the axial direction $z = x_3$. The resulting Navier-Stokes equations reduce then to the system

$$\begin{aligned} \nabla v + u_{3,3} &= 0 \\ u_3 v_{,3} + v.\nabla v - \nu\Delta v + \nabla p_t &= f \quad \text{in} \quad \Omega = \omega \times (0, z_{max}) \\ u_3 u_{3,3} + v.\nabla u_3 - \nu\Delta u_3 + p_{t,3} &= f_3. \end{aligned}$$

Above, $(v, u_3) = (u_1, u_2, u_3)$ denote the three dimensional velocity field, and the differential operators ∇, Δ denote the gradient and Laplacian with respect to the two-dimensional variables (x_1, x_2). Differentiation of a given function f with respect to the longitudinal direction $x_3 = z$ is denoted by $f_{,3}$. In addition, we suppose that the pressure gradient along the x_3 direction is quasi independent of the two-dimensional variables (x_1, x_2). This amounts to assume that the total pressure field is of the form

$$p_t = \Lambda(x_3) + p(x_1, x_2, x_3)$$

with

$$\Lambda'(x_3) >> \frac{\partial p}{\partial x_3}.$$

The equations reduce then to the Parabolised Navier-Stokes equations (PNS)

$$\begin{aligned} \nabla v + u_{3,3} &= 0 \\ u_3 v_{,3} + v.\nabla v - \nu\Delta v + \nabla p &= f \quad \text{in} \quad \Omega = \omega \times (0, z_{max}) \qquad (1) \\ u_3 u_{3,3} + v.\nabla u_3 - \nu\Delta u_3 + \Lambda_{,3}(z) &= f_3. \end{aligned}$$

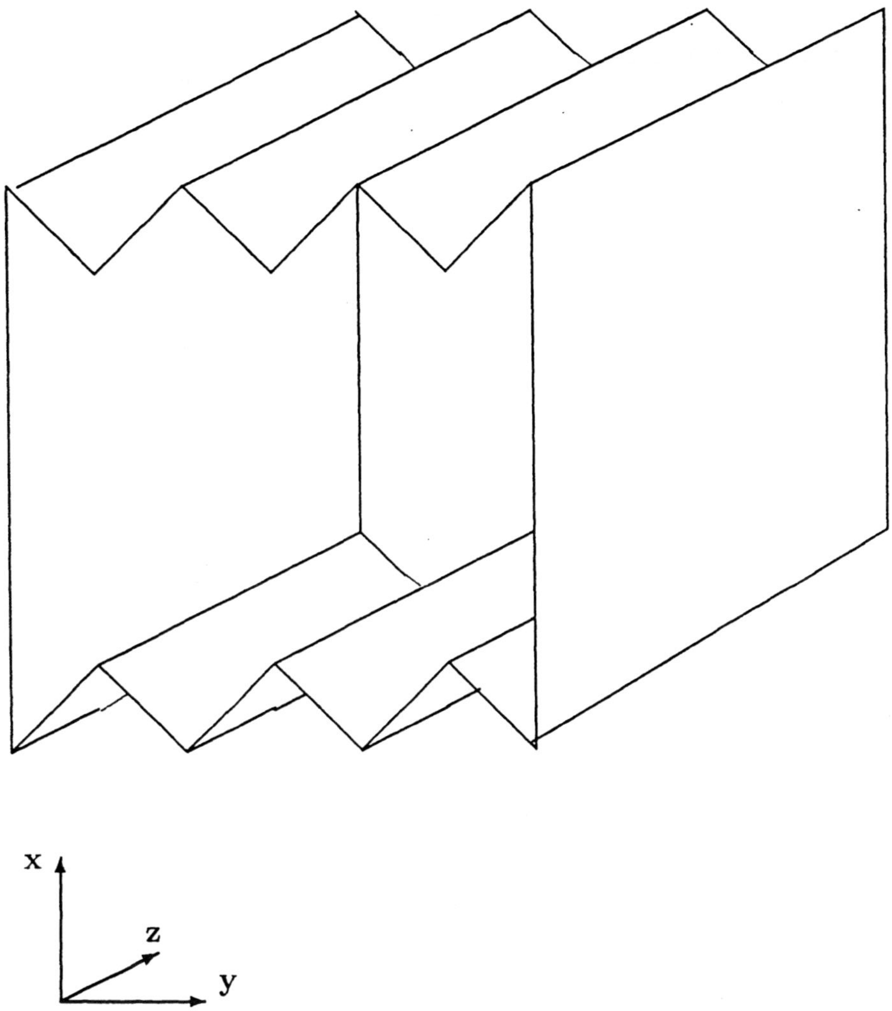

Figure 1 General geometry

2 Boundary Conditions

For any given cross section, i.e. for a given z, using standard symmetry and periodicity arguments, the boundary conditions take the form (Fig. 2-b) :

$$\begin{aligned} u_3 = 0, v = 0 \qquad & \text{on} \qquad \Gamma_1, \\ v.n = 0, \frac{\partial v.\tau}{\partial n} = \frac{\partial u_3}{\partial n} = 0 \qquad & \text{on} \qquad \Gamma - \Gamma_1. \end{aligned} \tag{2}$$

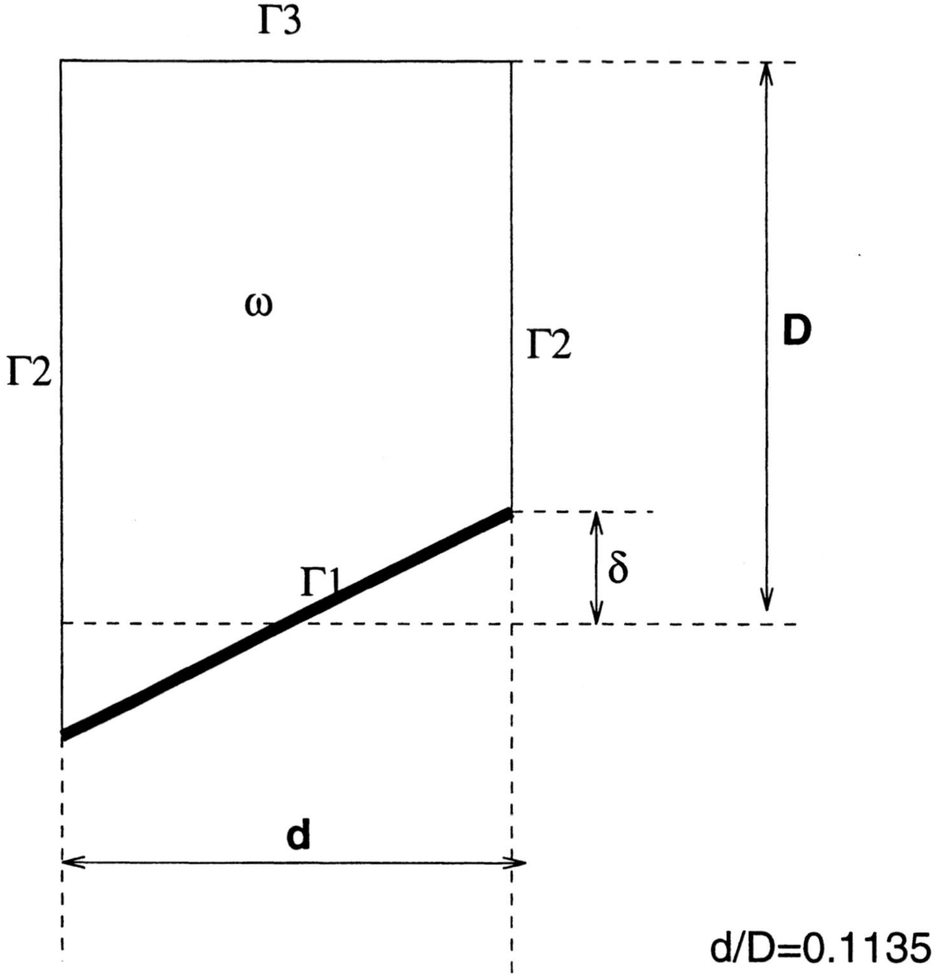

Figure 2 Cross section for a half riblet and definition of the boundaries

Due to these boundary conditions, by integrating the incompressibility condition over a given cross section,

$$\nabla . v = -u_{3,3} \tag{3}$$

we have

$$\int_{\Omega} u_{3,3} = \int_{\Omega} \nabla . v = \int_{\partial\Omega} v.n = 0. \tag{4}$$

By integration along x_3, we obtain a global compatibility constraint to be imposed on the axial velocity u_3, which writes

$$\int_\Omega u_3 = \int_\Omega u_3^0. \tag{5}$$

This equation must be added as a constraint to the PNS equation (1). Actually the multiplier of this constraint is precisely the pressure term $\Lambda_{,3}$ which appears in the third equation of (1). Problem (1)-(5) has therefore the structure of a classical well-posed mixed problem.

3 Discretisation in z

In view of the PNS equations (1), it seems perfectly natural to treat the axial variable $z = x_3$ as the standard time variable in a two-dimensional time evolution problem, to treat all derivatives in $z = x_3$ by a first order backward Euler discretisation scheme, and to treat all convection terms in v by the method of characteristics.

For this purpose, for any given point $x \in \omega$, for any given cross section z, we compute the characteristic curve χ^z arriving at point x in section z by solving the first order differential equation

$$\frac{\partial \chi^z(x;\tau)}{\partial \tau} = \frac{v(\chi^z,\tau)}{u_3(\chi^z,\tau)}, \tag{6}$$

$$\chi^z(x,0) = x. \tag{7}$$

We then approximate all convection terms on the cross section $n+1$ by the first order upwind formula

$$(u_3 f_{,3} + v\nabla f)(x, z^{n+1}) = u_3^n \left(\frac{f^{n+1} - f^n o \chi^n(x, -\Delta z)}{\Delta z}\right). \tag{8}$$

With this choice, the PNS equations written on a given section $n+1$ take the form

$$u_3^n \left(\frac{u_3^{n+1} - u_3^n o \chi^n(-\Delta z)}{\Delta z}\right) - \nu \Delta u_3^{n+1} + \lambda^{n+1} = f_3^{n+1}, \tag{9}$$

$$\int_\omega u_3^{n+1} = \int_\omega u_3^0, \tag{10}$$

$$\nabla . v^{n+1} = -\frac{u_3^{n+1} - u_3^n}{\Delta z}, \tag{11}$$

$$u_3^n \left(\frac{v^{n+1} - v^n o \chi^n(-\Delta z)}{\Delta z}\right) - \nu \Delta v^{n+1} + \nabla p^{n+1} = f^{n+1}. \tag{12}$$

Problem (9)-(12) is now a sequence of perturbed two-dimensional Stokes problem which can be solved by standard finite element methods.

4 Discrete Variational Problem

We now introduce a standard finite element discretisation of this incompressible Stokes problem in which

- velocities at each cross section are approximated by second order quadrilateral finite elements $Q2$;
- pressure fields at each cross section are discretized by discontinuous $P1$ finite elements.

We therefore introduce a given triangulation $\mathcal{T}$ in regular quadrilaterals, together with the functional spaces

$$P_1 = \{p = p_1 x + p_2 y + p_3, \forall T \in \mathcal{T}\}$$
$$Q_2 = \{p = a_0 + a_1 x + a_2 y + a_3 x^2 + a_4 xy + a_5 y^2 + a_6 x^2 y + a_7 xy^2 + a_8 x^2 y^2, \forall T \in \mathcal{T}\}$$
$$U = \{u \in C(\omega), u|_T \in Q_2\}$$
$$V = \{v \in U^2\}$$
$$\mathring{\mathcal{U}} = \{u \in U, \text{u satisfies the boundary conditions}\}$$
$$\mathring{\mathcal{V}} = \{v \in V, \text{v satisfies the boundary conditions}\}$$
$$V_p = \{\varphi; \varphi|_T \in P1, \forall T \in \mathcal{T}\},$$
$$V_0 = \{v \in \mathring{\mathcal{V}}, \int_T \nabla.vp = 0, \forall p \in V_p\}.$$

With this choice, the fully discrete Parabolized Navier-Stokes equation reduce to the sequence of uncoupled variational problems :

Let the inflow (v^o, u_3^o) be given. Then, for for each section $n \geq 0$ and knowing (v^n, u_3^n),

Step 1 : Calculate the characteristics χ at each Gaussian integration point x of the finite element triangulation

$$\frac{\partial \chi^{n+1}(\tau)}{\partial \tau} = \frac{v^n(\chi)}{u_3^n(\chi)}, \tag{13}$$
$$\xi^{n+1}(0) = x, \tag{14}$$

and set

$$\chi^n = \xi(., -\Delta z);$$

Step 2 : Calculate u_3^{n+1} by solving the constrained Q_2 finite element problem

$$\int_\omega u_3^n u_3^{n+1} \varphi + \nu \Delta z \int_\Omega \nabla u_3^{n+1} \nabla \varphi = \int_\Omega u_3^n u_3^n o \chi^n \varphi + \Delta z \int_\Omega (\lambda^{n+1} + f_3^{n+1}) \varphi,$$
$$\forall \varphi \in \mathring{\mathcal{U}}, u_3^{n+1} \in \mathring{\mathcal{U}},$$
$$\int_\omega u_3^{n+1} = \int_\omega u_3^n, \lambda^{n+1} \in R. \tag{15}$$

Step 3 : Calculate an extension $\tilde{v} \in \mathring{\mathcal{V}}$ of the gradient $u_{3,3}$ such that

$$\int_\omega div \tilde{v} \varphi_p = \int_\omega \frac{u_3^n - u_3^{n+1}}{\Delta z} \varphi_p, \forall \varphi_p \in V_p. \tag{16}$$

Step 4: Calculate the cross vector field

$$v^{n+1} = \bar{v} + \tilde{v} \tag{17}$$

by solving the variational system

$$\begin{aligned} \int_\omega u_3^n \bar{v} \psi \quad &+ \quad \nu \Delta z \int_\omega \nabla \bar{v} \nabla \psi \\ &= \int_\omega u_3^n v^n o \chi^n \psi - \int_\omega u_3^n \tilde{v} \psi - \nu \Delta z \int_\omega \nabla \tilde{v} \nabla \psi + \int_\Omega f^{n+1} \psi, \\ &\forall \psi \in V_0, \bar{v} \in V_0; \end{aligned} \tag{18}$$

Step 5: Calculate the cross pressure field by solving

$$\begin{aligned} \int_\omega p^{n+1} \nabla . \phi \quad &= \quad \frac{1}{\Delta z} \int_\Omega u_3^n (v^n o \chi^n \phi - v^{n+1} \phi) - \nu \int_\Omega \nabla v^{n+1} \nabla \phi \\ &+ \int_\Omega f^{n+1} \phi \\ &\forall \phi \in \mathring{\mathcal{V}}, p^{n+1} \in V_p. \end{aligned} \tag{19}$$

Remarks : The total pressure is given by

$$p_{total} = \Lambda(z) + p(x_1, x_2), \quad \Lambda'(z) = \lambda(z).$$

Problem (15) is linear and reduces to a standard linear system after expansion of u_3 in a nodal finite element basis of $\mathring{\mathcal{U}}$.

Problem (18) is linear, and reduces to a linear system after development of the unknown velocity in a finite element basis of discrete divergence free nodal functions. The resulting systems are solved by a direct Choleski solver.

In abstract form, we also rewrite this system of uncoupled equation as

$$a(\vec{v}, \vec{\varphi}) = 0, \forall \vec{\varphi}.$$

5 Code Validation

The first validation of the code is based on the following three test cases

Test 1 : Poiseuille

We consider the flow between two flat plates. Because of symmetry, the cross section reduces to its lower half section.

This section is a 1×0.25 rectangle. Imposed boundary conditions are :

$$\begin{aligned}
\Gamma_1 &: \quad v = (u_1, u_2) = (0,0), u_3 = 0, \\
\Gamma_2 &: \quad \frac{\partial u_1}{\partial x_2} = 0, u_2 = 0, \frac{\partial u_3}{\partial n} = 0, \\
\Gamma_3 &: \quad u_1 = 0, \frac{\partial u_2}{\partial x_1} = 0, \frac{\partial u_3}{\partial n} = 0.
\end{aligned}$$

The inflow was either a Poiseuille profile $\vec{v} = (0, 0, 1 - x^2)$ (test 1.a), or a uniform flow $\vec{v} = (0,0,1)$ (test 1.b) . Three meshes were used. All meshes have the same number of elements (400 quadrangles, 1071 nodes), but they correspond to three different spacings.

i) The first mesh is uniform. Corresponding results are prefixed by $KR400$.

ii) The second mesh is refined in the central part $x_1 = 1$. Corresponding results are prefixed by $KR400R1$.

iii) The third mesh is refined in the lower part $x_1 = 0$. Corresponding results are prefixed by $KR400R2$.

Test 1.a : Even with $Re = 4000$ and after 2000 iterations, the cross velocity field $v = (u_1, u_2)$ remains at level 10^{-14} for KR400, 10^{-14} for KR400R1 and 10^{-13} and10^{-12} for KR400R2. Observe also that we recover $\frac{\partial p}{\partial z} = \lambda = -0.0005$ with an error less than 10^{-12}.

Cross profiles of u_3 are also compared with the Poiseuille flow in the next figures.

Test 1.b : Here, we have taken $Re = 400$ and we observe how the flow converges to a Poiseuille flow as it flows above the plate. We show profiles of the velocity component u_i, together with the convergence curves $|u_3 - u_{ex}|_{L^\infty}$ and λ for all three meshes.

At the final iteration, the cross velocity field is of order 10^{-7} with cross pressure gradient of order 10^{-6}, and the error on pressure is $\frac{\partial p_t}{\partial z} - \lambda \simeq 10^{-8}$.

Test 2 : Blasius boundary layer

We keep the same domain ω, and take as boundary conditions

$$\begin{aligned}
\Gamma_1 &: \quad v = (u_1, u_2) = (0,0), u_3 = 0 \\
\Gamma_2 &: \quad \frac{\partial u_1}{\partial x_2} = 0, u_2 = 0, \frac{\partial u_3}{\partial n} = 0 \\
\Gamma_3 &: \quad u_1 = 0, u_2 = 0, u_3 = 1.
\end{aligned}$$

We also keep the same meshes as in the first test, which are still referred as KRB400, KRB400R1 and KRB400R2.

The inflow is now given by

$$v = (u_1, u_2) = 0,$$

$$u_3 = \begin{cases} \text{PB if } 1-\delta \leq x_1 \leq 1, \\ 1 \text{ if not.} \end{cases}$$

Above, PB is the so-called Blasius profile approximated by sixth order polynomials

$$PB = 2\xi - 5\xi^4 + 6\xi^5 - 2\xi^6 \text{ with } \xi = \frac{1-x_1}{\theta},$$

where θ denotes the boundary layer thickness. Here, we have taken

$$\theta = 6.2\frac{z}{\sqrt{R_z}} \text{ avec } R_z = \frac{u_e z}{\nu} (= \frac{z}{\nu} \text{ in our case}),$$

with data

$$z_o = 350, \quad R_{z_o} = 490000 \times 350.$$

In particular, at the beginning of our computational domain, the boundary layer thickness is equal to

$$\theta_o = 0.1657.$$

We represent hereafter the cross velocity $v = (u_1, u_2)$, the cross pressure p and the profile of the axial velocity u_3 as compared to the sixth order polynomial expansion given for Blasius. These are given on the cross section $n = it = 1, 10, 100$.

Test 3 : Riblet

In this test, the computational domain is as described on Figure 1 with $D = 1, d = 2l = 0.1135D$.

The inflow velocity is the uniform field $\vec{v} = (0, 0, 1)$ and we have used a "time step" $\Delta z = 0.1$ with Reynolds number $Re = 400$.

We represent below the cross velocity field $v = (u_1, u_2)$ and the cross pressure field p_o at section $n = it = 1, 10, 100$ for the three proposed meshes. The velocity profile of u_3 is given at section $n = it = 1, 100$.

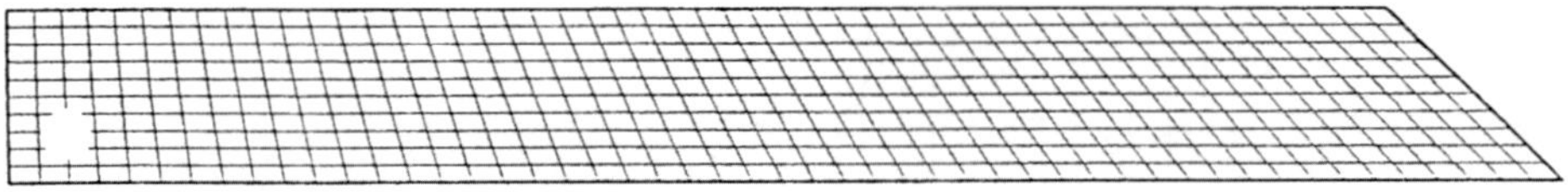

Figure 1: Uniform mesh

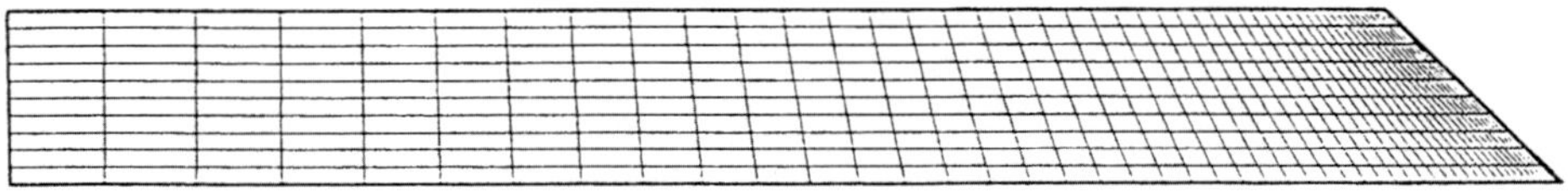

Figure 2: R1 mesh

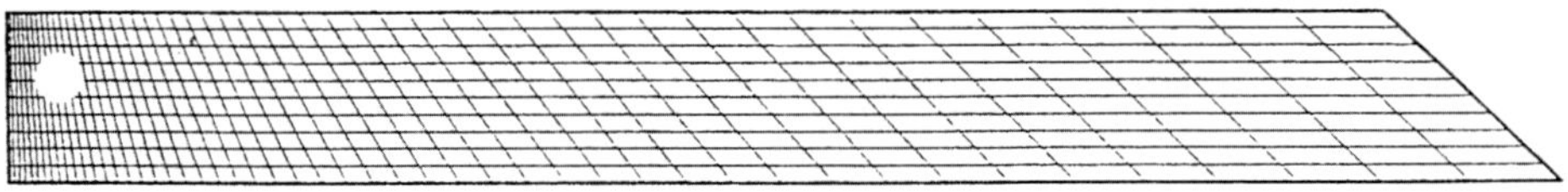

Figure 3: R2 mesh

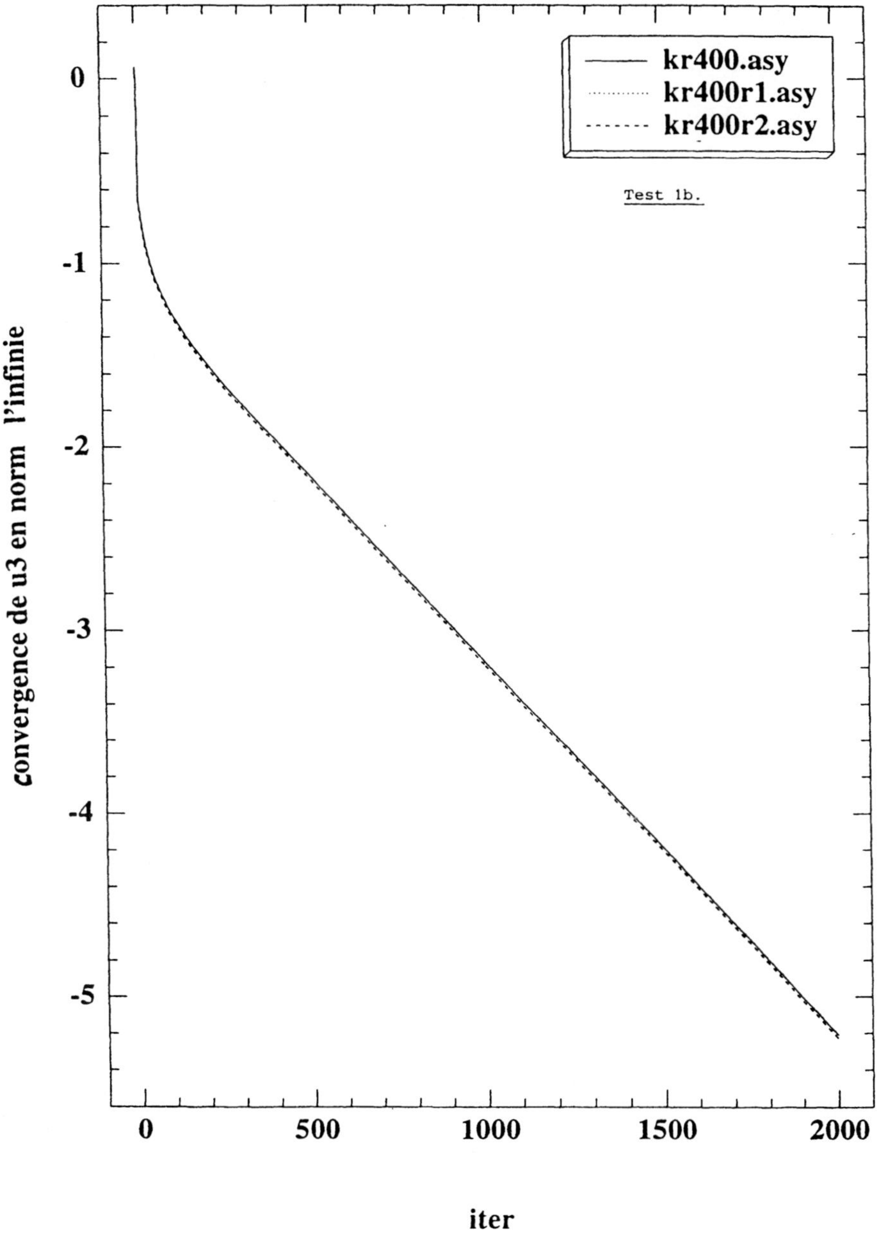

kr400.asy
kr400r1.asy
kr400r2.asy
Test 1b.
0
-1
-2
-3
-4
-5
convergence de u3 en norm l'infinie
0
500
1000
1500
2000
iter

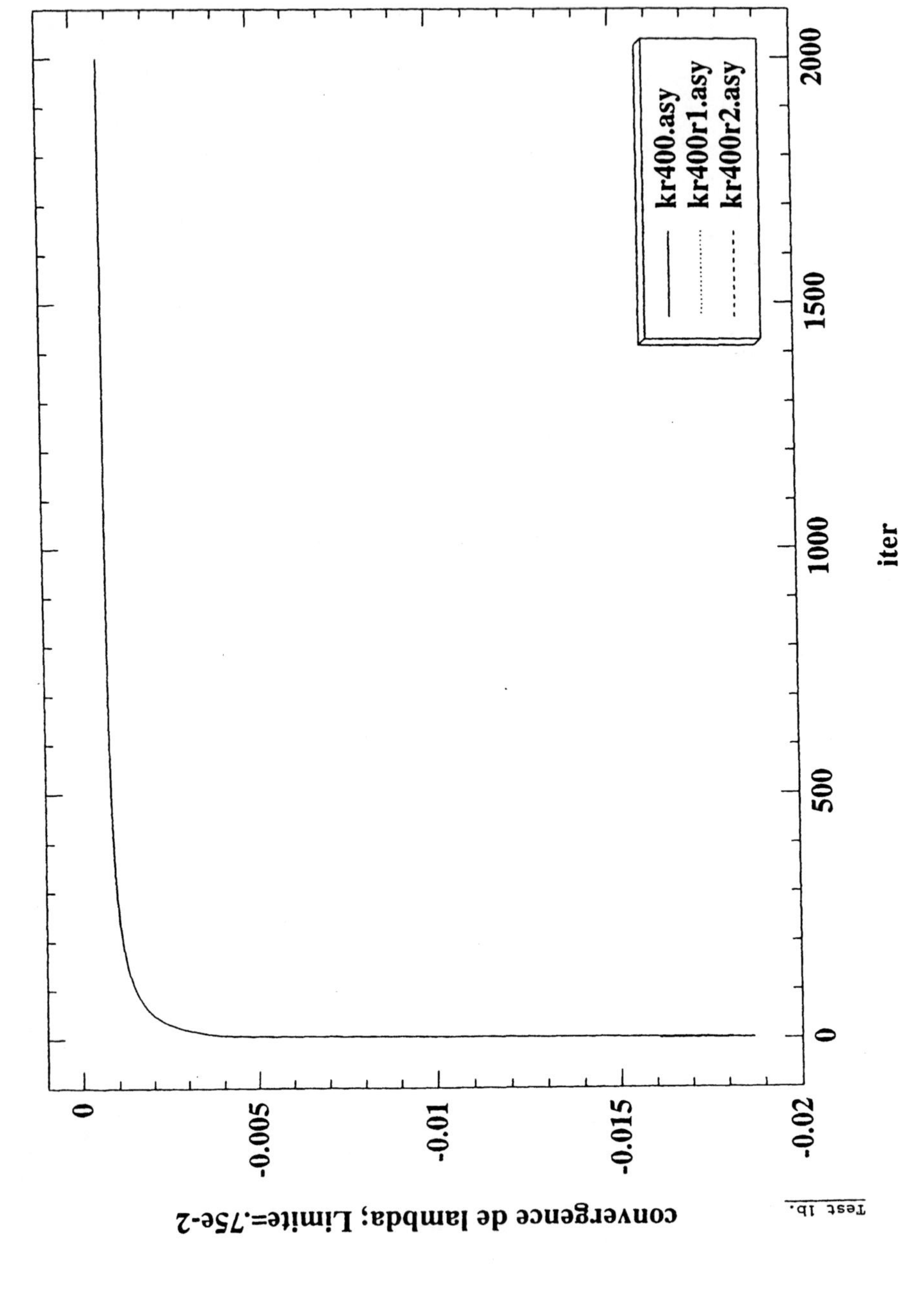
kr400.asy
kr400r1.asy
kr400r2.asy
0
500
1000
1500
2000
iter
0
-0.005
-0.01
-0.015
-0.02
convergence de lambda; Limite=.75e-2
Test 1b.

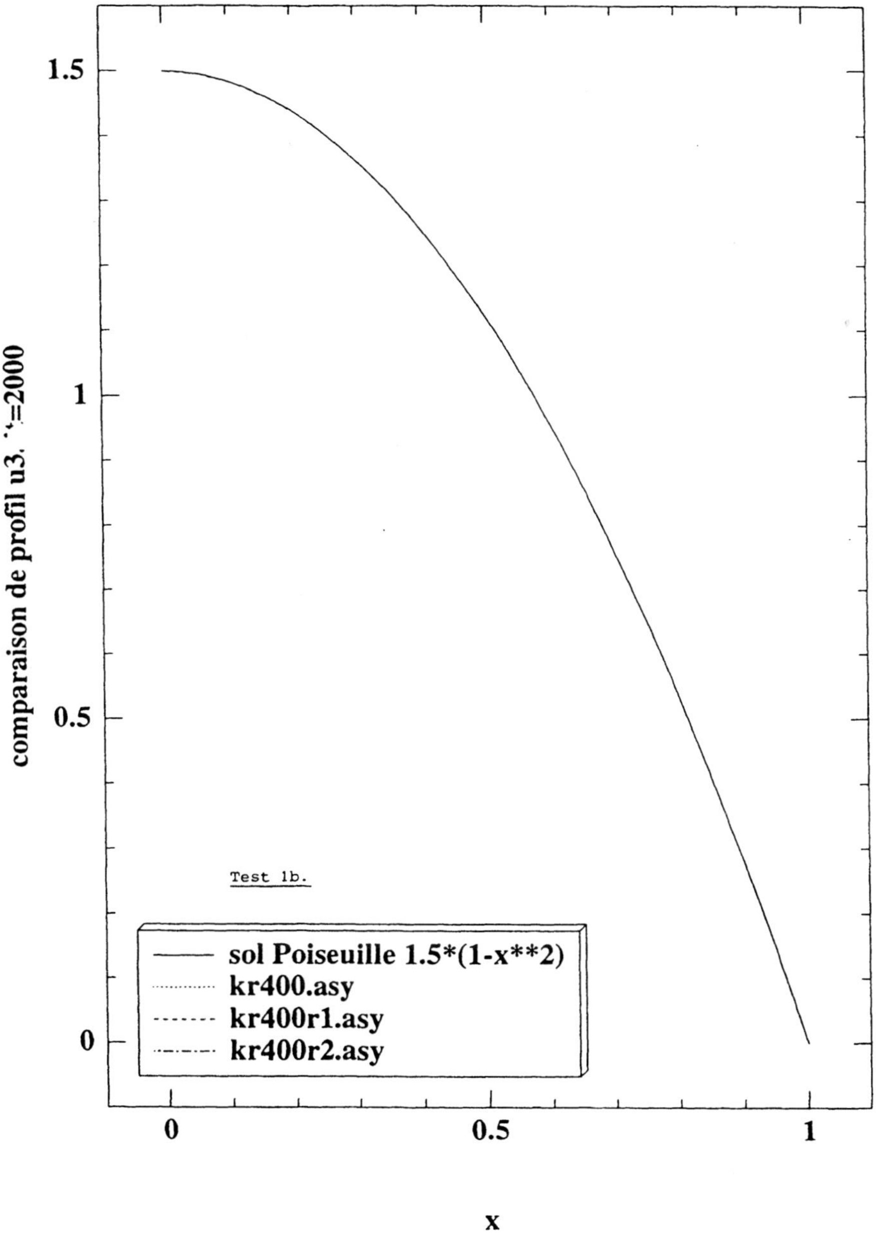

1.5
1
0.5
0
comparaison de profil u3. `‘=2000
Test 1b.
sol Poiseuille 1.5*(1-x**2)
kr400.asy
kr400r1.asy
kr400r2.asy
0
0.5
1
x

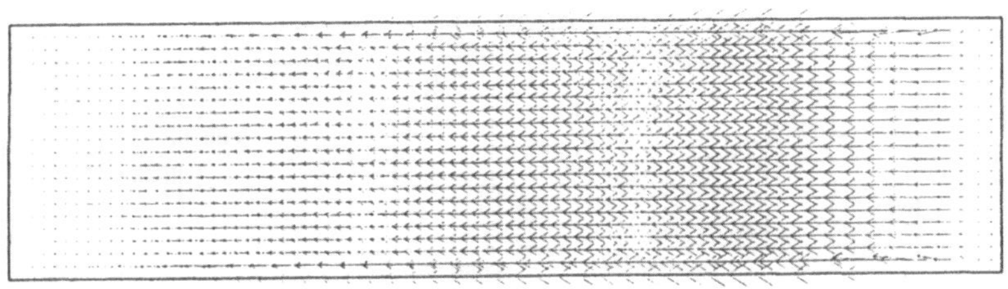

MODULEF : xiang
BLASIUS.CHAMP_V.IT=1
31/05/94
krb400.mail
krb400.coor
krb400_10v

451	POINTS
1701	NOEUDS
400	ELEMENTS
400	QUADRANGLES

EXTREMA DU CHAMP B :
.0000E+00 3.8501E-05
ECHELLE :
2.000 CM. = 3.8501E-

Test 2 : Blasius
KRB400
transverse solution at first step.

VITESSES

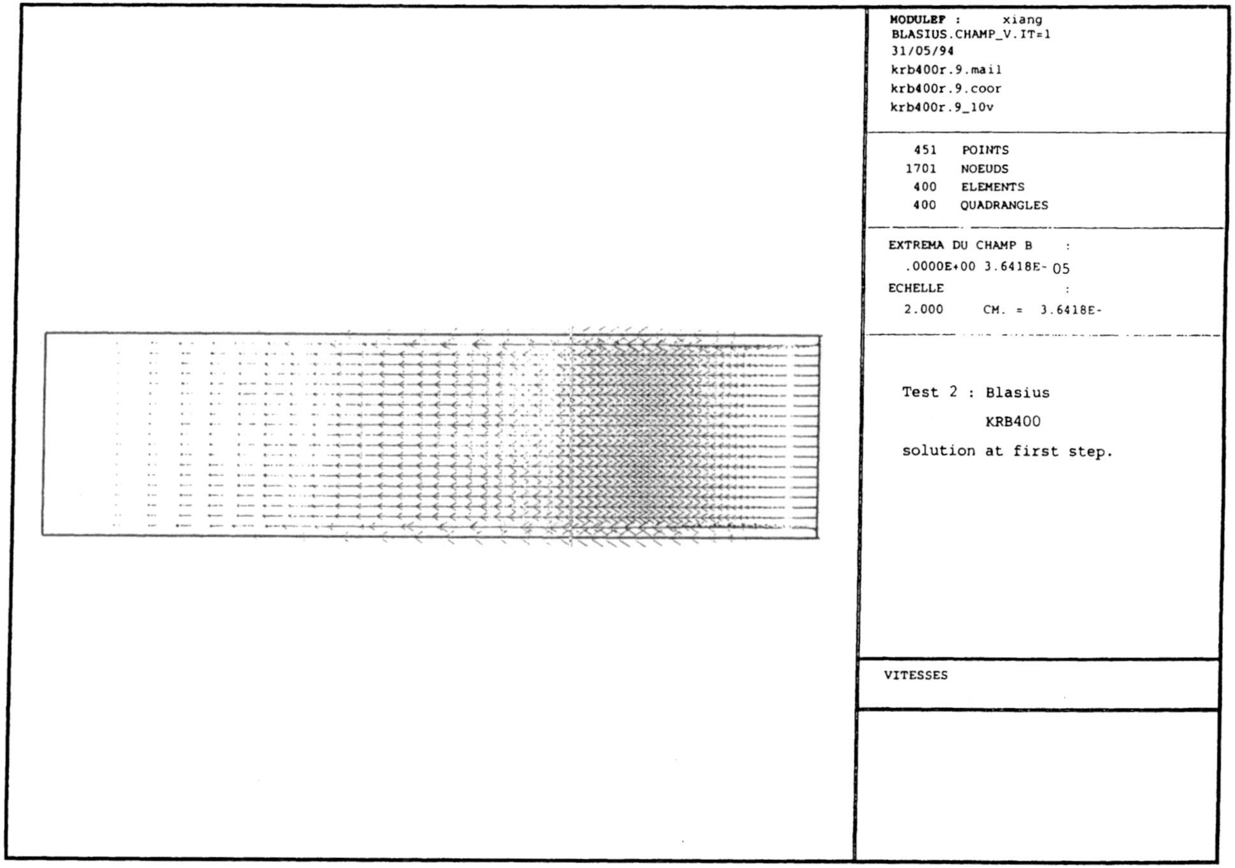
MODULEF : xiang
BLASIUS.CHAMP_V.IT=1
31/05/94
krb400r.9.mail
krb400r.9.coor
krb400r.9_10v
451 POINTS
1701 NOEUDS
400 ELEMENTS
400 QUADRANGLES
EXTREMA DU CHAMP B :
.0000E+00 3.6418E-05
ECHELLE :
2.000 CM. = 3.6418E-
Test 2 : Blasius
KRB400
solution at first step.
VITESSES

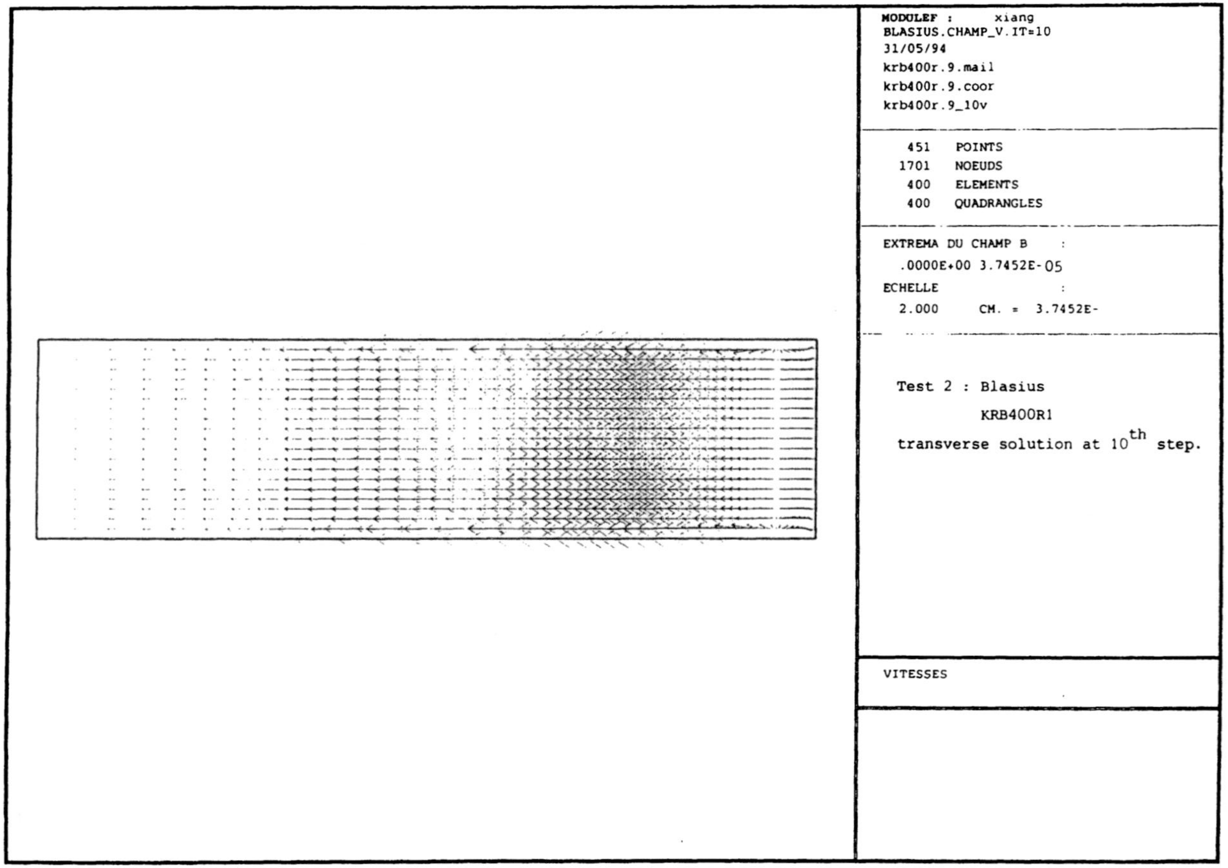
MODULEF : xiang
BLASIUS.CHAMP_V.IT=10
31/05/94
krb400r.9.mail
krb400r.9.coor
krb400r.9_10v
451 POINTS
1701 NOEUDS
400 ELEMENTS
400 QUADRANGLES
EXTREMA DU CHAMP B :
.0000E+00 3.7452E-05
ECHELLE :
2.000 CM. = 3.7452E-
Test 2 : Blasius
KRB400R1
transverse solution at 10th step.
VITESSES

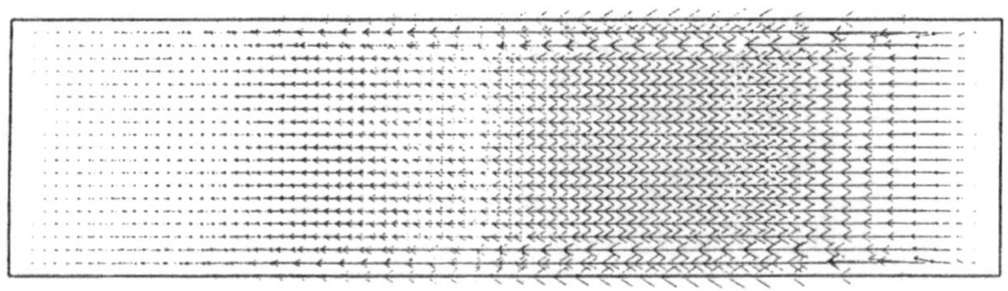

MODULEF : xiang
BLASIUS.CHAMP_V.IT=100
31/05/94
krb400.mail
krb400.coor
krb400_v

451 POINTS
1701 NOEUDS
400 ELEMENTS
400 QUADRANGLES

EXTREMA DU CHAMP B :
.0000E+00 5.5542E-05
ECHELLE :
2.000 CM. = 5.5542E-

Test 2 : Blasius
KRB400
transverse solution at 100th step.

VITESSES

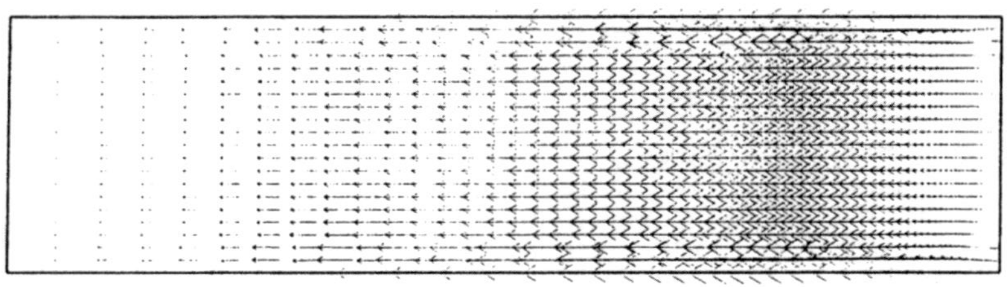

MODULEF : xiang
BLASIUS.CHAMP_V.IT=100
31/05/94
krb400r.9.mail
krb400r.9.coor
krb400r.9_v

451 POINTS
1701 NOEUDS
400 ELEMENTS
400 QUADRANGLES

EXTREMA DU CHAMP B :
.0000E+00 5.6450E- 05
ECHELLE :
2.000 CM. = 5.6450E-

Test 2 : Blasius
KRB400R1
transverse solution at 100th step.

VITESSES

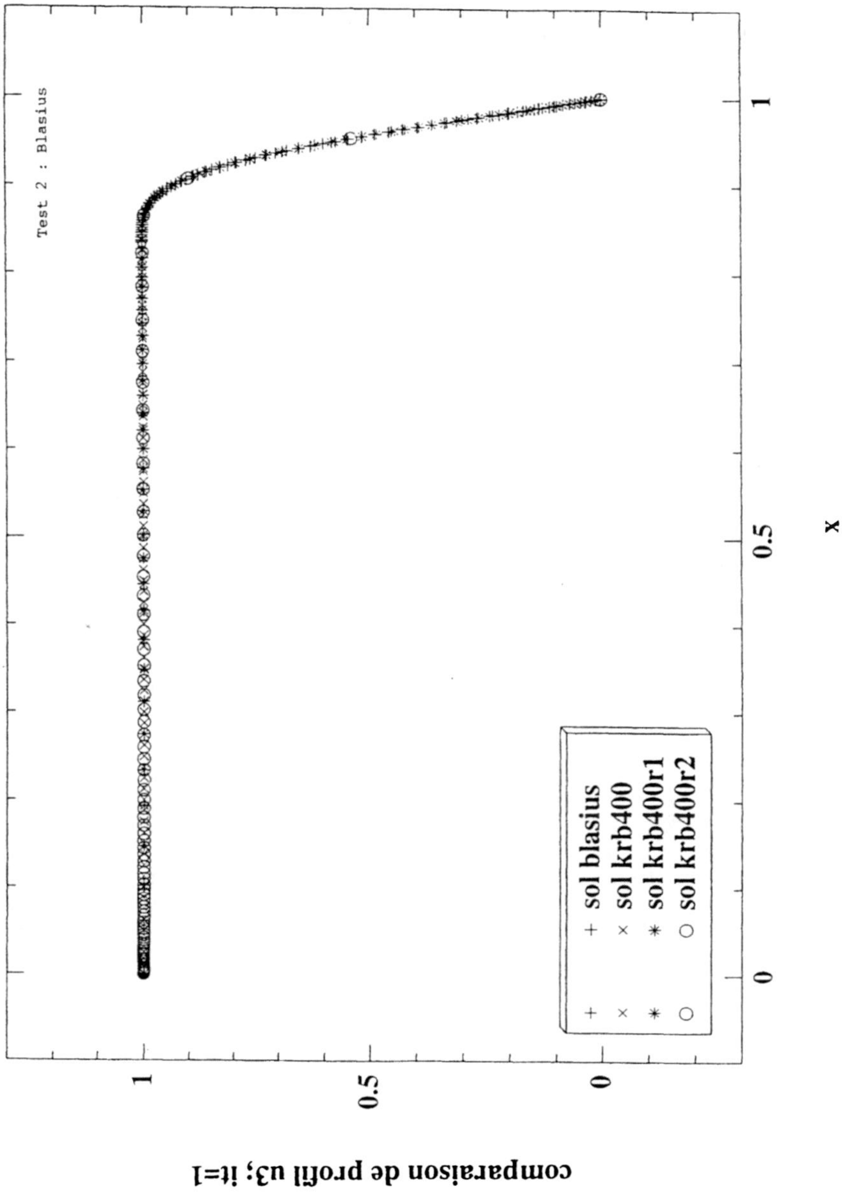
comparaison de profil u3; it=1
Test 2 : Blasius
sol blasius
sol krb400
sol krb400r1
sol krb400r2
0
0.5
1
0
0.5
1
x

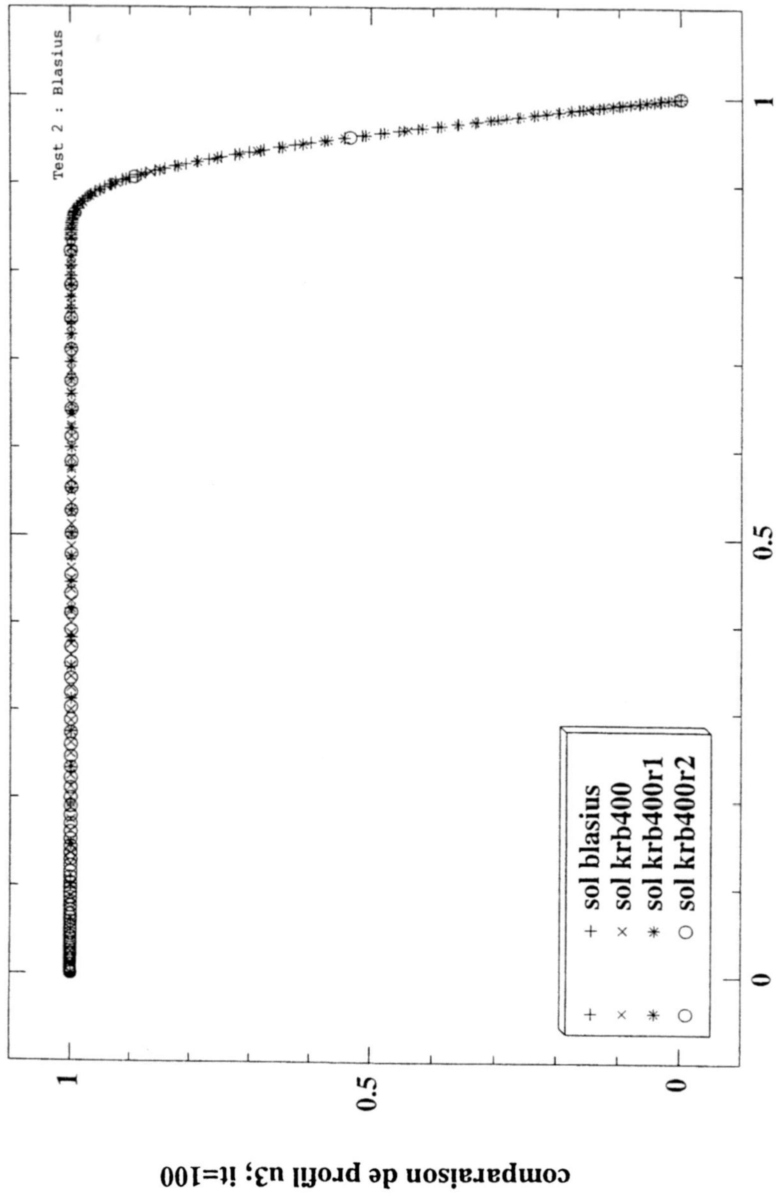
comparaison de profil u3; it=100
Test 2 : Blasius
sol blasius
sol krb400
sol krb400r1
sol krb400r2
x
0
0.5
1
0
0.5
1

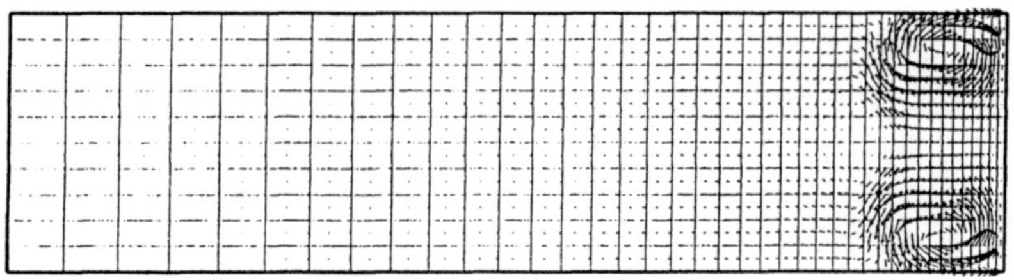

Test 3 : transverse velocity after first step.

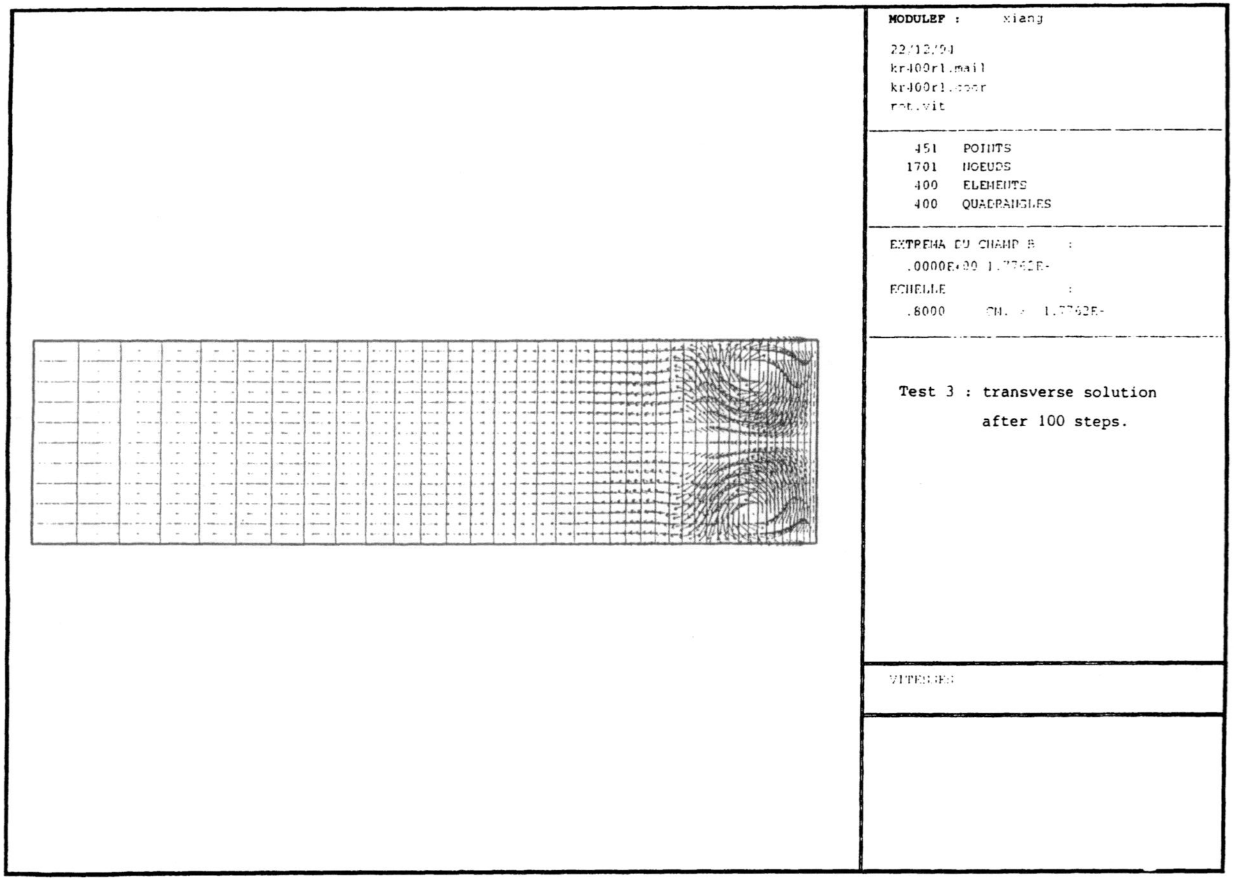

Test 3 : transverse solution after 100 steps.

6 Validation of the PNS assumptions

By assumption, PNS requires the longitudinal pressure gradient to be independent of x and y. In this paragraph, we check this assumption by the following fixed point iteration, operating on the corrected pressure $p_{t,n}(x,y)$ on cross section n :

- Initialisation

  $p_{t,1}(x,y) = p_1(x,y)$ as computed by PNS.

- Calculation of $p_{t,n}(x,y)$ from $p_{t,n-1}(x,y)$ by a fixed point iteration operating on the index m :

  - Initialisation
    $p^o_{t,n}(x,y) = p_n(x,y)$ as computed by PNS;
  - Updating :
    Add the source term

    $$\frac{p^{m-1}_{t,n}(x,y) - p_{t,n-1}(x,y)}{\delta z}$$

    in the u_3 equation and solve the resulting PNS equation on cross section n, yielding an updated value $p^m_{t,n}(x,y)$ of the pressure field.
  - Convergence test: stop the fixed point iterations in m when

    $$\sup_{x,y} \mid p^m_{t,n}(x,y) - p^{m-1}_{t,n}(x,y) \mid < \varepsilon.$$

This fixed point iteration changes the velocity by less than 10^{-4} in relative value. The assumption of $\frac{\partial P}{\partial z}$ being independent of x and y appears therefore to be perfectly validated in such situations.

III Adjoint State Equation

1 The Lagrangian

Our optimisation algorithm requires the evaluation of the gradient of the cost function. This gradient will be computed from an adjoint state. The calculation of the adjoint state is done at the discrete level, by reformulating our original minimisation problem (P_m) in a Lagrangian form. Introducing the Lagrange multiplier (the adjoint state) of the state equation, we can reformulate (P_m) as the problem of finding the critical points of the lagrangian

$$\mathcal{L}(r, \vec{v}, \vec{w}) = J(\vec{v}) - a(\vec{v}, \vec{w}).$$

By solving the equation $\frac{\partial \mathcal{L}}{\partial \omega} = 0$ with respect to the adjoint state $\vec{\omega}$, we obtain that the velocity $\vec{v}$ must be solution of the state equation. The differentiation with respect to $\vec{v}$ yields the adjoint state equation $\frac{\partial \mathcal{L}}{\partial v} = 0$. Then, if $\vec{v}$ and $\vec{\omega}$ are solutions of the state equation and adjoint state equations respectively, the total gradient of the cost function with respect to the control variable r is classically given by

$$\frac{dJ}{dr} = \frac{\partial \mathcal{L}(r, \vec{v}, \vec{\omega})}{\partial r}.$$

Let us develop our calculations. By construction, we have

$$J = \frac{\nu}{2} \int_{\Omega} |\nabla v|^2 + |\nabla u_3|^2.$$

After discretisation in z and using the same functional spaces as in the preceeding section , this writes

$$J = \frac{\nu_z}{2} \sum_{n=1}^{N-1} \{ \int_{\omega} |\nabla v^n|^2 + |\nabla u_3^n|^2 \} + \frac{\nu_z}{4} \int_{\omega} |\nabla v^N|^2 + |\nabla u_3^N|^2 + \frac{\nu_z}{4} \int_{\omega} |\nabla v^o|^2 + |\nabla u_3^o|^2.$$

Here

$$\nu_z = \Delta z . \nu$$

and v^o et u_3^o are the invariant initial data on the inflow cross section. Let us now rewrite our discrete variational state equation. Introducing

$$\tilde{x}^{n-1} = x - \frac{v^{n-1}}{u_3^{n-1}} \Delta z \approx \chi(-\Delta z)$$

the equation in u_3 at section n takes the form

$$\int_{\omega} u_3^{n-1}(u_3^n - u_3^{n-1}(\tilde{x}^{n-1}))\varphi^n + \Delta z \lambda^n \int_{\omega} \varphi^n + \nu_z \int \nabla u_3^n \nabla \varphi^n = \int_{\omega} f_3 \varphi^n, \forall \varphi^n \in \mathring{\mathcal{U}},$$

that is

$$\alpha_1^n(u_3, v, \lambda, f_3; \varphi) = 0, \forall \varphi.$$

The equation in average pressure is similarly

$$q^n(\int_{\omega} u_3^n - \int_{\omega} u_3^o) = 0, \forall q^n \text{ constant},$$

that is

$$\alpha_2^n(u_3; q) = 0, \forall q \in \mathbb{R}.$$

The equation in cross velocity writes

$$\int_\omega u_3^{n-1}(v^n - v^{n-1}(\tilde{x}^{n-1}))\varphi_t^n - \Delta z \int_\omega p^n \nabla.\varphi_t^n + \nu_3 \int_\omega \nabla v^n \nabla \varphi_t^n$$
$$= \int_\omega f\varphi_t^n, \forall \varphi_t^n \in \mathring{\mathcal{V}}$$

that is

$$\alpha_3^n(u_3, v, p, f; \varphi_t) = 0, \forall \varphi_t \in \mathring{\mathcal{V}}.$$

Finally, the equation in cross pressure is

$$\int_\omega (\Delta z \nabla.v^n + u_3^n - u_3^{n-1})\mu^n = 0, \forall \mu^n \in V_p,$$

that is

$$\alpha_4^n(u_3, v; \mu) = 0, \forall \mu \in V_p.$$

With the above notation, the Lagrangian takes the form

$$\mathcal{L} = J - \sum_{n=1}^{N}(\alpha_1^n + \alpha_2^n + \alpha_3^n + \alpha_4^n).$$

2 Adjoint equations

The adjoint state vector is now simply obtained by solving

$$\frac{\partial \mathcal{L}}{\partial u_3} = 0, \frac{\partial \mathcal{L}}{\partial \lambda} = 0, \frac{\partial \mathcal{L}}{\partial v} = 0, \frac{\partial \mathcal{L}}{\partial p} = 0.$$

Its solution corresponds then to the four adjoint fields

$$\vec{\omega} = (\varphi, q, \varphi_t, \mu).$$

The equation in φ is first obtained by differentiating $\mathcal{L}$ with respect to u_3. We have

$$\frac{\partial \alpha_1^n}{\partial u_3} = \int_\omega \delta u_3^{n-1}(u_3^n - u_3^{n-1}(\tilde{x}^{n-1}))\varphi^n$$
$$+ \int_\omega u_3^{n-1}\left(\delta u_3^n - [\delta u_3^{n-1}(\tilde{x}^{n-1}) + \nabla u_3^{n-1}(\tilde{x}^{n-1})(\frac{v^{n-1}}{(u_3^{n-1})^2}\delta u_3^{n-1}\Delta z)]\right)\varphi^n$$
$$+\nu_z \int \nabla \delta u_3^n \nabla \varphi^n,$$
$$\frac{\partial \alpha_2^n}{\partial u_3} = q^n \int_\omega \delta u_3^n,$$
$$\frac{\partial \alpha_3^n}{\partial u_3} = \int_\omega \delta u_3^{n-1}(v^n - v^{n-1}(\tilde{x}^{n-1}))\varphi_t^n$$

$$+\int_\omega u_3^{n-1}\Big[-\nabla v^{n-1}(\tilde{x}^{n-1})(\frac{v^{n-1}}{(u_3^{n-1})^2}\delta u_3^{n-1}\Delta z)\Big]\varphi_t^n,$$

$$\frac{\partial \alpha_4^n}{\partial u_3} = \int_\omega (\delta u_3^n - \delta u_3^{n-1})\mu^n.$$

Regrouping all terms related to the test function δu_3^n in the equation $\frac{\partial \mathcal{L}}{\partial u_3^n} = 0$, this equation reduces to the system of variational equations, $\forall n = 1, N-1$,

$$-\Big\{\int_\omega \delta u_3^n(u_3^{n+1} - u_3^n(\tilde{x}^n))\varphi^{n+1} + \int_\omega u_3^{n-1}\delta u_3^n\varphi^n -$$

$$\int_\omega u_3^n[\delta u_3^n(\tilde{x}^n) + \Delta z\delta u_3^n\frac{v^n}{(u_3^n)^2}\cdot\nabla u_3^n(\tilde{x}^n)]\varphi^{n+1} + \nu_z\int\nabla\delta u_3^n\nabla\varphi^n$$

$$+q^n\int_\omega \delta u_3^n + \int_\omega \delta u_3^n(v^{n+1} - v^n(\tilde{x}^n))\varphi_t^{n+1} - \int_\omega \Delta z\delta u_3^n\frac{v^n}{u_3^n}.\nabla v^n(\tilde{x}^n).\varphi_t^{n-1}$$

$$+\int_\omega(\delta u_3^n\mu^n - \delta u_3^n\mu^{n+1})\Big\} + \nu_z\int\nabla\delta u_3^n\nabla u_3^n = 0, \forall\delta u_3^n \in \mathring{\mathcal{U}}.$$

In summary, the adjoint φ^n is obtained by solving

$$\int_\omega u_3^{n-1}\delta u_3^n\varphi^n + \nu_z\int\nabla\delta u_3^n\nabla\varphi^n + q^n\int\delta u_3^n$$

$$= \nu_z\int\nabla\delta u_3^n\nabla u_3^n - \{\int_\omega \delta u_3^n u_3^{n+1}\varphi^{n+1}$$

$$+\Delta z\delta u_3^n\frac{v^n}{u_3^n}\cdot\nabla u_3^n(\tilde{x}^n)]\varphi^{n+1} - \int_\omega \delta u_3^n u_3^n(\tilde{x}^n)\varphi^{n+1} - \int_\omega u_3^n\delta u_3^n(\tilde{x}^n)\varphi^{n+1}$$

$$-\int_\omega \Delta z\delta u_3^n\frac{v^n}{u_3^n}.\nabla u_3^n(\tilde{x}^n)\varphi^{n+1} + \int_\omega \delta u_3^n v^{n+1}\varphi_t^{n+1} - \int_\omega \delta u_3^n v^n(\tilde{x}^n)\varphi_t^{n+1}$$

$$-\int_\omega(\mu^{n+1} - \mu^n)\delta u_3^n\}, \forall\delta u_3^n \in \mathring{\mathcal{U}}, \forall n < N,$$

and at section N :

$$-\{\int_\omega u_3^{N-1}\delta u_3^N\varphi^N + \nu_z\int\nabla\delta u_3^N\nabla\varphi^N + q^N\int\delta u_3^N + \int\mu^N\delta u_3^N\}$$

$$+\frac{\nu_z}{2}\int\nabla u_3^N\nabla\delta u_3^N = 0$$

$$\text{i.e.}\int_\omega u_3^{N-1}\delta u_3^N\varphi^N + \nu_z\int\nabla\delta u_3^N\nabla\varphi^N + q^N\int\delta u_3^N =$$

$$\frac{\nu_z}{2}\int\nabla u_3^N\nabla u_3^N - \int\mu^N\delta u_3^N, \forall\delta u_3^N \in \mathring{\mathcal{U}}.$$

The equation in q is obtained by differentiating with respect to λ. We have simply

$$\frac{\partial \alpha_1^n}{\partial \lambda} = \Delta z \delta \lambda^n \int_\omega \varphi^n$$

and thus the equation $\frac{\partial \mathcal{L}}{\partial \lambda} = 0$ writes simply

$$\int_\omega \varphi^n = 0, \forall n.$$

The equation in φ_t is obtained by differentiating with respect to the variable v. We have

$$\frac{\partial \alpha_1^n}{\partial v} = -\int_\omega u_3^{n-1} \nabla u_3^{n-1}(\tilde{x}^{n-1})(-\frac{\delta v^{n-1}}{u_3^{n-1}}\Delta z)\varphi^n$$

$$\frac{\partial \alpha_3^n}{\partial v} = \int_\omega u_3^{n-1}[\delta v^n - [\delta v^{n-1}(\tilde{x}^{n-1}) - \nabla v^{n-1}(\tilde{x}^{n-1})(-\frac{\delta v^{n-1}}{u_3^{n-1}}\Delta z)]]\varphi_t^n$$

$$+\nu_z \int_\omega \nabla \delta v^n \nabla \varphi_t^n$$

$$\frac{\partial \alpha_4^n}{\partial v} = \int_\omega \Delta z \nabla . \delta v^n \mu^n .$$

The equation $\frac{\partial \mathcal{L}}{\partial v} = 0$ writes now , for each test function δv^n and each section $(n = 1, N-1)$.

$$-\Big\{\int_\omega \Delta z \delta v^n . \nabla u_3^n(\tilde{x}^n)\varphi^{n+1} + \int_\omega u_3^{n-1}\varphi_t^n \delta v^n - \int_\omega u_3^n \delta v^n(\tilde{x}^n)\varphi_t^{n+1}$$

$$+\Delta z \delta v^n . \nabla v^n(\tilde{x}^n).\varphi_t^{n+1} + \nu_z \int_\omega \nabla \delta v^n \nabla \varphi_t^n$$

$$+\int_\omega \Delta z \nabla . \delta v^n \mu^n \Big\} + \nu_z \int \nabla \delta v^n \nabla v^n = 0.$$

In summary, the adjoint φ_t^n is obtained by solving

$$\int_\omega u_3^{n-1}\varphi_t^n \delta v^n + \nu_z \int_\omega \nabla \delta v^n \nabla \varphi_t^n + \Delta z \int_\omega \mu^n \nabla . \delta v^n$$

$$= \nu_z \int_\omega \nabla \delta v^n \nabla v^n - \Big\{\int_\omega \Delta z \delta v^n . \nabla u_3^n(\tilde{x}^n)\varphi^{n+1} - \int_\omega u_3^n \delta v^n(\tilde{x}^n)\varphi_t^{n+1}$$

$$+\Delta z \int_\omega \delta v^n . \nabla v^n(\tilde{x}^n)\varphi_t^{n+1}\Big\}, \forall \delta v^n \in \mathring{\mathcal{V}}, \forall n < N,$$

and at section N

$$-\{\int_\omega u_3^{N-1}\delta v^n \varphi_t^N + \nu_z \int_\omega \nabla \delta v^N \nabla \varphi_t^N + \Delta z \int \mu^N \nabla . \delta v^N\} + \frac{\nu_z}{2}\int \nabla \delta v^N \nabla v^N = 0$$

i.e. $$\int_\omega u_3^{N-1}\varphi_t^N \delta u^N + \nu_z \int_\omega \nabla \varphi_z^N \nabla \delta v^N + \Delta z \int \mu^N \nabla . \delta v^N = \frac{\nu_z}{2}\int \nabla \delta v^N \nabla v^N,$$

$\forall \delta v^N \in \mathring{\mathcal{V}}$.

The last multiplier is obtained by solving $\frac{\partial \mathcal{L}}{\partial p} = 0$. Since

$$\frac{\partial \alpha_3}{\partial p} = -\Delta z \int \delta p^n \nabla . \varphi_t^n$$

this equation writes

$$\int \delta p^n \nabla . \varphi_t^n = 0, \forall \delta p^n \in V_p, \forall n.$$

3 Practical algorithm for computing the adjoint

The calculation of the adjoint state now starts at section N by the following sequence of operations

- Solve in φ_t^N :

$$\int_\omega u_3^{N-1} \varphi_t^N \delta v^N + \nu_z \int_\omega \nabla \varphi_t^N \nabla \delta v^N = \frac{\nu_z}{2} \int \nabla \delta v^N \nabla v^N, \forall \delta v^N \in V_o.$$

(recalling that $\int \delta p^N \nabla . \varphi_t^N = 0, \forall \delta v^N \in V_o$).

- Calculation of the adjoint pressure μ^N :

$$\Delta z \int_\omega \mu^N \nabla . \delta v^N = \frac{\nu_z}{2} \int_\omega \nabla v^N \nabla \delta v^N - \{ \int_\omega u_3^{N-1} \varphi_t^N \delta v^N + \nu_z \int_\omega \nabla \varphi_t^N \nabla \delta v^N \},$$

$\forall \delta v^N \in \mathring{\mathcal{V}}$.

- Calculation of φ^N and q^N :

$$\int_\omega u_3^{N-1} \varphi^N \delta u_3^N + \nu_z \int_\omega \nabla \varphi^N \nabla \delta u_3^N + q^N \int_\omega \delta u_3^N =$$

$$\frac{\nu_z}{2} \int_\omega \nabla u_3^N \nabla \delta u_3^N - \int_\omega \mu^N \delta u_3^N, \forall \delta u_3^N \in \mathring{\mathcal{U}},$$

$$\int_\omega \varphi^N = 0.$$

Next, $\varphi^{n+1}, \varphi_t^{n+1}, \mu^{n+1}, q^{n+1}$ being known, we iteratively calculate $\varphi^n, \varphi_t^n, \mu^n, q^n$ as follows

1. Calculation of φ_t^n

$$\int_\omega u_3^{n-1} \varphi_t^n \delta v^n + \nu_z \int \nabla \delta v^n \nabla \varphi_t^n$$

$$= \nu_z \int \nabla \delta v^n \nabla v^n - \left\{ \Delta z \int_\omega \delta v^n . \nabla u_3^n(\tilde{x}^n) \varphi^{n+1} + \Delta z \delta v^n \nabla v^n(\tilde{x}^n) . \varphi_t^{n+1} \right.$$

$$\left. - \int_\omega u_3^n \delta v^n(\tilde{x}^n) \varphi_t^{n+1} \right\},$$

$$= F, \forall \delta v^n \in V_o.$$

2. Calculation of the adjoint pressure $\mu^n \in V_p$

$$\Delta z \int_\omega \mu^n \nabla.\delta v^n = F - \{\int_\omega u_3^{n-1}\varphi_t^n \delta v^n + \nu_z \int \nabla\delta v^n \nabla\varphi_t^n\}, \forall \varphi_t^n \in \mathring{\mathcal{V}}.$$

3. Calculation of φ^n and q^n

$$\int_\omega u_3^{n-1}\delta u_3^n \varphi^n + \nu_z \int_\omega \nabla\varphi^n \nabla\delta u_3^n + q^n \int_\omega \delta u_3^n =$$
$$\nu_z \int_\omega \nabla\delta u_3^n \nabla u_3^n + \int_\omega (\mu^{n+1} - \mu^n)\delta u_3^n$$
$$-\Big\{\int_\omega \delta u_3^n u_3^{n+1}\varphi^{n+1} - \int_\omega \delta u_3^n u_3^n(\tilde{x}^n)\varphi^{n+1} - \int_\omega \Delta z \delta u_3^n \frac{v^n}{u_3^n}.\nabla u_3^n(\tilde{x}^n)\varphi^{n+1}$$
$$- \int_\omega u_3^n \delta u_3^n(\tilde{x}^n)\varphi^{n+1} + \int_\omega \delta u_3^n v^{n+1}\varphi_t^{n+1}$$
$$- \int_\omega \delta u_3^n v^n(\tilde{x}^n)\varphi_t^{n+1} - \int_\omega \Delta z \delta u_3^n \frac{v^n}{u_3^n}.\nabla v^n(\tilde{x}^n)\varphi_t^{n+1}\Big\}, \forall \delta u_3^n \in \mathring{\mathcal{U}}$$
$$\int_\omega \varphi^n = 0.$$

4 Validation of the adjoint state

The adjoint state has no physical interpretation. Therefore, the adjoint state equation can only be validated by calculating the gradient of the cost function as predicted from the adjoint state and comparing the result to a parallel evaluation of this gradient by centered finite difference

$$J'(g) \approx \frac{J(g+\varepsilon) - J(g-\varepsilon)}{2\varepsilon}.$$

Observe that we should have an error given by

$$J'(g) - \frac{J(g+\varepsilon) - J(g)}{\varepsilon} \approx \varepsilon\, H(g)$$

where H is the Hessian of J.

We now consider the cost function

$$J(g) = \frac{\nu_z}{4}\{\int_\omega |\nabla v^o|^2 + |\nabla u_3^o| + \int_\omega |\nabla v^N|^2 + |\nabla u_3^N|^2\} + \frac{\nu_z}{2}\sum_{n=1}^{N-1}\int_\omega |\nabla v^n|^2 + |\nabla u_3^n|^2,$$

as a function of the imposed Dirichlet boundary condition $v = g$ on the bottom wall Γ_1, with g considered as the control variables r. The test problem is a standard Poiseuille flow problem, and the errors on the gradient are plotted on figure (3). As expected, the error between the finite difference and the analytic gradient (based on our adjoint state) is of order ε.

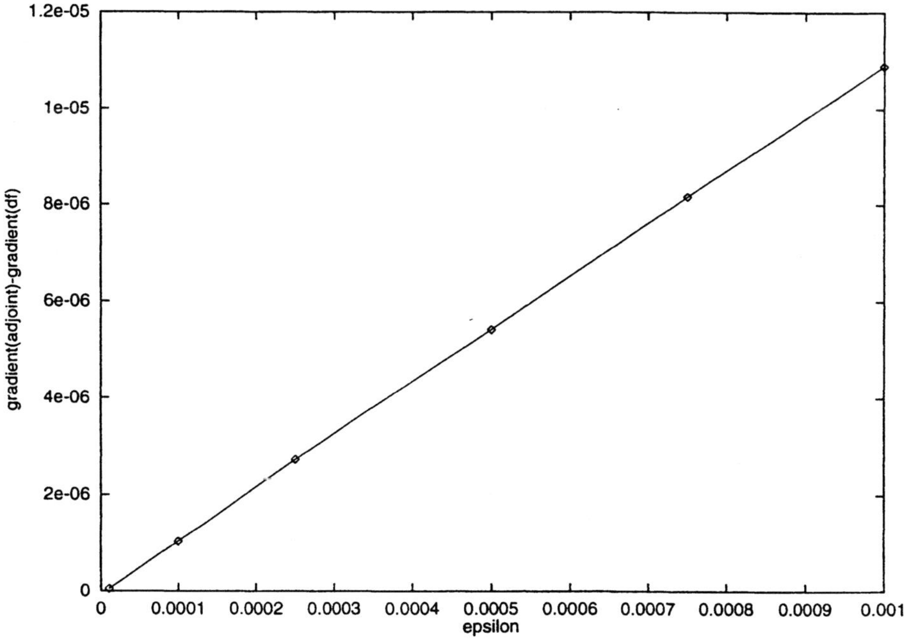

Figure 3

The next validation uses the same control variables, but with a cost function given by the L^2 norm of the vorticity $\nabla \wedge v$. The entrance flow is of Blasius type with superimposed cross vortices. In this case also, as indicated in figure (4), there is a very good agreement between the analytic gradient and its finite difference approximation.

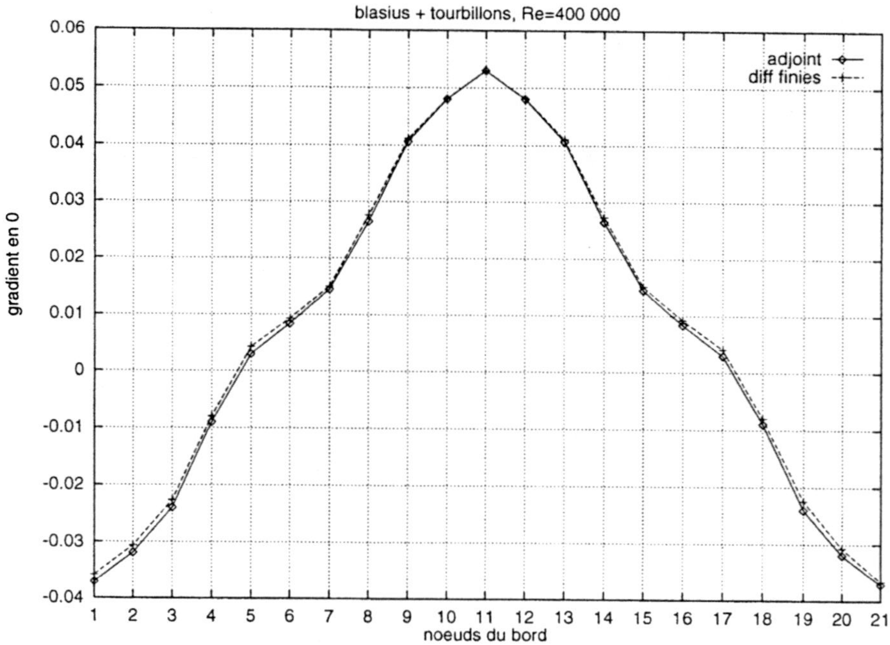

blasius + tourbillons, Re=400 000
adjoint
diff finies
gradient en 0
noeuds du bord
0.06
0.05
0.04
0.03
0.02
0.01
0
-0.01
-0.02
-0.03
-0.04
1
2
3
4
5
6
7
8
9
10
11
12
13
14
15
16
17
18
19
20
21

Figure 4

IV Shape optimisation

1 Reduction to a transpiration problem

At first order, changing the shape of the flow domain while conserving its volume can be identified to a change of Dirichlet boundary conditions. To be more specific, let $\vec{v}$ be a solution of the flow equation

$$a(\vec{v}, \vec{w}) = 0 \quad \text{on} \quad \Omega^i,$$

set on the initial domain Ω^i with boundary condition

$$\vec{v} = 0 \quad \text{on} \quad \Gamma_1^i.$$

Let any point $\vec{x}$ of the initial boundary Γ_1^i move of a given small displacement $d\vec{x}$. Let $\vec{v} + d\vec{v}$ be the solution of our initial equation on the same initial domain, but with the new boundary condition

$$\vec{v} + d\vec{v} = -\nabla \vec{v}.d\vec{x} \quad \text{on} \quad \Gamma_1^i.$$

At first order, we have on the new boundary

$$\begin{aligned} (\vec{v} + d\vec{v})(\vec{x} + d\vec{x}) &= (\vec{v} + d\vec{v})(\vec{x}) + \nabla(\vec{v} + d\vec{v}).d\vec{x} \\ &= -\nabla \vec{v}.d\vec{x} + \nabla \vec{v}.d\vec{x} + 0(|d\vec{x}|^2) \\ &= 0. \end{aligned}$$

Hence $\vec{v} + d\vec{v}$ is solution of the original flow problem set on the new domain $\Omega = \Omega^i + d\Omega$ with noslip boundary condition

$$\vec{v} + d\vec{v} = 0 \quad \text{on} \quad \Gamma_1^i + d\Gamma_1.$$

In other words, updating the shape of the domain Ω amounts to update the boundary conditions of the original problem by the quantity $-\nabla \vec{v}.d\vec{x}$.

In variational form, this means that we can locally replace the state equation for any shape configuration close to the initial shape Ω^i by the new equation

$$\begin{aligned} &a(\vec{v} + d\vec{v} - Tr^{-1}(\nabla \vec{v}.d\vec{x}), \vec{w}^i) = 0, \\ &\forall \vec{w}^i \in V^i, \vec{v} + d\vec{v} \in V^i, \end{aligned}$$

where the variational form a and the functional space V^i are associated to the fixed domain Ω^i and are considered as independent of the shape variables.

2 Calculation of the gradient

With this new notation, our Lagrangian becomes locally around the present shape Ω :

$$\mathcal{L}(\vec{x},\vec{v},\vec{w}) = J(\vec{x},\vec{v}) - a(\vec{v} - Tr^{-1}(\nabla\vec{v}.d\vec{x}),\vec{w}).$$

The total gradient of the cost function J with respect to $\vec{x}$ is then given as

$$\begin{aligned}\frac{dJ}{d\vec{x}}(\vec{x},\vec{v}(\vec{x})).d\vec{x} &= \frac{\partial J}{\partial\vec{x}}(\vec{x},\vec{v}).d\vec{x} \\ &\quad -a'_u(-Tr^{-1}(\nabla\vec{v}.d\vec{x}),\vec{w}) \\ &= \frac{\nu}{2}\int_{\Gamma_1}\int_0^1 |\nabla\vec{v}|^2(\vec{x}+sd\vec{x})\frac{dads}{\vec{x}.\vec{n}} \\ &\quad -a'_u(-Tr^{-1}(\nabla\vec{v}.d\vec{x}),\vec{w}).\end{aligned}$$

This formula is quite easy to program once the adjoint state has been computed as indicated in the previous section.

V Herskovits Interior Point Methods

Our original constrained optimisation problem is solved by using the interior point algorithm developed by Herskovits. In abstract form, after elimination of the state vector $\vec{v}$, our optimisation problem can be written as

$$\min_{\substack{z\in R^m \\ g_j(z)\leq 0}} j(z).$$

Its solution satisfies the optimality condition

$$\frac{dj}{dz}(\bar{z}) + \sum_{j=1}^{p}\bar{s}_j.\frac{\partial g_j}{\partial z}(\bar{z}) = 0 \in R^m,$$

$$\bar{s}_j.g_j(\bar{z}) = 0, \forall j = 1,p.$$

$$(\bar{z},\bar{s}) \in C = \{(z,s) \in R^m \times R^p, g(z) \leq 0, s \geq 0\}.$$

The idea of interior point algorithms is to solve such a problem by a second order quasi-Newton algorithm, modified so that each approximate solution satisfies all inequality constraints. The Herskovits algorithm achieves that in four steps

i) **Newton Prediction** Solve

$$Bdz^o + ds^o\frac{dg}{dz} = -\frac{dj}{dz} - s\frac{dg}{dz},$$
$$s\frac{dg}{dz}dz^o + ds^o g = -sg,$$

with B a BFGS approximation of the Hessian of the Lagrangian.

ii) **Deflexion** Compute a deflected direction pointing towards the set of admissible variables by

$$Bdz^1 + ds^1\frac{dg}{dz} = 0,$$
$$s\frac{dg}{dz}dz^1 + ds^1 g = -\omega s.$$

iii) **Descent** Compute a deflected direction of descent by setting

$$dz = dz^o + \rho dz^1,$$

where ρ is the maximum positive number such that

$$\frac{dj}{dz}.dz < \frac{1}{2}\frac{dj}{dz}.dz^o.$$

iv) **Line search** Find the best solution along the line dz (let us say by an Armijo type algorithm)

$$\begin{aligned} z &= z + tdz, \\ s_i &= s_i + ds_i + \varepsilon_i. \end{aligned}$$

VI Validation results

1 Problem Definition

We now present the final results that we have obtained in the ECARP validation workshop, for validating our global optimisation approach, combining the Herskovits algorithm of Section 5 for optimising the cost, and the PNS solver of Section 2 for solving the equation of state.

The Geometry

To test the efficiency of riblets we shall consider the flow in a semi-infinite channel with approximate rectangular cross section as shown in Figure (1) with two horizontal walls with riblets (the floor and the ceiling). The riblets are small and shaped like a saw tooth. The Reynolds number is based on the horizontal velocity in the centre of the channel w_o, on the inflow cross section. The Reynolds number, $Re = w_oD/\nu = 4200$ and the riblet spacing is to be given by $d/D = 0.1135$.

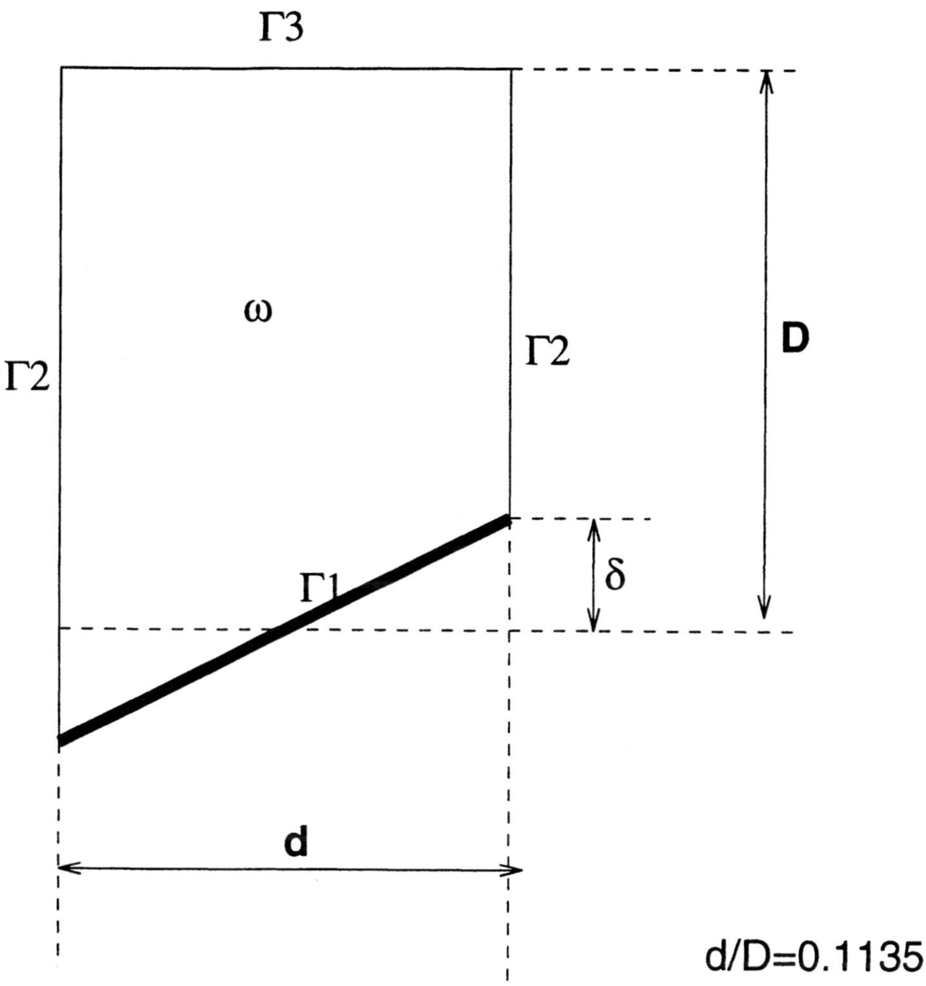

Figure 5 Cross section for a half riblet

In order to keep a constant cross section during the design, the height 2D of the riblet is the average height of the channel (measured at half slope) and not the maximum height.

The depth $\Delta = 2\delta$ of a tooth is the design parameter, which is to be optimized in the process.

Boundary Conditions

At inflow $\vec{u} = (0, 0, w_o)$ which is a constant velocity field entering the channel. On the surface of the riblets we impose a noslip boundary condition $\vec{u} = 0$.

Test Case 8.1. : Direct Simulation

The flow is laminar and stationary. Periodicity and symmetry allows one to reduce the computational domain to a half riblet, limited by two planes of symmetry further cut in half by the horizontal plane of symmetry so that only one riblet wall is considered. The Reynolds number is $Re = 4200$. The shape of the riblet is a saw tooth with fixed base size.

Output :

1) Wall shear stress as a function of the section 2) Velocity level plot for each 3 components at distances along the channel given by $L = 10D, L = 1000D$.

Test Case 8.2. : Optimisation

Riblets in general give increase in drag in laminar flow. However we solve the following optimisation problems when $L = 1000D$:

8.2.1. Show that, for the classes of riblets defined in 8.1, the riblet shape giving the least drag is of zero height, i.e. a flat plate.

8.2.2. Find the maximum height for which the drag is increased by at most 10% of the flat plate drag.

2 Direct simulation

The first results presented in the ECARP data base (inr-tc8.1-sol0, inr-tc8.1-sol05, inr-tc8.1-sol1) are the results of direct simulations run with imposed values of the tooth half depth δ ($\delta = 0, \delta = 0.05, \delta = 0.1$, respectively).

The mesh is constant in each cross section, and uses second order $Q2$ quadrilateral finite elements. The mesh contains 4141 nodes per section and is very much refined along the vertical direction, the first element height being equal to

$$h_{min} = 1/Re.$$

The tests are run from $z = 0$ to $z = 1000D$, with $\Delta z = 0.1$ for the first tenth of the riblet ($0 \leq z \leq 100D$) and $\Delta z = 0.5$ for the remaining part ($100D \leq z \leq 1000D$). Altogether, the simulation uses 7653800 velocity nodes, but, because of the PNS assumption, the calculation only requires several hours of CPU on a workstation. The three components of the velocity are stored in the ECARP data base for the 22 cross sections :

$$\begin{aligned}\frac{z}{D} &= 0, 10, 20, 30, 40, 50, 60, 70, 80, 90, 100, 100, 190, 280, \\ &\quad 370, 460, 550, 640, 730, 820, 910, 1000.\end{aligned}$$

The cross velocity at the second section and the isolines of the transversal velocity at the section $z = 100D$ are also represented on figures 6 and 7. The corresponding friction coefficients

$$C_f = \int_{\Gamma_1} \nu \frac{\partial u_3}{\partial n} d\Gamma$$

are computed at each section z, from $z = D$ to $z = 1000D$, and are represented for each value of δ on the file inr-te8.1-cf-coef.data. They are represented on figure 8.

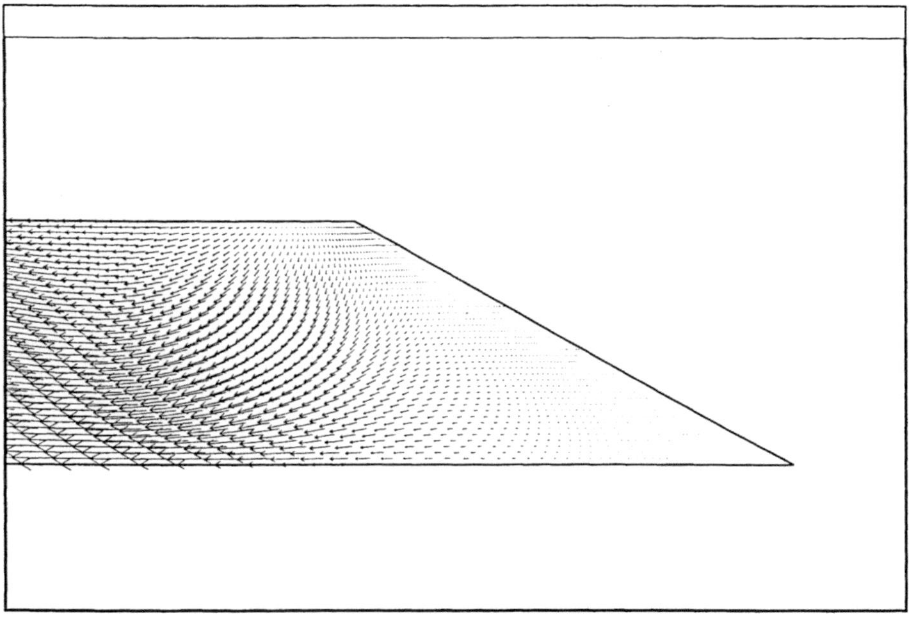

Figure 6 Cross Velocity next to the riblet entrance

As expected, the friction is minimal for the flat plate case. For $\delta = 0$, we have a standard "Blasius type" profile, the transverse flow being accelerated at the center line and slowed down at the wall until one reaches an asymptotic Poiseuille flow pattern. For $\delta = 0.05$, one can observe a similar pattern, combined with a shrinking (resp. dilatation) of the streamlines next to the vertices (resp. to the bottom) of the riblet. For $\delta = 0.1$, a transverse vortex appears at the beginning of the riblet, but rapidly fades away and disappears before $z = 10D$.

For further validation, the case $\delta = 0.05$ has been run for the half riblet with $\Delta z = 0.05$ between $z = 0$ and $z = 100D$, and for a double riblet configuration. The friction coefficient appears to be independent of Δz. The double riblet simulation gives a perfectly symmetric answer and therefore fully validates the symmetry assumptions used in the problem's definition.

Figure 7 Isolines of the transverse velocity at $z = 100D$.

3 Optimisation

We have observed in our direct simulations that all major effects on friction were governed by the first part of the riblet. Therefore, in our optimisation process, and in order to reduce simulation costs, we have reduced the computational domain to the strip

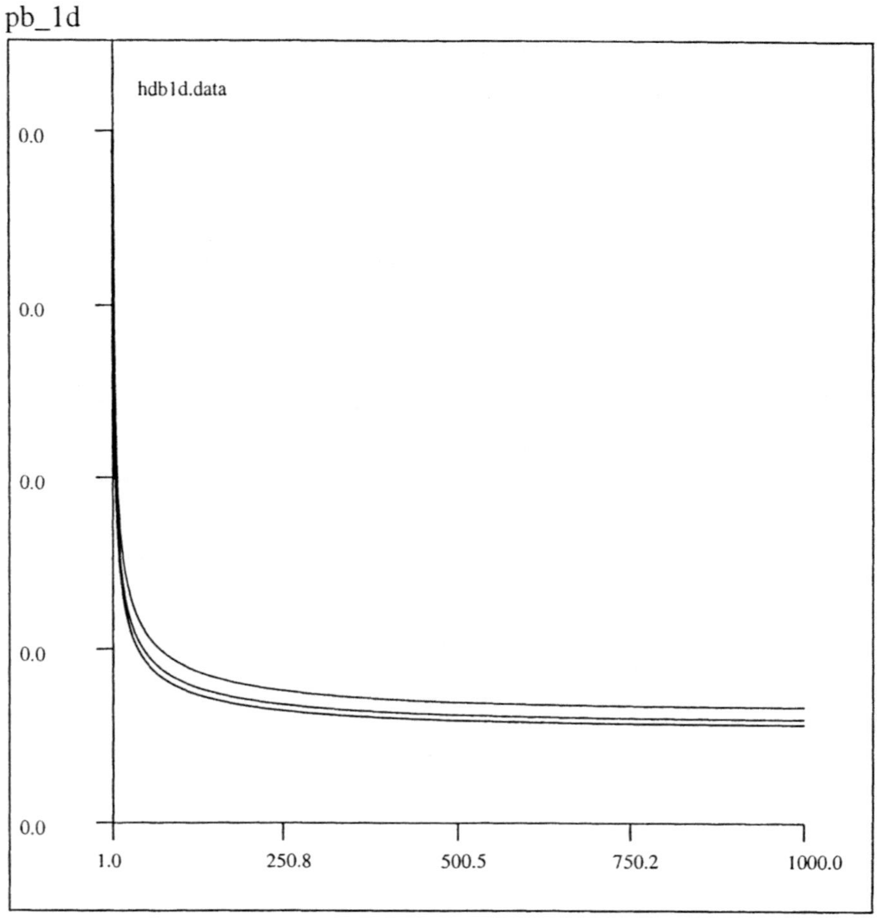

Figure 8 Values of C_f for the three depth.

$$0 \leq z \leq 10D$$

and run all tests with $\Delta z = 0.1$.

The original contract was to solve this optimisation problem using a simplified two-dimensional model. However, this turned out to be very artificial. Therefore, it was decided in the TOP UP contract to solve this optimisation problem using the full three-dimensional PNS model.

The first optimisation is a one dimensional constrained problem trying to find the optimal height $\delta \geq 0$ minimizing the total cost

$$J = \int_z C_f dz = \int_z \int_w \nu |\nabla \vec{v}|^2 dw dz.$$

The algorithm is the quasi Newton (BFGS) algorithm of Section 5 that we have initialized with the value $\delta = 0.1$. For simplicity, the gradient of the cost function was calculated by a first order finite difference formula.

As expected, the algorithm converges towards the flat plate configuration. The convergence is fast and occurs in 7 iterations, as indicated in the following table.

Table 1 Convergence of the direct optimisation

iter	log grad	δ	cost
1	-5.24291616828739	0.1	1.3504628e-03
2	-3.19393015943261	0.078287	1.2708938e-03
3	-3.16793537050826	2.8397994e-02	1.1687085e-03
4	-4.01703798038321	2.2261014e-02	1.1617886e-03
5	-3.78864701938956	-3.2028102e-03	1.1502200e-03
6	-5.02334411568821	2.4019958e-04	1.1499734e-03
7	-6.91963695249183	1.6893614e-04	1.1499725e-03

The second optimisation problem tries to find the maximum value of δ for which the drag is less than 1.1 times the drag of a flat plate :

$$J(\delta) = \int_z C_f dz \leq 1.1 * J(0).$$

Using the same Herskovits interior point algorithm as before, we obtain a solution $\delta = 7.6305\ 10^{-2}$ in six iterations. For this value, the drag constraint is saturated within a 10^{-10} accuracy (Table 2).

Table 2 Convergence of the Herskovits interior point algorithm for the inverse problem (optimisation of δ).

iter	δ	val.max constraint
1.0	2.0335981e-02	-1.0507274e-04
2.0	7.4844836e-02	-4.2366251e-06
3.0	7.5881480e-02	-1.2365828e-06
4.0	7.6243493e-02	-1.8012456e-07
5.0	7.6303915e-02	-3.1610285e-09
6.0	7.6304996e-02	-5.7011147e-10

VII Conclusion

The direct simulation of a laminar riblet channel can be achieved with great accuracy using the incompressible Parabolized Navier-Stokes solver developed for this ECARP contract. This solver gives detailed information on the flow structure (one may use 8 millions nodes at quite a reasonable cost) without having to solve the full three-dimensional Navier-Stokes problem. Moreover, a sensitivity analysis has been added to this solver, based on an adjoint state formulation, which enables us to compute at a reasonable cost the gradient of any objective function related to such flows.

For the Reynolds values which have been used, these simulations indicate that the optimal configuration is a flat plane. To confirm this statement, we have also run a few experiments with more complex cross sections, either convex or concave. We have always observed for our model an increase of drag when increasing the riblet depth. This increase is less dramatic for concave shapes. In any case, it appears that the efficiency of riblets cannot be properly explained by such laminar models operating on flat profiles.

The optimisation of the shape of the riblets has then been performed, either in a direct way (optimisation of the drag) or in an inverse way (maximisation of the depth with imposed drag). For both problems, the use of Herskovits interior point algorithm, combined with our full three-dimensional Parabolised Navier-Stokes solver for computing the state and the adjoint state vectors, was very efficient.

Bibliography

[1] CEA, J. (1971). Optimisation, théorie et algorithmes. *Dunod.*

[2] GILL, P., MURRAY, W., WRIGHT, M. (1981). Practical optimization. *Academic Press.*

[3] HECHT, F. (1983). Ecoulement laminaire derrière une marche : utilisation de base à divergence nulle en éléments finis. *Proceedings of the workshop on numerical analysis of laminar flow over a step, Bièvre, jan. 83.*

[4] HERSKOVITS, J. (1991). An interior points method for non-linear constrained optimization. *NATO/ASI conference on structural optimization, Berchtesgaden, Germany sept. 91.*

[5] PIRONNEAU, O. (1982). On the transport-diffusion algorithm and its application to the Navier-Stokes equations. *Num. Math. vol. 39, pp. 309-332, Springer Verlag.*

[6] SWEARINGEN, J.D., BLACKWELDER R.F. (1987). The growth and breakdown of streamwise vortices in presence of a wall. *Journal of Fluid Mech., vol. 182, pp. 255-290.*

9 Aerodynamic Design of a M6 Wing Using Unstructured Meshes

Alain Dervieux, Nathalie Marco and Jean-Michel Malé,
INRIA, Sophia Antipolis Cedex, France

Abstract
The ECARP M6 shape optimization problem is solved on unstructured meshes. A multilevel method relying on a gradient approach is built. The multilevel method is introduced either in an optimization loop or in a one-shot method. Its interest is demonstrated by an application to 3D aerodynamical shape optimization in which the shape variation is accounted for by a transpiration boundary condition.

9.1 Introduction

This contribution aims at experimenting a strategy for the Optimum Shape Design of aircrafts in complex flows.
The main characteristic of this strategy is to consider arbitrary shapes represented by **unstructured tetrahedrizations**.
The main option is to compute **gradients** rather than paying divided differences.
The optimization chain built in cooperation with a Dassault team and experimented here involved a set of new ingredients:
- hierarchical parametrization
- transpiration
- exact gradient (first-order), partly obtained by automated differentiation.

The first next section (Section 8.2) is devoted to recalling the main notions of Optimal Control methods relying on the adjoint state. Section 8.3 presents another strategy, the one-shot multi-level method allowing CPU savings. The 3D multilevel parametrization method is also briefly depicted. Section 8.4 gives the global Optimization strategy applied to the M6 Ecarp test-case.

9.2 Exact gradient approach for an airfoil flow

Let γ be a set of control parameters for the shape of an airfoil, and $W(\gamma)$ the corresponding flow variables, implicitly defined from the discretized steady Euler equation written as follows

$$\Psi(\gamma, W(\gamma)) = 0 \tag{1}$$

if the mesh is of fixed topology, with a deformation parametrized by γ, then the discrete flow variables W are a smooth function of γ.
Introducing a discrete cost function :

$$j(\gamma) = J(\gamma,\ W(\gamma)) \tag{2}$$

the gradient is given by :

$$j'(\gamma) = \frac{\partial J}{\partial \gamma}(\gamma,\ W(\gamma)) - < \Pi(\gamma), \frac{\partial \Psi}{\partial \gamma}(\gamma,\ W(\gamma)) > \tag{3}$$

where Π is an adjoint state, solution of the linear system :

$$\frac{\partial \psi^*}{\partial W}(\gamma,\ W(\gamma))\Pi(\gamma) = \frac{\partial J}{\partial W}(\gamma,\ W(\gamma)) \tag{4}$$

and the cost functional to be minimized is given as follows :

$$j(\gamma) = \sum (P_l(\gamma) - P_l^{target})^2 \tag{5}$$

where $P_l(\gamma)$ is the pressure variable from $W(\gamma)$, and where l belongs to a set of points on the airfoil wall.

In these conditions, the whole chain can be exactly differentiated, the gradient of j is expressed with an adjoint-state and a conjugate gradient algorithm can be applied to the minimization of j (see [1] for details). In practice, the Euler system (1) is assembled on the triangulations by means of the Van Leer first-order Flux Vector splitting ([2]).

9.3 One–shot Multilevel

We now consider a "one-shot" approach (denomination introduced by Ta'asan in [7] and [8]) which consists in solving simultaneously system (6).

$$\begin{cases} \Psi(\gamma, W) = 0 \\ \left(\dfrac{\partial \Psi^*}{\partial W}\right)(\gamma, W)\ \Pi = \dfrac{\partial J}{\partial W}(\gamma, W) \\ j'(\gamma) = \dfrac{\partial J}{\partial \gamma}(\gamma, W) - < \Pi, \dfrac{\partial \Psi}{\partial \gamma}(\gamma, W) > \\ \gamma^{\alpha+1} = \gamma^{\alpha} - \rho j'(\gamma^{\alpha}). \end{cases} \tag{6}$$

We then have three unknowns: W, Π, γ. W and Π are not implicit functions of γ, but W, Π, γ are solved simultaneously in a relaxation process that advances the three unknowns at each iteration consisting, typically, of :

$$\begin{pmatrix} \text{1 time iteration for } W, \\ \text{3 sweeps Jacobi for } \Pi, \\ \text{1 update of } \gamma^{\alpha+1} = \gamma^{\alpha} - \rho j'(\gamma^{\alpha}) \end{pmatrix}.$$

To get a better cost, the one–shot method is combined with a multilevel method : we solve the problem alternatively on the different levels. The cost functional is defined on a vector space E (fine level). Let us consider F a second space and P a linear continuous mapping from F to E; the algorithm is :

$$\begin{cases} \gamma^0 \text{ given} \\ \gamma^{\alpha+1} = \gamma^{\alpha} - \rho P P^* j'(\gamma^{\alpha}) \end{cases} \tag{7}$$

where P^* is the dual operator of operator P.

If γ is an element of a space F, this step is equivalent to a gradient step for the minimization of $j(\gamma)$ in the subspace $P(F)$.

Since we consider the parametrization for optimizing an aircraft in a 3D Euler flow, the parametrized shape is a 2D shell.
The shell is assimilated to a manifold $\gamma \in C^{\infty}$, that is smooth enough. Any deformation of the discrete manifold γ_h can be parametrized with multilevelling thanks to the agglomeration method [4]. Coarse parametrizations are obtained by combining a coarse "agglomerated" parametrization with an adequate choice of operator P. A new manifold $\gamma_h + PP^*\delta\gamma_h$ is generated, close to the initial one (details are given in [5]).

The resulting multilevel parametrization has the following characteristics :
- the computer code applies to any triangulated geometry,
- the number of parameters is adjustable,
- there is a mesh independent multilevel convergence for model cases.

9.4 Application to 3D Euler flows

9.4.1 Flow model and optimization loops

In this work, inspired by the approach used by Young et al ([3]), we are considering in a first phase the option of representing the shape modification by applying a transpiration condition; this means that the current shape is the combination of the mesh skin with a perturbation simulated by transpiration (introduced by G.D. Mortchelewicz in [6]), refered in the sequel as the "transpired perturbation".
We recall the transpiration condition for Euler flows:
Let us denote by *shell* the shape to be emulated by transpiration and by $\vec{n}_{shell}$ the normal of the shell. The slip boundary term of the flux $\Psi(W)$ is defined by :

$$\Psi(W)_{slip\ boundary} = W\ .\ q + \begin{pmatrix} 0 \\ p(W)\ .\ n_x \\ p(W)\ .\ n_y \\ p(W)\ .\ n_z \\ p(W)\ .\ q \end{pmatrix}$$

with : $q = \vec{V}\ .\ (\vec{n} - \vec{n}_{shell})$ where $\vec{V}$ is the velocity of the fluid.
The linearization (by differentiation) of the transpiration condition is straightforward and an adjoint state is easily computed.
The global method is essentially made of three loops (Figure 8). The external loop is an optional remeshing loop in which a new shape is derived from an old one updated by the transpired perturbation; then the medium loop is called. The medium loop is a multilevel gradient optimization loop in which the shape variable is taken into account in the transpired perturbation; the operator P is taken alternately equal to identity (fine level) or equal to the projection into one of the coarse levels (according to a V-cycle multilevel strategy). This loop

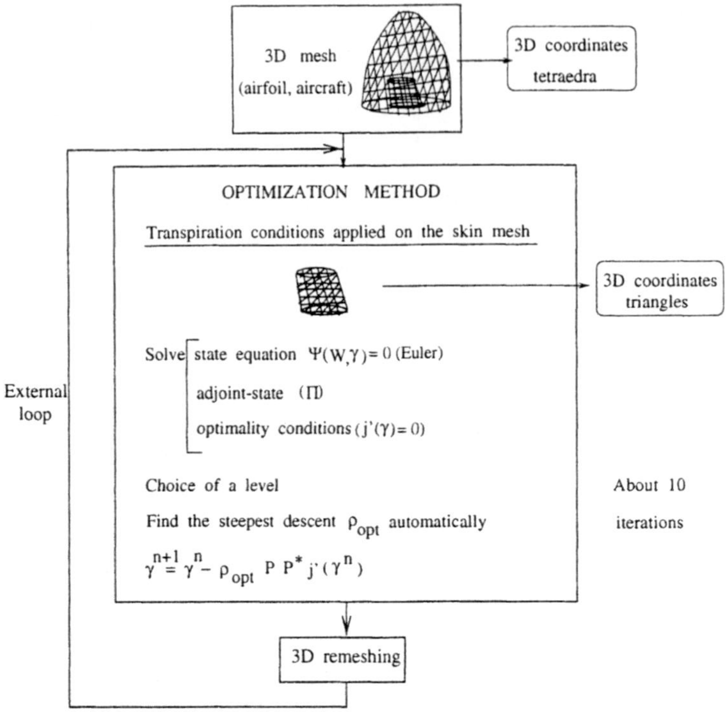

Figure 8 Organization of the optimization loop.

involves the evaluation of the gradient of the cost function through an adjoint state. The internal loop is the 1D local research of the steepest descent option.

9.4.2 Minimization of the shock induced drag on an M6 ONERA wing

The skin mesh of the simplified M6 wing that we considered is represented on Figure 9. The cost functional we use is of the form :

$$j(\gamma) = \omega_1(C_D - C_D^{target})^2 + \omega_2(C_L - C_L^{target})^2 + \omega_3 \int_\gamma (P - P^{target})^2 \, d\sigma$$

with $\omega_1 = 10$, $\omega_2 = 1$, $\omega_3 = 1$ and $C_D^{target} = 0$.

The initial conditions are defined by a farfield Mach number of 0.84 and an angle of attack of 3.06 degrees.

9.4.3 3D mesh with 2203 nodes and 814 nodes on the skin (Figure 10)

We present on Figure 11 the existing number of parameters on the different levels and the number of Jacobi relaxations used (for the flow and the adjoint-state) in the case of one optimization iteration and one One-Shot iteration.

Initial drag and lift coefficients, C_D^o and C_L^o , are defined by :

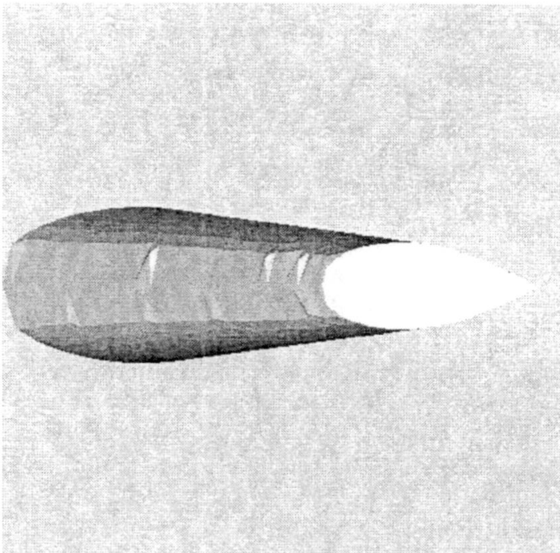

Figure 9 Initial shape.

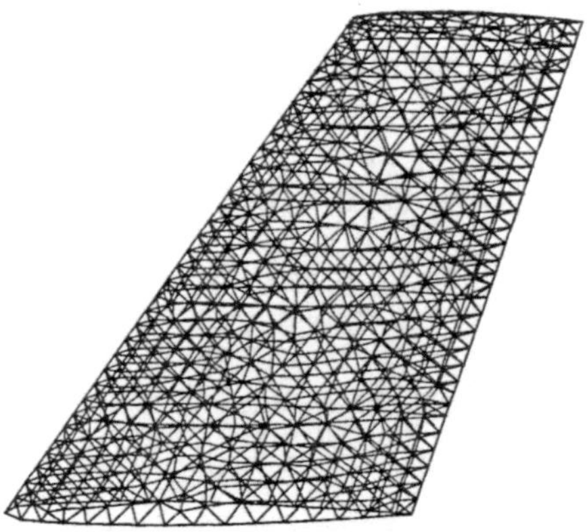

Figure 10 814-node skin mesh.

$$C_D^o = C_D^{M6} = 2.7654e - 02 \quad \text{and} \quad C_L^o = C_L^{M6} = 0.128439.$$

The initial value of the cost functional is equal to $j^o = 7.64e - 03$.

Before applying the multilevel method to the optimization problem, we work

Level 1	:	814 parameters
Level 2	:	201 parameters
Level 3	:	59 parameters
Level 4	:	20 parameters

Optimization method : 100 Jacobi relaxations for a flow.
180 Jacobi relaxations for the adjoint-state
About 5 flows for the evaluation of the optimal step

One-Shot method : 8 Jacobi relaxations for a flow
8 Jacobi relaxations for the adjoint-state
Optimal step fixed at each level

Figure 11 Presentation of the different levels and the number of Jacobi relaxations for the evaluation of the flow and the adjoint-state in the case of an optimization method and a One-Shot method.

with all the parameters (fine level). The optimal step has been evaluated at each iteration by a 1D minimization method. After 10 iterations (and 23 minutes of CPU time), the drag coefficient C_D was reduced by 59%. However, the obtained shape presents lots of roughnesses (figure 12).

Figure 12 Shape of the wing after 10 iterations on the finest level.

For applying the multilevel method, we have used a V-cycle strategy by alter-

nating the levels between the level 1 and the level 4 (20 parameters).
- First of all, an optimization method was used. The optimal shape was evaluated at each generation with a 1D minimization method. After 9 iterations (and 17 minutes of CPU time), we got :

$$C_D^9 = 0.43\ C_D^o \quad \text{et} \quad C_L^9 = 0.97\ C_L^{target}$$

and the value of the cost functional was : $j^9 = 2.75e - 03$.
41 cost functionals have been evaluated. This is equivalent to 5720 relaxations for reducing the drag of a factor about 2.

- The One-Shot method (**1** time iteration involving **8** Jacobi relaxations for the flow, **8** Jacobi relaxations for the adjoint-state) was combined with the multilevel method. Equivalent results where obtained after 3 minutes CPU and 9 One-Shot iterations. Globally, only 144 Jacobi relaxations where needed ! The new shape is depicted on Figure 13.

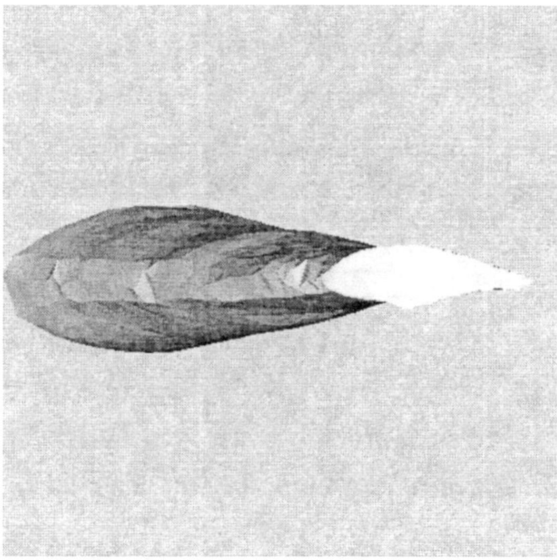

Figure 13 Optimized shape.

9.4.4 3D mesh with 15460 nodes and 3219 nodes on the skin (Figure 14)

Figure 15 presents the number of parameters (or control points) for each level.

On the finest level, the same rough shape as in the previous study was obtained. We restrict the optimization to the variables of the fourth level.

The skin mesh of the M6 wing involves 3819 nodes.

$$C_D^0 = 1.761e - 02 \quad \text{and} \quad C_L^0 = C_L^{target} = 0.145\ .$$

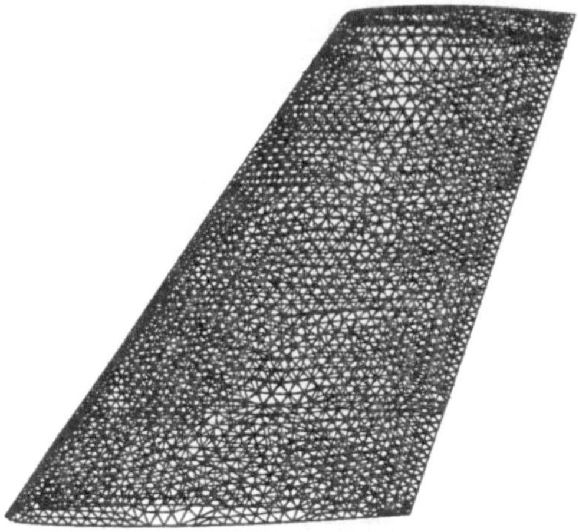

Figure 14 3219 nodes on the skin.

Level	1 :	3219 parameters
Level	2 :	814 parameters
Level	3 :	201 parameters
Level	4 :	59 parameters
Level	5 :	20 parameters

Optimization method :	100 Jacobi relaxations for a flow
	180 Jacobi relaxations for an adjoint-state
	About 5 flows for the evaluation of the optimal step

Figure 15 Presentation of the different levels and the number of Jacobi relaxations needed for a flow and the evaluation of the adjoint-state for one optimisation iteration. The One-Shot method was not applied.

We have used a sawtooth V-cycles strategy between the fourth level, with 59 parameters and the finest level. After 10 optimization iterations, the obtained drag is $C_D^{10} = 1.003e - 02$ and the obtained lift is $C_L^{10} = 0.1421$, i.e :

$$C_D^{10} = 0.57\, C_D^0 \quad \text{and} \quad C_L^{10} = 0.98\, C_L^0 .$$

CPU time is about 5 hours on a DEC station.
Figure 16 presents the final shape. On Figure 17, we observe a reduction of the shock on the wing.

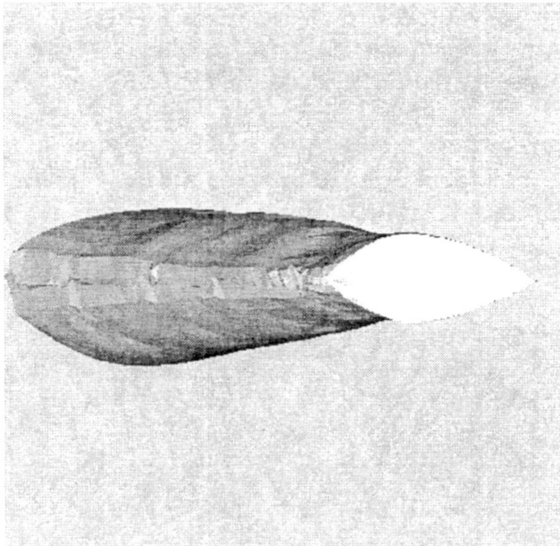

Figure 16 Final shape after 10 optimization iterations.

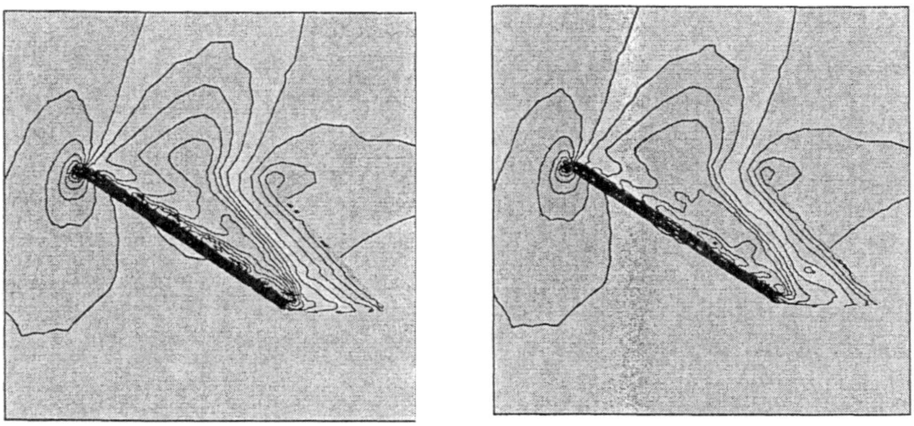

Figure 17 Reduction of the initial λ-shock after 10 optimization iterations.

9.5 Conclusion

The optimization of aerodynamical shapes is entering in an era of maturity. Complex CAO shapes will be more and more easily transformed in skin meshes from which advanced unstructured meshes generators (Voronoi, Advancing Fronts, Reconnectors) will automatically produce volumic meshes. Then Shape Optimizers of the type of the code presented in this work will be able to optimize the shape of a complex aircraft, taking into account the interaction between wing, body, pylon, nacelle, winglets, etc.

Although still in preliminary development phases, the presented method is able to give simplified answers for complex geometries. For reaching the above more ambitious goals, several numerical improvements and complements are to be produced:
- faster flow solver (multigrid ?),
- better gradients (for second-order approximation),
- smarter optimization algorithms taking into account constraints,
- complete optimization loop by remeshing.

Acknowledgements:
The contribution of Inria-Sophia to the ECARP project has got the best benefit from the helpful cooperation with Polytechnical University of Catalunya, of National Technical University of Athens, of Dassault-Aviation, and from fruitful discussions with the partner of the ECARP project.

Bibliography

[1] F. BEUX, A. DERVIEUX, M.-P. LECLERCQ, and B. STOUFFLET. Techniques de contrôle optimal pour l'optimisation de forme en aérodynamique avec calcul exact du gradient, 1994. Revue Scientifique et Technique de la Défense.

[2] L. FEZOUI and B. STOUFFLET. A Class of Implicit Upwind Schemes for Euler Simulations with Unstructured Meshes. *J. Comput. Phys.*, 1991.

[3] W. P. HUFFMAN, R. G. MELVIN, D. P. YOUNG, F. T. JOHNSON, J. E. BUSSOLETTI, M. B. BIETERMAN, and CRAIG L. HILMES. Practical Design and Optimization in Computational Fluid Dynamics. *AIAA Paper 93-3111*, 1993.

[4] M.-H. LALLEMAND. *Schémas décentrés multigilles pour la résolution des équations d'Euler en éléments finis.* PhD thesis, Université de Marseille, 1988.

[5] N. MARCO and A. DERVIEUX. Agglomeration Method applied to the Hierarchical Parametrization of a Skin Mesh in 3-D Aerodynamics. In *Contributions to 12th month of European Project ECARP*, April 1994.

[6] G.D. MORTCHELEWICZ. Résolution des équations d'Euler tridimensionnelles instationnaires en maillages non structurés. *La Recherche Aérospatiale*, (6):17–25, Novembre-Décembre 1991.

[7] S. TA'ASAN. One Shot Methods for Optimal Control of Distributed Parameter Systems. I : Finite Dimensionnal Control. ICASE Report 91-2, 1991. NASA Contractor Report 187497.

[8] S. TA'ASAN and G. KURUVILA. Aerodynamic Design and Optimization in One-Shot. AIAA Paper 92-0025, 1992.

10 Part I. Single and two-point airfoil design using Euler and Navier-Stokes equations Th. E. Labrujère, NLR, Holland

10.1 Test case TE2

1 Methodology

The method used for treating this test case is the NLR residual-correction method, described in section III.4.11., setting the number of design conditions equal to one.

2 Numerical results

The NLR residual-correction method has been applied to the present test case using a Navier Stokes solver as analysis code in the outer loop.

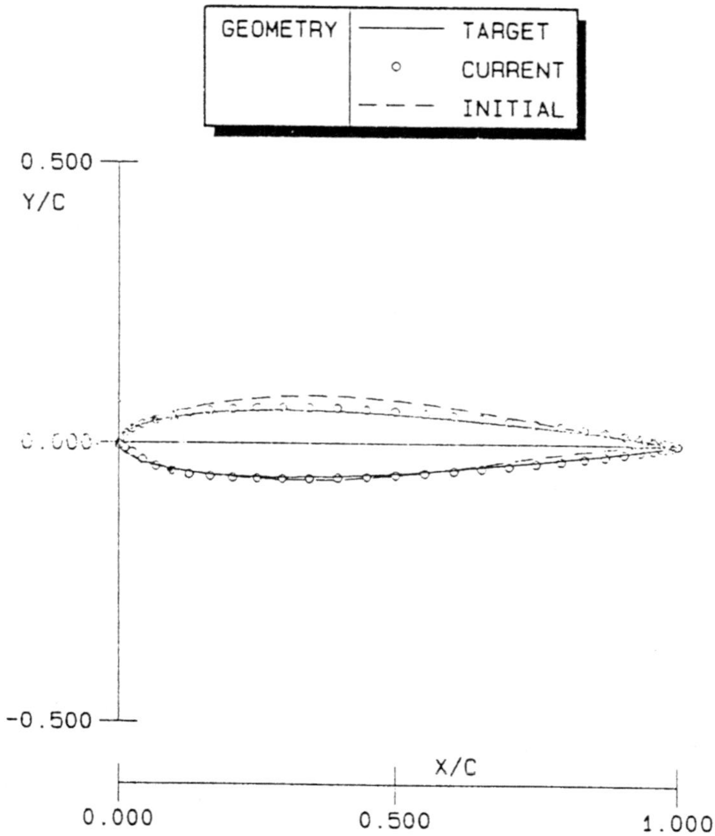

Figure 1 Resulting geometry

In agreement with the test case definition a target pressure distribution has been calculated by applying the Navier Stokes solver to the NACA0012 airfoil with subsonic onset flow conditions $\alpha = 3.0°$, Mach = 0.3 and $Re = 3 * 10^6$.

The NACA63215 airfoil was used as initial guess for the geometry.

The "designed" geometry is depicted in Figure 1. The associated pressure distribution is depicted in Figure 2. These figures show still rather large discrepancies between targets and obtained results, mainly in the leading edge region. The reason is that the iteration process was stopped after 10 iterations (see Figure 3) because of convergence problems near the leading edge of the airfoil. This

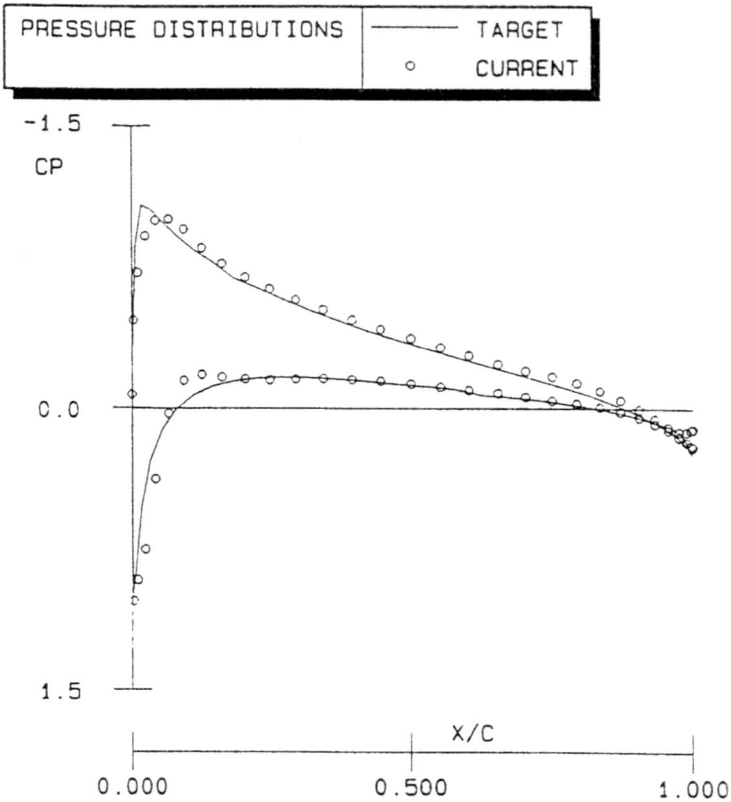

Figure 2 Resulting pressure distribution

lack of convergence seems to be due to an inconsistency in the determination of the the equivalent incompressible velocity in the stagnation point region. The results demonstrate however that the residual-correction concept usede is applicable with a Navier Stokes code for analysis, as soon as the problems in the stagnation point region are solved.

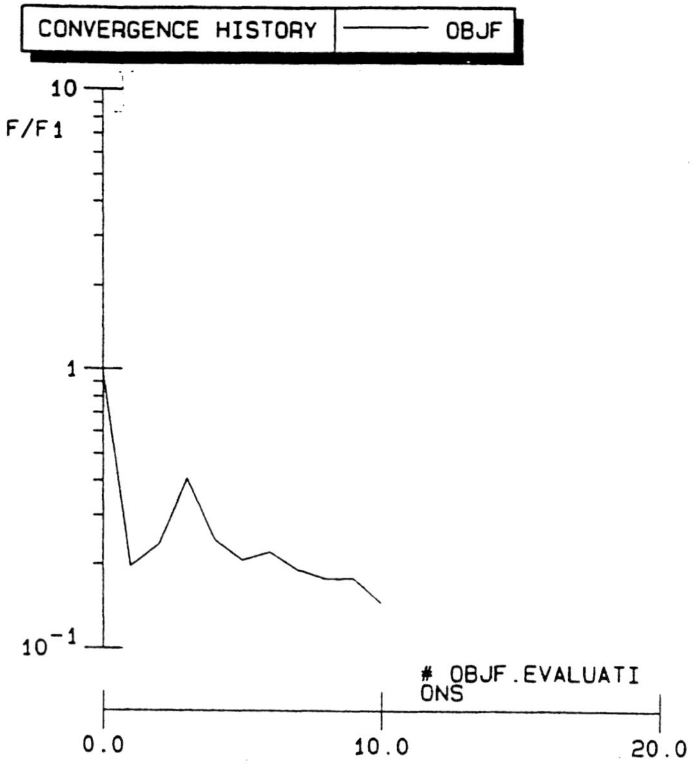

Figure 3 Convergence history

10.2 Test case TE4

1 *Abstract*

Multi-point airfoil design is related to the fact that aircraft have to operate under a number of quite different conditions. This implies in general that the aerodynamic design process must be able to deal with different requirements for different flow conditions.

Within the ECARP project "Optimum Design in Aerodynamics" NLR's task has been to develop and explore an algorithm for multi-point aerodynamic airfoil design for subsonic/ transonic flow conditions. Such an algorithm should lead to one single airfoil shape, optimized with respect to a priori specified requirements for a number of flow conditions and satisfying a number of constraints on aerodynamic and geometric characteristics. The algorithm described below is based on the residual-correction approach and offers in principle the possibility to incorporate different types of constraints. The algorithm has however been applied to the test case without any constraints.

2 *Problem definition*

The design problem considered can be defined as the minimization of an objective function of the form

$$F = \sum_{i=1}^{N} [W^i F^i(\vec{X})], \tag{1}$$

where the summation is over N operating conditions and the vector $\vec{\mathbf{X}}$ contains the design variables. Each function F^i attains its minimum when the design requirements for the corresponding operating condition i are fulfilled. The W^i are weight factors that balance the requirements for the different operating conditions.

In inverse airfoil design the design requirements are specified in terms of target pressure distributions. For potential flow this implies specification of target tangential velocity distributions, and thus the objective function takes the form

$$F\{\alpha_1, \alpha_2, .., \alpha_N, x(t), z(t)\} = \sum_{i=1}^{N} [W^i \int_0^1 \{V_t^i(\bar{t}) - V_{tar}^i(\bar{t})\}^2 d\bar{t}], \tag{2}$$

where α_i is the angle of attack associated with operating condition i, x and z are the coordinates of the airfoil contour $\bar{t}$ is , t is the arclength measured along the airfoil contour, $\bar{t}$ is the fractional arclength and V_t is the tangential velocity.

Basic principle
The NLR residual correction method is based on the assumption that it is possible to split the design process into two major steps, which can be iterated until satisfactory results have been obtained. In the first step the flow about the

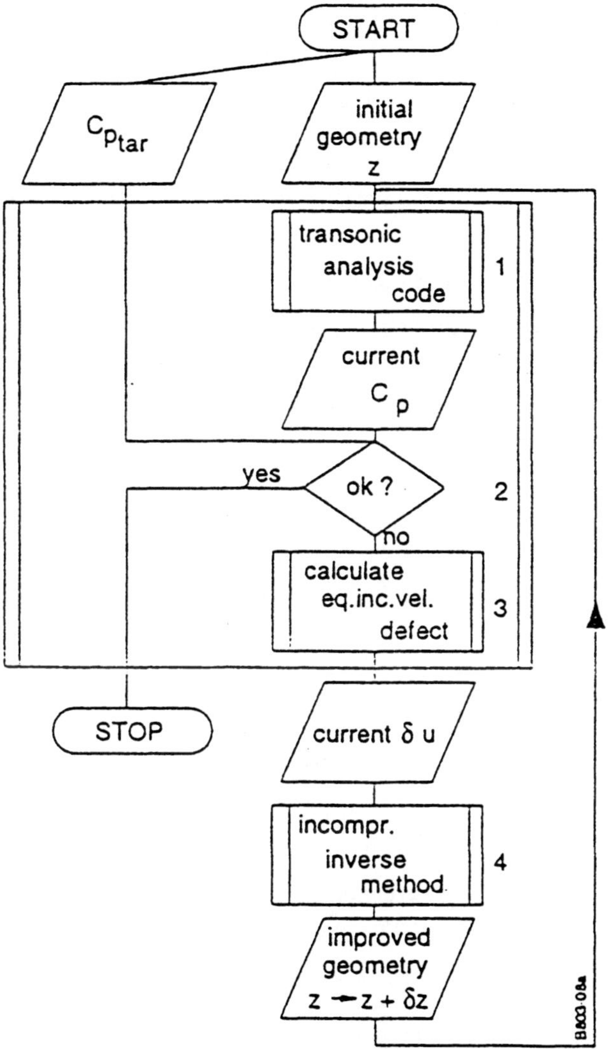

Figure 4 Residual correction method, outer loop

current estimate of the geometry is calculated for each operating condition by means of an analysis code applicable to each flow regime under consideration, thus giving the deviations (the residuals) from the specified targets.

In the second step, a geometry correction is calculated from the current residuals by means of an approximative procedure involving the specification of an equivalent incompressible multi-point inverse flow problem which is more amenable to fast computational methods. The computational process is initiated by specifying :

- the operating conditions in terms of lift (angle of attack), Mach number and Reynolds number,
- a target pressure distribution for each operating condition,
- a starting airfoil with initial angles of attack,
- geometric constraints and/or aerodynamic constraints (e.g. pitching moment limitations),
- weight factors for each operating condition.

The computation proceeds by utilizing the following loop in an iterative fashion (see Figure 4) :

1. Calculate the flow about the current estimate of the airfoil geometry z=z[x] for each operating condition (or design point) considered and obtain the current pressure distribution Cp[t] on the airfoil surface.
2. Decide whether or not the current airfoil geometry z=z[x] needs further improvement, by comparing the current pressure distribution Cp[t] with the target pressure distribution $Cp_{tar}[t]$ and by considering the convergence history.
3. If further improvement of the current airfoil geometry z=z[x] is considered necessary, calculate the equivalent incompressible perturbation velocities, from the discrepancies between the target pressure distribution $Cp_{tar}[t]$ and the current pressure distribution Cp[t].
4. Calculate a new estimate for the airfoil geometry by solving an equivalent incompressible design problem as will be defined below.
5. Iterate the whole process until satisfactory results in terms of approximations of the pressure distributions and geometric requirements are obtained.

During the process, the weight factors mentioned above may be used to balance the different design requirements.

The inverse calculation consists of two major steps as indicated in Figure 5. In the first step the pressure-defect distribution is replaced by an equivalent subsonic perturbation velocity defect distribution as will be described below, applying for the transonic case a so-called pressure defect splitting technique. In the second step this equivalent velocity distribution is used for the determination of a new airfoil shape.

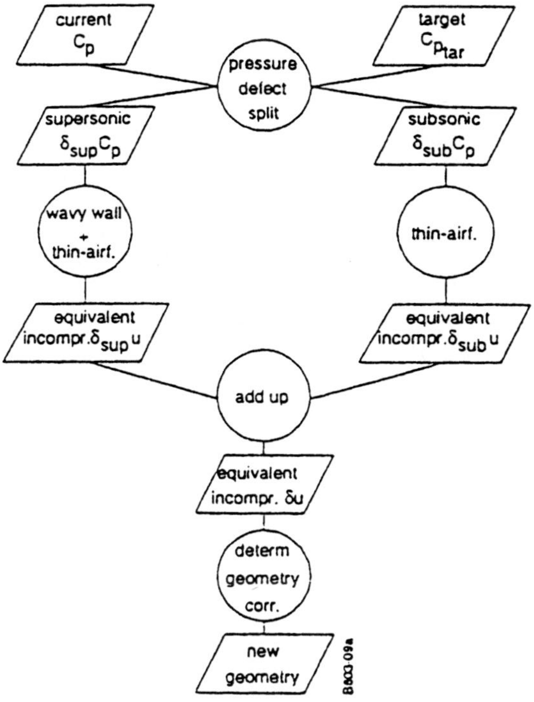

Figure 5 Inverse computational process

Equivalent incompressible velocity

The pressure defect splitting technique, which is used to distinguish between a subsonic and supersonic part of the pressure defect, is illustrated in Figure 6. The split made is based primarily on the assumption that subsonic thin-airfoil theory is applicable if the local actual velocity and the local target velocity are both subsonic, and that supersonic wavy-wall theory is applicable if both velocities are supersonic. In case these velocities are of a different nature, the sonic velocity is used as upper or lower limit. In [Fray et al84] a detailed description of the derivations is given.

The subsonic and supersonic parts of the pressure defect δCp are defined as

$$\delta_{sup}Cp = min(Cp_{tar}, Cp^{*}) - min(Cp, Cp^{*}), \qquad (3)$$

$$\delta_{sub}Cp = max(Cp_{tar}, Cp^{*}) - max(Cp, Cp^{*}), \qquad (4)$$

where Cp is the actual pressure coefficient, Cp_{tar} is the target pressure coefficient, and Cp^{*} is the pressure coefficient for sonic conditions.

The subsonic and supersonic parts of the pressure-defect distribution are each converted into incompressible perturbation velocity defects $\delta_{sub}u$ and $\delta_{sup}u$, from which the equivalent incompressible perturbation velocity defect is obtained as

$$\delta u = \epsilon(\delta_{sub}u + \delta_{sup}u), \tag{5}$$

where ϵ is a relaxation parameter.

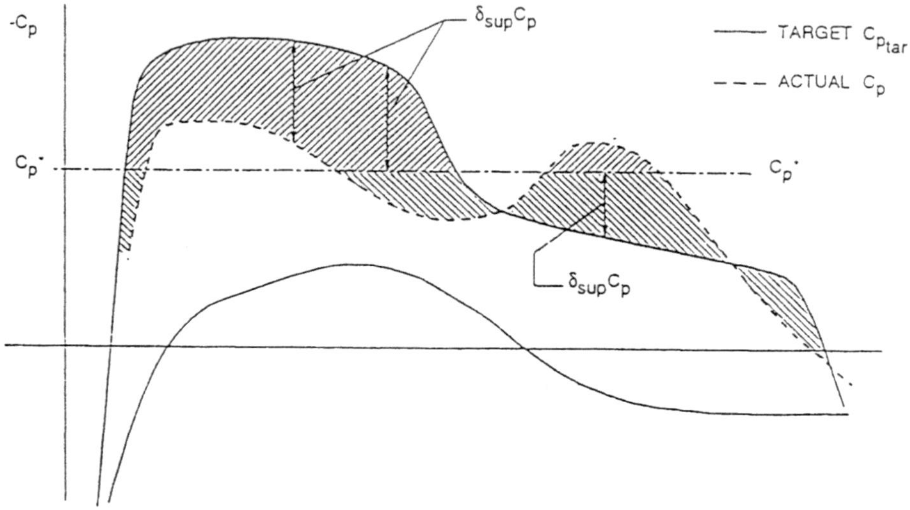

Figure 6 Pressure defect splitting

Equivalent multi-point incompressible design problem

A new estimate of the airfoil shape is obtained by solving an equivalent incompressible multi-point design problem. This involves the determination of a target velocity distribution for each design condition i according to the above defined methodology.

The requirements according to the Betz-Mangler constraints, associated with trailing edge closure and compatibility of target velocity and free stream velocity, are addressed by introducing a certain amount of freedom in the target velocity distributions. To this end, each specified target velocity distribution is augmented with auxiliary functions, containing three parameters which may be utilized to modify the target, simultaneously with solving the design problem such that the Betz-Mangler constraints will be fulfilled.

Following [Soemarwoto93], each target velocity is considered to have the form

$$V^i_{tar} = g^i[\bar{t}, p^i_1, p^i_2, p^i_3]. \tag{6}$$

Here, $\bar{t}$ is the fractional arclength measured along the airfoil contour, p^i_1 scales the tangential velocity level such that compatibility of the target velocity and the free stream velocity is established. A second degree of freedom (with a parameter

p_2^i) is used to adjust the velocity level near the trailing edge for compliance with trailing edge closure. The location of the leading edge stagnation point is used as a third degree of freedom in the target. One of the auxiliary functions chosen in [Soemarwoto93] achieves a coordinate transformation correlating the stagnation point location with the parameter p_3^i. If the target is specified such that all consistency constraints are implicitly fulfilled, the parameters should assume the values $p_1^i = 1$, $p_2^i = 0$, and $p_3^i = 0$, and the original target is not modified.

The algorithm considered here for solving the equivalent incompressible multi-point design problem is based on application of a first order panel method for the determination of the flow around an airfoil, utilizing piecewise constant doublet and source distributions on the airfoil contour.

According to this flow simulation, the total velocity potential is determined by

$$\phi^i = \phi_\infty^i + \phi_d^i + \phi_s^i \tag{7}$$

where the velocity potential of the undisturbed flow ϕ_∞^i is given by the inner product

$$\phi_\infty^i = \vec{q}_\infty^{\,i} . \vec{x}, \tag{8}$$

with

$$\vec{q}_\infty^{\,i} = \begin{bmatrix} cos(\alpha^i) \\ sin(\alpha^i) \end{bmatrix} \tag{9}$$

and where the magnitude of the onset flow velocity has been set equal to one.

The design problem may then be defined as the minimization of the functional

$$F = \sum_{i=1}^{N} [W^i F^i(\vec{X})] \tag{10}$$

with

$$\begin{aligned} F^i &= \int_0^1 \{V_t^i(\bar{t}) - V_{tar}^i()\}^2 d\bar{t} = \\ &= \int_0^1 \{\frac{\partial \phi^i}{\partial t} - V_{tar}^i\}^2 d\bar{t} \end{aligned} \tag{11}$$

under the condition

$$(\phi_d^i + \phi_s^i)^- = 0, \tag{12}$$

where the minus sign denotes application to the inner side of the airfoil contour, which is equivalent to the Neumann condition of zero normal velocity on the airfoil contour, and applying an appropriate Kutta condition at the trailing edge. A discretized form of equation(11) is obtained by evaluating the integrand at the midpoints of the panels and keeping it constant along each panel. Solution of the design problem then involves the minimization of the discretized functional

$$F = \sum_{i=1}^{N} W^i F^i =$$
$$= \sum_{i=1}^{N}[W^i \sum_{k=2}^{NC}(\frac{\partial \phi^i}{\partial t} - V^i_{tar})^2_k \Delta \bar{t}_k], \qquad (13)$$

under the condition

$$[\phi^i_d + \phi^i_s]^-_k = 0, \forall k, \qquad (14)$$

and applying the Kutta condition at the trailing edge.

Here, the indices k refer to the panel midpoints; k=2 denotes the first midpoint near the trailing edge on the airfoil lower side; k=NC denotes the last midpoint near the trailing edge on the airfoil upper side.

Considering the minimization of the functional, obvious design variables are the angles of attack α^i and the parameters p^i_k in the target velocities. With respect to the representation of the airfoil contour, x as well as z may be considered as unknown functions that are to be determined during the minimization. It has been assumed before that the velocity distribution is specified as a function of the fractional arclength. So,the prescribed velocity will be attained on the airfoil contour at given values of the fractional arclength. Therefore, because of the fact that the contour coordinates x and z are connected through the arclength, only one of these functions is actually required as design variable. However, it has appeared that a greater flexibility for modification of the airfoil contour is obtained when both x and z are considered as design variables while prescribing the arclength. Thus, the functional will be minimized with respect to the design variables α^i, x, z and the parameters p^i_k.

The design problem as formulated above may be solved in the least squares sense by considering the minimization of the augmented functional

$$Q = F + H =$$
$$= \sum_{i=1}^{N}[W^i F^i + H^i] =$$
$$= \sum_{i=1}^{N}[W^i \int_0^1 \{\frac{\partial \phi^i}{\partial t} - V^i_{tar}\}^2 d\bar{t} + \int_0^1 \{(\phi^i_d + \phi^i_s)^-\}^2 d\bar{t}]. \qquad (15)$$

This functional will attain its minimum (zero) for all operating conditions when the design requirements are fulfilled (first term vanishes) and when the Neumann condition of zero normal velocity on the airfoil contour is satisfied (second term vanishes).

During the investigations based on this approach to the solution of the design problem it appeared to be necessary to have some control over the values of the parameters p_k^i. Also, the designer obviously wishes to keep the deviations from his targets as small as possible. Therefore, the functional (15) is augmented further (in discretized form) to

$$\begin{aligned} Q \;=\; & \sum_{i=1}^{N}[W^i \sum_{k=2}^{NC}\{\frac{\partial \phi^i}{\partial t} - V_{tar}^i\}^2 \Delta \bar{t}_k + \\ & + \sum_{k=2}^{NC}\{(\phi_d^i + \phi_s^i)_{\bar{k}}^-\}^2 \Delta \bar{t}_k + \\ & + \{W_{p1}^i (p_1^i - 1)\}^2 + (W_{p2}^i p_2^i)^2 + (W_{p3} p_3^i)^2]. \end{aligned} \quad (16)$$

The weight factors in the last term enable the designer to control the deviation of the free parameters from their ideal values (1,0,0).

The functional Q has essentially the form

$$Q = \sum_{n=1}^{NF}[f_n(\vec{z})]^2, \quad (17)$$

where $\vec{z}$ is the vector of design variables and where the f_n form together a vector $\vec{f}$. This type of minimization problem can efficiently be solved by means of the method described in [Fletcher68].

The associated computational algorithm proceeds as follows :

1. Given $\vec{z} = \vec{z}^{(m)}$, compute the vector of residuals $\vec{f}^{(m)}$ and the Jacobian $J^{(m)}$; the Jacobian is derived by differentiating analytically the residuals f_n with respect to the design variables.

2. Determine a search direction by computing $\vec{s}^{(m)} = -J^{(m)+}.\vec{f}^{(m)}$, where $J^{(m)+}$ is the generalized inverse of $J^{(m)}$.

3. Set $\vec{z}^{(m+1)} = \vec{z}^{(m)} + \lambda \vec{s}^{(m)}$, and determine $\lambda > 0$ such that $Q[\vec{z}^{(m+1)}] < Q[\vec{z}^{(m)}]$.

4. If preset conditions with respect to the variation of $\vec{z}$, or with respect to the residuals f_n are fulfilled, convergence is considered to be attained; then terminate the iteration process, otherwise repeat from 1.

Bibliography

[Fray et al84] *Fray, J.M.J., Slooff, J.W., Boerstoel, J.W. and Kassies, A., (1984)* :, Design of transonic airfoils with given pressure, subject to geometric constraints, **NLR TR 84064 U.**

[Soemarwoto93] *Soemarwoto, B.I., (1993) :*, Robust inverse shape design in aerodynamics , **NLR TP 93432 L.**

[Fletcher68] *Fletcher, R., (1968) :*, Generalized inverse methods for the best least squares solution of systems of non-linear equations, **Comp.J 10 pp 392-399.**

4 *Numerical results*

The results shown here, have been obtained by applying the residual-correction method described above in two different ways. The two-point design problem was treated using an incompressible potential flow solver as well as an Euler solver as analysis code.

Incompressible potential flow

In the case of incompressible potential flow two target pressure distributions were obtained by analyzing the "High-lift" airfoil for an angle of attack of $\alpha_1 = 9.0°$ and the "Low-drag' airfoil for an angle of attack of $\alpha_2 = 1.0°$. Starting with the NACA4412 airfoil as initial guess for the geometry to be designed, a straightforward solution to the two-point design problem ($W^1 = 0.5$ and $W^2 = 0.5$) was obtained within 10 iteration steps (see Figure 7).

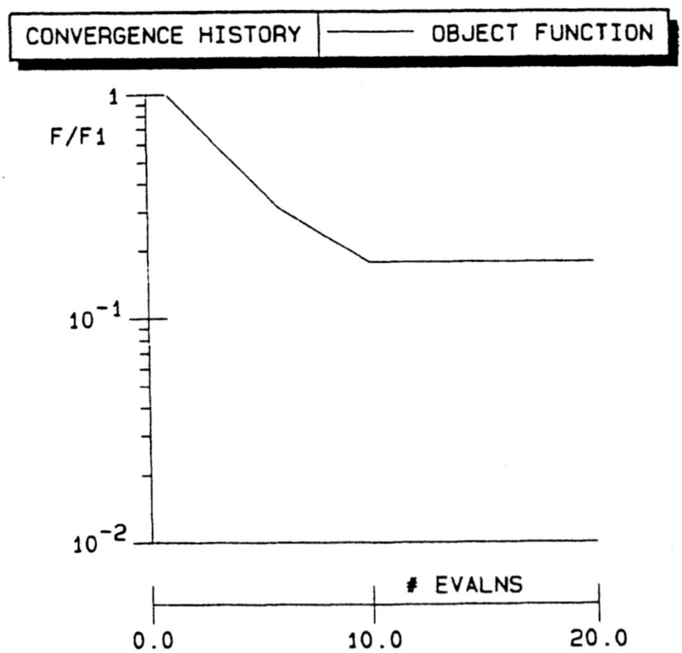

Figure 7 Test case TE4 : Convergence history, Incompressible Potential flow

A comparison of the resulting airfoil geometry with that of the "High-lift" and "Low-drag" airfoils is given in Figure 8. It may be observed that the upper part of the resulting airfoil is quite similar to the "High-lift" airfoil in the leading edge region while there is some resemblance to the "Low-drag" airfoil in the trailing edge region. The lower side of the resulting airfoil is closest to the "Low-drag" airfoil.

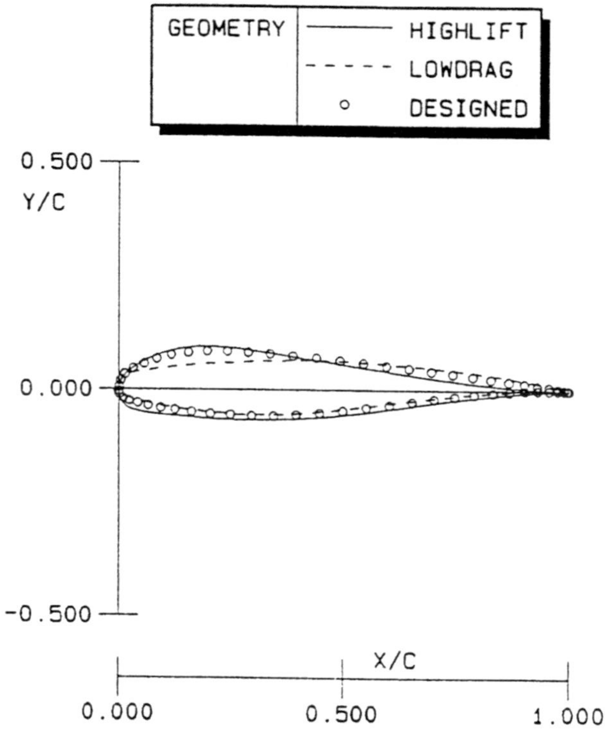

Figure 8 Test case TE4 : Resulting geometry, Incompressible Potential flow

The resulting pressure distributions are shown in Figures 9 and 10 in comparison with the "High-lift" and "Low-drag" targets. It may be observed that the deviation of the obtained pressure distributions from the targets is smallest at the airfoil lower side. The discrepancies are largest for the "Low-drag" flow condition (Figure 10). Altogether the result may definitely be seen as a compromise between two incompatible targets.

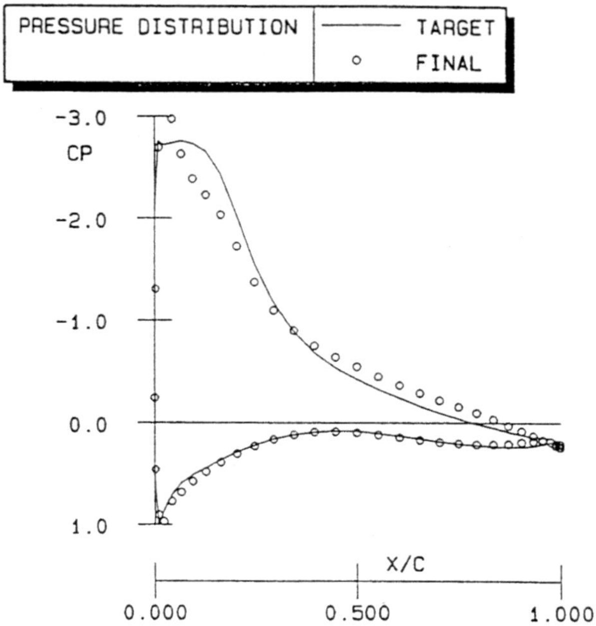

Figure 9 Test case TE4 : Pressure distribution, design condition 1 High-lift, Incompressible Potential flow

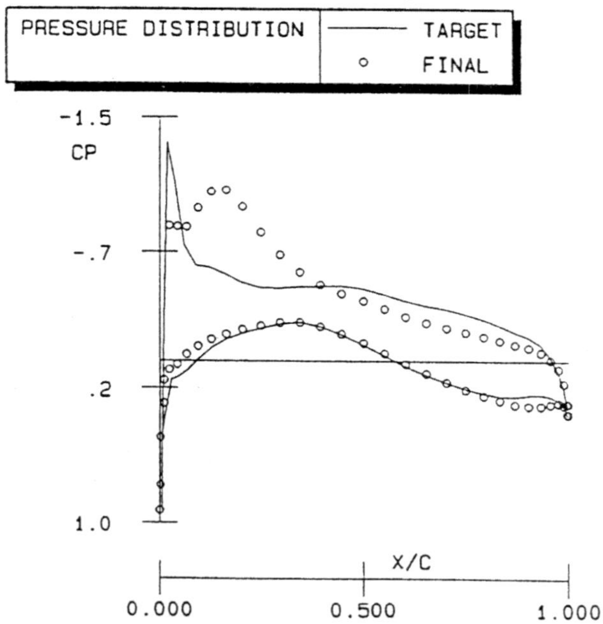

Figure 10 Test case TE4 : Pressure distribution, design condition 2 Low-drag, Incompressible Potential flow

Euler flow
In the case of the Euler flow two target pressure distributions were obtained by performing analysis computations for the "High-lift" airfoil at $\alpha_1 = 9.0°$ and $Ma_1 = 0.2$ and for the "Low-drag" airfoil at $\alpha_2 = 1.0°$ and $Ma_2 = 0.77$.

Again starting with the NACA4412 airfoil as initial guess for the geometry to be designed, the two-point design problem ($W^1 = 0.5$ and $W^2 = 0.5$) was treated. The residual-correction design process proceeded until the seventh iteration step (see Figure 7)when an unrealistic airfoil shape was obtained (distortion of the trailing edge region) giving rise to divergence of the computational process. As a consequence, fully converged results cannot be presented here.

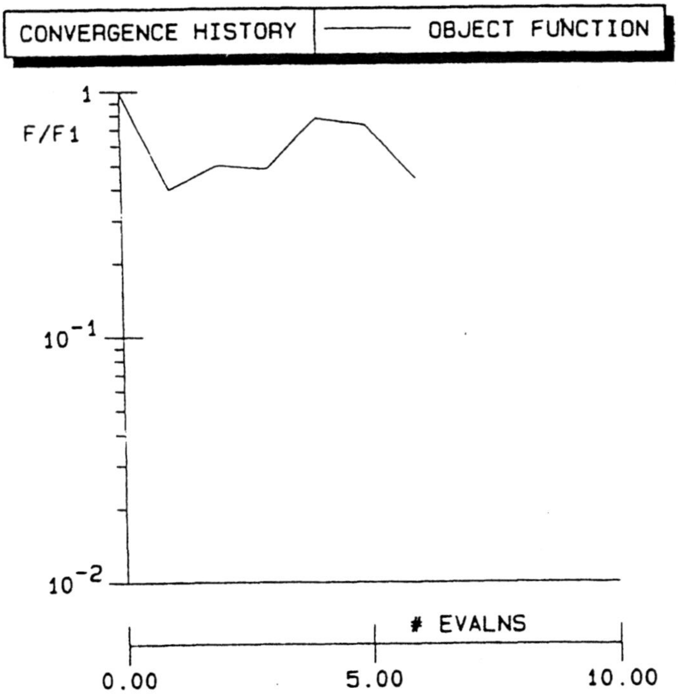

Figure 11 Test case TE4 : Convergence history, Euler flow

However, Figures 12 through 14 show qualitatively completely similar results as obtained for incompressible potential flow. Again, a compromise has been obtained. Both geometry and pressure distributions show features of the different targets.

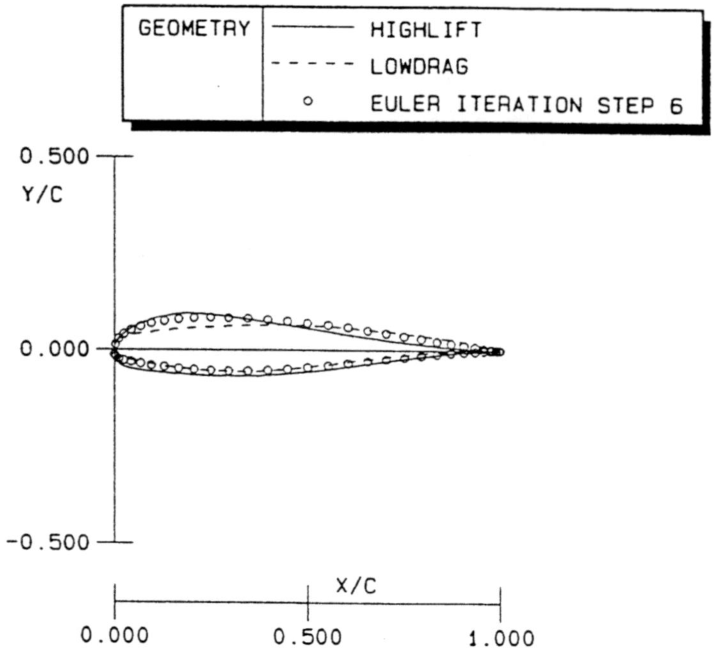

Figure 12 Test case TE4 : Resulting geometry, Euler flow

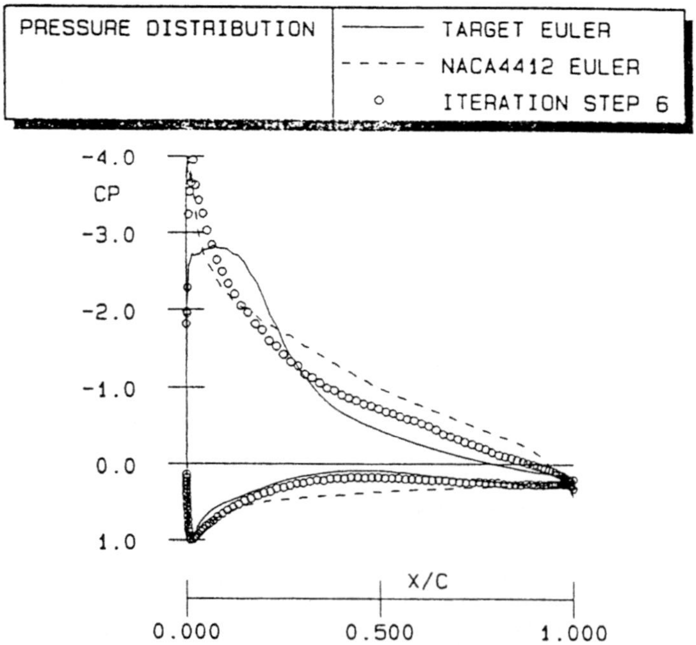

Figure 13 Test case TE4 : Pressure distribution, design condition 1 Ma = 0.2, Euler flow

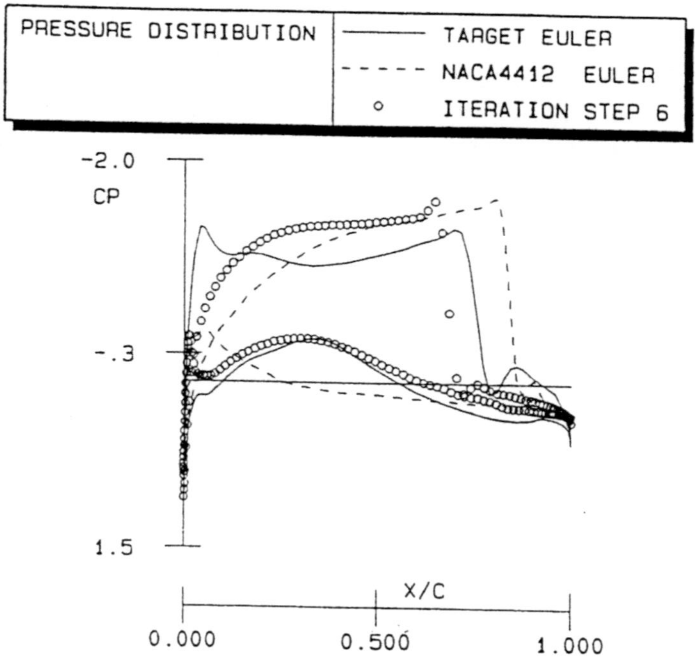

Figure 14 Test case TE4 : Pressure distribution, design condition 2 Ma = 0.77, Euler flow

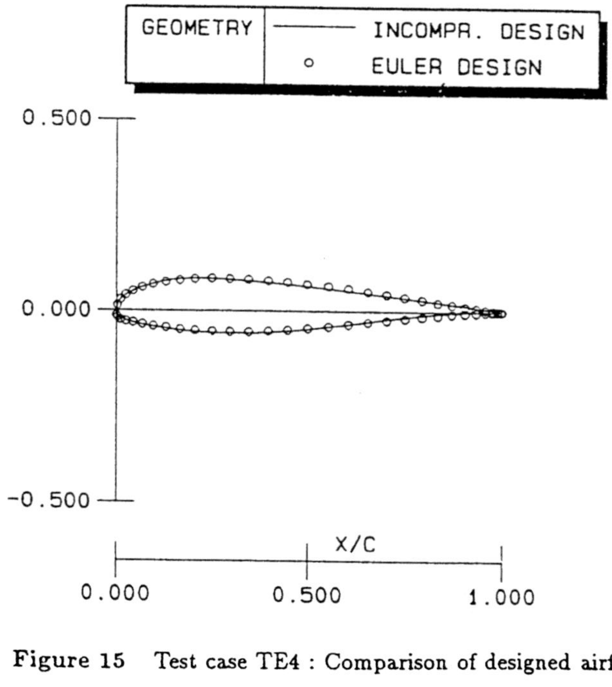

Figure 15 Test case TE4 : Comparison of designed airfoils

Comparison of "Incompressible potential flow results" and "Euler flow results"

Notwithstanding the lack of convergence of the computational process for Euler flow, a comparison of both airfoils obtained appeared to be very instructive. First of all, there is a close agreement between the shapes of the airfoils. Differences are visible only in the leading edge region in Figure 15.

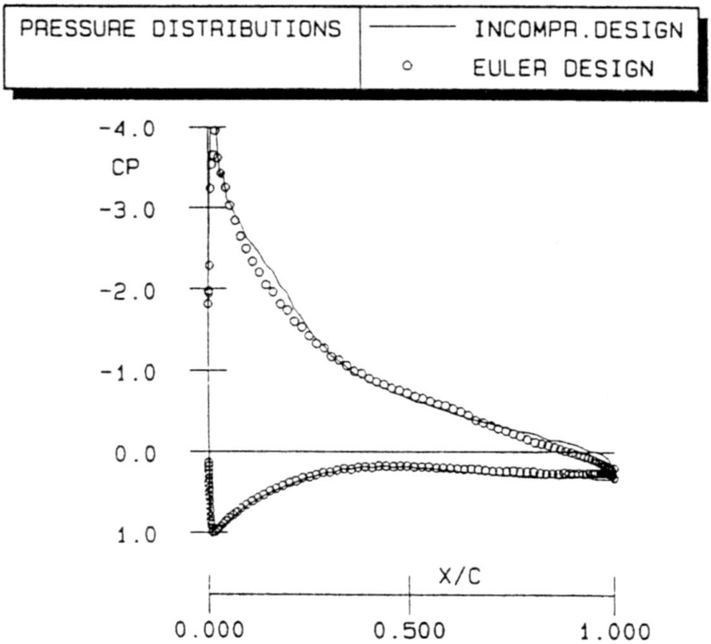

Figure 16 Test case TE4 : Comparison of designed airfoils, design condition 1 Ma = 0.2 , Euler computation

Figures 16 and 17 show the pressure distributions obtained from analysis by means of the Euler solver for both airfoils designed for the "High-lift" flow condition (Ma = 0.2) and the "Low-drag" flow condition (Ma = 0.77). Obviously, the two airfoils show a very similar behaviour at either flow condition.

Navier Stokes flow

Attempts to apply the NLR residual-correction method to the two-point design problem using a Navier Stokes code for analysis (as for test case TE2) have been reported at the Ecarp workshop, but acceptable results were not obtained. Partly, this will be due to convergence problems similar to those described for test case TE2 (see section III.2.11). But the problems may also partly be due to the fact that geometry modifications easily led to airfoils with separated flow regions at the specified transonic flow condition.

The latter fact may be illustrated by means of Figures 18 and 19. These figures present the Navier Stokes pressure distributions on the "Euler designed" airfoil in comparison with the Navier Stokes solutions for the "High-lift" airfoil (first target for Navier Stokes design) and the "Low-drag" airfoil (second target for Navier Stokes design). First of all it may be noticed that the differences between

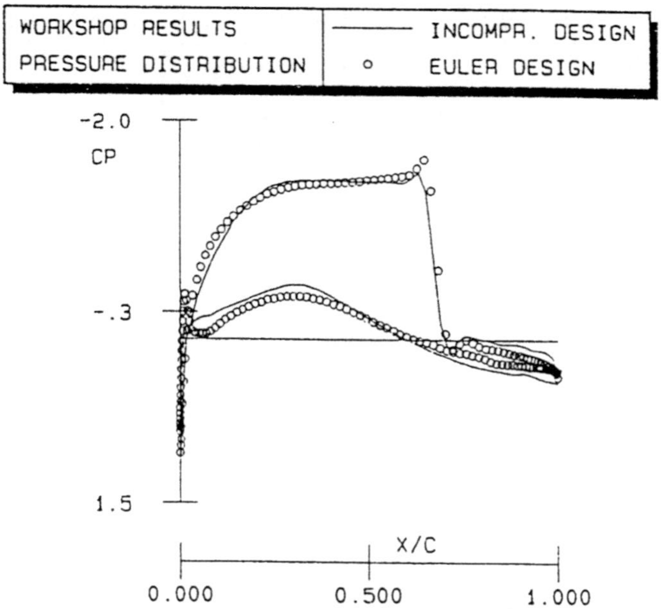

Figure 17 Test case TE4 : Comparison of designed airfoils,design condition 2 Ma =0.77 , Euler computation

obtained and target pressure distributions are of a quite similar nature as in Figures 13 and 14. Further, it may be observed that a useful airfoil is obtained for the subsonic "High-lift" design condition. But it should also be noticed that at the transonic flow condition there is a flow separation at the airfoil upper side after the shock near the trailing edge. As a consequence the designed airfoil will be not useful for the transonic design condition.

4.11.5 Conclusions

Notwithstanding the fact that the computational process did not converge when considering Euler and Navier Stokes flow ,valuable information has been obtained with respect to the multi-point design problem. The feasibility of solving the multi-point design problem on the basis of prescribed target pressure distributions has been demonstrated.

It has been shown that the result of the computations is practically independent of the complexity of the flow similutation when it is attempted to obtain an airfoil that combines the aerodynamic characteristics of two given highly different airfoils. This implies that, in cases similar to the one treated here, it should be possible to determine a good estimate of the airfoil to be designed by means of a simple incompressible potential flow design.

It may also be concluded that taking into account viscous flow effects will in general require the application of aerodynamic constraints, e.g. in order to avoid that the computations will lead to airfoils which show separated flow regions at the design conditions. In that case, of course, the resulting airfoil may differ to a much larger extent from an airfoil obtained with inviscid flow simulations.

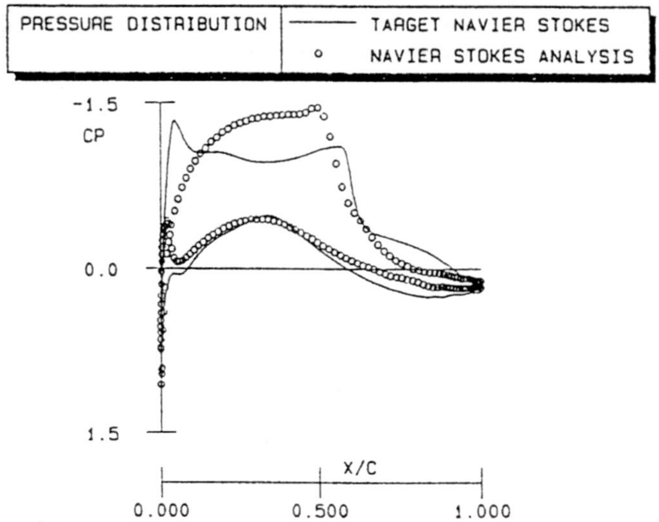

Figure 18 Test case TE4 : Analysis of designed airfoil, design condition 1 Ma = 0.2 Re=5*10^6, Navier Stokes computation

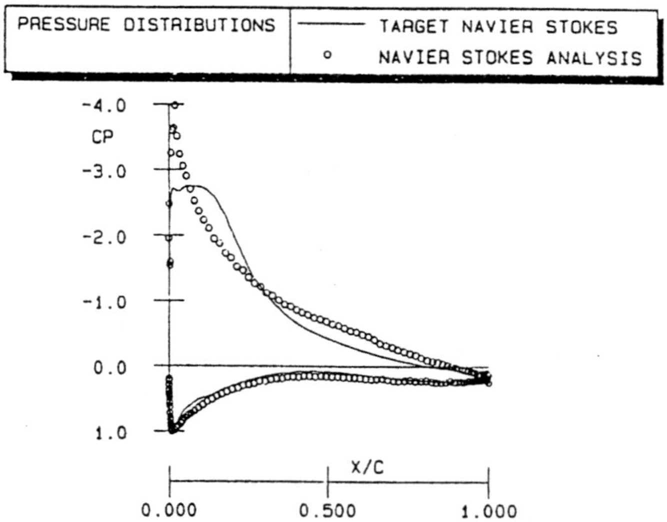

Figure 19 Test case TE4 : Analysis of designed airfoil, design condition 2 Ma =0.77 Re=1*10^7, Navier Stokes computation

10 Part II. Application of genetic algorithms to the design of airfoil pressure distributions

H. Kuiper, A. J. van der Wees, C. F. W. Hendriks and Th. E. Labrujère, NLR, Amsterdam, Holland

Summary

NLR participates in the ECARP (European Computation Aerodynamics Research Project) "Optimum Design" project, developing an algorithm for multi-point airfoil design. Within the ECARP project a test case has been defined for two-point design with the aid of well-specified target pressure distributions. It appears that optimization of pressure distributions requires the application of efficient non-gradient-based optimization algorithms. The investigation described in this report has been carried out in order to examine the possibilities of genetic algorithms.

In this report the problem description is given, as well as how genetic algorithms can be used to optimize the problem. Next, the simulator is decribed which has been used to conduct some small-scale experiments. Finally, results from these simulations are presented.

Summarized, the results are:

- Genetic algorithms can be used successfully to optimize pressure distributions.
- Software in the public domain can be used to solve the optimization problem. Only some problem specific operators have been added.
- The current implementation of the genetic algorithm generates different minima depending on the starting point.
- The results suggest that the current choice of the initial population plays an important role in the outcome of the genetic algorithm.
- Although the genetic algorithm uses about 100 times more function evaluations than the application of the Simplex algorithm to the present constrained minimization problem, the efficiency of the genetic algorithm can be improved by improving the function formulation, by specifying more appropriate initial populations and by applying problem specific mutation and recombination operators.

1 Introduction

Given the task to apply Genetic Algorithms (GAs) to the design of airfoil pressure distributions (see Ref. 2) the following approach was used:

1. Study the problem.
2. Get a suitable GAs simulator, install and study it.
3. Sketch a GA approach to the problem, that is find suitable:
   - problem coding,
   - objective (fitness) function,
   - genetic operators.
4. The fitness function must be specified and implemented; it has to be found out cq. realised:
   - what the input and output is,
   - how the implementation can interface with the GAs simulator.
5. Prepare the GAs simulator for the given coding, operators and fitness function.
6. Plan and perform first tests.
7. Evaluate and report results from first tests, and:
   - decide if knowledge (heuristic) based operator(s) should be added to the GA,
   - (if necessary) plan and perform subsequent tests.

In this report, the results from the study are described. Chapter 2 contains a short problem description. Chapter 3 describes the objective function used to evaluate candidate solutions. Next, various aspects of the type of GA used are described. The starting points for the GA optimization are described in chapter 5. The simulations itself, and the results, are described in chapter 7. The report concludes with conclusions and recommendations. Some basic knowledge of genetic algorithms and pressure distributions is required.

2 Target pressure distributions for airfoil design

Many airfoil and wing design methods are based on the solution of a so-called inverse problem. An inverse problem involves the determination of the shape of an airfoil or wing such that on its contour an a priori prescribed pressure distribution exists at the flow condition considered. In airfoil design, using these type of methods, the basic idea is that the designer will be able to translate his design requirements in terms of aerodynamic quantities such as lift, drag, pitching moment into a properly defined target pressure distribution.

Though skilful designers are capable of producing successful designs, the design efficiency can be improved by providing the designer with tools for target pressure specification. Such a tool has been developed at NLR, based on the solution of an optimization problem (see Ref. 1). This problem involves the optimization of an airfoil pressure distribution, subject to constraints on aerodynamic and geometric characteristics, using a simple parametric representation of the pressure distribution as depicted in figure 1. Here, the solution of a similar problem is considered. The characteristic pressure distribution is defined by eight coordinates, where linear interpolation

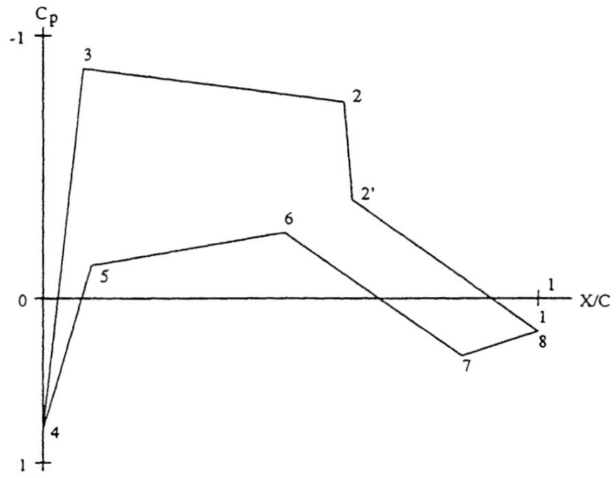

Fig. 1 Schematic representation of pressure distributions

is chosen between these coordinates to simplify the approach. When there is no shock, coordinates 2 and 2′ coincide; the shock jump between these two coordinates is determined by the local Mach number at 2. For a given free stream Mach number the coordinates 4 (stagnation pressure) and 1 and 8 (trailing edge pressures) are considered to be fixed. This leaves level and position of the coordinates 2, 3, 5, 6 and 7 (ten design variables) free to represent a large class of (simplified) pressure distributions. Figure 2 is given to get a global idea about the relation between a pressure distribution as depicted in figure 1 and an airfoil where the numbers in both figures correspond with each other.

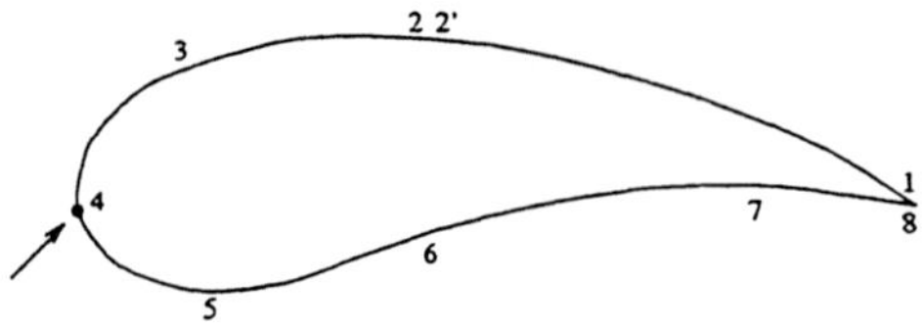

Fig. 2 Relation between airfoil and pressure distribution

In order for the genetic algorithm to evaluate various solutions, an objective function is required, which rates the quality of the solution. This objective function is described in the next chapter.

3 The objective function

An already implemented objective function will be used to evaluate the candidate pressure distributions (see ref. 2). In table 1 the structure of the input file for the objective function is given, where $(x, c_p)_i$ refers to coordinate i in figure 1.

Table 1 Structure of input file for objective function

Input	Explanation	Input	Explanation
N	Number of grid points	$x(13, 14)$	(x, c_p) of coordinate 6
$XT(i)$	X coordinates of grid	$x(15, 16)$	(x, c_p) of coordinate 7
$YT(i)$	Y coordinates of grid	$x(17, \ldots, 34)$	not active with linear interp.
$ST(i)$	arc length of grid	$x(35, 36)$	shock width upper & lower
$cps(i)$	initial pressure coefficient	$x(37)$	required thickness
$x(1)$	mode (not used)	$x(38)$	weight on thickness
$x(2)$	method (not used)	$x(39)$	required lift
$x(3)$	free stream Mach number	$x(40)$	weight on lift
$x(4)$	free stream Reynolds number	$x(41)$	weight on upper separation
$x(5, 6)$	transition point upper & lower	$x(42)$	weight on lower separation
$x(7, 8)$	(x, c_p) of coordinate 2	$x(43)$	weight on buffet onset criteria
$x(9, 10)$	(x, c_p) of coordinate 3	$(1.0, cps(1))$	(x, c_p) of coordinate 1 (fixed)
$x(11, 12)$	(x, c_p) of coordinate 5	$(0.0, cpstag)$	(x, c_p) of coordinate 4 (fixed)
		$(1.0, cps(N))$	(x, c_p) of coordinate 8 (fixed)

In table 2 the structure of the output file from the object function is given.

These structures have to be used when realising an interface between the already implemented objective function and the optimizer.

In ref. 2 the following function is used as objective function, which has to be minimized:

$$
\begin{aligned}
f(28) \;=\; & 100 \cdot f(19) && \text{drag coefficient} \\
+ \; & x(38) \cdot |x(37) - f(7)| && \text{penalty thickness} \\
+ \; & x(40) \cdot |x(39) - f(18)| && \text{penalty lift coefficient} \\
+ \; & x(41) \cdot |f(24)| && \text{penalty separation upper} \\
+ \; & x(42) \cdot |f(25)| && \text{penalty separation lower.}
\end{aligned}
$$

Where $x(38) = 1000, x(40) = 100, x(41) = 10$ and $x(42) = 10$ have been chosen, and it is stated that this function was adapted according to the needs of the optimization method. In ref. 2

Table 2 Structure of the output file from objective function

Output	Explanation
$f(1)$	$\|cps - cpt\|$; cpt defined by input points
$f(2,3)$	spare
$f(4,5,6)$	$(C_p)_{min}, (C_p)_{nose}, (C_p)_{t.e.}$
$f(7)$	current thickness according to cpt
$f(8)$	buffet onset – parameter
$f(9)$	x-position for max. buffet onset parameter
$f(10,11,12,13)$	$(fr, xr)_{upper}; (fr, xr)_{lower}$: roughness parameters/positions
$f(14,15,16,17)$	$(fs, xs)_{upper}; (fs, xs)_{lower}$: laminar in stability parameters/positions
$f(18)$	cl,lift coefficient
$f(19)$	cd, drag coefficient
$f(20)$	cm, moment coefficient
$f(21)$	cl/cd
$f(22,23)$	$(X_t)_{upper}, (X_t)_{lower}$: transition location
$f(24,25)$	separation parameters $\|cpt(1) - cpt(isep)\|$
$f(26,27)$	not active in this version
$f(28)$	objective function

the objective function evolved into the above given function because experiments showed it was best to put equality contraints on thickness, lift and no separation into a penalty function within the objective function.

4 GA approach to target pressure optimization

4.1 Rationale for using GAs

NLR participates in the ECARP project "Optimum Design" developing an algorithm for multi-point airfoil design. The algorithm is based on a residual-correction method, involving the optimization of an objective function defined in terms of target pressure distributions. Gradient-based optimization methods perform efficiently solving this problem.

Within the ECARP project a test case has been defined for a two-point design with the aid of well-specified target pressure distributions. In practical airfoil design, however target pressure distribution specification may require the application of a tool as described in Ref. 1. It has appeared that optimization of pressure distributions then requires the application of efficient non-gradient-based optimization algorithms (see Ref. 2). The present investigation has been carried out in order to examine the possibilities of genetic algorithms in this respect.

4.2 Problem encoding

As mentioned in chapter 2 the coordinates 2, 3, 5, 6 and 7 in figure 1 are the (design) variables. A natural and logic choice for the representation of a pressure distribution in a genetic search string (a so called "chromosome" used by GAs to manipulate with) is:

$$x_1 y_1 x_2 y_2 x_3 y_3 x_4 y_4 x_5 y_5.$$

In the "chromosome" x_i and y_i denote the (x, c_p) of coordinate i, see also table 1. For convenience, the coordinates have been renumbered as follows:

part of chromosome	coordinate in pressure distribution
$x_1 y_1$	2
$x_2 y_2$	3
$x_3 y_3$	5
$x_4 y_4$	6
$x_5 y_5$	7

To encode the chromosomes for use within a GA simulator, the choice is between binary and non-binary encoding. Although non-binary codings can be more practical, the theoretical properties of binary encoding are better understood. In ref. 3 it is concluded that floating point representations can be both practical and efficient. For the experiments, a floating point representation is used.

4.3 Fitness function

The implementation of the objective function as described in Ref. 2 will be used as fitness function within the genetic algorithm. Advantages of this approach are that it is not necessary to look for a fitness function and to implement it, and it makes comparison of the results with previous studies easier.

A (major) drawback could be that the objective function was specifically designed for another optimization method, which could negatively influence the performance of the genetic algorithm. This could result in a biased comparison of methods.

The fitness function should also be modified such that only valid solutions (see section 5) will be fully evaluated, as the fitness function was not originally designed to handle all possible input values. Through recombination and mutation the GA will generate invalid solutions. These invalid solutions should receive a minimal fitness value to prevent them to propagate through the population.

4.4 Genetic operators

The first test phase will be done using the three basic genetic operators: selection, crossover and mutation. The crossover and mutation operators will be enhanced with some problem specific knowledge to be described below.

Besides using the below described genetic operators, it is also possible to add operators exploiting available heuristic knowledge or other optimizations techniques, for example local search. Another possibility is to realise a hard contraint in an operator, where an operator assigns the lowest possible fitness to those chromosomes that do not obey the constraint.

4.4.1 The selection operator

The selection operator will be the classical fitness proportional random selection. This process of selection can be imagined as the spinning of a wheel of fortune to point out the chromosomes that will make it into the next generation, where each chromosome gets a part on the wheel proportional to it's fitness. Scaling of the fitness value can be applied in order to ensure that solutions close the overall optimum.

4.4.2 The crossover operator

Although there are a number of variants of the crossover operator, one has been chosen that is most suitable for the continuous variables used in the chromosomes. This operator, called *intermediate recombination* (see Ref. 4) is based on the principle of averaging:

$$\begin{aligned} x_i^a &:= x_i^a + \lambda(x_i^b - x_i^a) \qquad & y_i^a &:= y_i^a + \lambda(y_i^b - y_i^a) \\ x_i^b &:= x_i^b + \lambda(x_i^a - x_i^b) \qquad & y_i^b &:= y_i^b + \lambda(y_i^a - y_i^b). \end{aligned}$$

Where $\lambda \in [0, 1]$, a and b are the parents for the operation.

In the given context, this crossover operator brings variation into the population, which otherwise would only be done by the mutation operator (see section 4.4.3). Also, this crossover explores the search space inbetween two solutions. The fraction λ indicates a difference of sensitivity for changes of the parameters. If this sensitivity cannot be quantified, these fractions could be chosen randomly (resulting in *random intermediate recombination*).

4.4.3 The mutation operator

The mutation operator supplies the noise to avoid that the optimization process gets stuck in a local optimum. On one hand it must supply enough noise to get out of local optima and on the other hand it may not add too much noise, in order to keep search directed towards a global optimum. Too much noise (mutation) can lead to pure random search.

In case of a binary coding bits are flipped with a certain probability by the mutation operator. For the proposed floating point representation an alternative has to be found. A possibility is to mutate x_i with a certain probability by: $x_i := x_i + \xi_i$. Where $\xi_i \in [-\varepsilon_i, \varepsilon_i]$, and ε_i small. This approach can also be used for y_i. In first instance fixed mutation intervals shall be used where $0.0 \leq \varepsilon_i \leq 0.1$.

The quality of the pressure distribution reacts differently on changes in different coordinates of figure 1. Therefore it might be necessary to use different values for ε_i for each of the design variables.

5 Planning of first tests

The test problem has been adopted from ref. 2, chapter 6, where a pressure distribution of an airfoil is optimized for a laminar flow condition. The freestream Mach number is 0.65 and the Reynolds number 15 million. Initial pressure distributions to start the optimization process with have been taken from subsection 6.4.4 of ref. 2:

	Initial pressure distributions	
Coordinates	B'	C
(x_1, c_p)	(1.000, 0.200)	(1.000, -0.200)
(x_2, c_p)	(0.740,-1.000)	(0.8941, -0.9942)
(x_3, c_p)	(0.030,-0.610)	(0.03716,-0.624)
(x_4, c_p)	(0.000, 1.100)	(0.000, 1.100)
(x_5, c_p)	(0.137,-0.270)	(0.2246, -0.3974)
(x_6, c_p)	(0.605,-0.650)	(0.4542, -0.8225)
(x_7, c_p)	(0.900, 0.205)	(0.900, 0.205)
(x_8, c_p)	(1.000, 0.205)	(1.000, 0.205)

Figure 3 shows a depiction of these starting points.

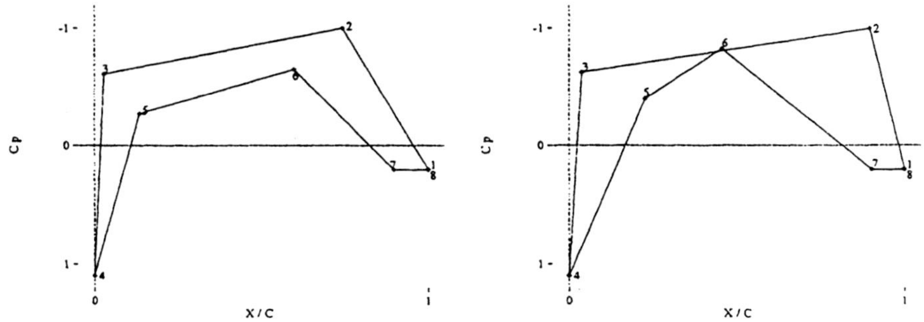

a) Starting point B'.　　*b) Starting point C.*

Fig. 3 Starting points for the optimization process

Constraints for the design variables are:

$$0.02 \leq x_i \leq 1.0 \qquad c_{p,i} \leq 1.0 \quad i = 2, 3, 5, 6$$
$$0.85 \leq x_7 \leq 0.95 \qquad 0.1 \leq c_{p,7} \leq 0.25.$$

The latter condition on $(x_7, y_{p,7})$ requires that the pressure distribution has "rear loading" at the trailing edge. Although the constraint $x_4 < x_5 < x_7 < x_8$ and $x_4 < x_3 < x_2 < x_1$ has not yet been prescribed explicitly in any optimization study for the problem considered, it will be used with the GA, in order to prevent non-valid solutions to be evaluated (other optimization methods presumably do not generate solutions that violate this constraint).

The following weights for the penalty function in the objective function will be taken (as in subsection 6.4.4 of ref. 2):

$x(37)$	=	0.18	required thickness
$x(38)$	=	1000.00	thickness weight
$x(39)$	=	0.50	required lift
$x(40)$	=	100.00	lift weight
$x(41)$	=	10.00	upper separation weight
$x(42)$	=	10.00	lower separation weight

The starting points have the following features:

	Initial pressure distribution		"Target"
Input variables	B'	C	values
cd-value	$35.1e-4$	$36.6e-4$	min
cl-value	.4627	.5480	.50
thickness	.1648	.1713	.18
separation	no	yes	no
objective function	19.23	16.13	min

Consequently, both starting points have a thickness which is too low. Moreover, starting point B has a too low cl-value, and starting point C has separation at the trailing edge. The objective function values are about equal. There exists a feasible solution (appropriate cl-value and thickness with no separation) which in ref. 2 only the Simplex method could find. The cd-value of this feasible solution is $34.75e{-}4$. In ref. 2 the optimization package MINOS, which makes use of derivatives in the optimization process, could not find a feasible solution. It is speculated in ref. 2 that the surface spanned by the feasible solutions is probably very complex and irregular and therefore not suitable for gradient-based optimization methods.

6 Initial experiments

6.1 Selection of simulator

In order to investigate whether GAs are suitable for the problem described, a small number of simple experiments have been conducted. For this purpose, a number of public domain GA simulators have been studied.

The GENEsYs simulator has been selected (see ref. 5). This simulator provides various selection, mutation and recombination operators. Also, usage of non-binary variables is supported (although they are encoded as binary strings).

Two major additions are made to the simulator: a new mutation operator (as described in section 4.4.3) and a new input facility to use an initial population in non-binary format (i.e. a text file containing floating point numbers – the original facility expected binary strings).

The FORTRAN code which implements the fitness functions has also been modified slightly. The generation of output is reduced for performance reasons, and coding is modified to facilitate calling the fitness function from a "C"-environment.

6.2 General parameter settings

The following settings are used during all simulations:

- Population size: 100
- Proportional selection scheme (see section 4.4.1)
- Mutation scheme as described in section 4.4.3, with $\varepsilon_i = 0.1$, and mutation probability of 0.005 (which means for each *variable* there is a probability of 0.005 that it will be mutated)
- Random intermediate recombination (thus λ is chosen at random), with a crossover application rate of 0.6 (which means that 60% of the population will be recombined for each new generation)
- Scaling of fitness values
- Maximum of 100000 function evaluations per simulation
- Elitist strategy: the best performing solution will always survive into the next generation, therefore ensuring the best solution sofar will be saved.

Three experiments are performed using these parameters. Only the initial population is varied for each of the experiments.

6.3 Results for starting point B'

In the first experiment an initial population of 50% random members and 50% members representing starting point B' (see Fig. 3) is used.

After the first simulation run, the best solution found has an objective function value of .456. In order to investigate whether a better solution could be found, a second simulation run is started using the final population of the first simulation. This results in a value of .445. Obviously, hardly any progress is made in the second simulation. The final solution is shown in figure 4. The corresponding cd-value is 27.3e-04.

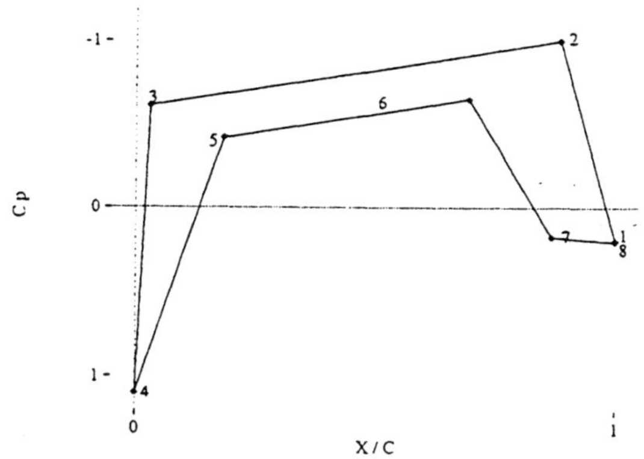

Fig. 4 Solution when starting from B'

6.4 Results for starting point C

In the second experiment an initial population of 50% random members and 50% members representing starting point C (see Fig. 3) is used.

After the first simulation run, the best solution found has an objective function value of .381. In order to investigate whether a better solution could be found, a second simulation run is started using the final population of the first simulation. This resulted in a value of .369. Again, hardly any progress is made in the second simulation. The final solution is shown in figure 5. The corresponding cd-value is 34.3e-04.

6.5 Results for starting points B' and C

Finally, an experiment is done using an initial population consisting of 50% members representing starting point B' and 50% representing point C. After one simulation run, the best solution found has an objective function value of .312. This solution is shown in figure 6. It is obvious

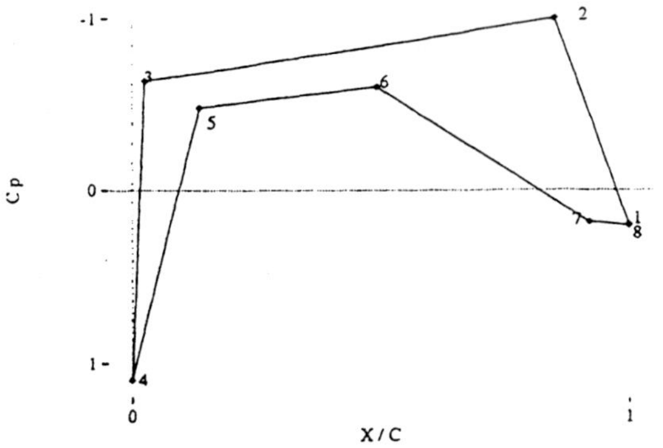

Fig. 5 Solution when starting from C

from an aerodynamic point of view, however, that something is wrong with the solution found. Apparently a difficulty has been found in the implementation of the object function, since the solution presented cannot exist without separation at the trailing edge. Whence the solution found is not feasible, although the implementation of the object function fails to signal the constraint violation. The solution found is therefore discarded.

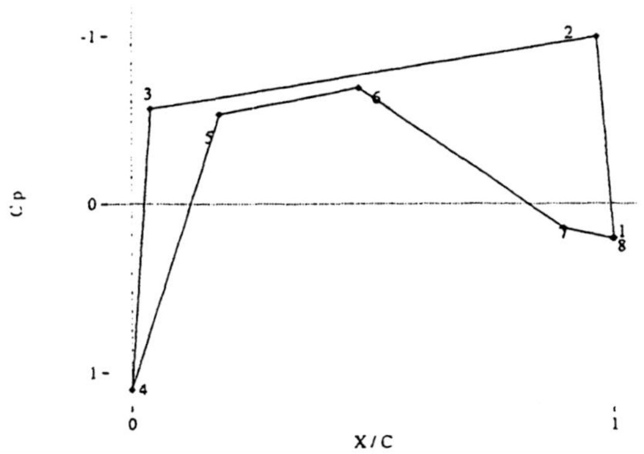

Fig. 6 Solution when starting from B′ and C

From the convergence history of this simulation shown in figure 7, it can be observed that the fitness values reach an extremum after approximately 100000 function evaluations (maximum per simulation).

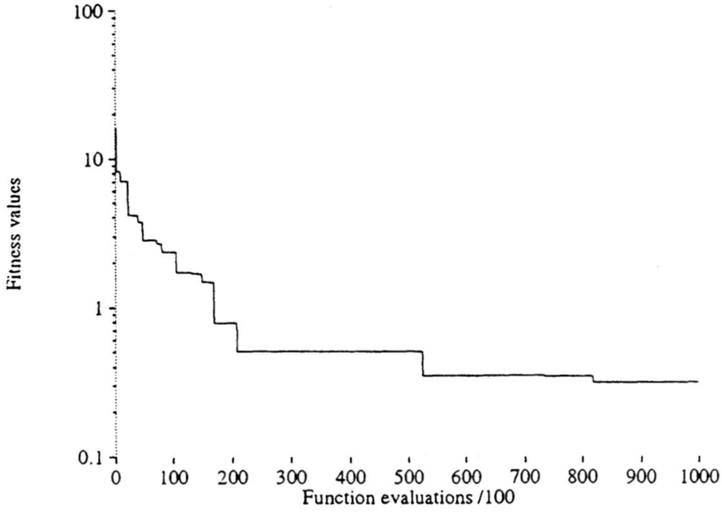

Fig. 7 Convergence history of solution shown in Fig. 6

7 Conclusions and recommendations

The present study deals with the application of a genetic algorithm to the constrained minimization of the airfoil drag coefficient; the constraints are:

- lift coefficient = 0.5
- minimum pressure coefficient $\geq$ -1.01
- maximum thickness to chord ratio = 0.18
- no boundary layer separation.

The Reynolds number is taken equal to 15 million, and the laminar-turbulent transition is modelled by the method of Granville (ref. 6).

From the results of this study it is concluded that:

1. Software available in the public domain could be used to solve the optimization problem using genetic algorithms. A problem-specific mutation operator is added to the software (see section 6.1).
2. The genetic algorithm generates different minima depending on the starting point; these minima do satisfy the constraints to a good approximation. The following minima are found:

Genetic algorithm	Starting from B'	Starting from C
Minimum cd-value	0.00273	0.00343

   In ref. 2 the simplex method is applied to the same, constrained minimization problem. The following results are reported:

Simplex algorithm	Starting from B'	Starting from C
Minimum cd-value	No minimum found	0.00348

   The small discrepancy between the minimum cd-value when starting from C is due to the approximate treatment of the constraints.
3. The pressure distributions found, starting from B' and C with the genetic algorithm and starting from C with the Simplex method are plotted in figure 8.
   The differences between the pressure distributions obtained are small at the airfoil upper surface. The differences are significantly larger at the lower surface. The corresponding laminar-turbulent transition locations at the lower side of the airfoil are as follows:

	Transition location in terms of x/c		
Algorithm	at initial iteration	at last iteration	minimum *cd*-value
Genetic from B'	0.62	0.70	0.0027^3
Genetic from C	0.48	0.52	0.0034^3
Simplex from C	0.48	0.52	0.0034^8

The differences in minimum *cd*-value correlate with the laminar-turbulent transition point location.

The different transition point locations found with the genetic algorithm, starting from B' and C, correlate with the different transition point locations at the two starting points. This observation suggests that the current choice of the initial population plays an important role in the outcome of the genetic algorithm.

4. The genetic algorithm uses about 100 times more function evaluations than the application of the Simplex method reported in Ref. 2. The Simplex method gives better convergence within a fixed number of iterations. The efficiency of the present application of genetic algorithms can be improved as follows:
   - by improving the formulation of the objective function; i.e. the problem constraints might be reformulated to better suit genetic algorithms, because currently it is necessary to adjust more than one parameter at a time in order to get a new solution that is still within the constraints;

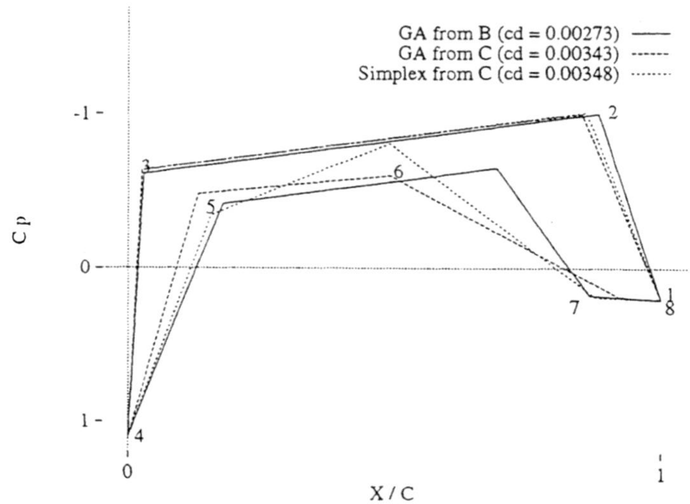

Fig. 8 Comparison of pressure distributions

- by specifying a more appropriate initial population, either by removing the randomness from the initial population (using known valid solutions as starting point, as is done when starting from BC'), or by first generating (randomly) an initial population satisfying the constraints;
- by applying problem specific mutation and recombination operators, since the convergence of the genetic algorithm depends highly on only the mutation operator.

8 References

1. J.A. van Egmond, *Numerical optimization of target pressure distributions for subsonic and transonic airfoil design*, AGARD CP 463, 1989.
2. J.P. van Wageningen and A.J. van der Wees, *Selection of an optimization for highly nonlinear problems. Evaluation of MINOS for some standard testproblems and an aerodynamic design problem*, NLR TR 91336 L, 1991.
3. C.Z. Janikow and Z. Michalewicz, *An experimental comparison of binary and floating point representations in genetic algorithms*, Proceedings of the fourth international conference on genetic algorithms, Morgan Kaufmann Publishers Inc., 1989.
4. H-P. Schwefel, *Numerical Optimization of Computer Models*, Wiley, Chichester, 1981.
5. Th. Bäck, *A user's guide to GENEsYs 1.0*, University of Dortmund, 1992.
6. P.S. Granville, *The calculation of the viscous drag of bodies of revolution*, David Taylor Model Basin Rep. 849, 1953.

11 Design and Optimization Aspects of 2D and Quasi-3D Configurations Using an Inverse Euler Solver

P. Chaviaropoulos, V. Dedoussis and K. D. Papailiou, NTUA, Athens, Greece

Abstract

This paper presents the extension of a 2D single-pass inverse design method to quasi-3D multi-block configurations. The flow is considered to be axially symmetric, compressible and rotational due to inflow stagnation temperature, entropy and/or swirl non-uniformities. The method is based on the potential function/streamfunction formulation. Clebsch decomposition is used to model the effect of rotationality. The present inverse inviscid approach can be used for viscous design. This is achieved via a weak viscous-inviscid interaction technique and by optimizing, in terms of drag, the "target pressure" distribution. Within the framework of the ECARP-Optimum design Workshop, the method has been applied to three test cases concerning a duct, an airfoil and an afterbody engine configuration.

I Introduction

A central issue in applied aerodynamics is the problem of determining the shape of the walls of an aerodynamic component on which the pressure (or velocity in inviscid flows) distribution is prescribed. This inverse problem is usually referred to as the target pressure one. Over the years quite a few methods have been developed for its solution. The cornerstone of inverse approaches is the method developed by Stanitz [Stanitz53] in which the governing equations are transformed employing the potential function and the stream function as natural body-fitted coordinates.

Potential/stream function inverse methods were developed for potential flows only. Rotational flows have received only limited attention mainly because of the brakedown of the concept of the potential function as such. The difficulty can be circumvented using the Clebsch formulation to decompose the velocity vector into a gradient-type potential part and another rotational part. The present authors [Dedoussis93] used this technique to solve the rotational 2D inverse problem for internal flow configurations. Koumandakis [Koumandakis94] extended the above method for the design of axisymmetric annular ducts. This latter work is further extended here, to tackle multi-block quasi-3D flows. A typical example of such a flow, which has been included in the workshop test cases, is the one encountered in the afterbody region of aircraft engines.

In the present method the main governing equation is expressed in terms of the magnitude of the meridional velocity component. It is derived using the defining relations of the potential and the stream function and employing differential geometry principles for the mapping-transformation of the physical space on the natural one. This nonlinear partial differential equation for the velocity is solved in conjunction with a transport equation for the drift function. The numerical integration of the equations is carried out on an auxiliary computational domain and the geometry sought is determined through the integration of Frenet equations along the computational grid lines. A difficulty inherently associated with axisymmetric flow conditions is that the velocity equation is coupled with the radial physical coordinate. This implies that the flow and geometry calculations are coupled. Complete discussion of the theoretical aspects of the method are included in [Koumandakis94] and [Chaviaropoulos94] and are briefly repeated here for the sake of completeness.

Inverse problem formulations assume an *a priori* specified target pressure distribution. This target pressure distribution is directly related to the behaviour of the boundary layers developing on the surface of the aerodynamic component. Considering that viscous effects are not dominant at the design point, one may take them into account through a weak viscous-inviscid interaction scheme. In this case, the target pressure optimization problem can be decoupled from the shape optimization. "Good" target pressure distributions may be, then, obtained using a simple boundary layer optimizer which minimizes some appropriate aerodynamic cost function, the viscous drag for instance, under certain constraints. This approach is applied for optimizing the convergent-divergent symmetric duct of the workshop.

The above mentioned technique can also be used with quite success in strong viscous-inviscid interaction cases, if the design shape is appropriatelly modified for the boundary layer displacement effect. This is implicitly demonstrated in the viscous flow airfoil reproduction test case of the workshop.

II Brief Description of the Simple Domain Inverse Euler Solver

1 Assumptions and Basic Equations

The method concerns steady, subsonic, inviscid and adiabatic flows of a perfect gas with axial symmetry. Although gradients in the peripheral direction vanish, non-zero peripheral velocity component V_u is assumed. The entropy as well as the total enthalpy level of different stream surfaces (or better meridional plane streamlines) may be different.

The natural coordinates employed in the method are the potential ϕ and the stream function ψ on the meridional plane.

A stream function may be defined as:

$$\hat{\rho}\vec{V}_m = \nabla_m \psi \times \vec{k} \quad , \quad \hat{\rho} = r\rho \tag{1}$$

where $\vec{k}$ is the unit vector in the peripheral direction. The definition of the stream function in axisymmetric flows is analogous to the usual two-dimensional one with the exception that the density is replaced by the modified density $\hat{\rho}$.

Clebsch formulation is used to decompose the meridional velocity vector to an irrotational and a rotational part. The rotational part is expressed in terms of the meridional gradient of the stream function. The Clebsch decomposition of the meridional velocity vector reads:

$$\vec{V}_m = \nabla_m \phi + \alpha \nabla_m \psi. \tag{2}$$

From the momentum and energy equations it may be seen that h_t, s and rV_u are conserved along meridional streamlines. It is evident, therefore, that their meridional gradients are parallel to the stream function gradient. Introducing the corresponding proportionality coefficients λ, μ and ν the following transport equation is deduced that

$$\vec{V}_m.\nabla_m \alpha = -\lambda + \mu T + \nu \frac{V_u}{r} \tag{3}$$

where T is the temperature and coefficient α is usually termed as the drift function.

2 Governing Equations

2.1 Velocity Equation

An equation for the magnitude of the meridional velocity component is obtained from the zero-curvature metrics compatibility condition, which has to be satisfied by any parametrization of the physical space, including the (ϕ, ψ) natural coordinates one. Referring to our previous work [Chaviaropoulos94] the relevant velocity equation reads:

$$a(lnV_m)_{\phi\phi} + b(lnV_m)_{\phi\psi} + c(lnV_m)_{\psi\psi} + d(lnV_m)_{\phi} + e(lnV_m)_{\psi} = f, \tag{4}$$

where subscripts ϕ and ψ indicate corresponding partial derivatives. Coefficients a,b,...,f are functions of the modified density and the drift function. The thermodynamic density is computed via the state equation in terms of stagnation thermodynamic quantities which are conserved along the streamlines.

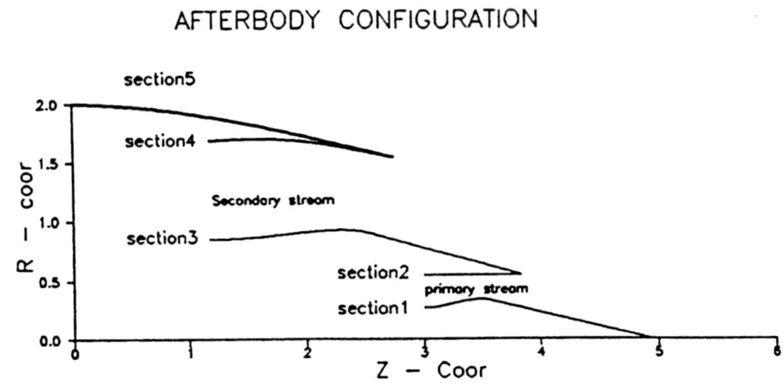

Figure 1 Typical by-pass afterbody configuration

2.2 Drift Function Equation

The transformation of the drift function equation in the natural coordinates space yields

$$V_m{}^2 \alpha_\phi = -\lambda + \mu T + \nu \frac{V_u}{r}. \tag{5}$$

2.3 Geometry Computation

In the previous sections it has been emphasized that the flowfield and the geometry solutions are inherently coupled via the radial coordinate. The calculation of the geometry as well as the integration of the flowfield equations are carried out on an auxiliary computational grid. The geometry is determined by integrating the generalized Frenet equations along the computational grid lines [Chaviaropoulos94] .

III Extension to Afterbody (Multi-Domain) Configuration

1 Preliminaries

The vast majority of modern military and commercial aircraft engines are by-pass rather than simple engines. In by-pass engines the thrust is effected via the combination of more than one backward directed jets. Usually, the thrust generating jet is composed of two simple jets. These jets, as well as the afterbody of a typical modern aircraft by-pass engine configuration, is shown schematically in figure 1.

The primary stream is the high momentum (and energy) stream of flow. This part of the flow comes directly out of the combustion chamber and the turbine of the engine. Since this part of the flow is the one with the highest energy content is appropriately termed as "hot". Compared to the "cold" stream which is associated with the free stream flight conditions, the "hot" stream has increased total temperature (energy) and entropy (flow losses in the turbomachine, combustion). Before leaving the engine this part of the flow has to be accelerated as much as possible in the primary nozzle, so that the exiting stream to have high momentum and thus produce a high thrust. Modern afterbody configurations accelerate the primary stream up to the supersonic flow regime using a convergent-divergent primary nozzle. The expansion of the flow at the nozzle exit creates a strong shock wave there, which increases the drag of the device considerably. One of the design aims, therefore is to reduce the shock drag by avoiding, for example, the formation of strong normal shocks. The secondary, "warm", stream of air which is processed only by the fan of the engine, serves to prohibit the uncontrolled expansion of the hot stream to the "cold" surrounding free-stream air and align, this way, the thrust vector with the axis of the machine. This secondary stream is also accelerated in the secondary nozzle up to the high subsonic flow regime and considerably contributes to the total thrust.

Due to the different level of total thermodynamic properties of the different streams attained via the processing of the air by different parts of the engine upstream, the flow has a strongly rotational character even if the viscous effects are neglected. In that respect, the simplest flow model which may simulate the flow quantities in (and downstream) the afterbody with acceptable accuracy is that of the Euler equations. Considering that the flow is inviscid the total properties differences existing along the inflow boundaries are convected without any attenuation downstream. There is obviously a discontinuity of the total properties along the lines separating the primary, secondary and free streams. These discontinuities imply the existence of vorticity sheets along these lines which emerge from the corresponding trailing edges of the afterbody.

2 Numerical Aspects

The analysis and the design of afterbody configurations was based on two codes, a direct and an inverse target pressure one. The basic features of the latter have been described in the previous sections. Both methods may treat rotational flows with strong incoming vorticity gradients and may operate in the transonic flow regime using the artificial density concept. The direct code applies the vorticity-streamfunction formulation for the kinematic field and solves the total enthalpy and entropy transport equations for the thermodynamic field [Kiousis92]. Both codes use an implicit multi-domain solver which treats the three streams in a concurrent manner.

In order to increase the flexibility of the numerical simulation, the afterbody configuration is split into three individual computational domains, corresponding to the "hot", "warm" and "cold" regions. A structured grid is generated into each sub-domain using simple interpolation schemes. The number of grid points in both the longitudinal and transverse directions may be different in each sub-domain. The accuracy of the computation is improved by selecting uniform grid spacing in the transverse direction in the primary and secondary streams and by stretching the grid in the near-trailing edge regions. The same grid generation strategy is used for both the direct (grid on the meridional plane) and the inverse (grid on the natural plane) solvers. The governing equations are discretized using centered finite-difference/finite-volume schemes. The numerical integration of the discretized equations is performed using an implicit scheme based on the preconditioned linear restarting GMRES [Saad83] algorithm. An incomplete LU preconditioner [Zedan83] is employed for both the direct and the inverse computations. The residuals of the governing equations are computed in each individual subdomain and, thenafter, stored in a global array. In order to assure the implicit treatment of the subdomain interfaces the LU preconditioner uses a global enumeration acting on the complete grid structure. This procedure is facilitated by the fact that the governing elliptic type equations have Dirichlet boundary conditions (stream function equation for the direct solver, V-equation for the inverse solver) along the solid walls. In that respect, each interface grid line (including the corresponding wall section) is stored only once in the global array.

Numerical experimentation has shown that the thermodynamic total properties discontinuities along the stream interfaces cause non-physical Mach number overshoots. For this reason these discontinuities are smoothed out by adding a second derivative term in the stagnation enthalpy, entropy and swirl transport equations. This second derivative term which is added in the transverse direction simulates to some extend the natural diffusion process of the real viscous flow. This strategy of smoothing the thermodynamic properties used in both the direct and the inverse codes.

It should be emphasised that the inverse target pressure approach proposed, is a single-pass one. In that respect, the cost associated with the inverse -design- computation is almost equivalent to the cost of the direct -analysis- code.

IV Results and Discussion

1 Afterbody Reproduction Test Case

Defining the afterbody reproduction test case, special attention has been paid in specifying realistic inflow conditions for both the primary and secondary streams. The kinematic and thermodynamic inflow quantities, which are used as boundary conditions, are typical for 10,000 ft altitude flight. Comparison of the computed temperature and pressure inflow values for the three streams indicate that the entropy level of the hot stream is quite higher than that of the cold stream. This is also true for the total temperature level.

The flow field was computed for the specified boundary conditions, using a

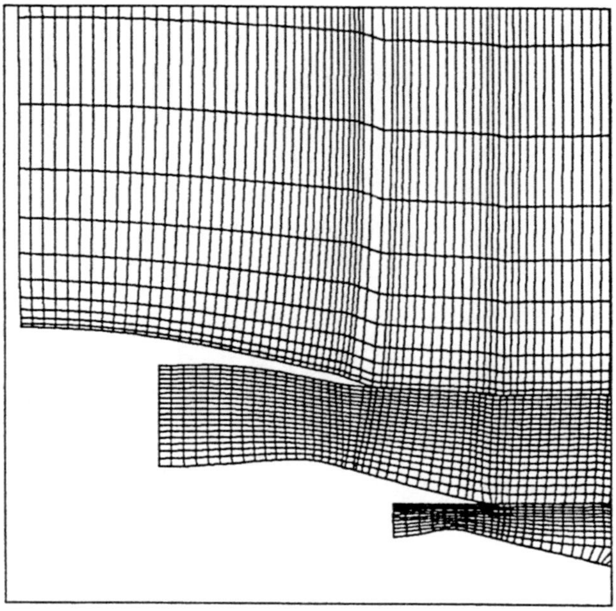

Figure 2 Structured grid derived from the inverse computation

direct Euler solver which employes the vorticity-stream function formulation. The Mach number distribution is characterized by a strong normal shock which is formed in the near exit region of the primary stream.

Using the surface pressure distribution of the direct solver as the target pressure boundary condition, the geometry was reproduced by the multi-block axisymmetric inverse solver. The calculated streamlined structured grid in the region of interest, is shown in figure 2.

The iso-mach contours of the inverse calculation are shown in figure 3. The presence of the strong normal shock in the near-exit section of the "hot" stream is evident.

The comparison between the reproduced and the original surface geometry of the axisymmetric afterbody sections is presented in figure 4. One may comment that the accuracy of the reproduction is quite satisfactory, considering that the final result accumulates the error of both the direct and the inverse solvers.

2 Viscous Drag Minimization Test Case

The viscous drag minimization problem is treated for a symmetric 2D convergent-divergent duct. The inflow and throat areas are kept fixed while optimizing the duct's wall geometry. The inflow Mach number is 0.2 and the Reynolds number is $2 * 10^6$ per unit length. The flow is considered to be fully turbulent. The viscous drag (pressure and skin friction) is estimated using the Squire and Young

Figure 3 Iso-mach contours calculated by the inverse solver

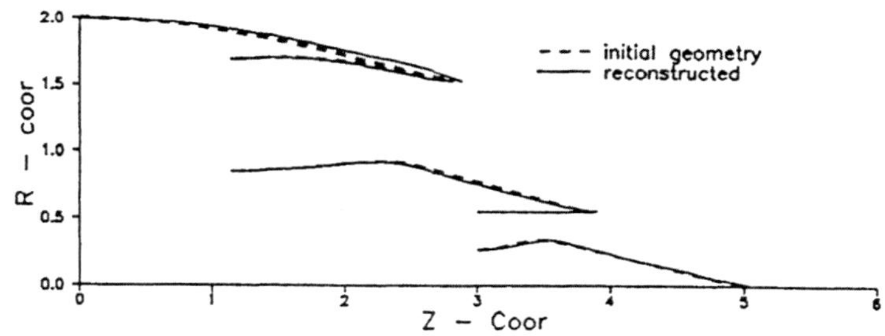

Figure 4 Comparison of initial and reproduced geometries

[Squire38] formula.

The optimization problem is treated in two steps. First the wall pressure distribution is optimized in the area of interest using an integral boundary layer solver (Head's lag-entrainment method, [Cebeci77]) coupled to a simplex based optimizer [Press89]. Starting from the pressure (or external velocity) distribution which corresponds to the initial (sinusoidal) geometry its values at the inlet outlet and throat sections, as well as the inlet and outlet slopes are kept fixed through constraints within the optimization process. These constraints are introduced in the cost function through penalties. The wall velocity distribution is parametrized using a small number of free parameters N combined with cubic

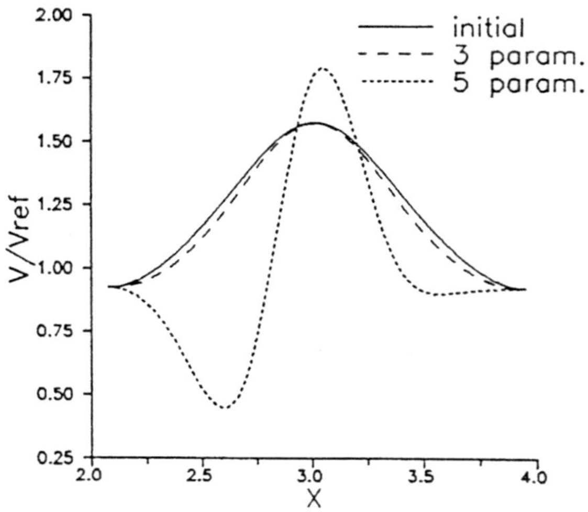

Figure 5 Initial and optimized velocity distributions

spline interpolation. For both examined cases (N=3 and N=5) the optimizer needed an order of one hundred boundary layer solutions before converging to the minimum drag value. This costs only a few CPU minutes on a PC-486 clone personal computer since one boundary layer pass is very rapid. The drag of the initial geometry was estimated to 20 drag counts. The N=3 optimization reduces the drag to 19 counts without significantly changing the upper wall geometry. This means that the initial geometry is already a good choice. When the number of free parameters is increased (N=5) a wavy optimized velocity distribution is obtained and the character of the flow is significantly changed. The optimized velocity is characterized by an early deceleration which draws the flow near separation, an accelerating part up to the throat velocity value and a steep deceleration after that, followed by a last relaxing part. The estimated drag drops to 15 counts but it is not evident that the Squire and Young formula is valid for such pressure distributions. The optimized external velocity distributions for 3 and 5 free parameters are compared to the initial velocity in figure 5. The boundary layer form factor H which correspond to the above three cases are compared in figure 6.

Once the optimum pressure distribution is obtained the inverse inviscid 2D solver [Dedoussis93] provides the corresponding optimized geometry using the optimal pressure distribution as the target pressure. A 141x41 grid was used in both direct and viscous computations. The resulting geometries for N=3 and 5 are compared to the initial one in figure 7.

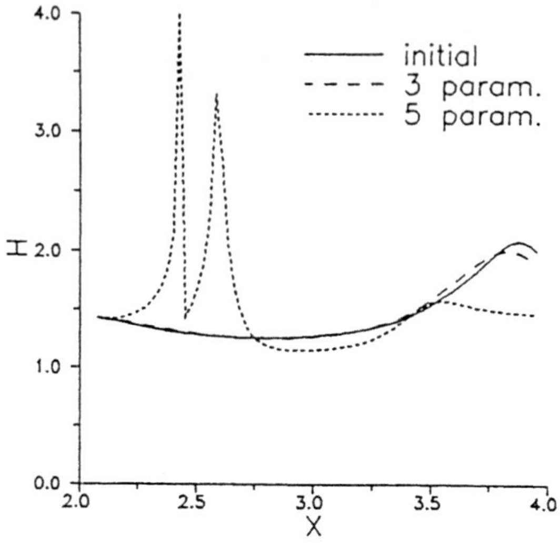

Figure 6 Boundary layer shape factor along the duct axis

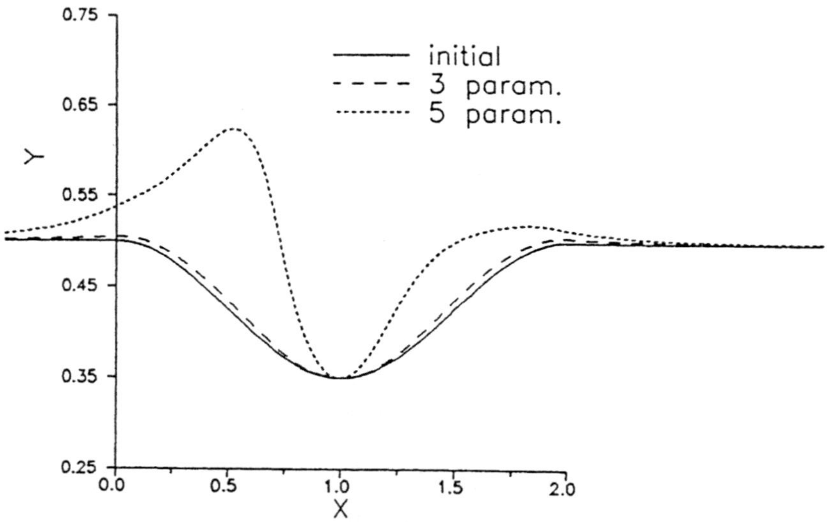

Figure 7 Initial and optimized duct geometries

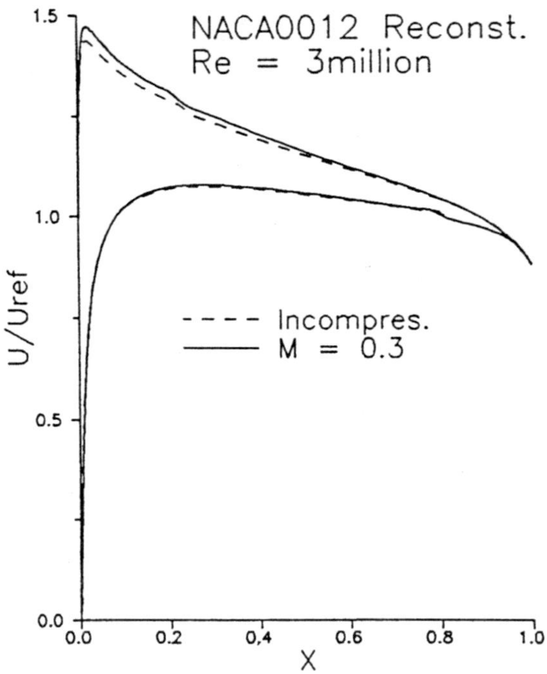

Figure 8 Normalized airfoil velocity distributions

3 Viscous Airfoil Reproduction Test Case

The viscous airfoil reproduction problem is treated for the NACA0012 symmetric profile. The inflow Mach number is 0.3 the incidence is 3 degrees and the Reynolds number based on chord is $3 * 10^6$. The flow is considered turbulent with free transition. Two direct computations, an incompressible and a compressible one, were performed using a viscous-inviscid interaction method. The converged external velocity distributions are presented in figure 8.

For the compressible case the lift coefficient is 0.3521 and the drag 58.2 counts. The computed displacement thickness distributions of the suction and pressure side boundary layer are presented in figure 9.

The calculated velocity distribution is given as the target velocity data to the inverse inviscid airfoil code of [Chaviaropoulos93]. The inverse problem is then solved on a 160x31 grid with 140 nodes on the profile. Since the viscous target velocity does not correspond to a closed profile, in the inviscid sense, the inverse code provides an open shape. However, if one subtracts the boundary layer displacement thickness from the open shape he must obtain the initial profile. The results of this exercise are presented in figure 10, where the initial

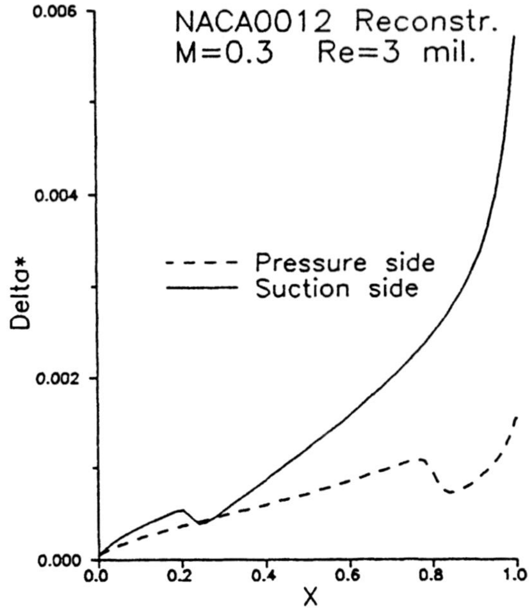

Figure 9 Displacement thickness distribution along the NACA0012 profile

and the reproduced profiles before and after the subtraction of the displacement thickness are compared.

V Conclusions

The development of an inverse method which can be applied to compressible, potential or rotational 2D or quasi-3D configurations has been briefly described. The method is based on the potential function/stream function formulation and employs Clebsch's approach to take into account rotational effects and it is capable of handling both simple and multi-block topologies. Using a prescribed (inviscid) target pressure distribution the corresponding shape is determined in a single-pass manner. Viscous effects can be accomodated via a weak viscous-inviscid interaction scheme. Drag minimization problems can be also tackled through boundary layer optimization. Computations have been carried out for three of the Optimum Design Workshop test cases. The satisfactory results indicate the efficiency of the proposed inverse approach.

Acknowledgment
This work constitutes the contribution of the Laboratory of Thermal Turboma-

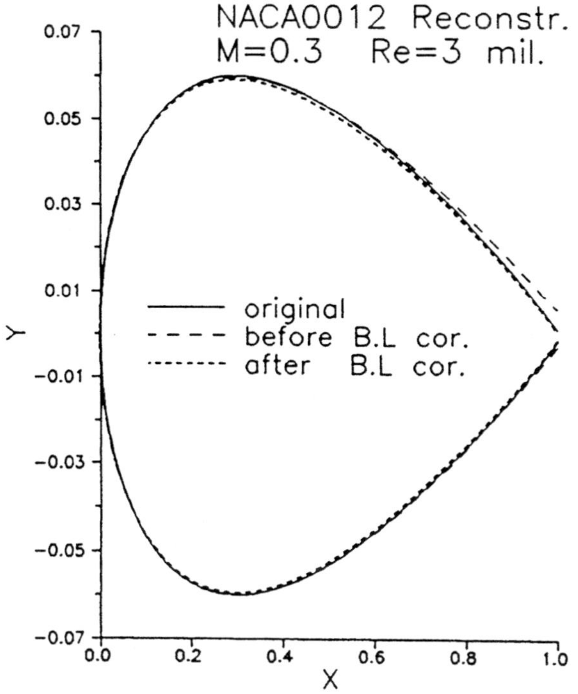

Figure 10 Comparison of initial and reproduced profile geometries

chines of NTUA in the "Optimum Design" Workshop held during the last phase of the ECARP Project.

Bibliography

[Cebeci77] *Cebeci, T., and Bradshaw, P., (1977) :* Momentum Transfer in Boundary Layers, Hemisphere - McGraw-Hill.

[Chaviaropoulos93] *Chaviaropoulos, P., Dedoussis, V. and Papailiou, K.D., (1993) :* Compressible Flow Airfoil Design Using Natural Coordinates, Computer Methods in Applied Mechanics and Engineering **110** , 131-142.

[Chaviaropoulos94] *Chaviaropoulos, P., Dedoussis, V. and Papailiou, K.D., (1994) :* Single-Pass Method for the Solution of Inverse Potential and Rotational Problems. Part I: 2-D and Quasi 3-D Theory and Application, AGARD-R-803 Optimum Design Methods for Aerodynamics, 1.1-1.19.

[Dedoussis93] *Dedoussis, V., Chaviaropoulos, P., and Papailiou, K.D., (1993) :* Rotational Compressible Inverse Design Method for Two-Dimensional, Internal Flow Configurations, AIAA Journal **31**, No. 3 , 551-558.

[Kiousis92] *Kiousis, P., Chaviaropoulos, P., and Papailiou, K.D. (1992) :* Meridional Flow Calculation Using Advanced CFD Techniques, ASME Paper 92-GT-325.

[Koumandakis94] *Koumandakis, M., Dedoussis, V., Chaviaropoulos P., and Papailiou, K.D., (1994) :* Design of Axisymmetric Channels with Rotational Flow, AIAA Journal of Propulsion and Power **10** , No. 5 , 729-735.

[Press89] *Press, W.H., Flannery, B.P., Teukolsky, S.A., and Vetterling, W.T., (1989) :* Numerical Recipies. The Art of Scientific Computing, Cambridge Univ. Press.

[Saad83] *Saad, Y., and Schultz, M.M., (1983) :* GMRES: A Generalized Minimal Residual Algorithm for Solving Nonsymmetric Linear Systems, Res. Rep. YALEU/DCS/RR-254.

[Squire38] *Squire, H.B., and Young, A.D., (1938) :* The Calculation of the Profile Drag of Aerofoils, ARC R & M No 1838.

[Stanitz53] *Stanitz, J.D., (1953) :* Design of Two-Dimensional Channels with Prescribed Velocity Distributions Alonng the Channel Walls, NACA Report 1115.

[Zedan83] *Zedan, M., and Schneider, G.E., (1983) :* A Three-Dimensional Modified Strongly Implicit Procedure for Heat Conduction, AIAA Journal **21**, No. 2, 295-303.

12 Optimum Aerodynamic Shape Design Including Mesh Adaptivity

G. Bugeda and E. Oñate, UPC, Barcelona, Spain

Abstract

This paper presents a methodology for solving shape optimization problems in the context of fluid flow problems including adaptive remeshing. The method is based on the computation of the sensitivities of the geometrical design parameters, the mesh, the flow variables and the error estimator to project the refinement parameters from one design to the next. The efficiency of the proposed method is checked out in a 2D optimization problem using a full potential model coupled with a boundary layer model.

12.1 Introduction

The increasing complexity of fluid flow problems to be analyzed with the finite element method makes necessary the use of meshes with an adequate sizing of the elements adapted to the flow features. This is necessary to capture the flow complexities (shock waves and boundary layers), but it increases considerably the computational cost of the analysis.

The high cost of the analysis of a complex flow can be reduced if an adaptive remeshing strategy is used. In this case small elements are used only in the zones where the flow is complex whereas bigger elements are used in the rest of the domain. A general scheme of this strategies is shown in Figure 1. The basic requirements of this strategy are the following:

- 1. A non structured mesh generator able to control the sizes of the elements everywhere.
- 2. An a posteriori error estimator or indicator to asses where smaller elements are needed.
- 3. An optimality criteria for the definition of the characteristics (elements sizing) of an optimal mesh for a given CPU cost, or a given precision.

The adaptive remeshing procedures are typically based on a series of successive analysis of the flow problem using meshes with increasing quality. This is possible because the characteristics of the flow problem are the same during the successive analyses. This procedure is much more complex if this characteristics are continuously changing as in the case of an optimization problem. Here a succession of different designs are obtained and the characteristics of the flow can change from one design to the next. An "optimal" mesh for the analysis of one design can become inadequate for the next one. This can be crucial if the location of flow discontinuities (i.e. shocks) changes from one design to the next.

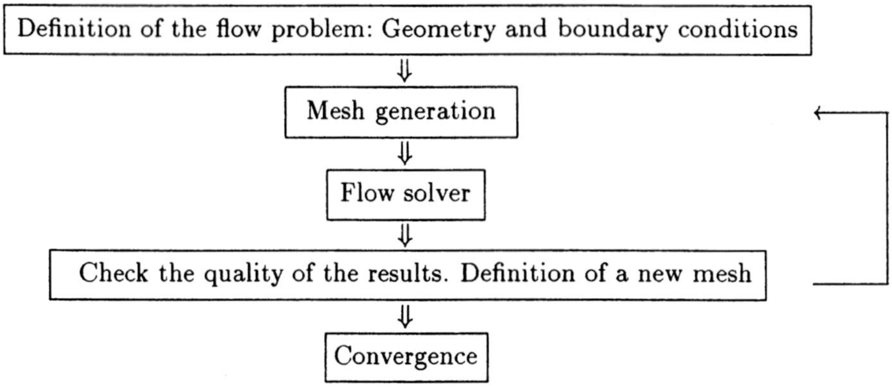

Figure 1 Adaptive remeshing scheme

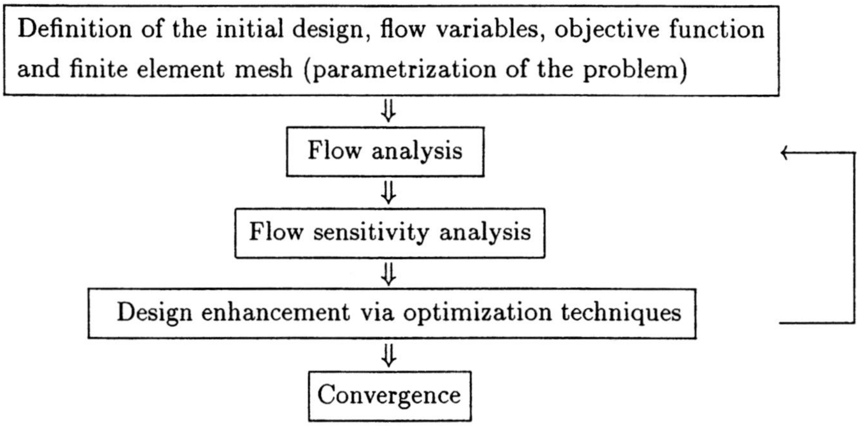

Figure 2 Classical optimization approach

Traditionally optimization problems are solved iteratively following the scheme reflected in Figure 2. In this the same mesh is properly adapted for the analysis of each geometry. There is not any control on the quality of the results which can lead to convergence problems and also to bad final results. Not only the quality of the analysis but also the quality of its sensitivities depend on the quality of the meshes.

To insert an adaptive remeshing loop inside the optimum design iterative process is an obvious possibility. The problem is that in this case the CPU cost of the problem grows proportionally to the number of optimization iterations multiplied by the number of adaptive remeshings for each design.

The problem of an adaptive remeshing process is that the definition of a good mesh requires a previous analysis using a given mesh. In the iterative

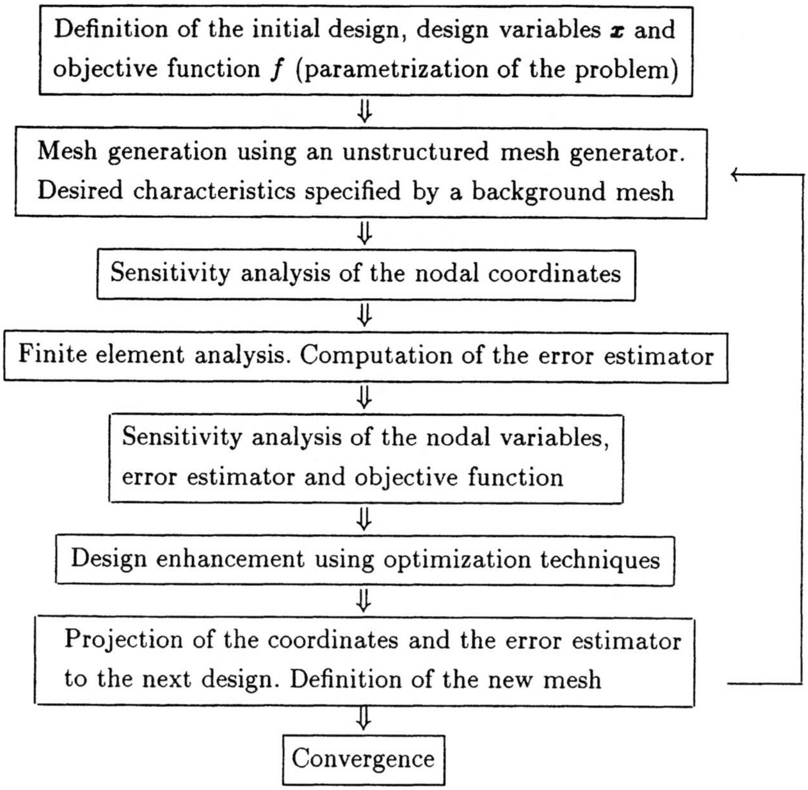

Figure 3 Proposed methodology

resolution of an optimum design problem different geometries are obtained and analyzed, but the fluid flow characteristics are different for each one. The authors have recently proposed a methodology where the information required for each adaptive remeshing is taken from the analysis of the previous design. Both the mesh parameters and the error estimator are "projected" to the next enhanced design. This allows to define "a priori" refined meshes which ensures improved quality of the results in the analysis of each design geometry. The general scheme of this methodology is shown in Figure 3. The main features of this methodology are discussed in the following sections.

12.2 Parametrization of the Problem

Each design geometry is represented by using "definition points" which specify some interpolation curves. The curves used here are parametric B-splines. The general expression of a closed B-spline for q points is (see Ref. [Beux94]):

$$\boldsymbol{r}(t) = \sum_{l=0}^{q} \boldsymbol{r}_l N_{4,l+1}(t) \tag{1}$$

where $\boldsymbol{r}(t)$ is the position vector depending on a parametric variable t. The coordinates of the definition points are recovered using $t = 0, 1, 2, \ldots$ The curve is expressed as a linear combination of $q + 1$ normalized fourth order (cubic) B-splines. The degree of continuity of a cubic B-spline is C^2. The $\boldsymbol{r}_l$ coefficients are the coordinates of the so called polygon definition points and they are found by using the coordinates of the definition points and some additional conditions about slopes and curvatures to build up a linear system of equations:

$$\boldsymbol{V} = \boldsymbol{N}\boldsymbol{R} \tag{2}$$

where $\boldsymbol{V}$ is a vector containing the imposed conditions at the definition points (coordinates, slopes, curvatures, etc.), $\boldsymbol{N}$ is a matrix containing some terms corresponding to the values of the polynomials $N_{4,l+1}(t)$ that define each B-spline evaluated at the definition points, and vector $\boldsymbol{R}$ contains the coefficients $\boldsymbol{r}_i$ to be computed. Details of this process can be found in [Beux94].

The first and second order sensitivities of $\boldsymbol{R}$ along a direction $\boldsymbol{s}$ in the design variable space are given by:

$$\begin{aligned} \frac{\partial \boldsymbol{R}}{\partial \boldsymbol{s}} &= \boldsymbol{N}^{-1}\left(\frac{\partial \boldsymbol{V}}{\partial \boldsymbol{s}} - \frac{\partial \boldsymbol{N}}{\partial \boldsymbol{s}}\boldsymbol{R}\right) \\ \frac{\partial^2 \boldsymbol{R}}{\partial \boldsymbol{s}^2} &= \boldsymbol{N}^{-1}\left(\frac{\partial^2 \boldsymbol{V}}{\partial \boldsymbol{s}^2} - \frac{\partial^2 \boldsymbol{N}}{\partial \boldsymbol{s}^2}\boldsymbol{R} - 2\frac{\partial \boldsymbol{N}}{\partial \boldsymbol{s}}\frac{\partial \boldsymbol{R}}{\partial \boldsymbol{s}}\right). \end{aligned} \tag{3}$$

The derivatives of $\boldsymbol{V}$ with respect the coordinates of the definition points chosen as design variables can be easily computed. Vectors $\partial \boldsymbol{R}/\partial \boldsymbol{s}$ and $\partial^2 \boldsymbol{R}/\partial \boldsymbol{s}^2$ will contain the terms $\partial \boldsymbol{r}_i/\partial \boldsymbol{s}$ and $\partial^2 \boldsymbol{r}_i/\partial \boldsymbol{s}^2$, respectively (see Ref. [Bugeda95]).

Finally, the sensitivities of the coordinates of any point on the interpolation curve corresponding to a constant value of t are obtained by:

$$\begin{aligned} \frac{\partial \boldsymbol{r}(t)}{\partial \boldsymbol{s}} &= \sum_{l=0}^{q} \frac{\partial \boldsymbol{r}_i}{\partial \boldsymbol{s}} N_{4,l+1}(t) \\ \frac{\partial^2 \boldsymbol{r}(t)}{\partial \boldsymbol{s}^2} &= \sum_{l=0}^{q} \frac{\partial^2 \boldsymbol{r}_i}{\partial \boldsymbol{s}^2} N_{4,l+1}(t). \end{aligned} \tag{4}$$

Last expression is used to compute the sensitivities of the nodal points placed at the boundary of the finite element mesh.

12.3 Mesh Generation and Sensitivity Analysis

In order to control the sizes of the elements a non structured mesh generation algorithm should be chosen. In this work the well known advancing front method has been chosen [Peraire89]. The characteristics of the desired mesh are specified via a background mesh over which nodal values of the size parameter δ are defined and interpolated via the shape functions. The background mesh for the first design has to be specified by the user. For subsequent designs the background mesh is taken to coincide with the mesh projected into this design from the previous one. This projection process will be described later.

Once the sensitivities of the coordinates of each boundary node are known, it is also possible to compute the sensitivities of the coordinates of each internal nodal point (mesh sensitivities). These sensitivities are necessary to asses how the mesh evolves when the design variables change.

There are many different ways to define the evolution of the mesh in terms of the design variables. It is possible to consider a simple analogous elastic medium defining the mesh movement. This is the bases of the so called "spring analogy" where each element side is regarded as a spring connecting two nodes. The force induced by each spring is proportional to its length and the boundary nodes are considered as fixed. The solution of the equilibrium problem in the spring analogy is simple but expensive and it involves to solve a linear system of equations with two degrees of freedom per node.

In this work the spring analogy problem has been solved iteratively using a simple Laplacian smoothing approach. This technique is frequently used to improve the quality of non-structured meshes. It consists on the iterative modification of the nodal coordinates of each interior node by placing it at the center of gravity of adjacent nodes. The expression of the new nodal position vector $\boldsymbol{r}_i$ for each iteration is given by:

$$\boldsymbol{r}_i = \frac{\sum_{j=1}^{m_i} \boldsymbol{r}_j}{m_i} \tag{5}$$

where $\boldsymbol{r}_j$ are the position vectors of the m_i nodes connected with the i-th node.

The solution of the spring analogy problem with a prescribed error tolerance requires to check quality of the solution after each smoothing cycle. Taking into account that the described iterative process is only a way to obtain mesh sensitivities, rather than the solution of the equilibrium problem itself, rigorous convergence conditions are not needed. For this reason the number of smoothing cycles to be applied can be fixed a priori. This allows to substantially decrease the CPU time of the mesh sensitivity analysis compared with that required for the full resolution of the spring analogy equilibrium equations. In the examples presented later we have checked that 50 "smoothing" iterations are enough to ensure a good quality of results.

The first-order and higher-order mesh sensitivities along any direction of the design variables space, $\boldsymbol{s}$, are obtained by differentiating eq. (5) with respect to $\boldsymbol{s}$ for each cycle, i.e.

$$\frac{\partial \boldsymbol{r}_i}{\partial \boldsymbol{s}} = \frac{\sum_{j=1}^{m_i} \frac{\partial \boldsymbol{r}_j}{\partial \boldsymbol{s}}}{m_i}$$
$$\frac{\partial^2 \boldsymbol{r}_i}{\partial \boldsymbol{s}^2} = \frac{\sum_{j}^{m_i} \frac{\partial^2 \boldsymbol{r}_j}{\partial \boldsymbol{s}^2}}{m_i} . \tag{6}$$

Last equations provide the sensitivity analysis of the coordinates of the all the internal nodes. There is a loop over all the internal nodes corresponding to each smoothing cycle. The sensitivities of each node are obtained from the sensitivities of the adjacent nodes. Before the first cycle only the boundary nodes have non null sensitivities, obtained from equations (4).

12.4 State equations for the full potential solver and its sensitivity analysis

The flow model considered in this work is the steady state solution of the following equations [Pironneau88]:

$$\frac{\partial \rho}{\partial t} + \nabla(\rho \boldsymbol{u}) - \frac{\partial}{\partial \boldsymbol{s}}(\delta s \frac{\partial F}{\partial \boldsymbol{s}}) = 0 \tag{7}$$

where the velocities $\boldsymbol{u}$ are computed as the gradient of a potential $\boldsymbol{u} = \nabla \phi$ and $\boldsymbol{s}$ is the unit vector on the flow direction and the upwinding term F is equal to

$$\begin{cases} \rho u - \rho_* c_* & \text{if M} > 1 \\ 0 & \text{otherwise} . \end{cases} \tag{8}$$

The star stands for critical values. A Dirichlet boundary condition must also be added in order to define zero potential level. The inlet part $\Gamma_{I\infty}$ of the far field boundary is chosen for this condition.

The weak formulation of eq. 7 must take into account the lift. Thus, a potential jump is taken on the wake Σ by considering two dissociate lines Σ^+ and Σ^- which replace Σ. Σ^+ joins the upper surface of the airfoil with the far field boundary and Σ^- the lower part of the boundary with it. Then, the weak formulation is formed in the domain Ω' whose boundary is $\Gamma_\infty \bigcup \Gamma_c \bigcup \Sigma^+ \bigcup \Sigma^-$. By multiplying 7 by a test function $w \in H^1$ and integrating by parts we obtain the weak formulation: find $\phi \in V_o(\Omega')$ such that $\forall w \in V_0(\Omega')$

$$\int_{\Omega'} \frac{\partial \rho}{\partial t} w d\Omega - \int_{\Omega'} \rho \boldsymbol{u} \nabla w d\Omega + \int_{\partial_{\Omega'_o}} \rho \boldsymbol{u} \boldsymbol{n} w d\gamma - \int_{\partial_{\Omega'}} \boldsymbol{s} \nabla (\delta s \frac{\partial F}{\partial \boldsymbol{s}}) w d\Omega = 0 . \tag{9}$$

In order to solve 9, an implicit Euler's scheme is used. Let us suppose that the potential ϕ^k is known at the kth step. Then, the solution at the $(k+1)^{th}$ step is given by the equation:

$$\int_{\Omega'} \frac{\rho^{k+1} - \rho^k}{t^{k+1} - t^k} w d\Omega - \int_{\Omega'} \rho^{k+1} \boldsymbol{u}^{k+1} \nabla w d\Omega + \\ \int_{\partial_{\Omega'_o}} \rho^{k+1} \boldsymbol{u}^{k+1} \boldsymbol{n} w d\gamma - \int_{\partial_{\Omega'}} \boldsymbol{s} \nabla (\delta s \frac{\partial F}{\partial \boldsymbol{s}}) w d\Omega = 0 \tag{10}$$

which can be rewritten as

$$G(\phi^{k+1}) \equiv \int_{\Omega'} \frac{\rho^{k+1} - \rho^k}{t^{k+1} - t^k} w d\Omega + J(\phi^{k+1}) = 0, \forall w \in V_0(\Omega'). \tag{11}$$

The velocity at the kth tiem step is equal to $\boldsymbol{u}^{k+1} = \boldsymbol{u}^k + \nabla(\phi^{k+1} - \phi k)$ and the fluid density is a function of $|\boldsymbol{u}^{k+1}|$. Therefore, eq. 11 gives ϕ^{k+1} since ϕ^k is known. The space discretization is done by P^1 finite elements. The solution of the steady state of eq. 11 is obtained by using the techniques described in [Pironneau88].

12.4.1 Sensitivity analysis of the full potential solver

The sensitivity analysis of all the magnitudes involved in eq. 10 can be obtained by direct derivation of its integral expressions. The first step will be the computation of the sensitivities of the potential ϕ with respect to any design variable x.

The term J of 11 depends of the potential ϕ and the design variables x. In the steady state it must hold:

$$J_i(x, \phi) = 0, i = 1, ..., n \tag{12}$$

where n is the number of trial functions used for the discretization of ϕ. This expressions can be derivated with respect the design variable x:

$$\frac{\partial J_i}{\partial x} + \frac{\partial J_i}{\partial \phi_j} \frac{\partial \phi_j}{\partial x}, \quad i, j = 1, ...n . \tag{13}$$

Last equations allow to obtain the sensitivities of the potential value at each node ϕ_j with respect to x:

$$\frac{\partial \phi_j}{\partial x} = - \left[\frac{\partial J_i}{\partial \phi_j} \right]^{-1} \frac{\partial J_i}{\partial x} . \tag{14}$$

Last expression involves the resolution of a linear system of equations.

After the computation of the sensitivities of ϕ, the sensitivities of any other magnitude (velocities, density, pressure, objective function, etc) can be obtained by direct derivation of its expressions in terms of the potential and the design variables.

12.5 Boundary Layer and sensitivity analysis

The boundary layer model that has been coupled with the full potential equations is the Head's method which is described in reference [Cebeci77]. It is an integral method for calculating two-dimensional turbulent boundary layers in pressure gradient which uses the momentum-integral equation, a formula relating the skin friction c_f to the Reynolds number and a profile shape parameter, and an ODE for the rate of change of profile shape parameter with x.

Head assumes that the dimensionless entrainment velocity, v_E/v_e is a function of the shape factor H_1 (see [Cebeci77]):

$$\frac{u_E}{u_e} = \frac{1}{u_e}\frac{d}{dx}[u_e(\delta - \delta^*)] = F(H_1) \tag{15}$$

where u_e is the external inviscid velocity, δ is the boundary layer thickness, δ^* is the displacement thickness and H_1 is defined in terms of the momentum thickness θ by:

$$H_1 = \frac{\delta - \delta^*}{\theta}. \tag{16}$$

After substitution of 16 in 15 we obtain

$$\frac{d}{dx}(u_e \theta H_1) = u_e F. \tag{17}$$

Head assumes that H_1 is related to the shape factor H by

$$H_1 = G(H). \tag{18}$$

The functions F and G are determined from experiment. A best fit to several tests of experimental data (see [Cebeci77]) is:

$$\begin{aligned} F &= 0.0306(H_1 - 3.0)^{-0.6169} \\ G &= \begin{cases} 0.8234(H - 1.1)^{-1.287} + 3.3 & H \leq 1.6 \\ 1.5501(H - 0.6778)^{-3.064} + 3.3 & H \geq 1.6\,. \end{cases} \end{aligned} \tag{19}$$

If we write the momentum integral equation as

$$\frac{d\theta}{dx} + (H+2)\frac{\theta}{u_e}\frac{du_e}{dx} = \frac{c_f}{2} \tag{20}$$

we see that it has three unknowns: θ, H, and c_f for a given external velocity distribution. Equation 17, with F, H and G defined by eqs. 18, 19, and 20 provides a relationship between θ and H. Another equation relating c_f to θ and/or H is needed. Head used the c_f law given by Ludwieg and Tillmann:

$$c_f = 0.246 \ \ 10^{-0.678H} {R_\theta}^{-0.268} \tag{21}$$

where $R_\theta = u_e\theta/\nu$ is the Reynolds number based on the momentum thickness and the kinematic viscosity ν. The system of equations 17 to 21, which includes two ODEs, can be solved numerically for a specific external velocity distribution to obtain the boundary layer development.

This method, like most integral methods, uses the shape factor H as the criterion for separation. Equation 21 predicts $c_f = 0$ only if H tends to infinity. It is not possible to give an exact value of H corresponding to separation, and a range between 1.8 and 2.4 is commonly quoted. The difference between the lower and upper limits of H makes only little difference in locating the separation point, since close to separation dH/dx is large.

After the resolution of the system of ODEs the Squire and Young formula is used for an estimation of the total drag:

$$c_d = 2\frac{\theta}{L}\frac{u_e}{u_E}^{0.5(H+2.5)} \tag{22}$$

where L is the total length of the boundary layer.

Reference [Cebeci77] presents a Fortran program for predicting the turbulent boundary layer development on two-dimensional bodies by Head's method. The system of equations is integrated by a Runge-Kutta scheme for a given set of initial conditions. This program has been implemented and tested obtaining good results.

The sensitivity analysis of the Head's method has been obtained by direct differentiation of all the expressions appearing in the Fortran code. This task has been developed with the help of the MATHEMATICA program which has helped for the differentiation of the most comple expressions. The MATHEMATICA program can provide, in Fortran, the derivative of an expression with respect to any of its symbols in terms of all the symbols that appear in the expression and its derivatives. The new code containing all the differentiated expressions provides the drag and its sensitivities with respect the design variables.

12.6 Error Estimator and its Sensitivity Analysis

An "energetic" norm of the potential ϕ will be used as the reference magnitude which will control the adaptivity process. This norm is defined in terms of the exact velocities field $\boldsymbol{u}$ as:

$$U = ||\phi||^2 = \int_\Omega \boldsymbol{u}^T \boldsymbol{u} d\Omega \, . \tag{23}$$

The discretization process provides an approximated potential $\hat{\phi}$ with an associated velocities field $\hat{\boldsymbol{u}}$ whose "energetic" norm will be

$$||\hat{\phi}||^2 = \int_\Omega \hat{\boldsymbol{u}}^T \hat{\boldsymbol{u}} d\Omega \, . \tag{24}$$

Last norm will contain an error that for elliptic problems can be computed using the expression:

$$||e||^2 = \int_\Omega (\boldsymbol{u} - \hat{\boldsymbol{u}})^T (\boldsymbol{u} - \hat{\boldsymbol{u}}) d\Omega. \tag{25}$$

For non elliptic problems last expression can be assumed to be an approximation to the error in the "energetic" norm.

Of course, expression 25 can not be computed if the exact solution ϕ is not known. Then, the exact velocities $\boldsymbol{u}$ are replaced by an approximated velocities field $\bar{\boldsymbol{u}}$ which is supposed to be of better quality than the solution $\hat{\boldsymbol{u}}$. This replacement provides us the value of the error estimator η^2 that can be computed element by element.

$$\eta^2 = \int_\Omega (\bar{\boldsymbol{u}} - \hat{\boldsymbol{u}})^T (\bar{\boldsymbol{u}} - \hat{\boldsymbol{u}}) d\Omega = \sum_e \int_{\Omega_e} (\bar{\boldsymbol{u}} - \hat{\boldsymbol{u}})^T (\bar{\boldsymbol{u}} - \hat{\boldsymbol{u}}) d\Omega. \tag{26}$$

There are some possibilities for the obtainment of the improved velocities field $\bar{\boldsymbol{u}}$. In this work the Superconvergent Patch Recovery (SPR) technique has been used. This is a recent technique developed by Zienkiewicz and Zhu in [Zienkiewicz92] where the value of $\bar{\boldsymbol{u}}$ for each node is obtained by a local least square fit of a polynomial expansion set using the high accuracy sampling points inside the elements connected with the node. This procedure does not involve the resolution of any global system of equations and it seems to provide good results.

The recovered velocities field will be expressed as $\bar{\boldsymbol{u}} = \boldsymbol{N}\bar{\boldsymbol{u}}^*$ where $\boldsymbol{N}$ are the chosen potential interpolation functions and $\bar{\boldsymbol{u}}^*$ are the recovered nodal values of the velocities. In the recovery process the nodal values $\bar{\boldsymbol{u}}^*$ are assumed to belong to a polynomial expansion $\boldsymbol{u}_p^*$ of the same complete order p as that present in the basis functions $\boldsymbol{N}$ and which is valid over an element patch surrounding the particular assembly node considered. Such a "patch" represents the union of elements containing this vertex node. This polynomial expansion will be used for each component of $\boldsymbol{u}_p^*$ and we write simply:

$$\boldsymbol{u}_p^* = \boldsymbol{P}\boldsymbol{a} \tag{27}$$

where $\boldsymbol{P}$ contains the appropriate polynomial terms and $\boldsymbol{a}$ is a set of unknown parameters. For the P^1 finite elements used in this work we have $\boldsymbol{P} = [1, x, y]$.

The determination of the unknown parameters $\boldsymbol{a}$ of the expansion 27 is best made by ensuring a least square fit of this to the set of high accuracy sampling points existing in the patch. This sampling points are the baricenters of the triangles forming the patch. To do this we minimize:

$$F(\boldsymbol{a}) = \sum_{i=1}^{n} (\hat{\boldsymbol{u}}(x_i, y_i) - \boldsymbol{u}_p^*(x_i, y_i))^2 = \sum_{i=1}^{n} (\hat{\boldsymbol{u}}(x_i, y_i) - \boldsymbol{P}(x_i, y_i)\boldsymbol{a})^2 \tag{28}$$

where (x_i, y_i) are the coordinates of a group of n sampling points. The minimization condition of $f(\boldsymbol{a})$ implies that $\boldsymbol{a}$ satisfies:

$$\sum_{i=1}^{n} \boldsymbol{P}^T(x_i, y_i)\boldsymbol{P}(x_i, y_i)\boldsymbol{a} = \sum_{i=1}^{n} \boldsymbol{P}^T(x_i, y_i)\hat{\boldsymbol{u}}(x_i, y_i). \tag{29}$$

This can be solved in matrix form as:

$$\boldsymbol{a} = \boldsymbol{A}^{-1}\boldsymbol{b} \tag{30}$$

where

$$\begin{aligned} \boldsymbol{A} &= \sum_{i=1}^{n} \boldsymbol{P}^T(x_i, y_i)\boldsymbol{P}(x_i, y_i) \\ \boldsymbol{b} &= \sum_{i=1}^{n} \boldsymbol{P}^T(x_i, y_i)\hat{\boldsymbol{u}}(x_i, y_i). \end{aligned} \tag{31}$$

In order to get the sensitivities of the error estimator it is necessary to obtain the sensitivities of the recovered nodal velocities. These will be obtained by direct derivation of expressions 27, 30 y 31:

$$\frac{\partial \boldsymbol{u}_p^*}{\partial \boldsymbol{d}} = \frac{\partial \boldsymbol{P}}{\partial \boldsymbol{d}}\boldsymbol{a} + \boldsymbol{P}\frac{\partial \boldsymbol{a}}{\partial \boldsymbol{d}} \tag{32}$$

$$\frac{\partial \boldsymbol{P}}{\partial \boldsymbol{d}} = [0, \frac{\partial x}{\partial \boldsymbol{d}}, \frac{\partial y}{\partial \boldsymbol{d}}]. \tag{33}$$

The terms $\partial x/\partial \boldsymbol{d}$ and $\partial y/\partial \boldsymbol{d}$ can be obtained from the mesh sensitivity analysis. For the rest of terms it holds:

$$\frac{\partial \boldsymbol{a}}{\partial \boldsymbol{d}} = \boldsymbol{A}^{-1}\left[\frac{\partial \boldsymbol{b}}{\partial \boldsymbol{d}} - \frac{\partial \boldsymbol{A}}{\partial \boldsymbol{d}}\boldsymbol{a}\right] \tag{34}$$

$$\frac{\partial \boldsymbol{A}}{\partial \boldsymbol{d}} = \sum_{i=1}^{n}\left[\left(\frac{\partial \boldsymbol{P}}{\partial \boldsymbol{d}}\right)^T \boldsymbol{P}(x_i, y_i) + \boldsymbol{P}^T(x_i, y_i)\frac{\partial \boldsymbol{P}}{\partial \boldsymbol{d}}\right] \tag{35}$$

$$\frac{\partial \boldsymbol{b}}{\partial \boldsymbol{d}} = \sum_{i=1}^{n}\left[\left(\frac{\partial \boldsymbol{P}}{\partial \boldsymbol{d}}\right)^T \hat{\boldsymbol{u}}(x_i, y_i) + \boldsymbol{P}^T(x_i, y_i)\frac{\partial \hat{\boldsymbol{u}}}{\partial \boldsymbol{d}}\right]. \tag{36}$$

The sensitivities of the velocities at the baricenters of the elements $\hat{\boldsymbol{u}}$ will be obtained from the flow sensitivity analysis. The sensitivities of the recovered velocities $\bar{\boldsymbol{u}}$ will be obtained from:

$$\frac{\partial \bar{\boldsymbol{u}}}{\partial \boldsymbol{d}} = \boldsymbol{N}\frac{\partial \bar{\boldsymbol{u}}^*}{\partial \boldsymbol{d}} = \boldsymbol{N}\frac{\partial \boldsymbol{u}_p^*}{\partial \boldsymbol{d}}. \tag{37}$$

The sensitivities of the error estimator η^2 can be obtained by direct derivation of the integral expression 26. The sensitivities of an integral expression are computed after its transformation into the isoparametric domain $\boldsymbol{\xi}$ which shape does not depend on the design variables. The jacobian of this transformation $|\boldsymbol{J}|$ can be expressed in terms of the nodal coordinates, so that, it can also be differentiated in order to know the integral sensitivities. Using this technique the sensitivities of the integral expression of the error corresponding to each element e can be obtained by:

$$\begin{aligned}\frac{\partial \eta_e^2}{\partial \boldsymbol{d}} = \int_{\Omega_{\boldsymbol{\xi}}} \Bigg[\left(\frac{\partial \bar{\boldsymbol{u}}}{\partial \boldsymbol{d}} - \frac{\partial \boldsymbol{u}}{\partial \boldsymbol{d}} \right)^T (\bar{\boldsymbol{u}} - \boldsymbol{u})|J| + (\bar{\boldsymbol{u}} - \boldsymbol{u})^T \left(\frac{\partial \bar{\boldsymbol{u}}}{\partial \boldsymbol{d}} - \frac{\partial \boldsymbol{u}}{\partial \boldsymbol{d}} \right) |J| + \\ (\bar{\boldsymbol{u}} - \boldsymbol{u})^T (\bar{\boldsymbol{u}} - \boldsymbol{u}) \frac{\partial |\boldsymbol{J}|}{\partial \boldsymbol{d}} \Bigg] d\xi_1 d\xi_2\end{aligned} \tag{38}$$

where the sensitivity of the jacobian is:

$$\frac{\partial |\boldsymbol{J}|}{\partial \boldsymbol{d}} = |\boldsymbol{J}| tr \left(\boldsymbol{J}^{-1} \frac{\partial \boldsymbol{J}}{\partial \boldsymbol{d}} \right). \tag{39}$$

This technique allows to obtain the sensitivities of any other integral expression like 24.

12.7 Mesh Adaptivity

For the complete definition of the characteristics of a new mesh in the remeshing procedure it is necessary to use a mesh optimality criterion. In this work a mesh is considered as optimal when the error density is equally distributed across the volume, i.e. when $\frac{||e||_e^2}{\Omega_e} = \frac{||e||^2}{\Omega}$ is satisfied. The justification of this optimality criterion can be found in [Bugeda92] and [Oñate93].

The combination of the optimality criterion and the error estimation allows to define the new element sizes. Previously, it is necessary to define the limit of the allowable global error percentage γ as:

$$\gamma = 100\frac{||e||}{||\Phi||} \approx 100\frac{||e||}{\sqrt{||e||^2 + ||\hat{\Phi}||^2}}. \tag{40}$$

The desired error level for each element is:

$$||e||_e^d = \frac{\gamma}{100}\sqrt{(||\hat{\Phi}||^2 + ||e||^2)\frac{\Omega_e}{\Omega}}. \tag{41}$$

The new element sizes $\bar{h}_e$ can be computed in terms of the old ones h_e using the expression:

$$\bar{h}_e = \frac{h_e}{\xi_e^{1/p}} \tag{42}$$

where $\xi_e = \frac{\|e\|_e}{\|e\|_e^d}$ and p is the order of the shape function polynomials. For further details see [Bugeda92] and [Oñate93].

12.8 Design Enhancement and Definition of the New Mesh

The objective function sensitivities are used to get improved values of the design parameters by means of a standard minimization method. Depending on the optimization algorithm it may be necessary to use second order sensitivities. The design variables corresponding to the improved design will usually be found as:

$$\boldsymbol{d}^{k+1} = \boldsymbol{d}^k + \theta \boldsymbol{s}^k \tag{43}$$

where θ is an advance parameter. The direction of change $\boldsymbol{s}^k$ can be obtained using a BFGS Quasi Newton method or a GMRES method which only requires first derivatives of the objective function. The value of θ can be obtained by a line search procedure. One possibility is to use a second order sensitivity analysis in the $\boldsymbol{s}^k$ direction. The objective function can then be approximated along this direction using a second order Taylor expansion which minimization provides the value of θ.

Once the new design has been defined all the relevant variables for the adaptive remeshing strategy can be projected from the old design to the new one using the sensitivity analysis. Second order sensitivities can be used if required. The coordinates of the projected mesh are obtained using:

$$(x,y)^{k+1} \approx (x,y)^k + \theta\left(\frac{\partial x}{\partial \boldsymbol{s}}, \frac{\partial y}{\partial \boldsymbol{s}}\right) + \frac{1}{2}\theta^2\left(\frac{\partial^2 x}{\partial \boldsymbol{s}^2}, \frac{\partial^2 y}{\partial \boldsymbol{s}^2}\right). \tag{44}$$

For the case of incompressible potential equations the projections of the error estimator and the "energy" can be obtained from:

$$\|e\|^{2^{k+1}} = \|e\|^{2^k} + \theta\frac{\partial \|e\|^2}{\partial \boldsymbol{s}} + \frac{1}{2}\theta^2\frac{\partial^2 \|e\|^2}{\partial \boldsymbol{s}^2} \tag{45}$$

$$\|U\|^{2^{k+1}} = \|U\|^{2^k} + \theta\frac{\partial \|U\|^2}{\partial \boldsymbol{s}} + \frac{1}{2}\theta^2\frac{\partial^2 \|U\|^2}{\partial \boldsymbol{s}^2}\,. \tag{46}$$

These projections provide a good approximation of each of above values for the next design configuration previously to any new computations. In fact, the projected values provide the necessary information to perform a remeshing over the next design, **even before any new computation is attempted**. In that sense, we have changed an error estimator computed "a posteriori" into an "a priori" error estimator for the definition of a new mesh.

This projection is of paramount important since it allows to control the quality of the meshe for each new design. Only a single mesh is generated in every new design analysis step. Thus, the extra computational cost involved in the control of the mesh quality is very cheap.

The projected values are used to create the background mesh information needed to generate the mesh corresponding to the new design geometry. This operation closes the iterative process which will lead to the optimum design geometry after convergence.

Bibliography

[Beux94] *Beux, F., Dervieux, A., Leclercq, M. P. and Stoufflet, B.* Techniques de contrôle optimal pour l'optimisation de forma en aérodynamique avec calcul exact du gradient, Revue Scientifique et Technique de la Défense, 1994.

[Bugeda92] *Bugeda, G. and Oñate, E.* New adaptive techniques for structural problems, First European Conference on Numerical Methods in Engineering, Brussels, Belgium, September, 1992.

[Bugeda95] *Bugeda, G. and Oñate, E.* Optimum Aerodynamic Shape Design Including Mesh Adaptivity, International Journal for Numerical Methods in Fluids, bf 20, 915,924 (1995).

[Cebeci77] *Cebeci, T. and Bradshaw, P.*, Momentun transfer in boundary layers. Hemisphere Publication Corporation, 1977.

[Faux87] *Faux, I. D. and Pratt, M. J.* Computational Geometry for Design and Manufacture, Edited by Ellis Horwood Limited, 1987.

[Oñate93] *Oñate, E. and Bugeda, G.* A Study of Mesh Optimality Criteria in Adaptive Finite Element Analysis, Engineering Computations, **10**, 307-321 (1993).

[Peraire89] *Peraire, J., Morgan, K. and Peiró, J.* Unstructured finite element mesh generation and adaptive procedures for CFD., AGARD FDP: Specialist's Meeting, Loen, Norway (1989).

[Pironneau88] *Pironneau, O.* Méthodes des éléments finis pour les fluides, Masson 1988.

[Vossinis93] *Vossinis A.* Optimisation de forme d'aile d'avion. Doctoral Thesis. 1993.

[Zienkiewicz92] *Zienkiewicz, O.C., and Zhu, J. Z.* The superconvergent patch recovery and a posteriori error estimates. Part 1:the recovery technique, International Journal for Numerical Methods in Engineering, **33**, 1331-1364, 1992.

12.9 Numerical Results for the TE1 test case

The TE1 test case has been solved by using the methodology previously described. Figure 4 shows the initial shape of the nozzle and the mesh used for the analysis of this initial design. Linear triangles (P1 elements) have been used for the discretization of the domain. The limit of the allowable global percentage of error γ (see expression 40) has been fixed to a 0.02%.

The objective function is the total drag of the nozzle that has been obtained by using the Square and Young formula (see expression 22). The velocities obtained from the full potential model have been used for the computation of the boundary layer model by using the Head's method.

Four design variables have been used for the definition of the shape of each design. This variables are y coordinates of four points placed on the upper surface. The x coordinates of this points are 0.333, 0.667, 1.333 and 1.667. The y coordinate corresponding to x=1.000 is 0.35 as specified in the definition of the test case.

After 100 iterations no significant improvements have been obtained in the shape of the nozzle. Figure 5 shows the shape of the final design. Figure 6 shows a detail of the mesh that has been used for the analysis of this final design. As it can be observed, a big concentration of elements has been produced in the zone of the nozzle.

Figures 7 and 8 show the evolution of the objective function and the norm of its gradient during the iterative process. It can be observed how after the initial 40 iterations the objective function is not minimized. Nevertheless, the gradient norm is being minimized until the 80 iteration. The oscillations that are present in the evolution of both the objective function and its gradient norm are probably due to the change of the mesh for each different design. This oscillations are small if a very small global percentage of error γ is prescribed. Bigger values of γ produce cheaper meshes, but the oscillations in the iterative process are bigger and the convergence is worst. Figure 9 shows how the global percentage of error is maintained below the 0.02 value prescribed for the generation of the mesh for each design.

Figure 10 shows the C_p distribution in the area of the nozzle. This distribution is not smooth as it can be expected from Figure 5. Also, Figure 11 shows a non smooth distribution of the H form factor of expression 15.

Finally, Figure 12 shows the isolines of the Mach number distribution for the final design.

12.9.1 Analysis of the results

The following conclusions can be obtained from the analysis of the results obtained for the TE1 test case.

- The results obtained by using the presented methodology including mesh adaptivity show a good behaviour of the method. The obtained final optimum shape is unexpected, but it is in accordance with the final solutions obtained by using other methods. The use of different parametrizations of the problem with a bigger number of design variables produces similar final results. The reason of this unexpected optimum shape seems to be the use of the Square and Young expression that can provide unrealistic results for this type of geometry.

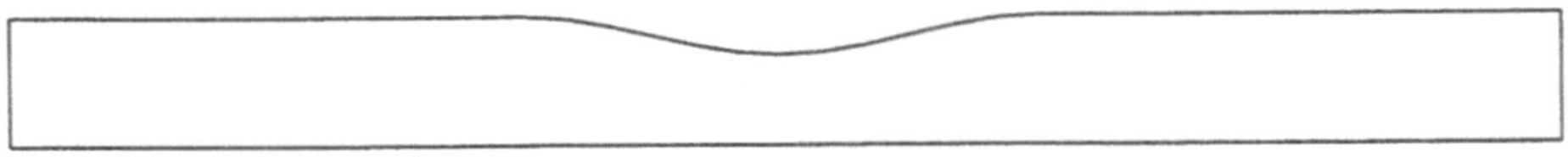

Figure 4 Initial shape and initial mesh

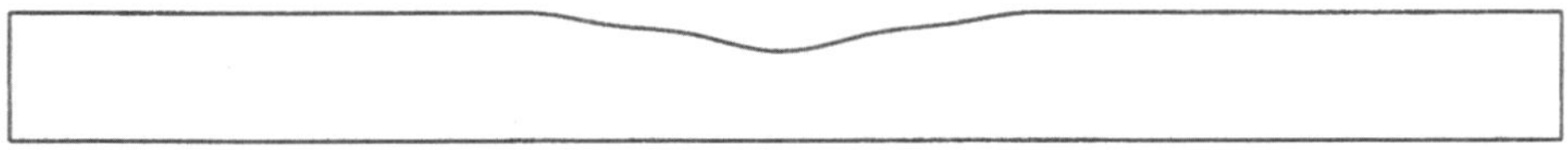

Figure 5 Final shape

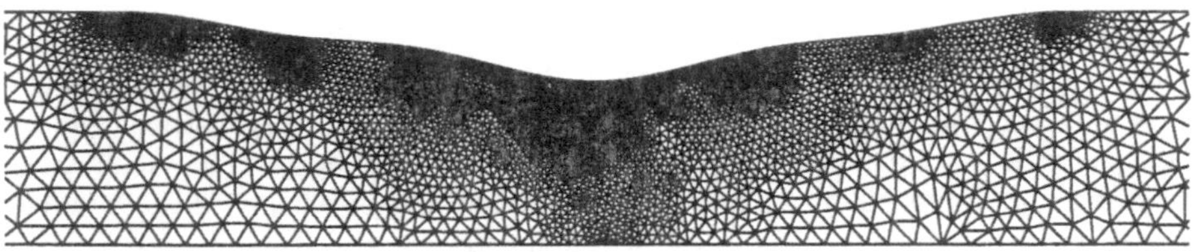

Figure 6 Detail of the final mesh

- The quality of the meshes used for the analysis of each design is well controlled and the good quality of the results obtained for each design is assured.
- In order to get a good convergence of the iterative process it is necessary to use a very strict limit in the global percentage of error.

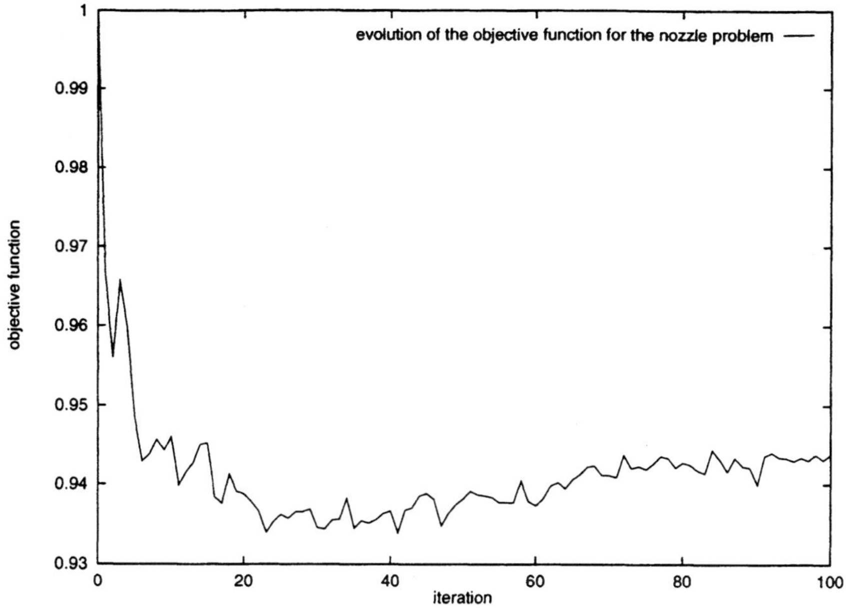

Figure 7 Evolution of the objective function

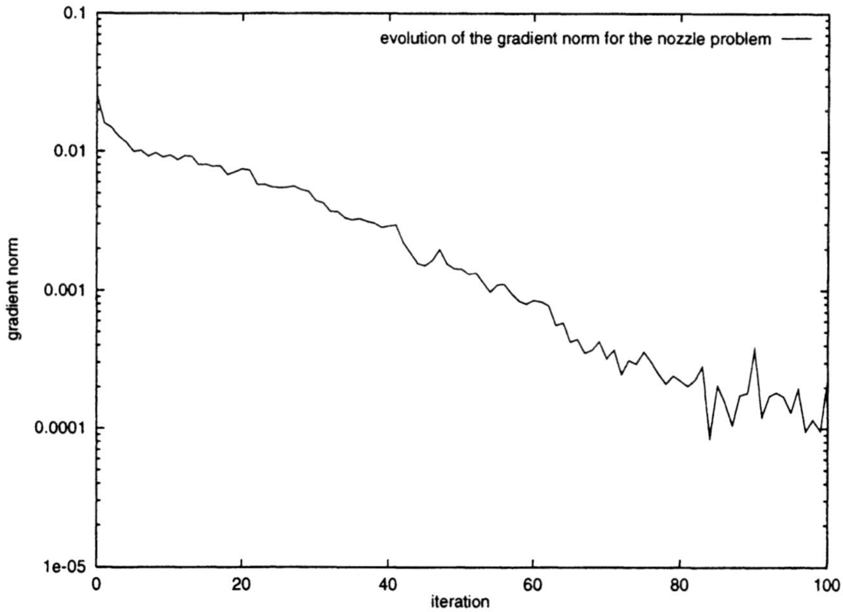

Figure 8 Evolution of the norm of the gradient of the objective function

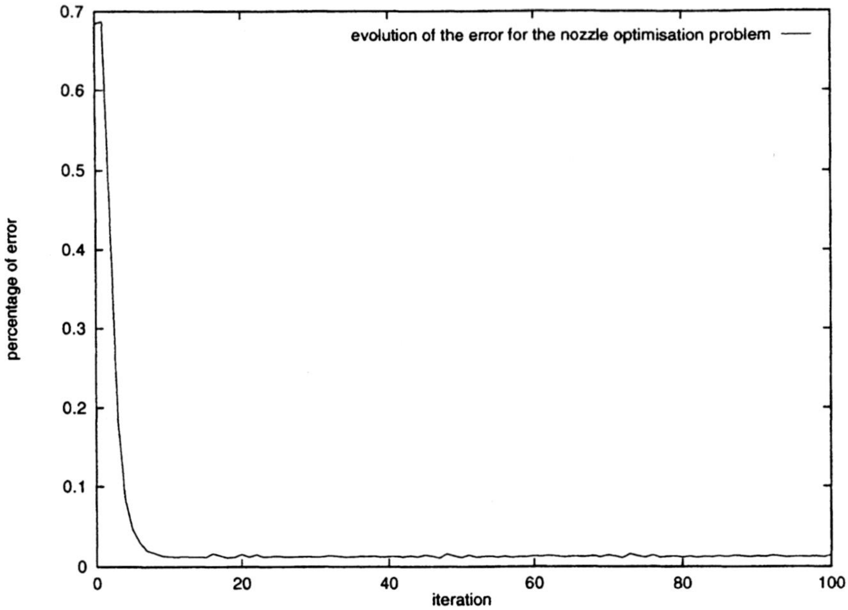

Figure 9 Evolution of the global percentage of error

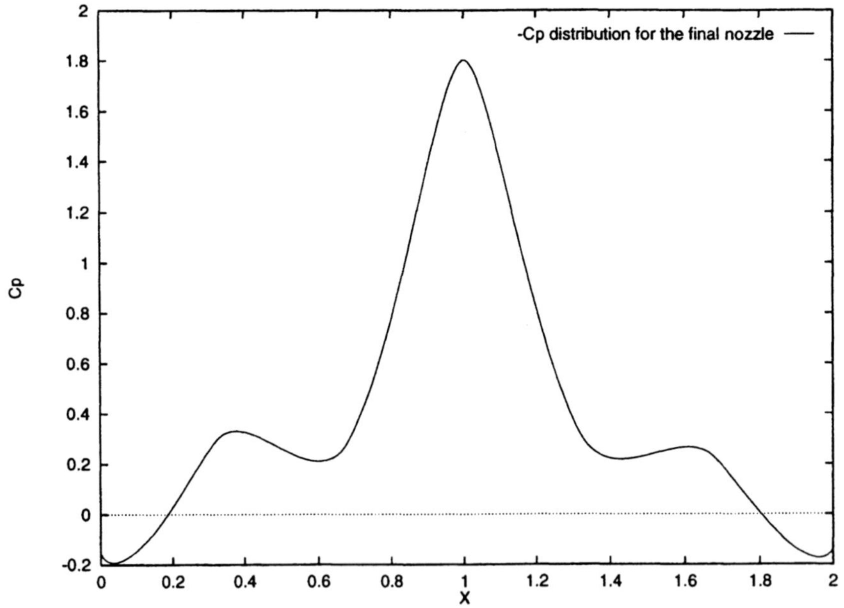

Figure 10 $-C_p$ distribution in the area of the nozzle

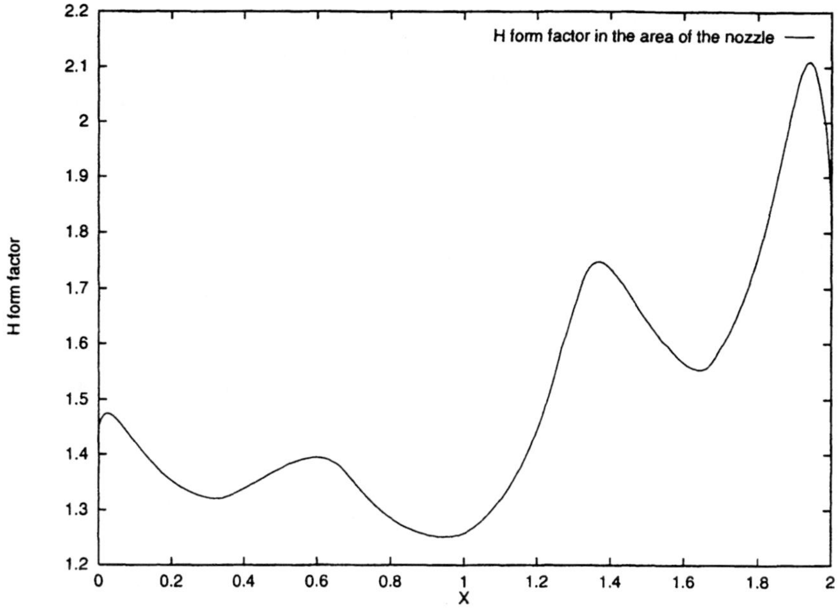

Figure 11 *H* form factor in the area of the nozzle

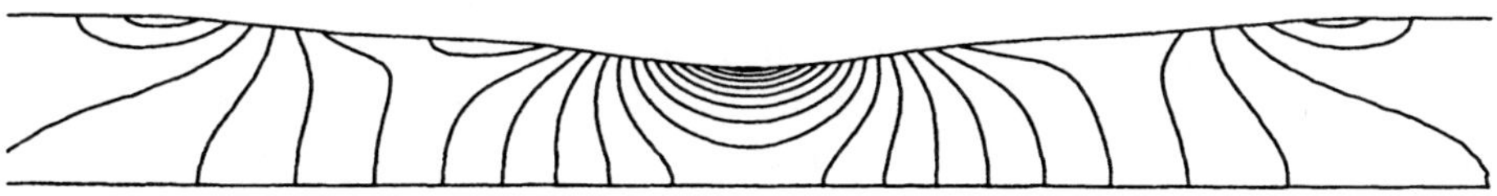

Figure 12 Isolines of the Mach numer obtained for the final design

12.10 Numerical Results for the TE2 test case

The methodology that has been used for the resolution of the test case TE2 by UPC is exactly the same as described for the tet case TE1. The TE2 test case is an inverse problem where the difference between a target $C_{p_{target}}$ distribution and the C_p distribution corresponding to a given design is used as an objective function. Due to the thin boundary layer obtained at $R_e = 3x10^6$ and assuming that the pressure distribution is constant across the boundary layer a week coupling has been used between the inviscid full potential model and the boundary layer equations. The procedure for this coupling is as follows:

- The velocities are computed by using the full potential model.
- The boundary layer equations are solved by using the inviscid velocities around the profile. This provides the boundary layer thickness at each point of the profile.
- For each point of the profile the C_p coefficient is evaluated at a distance equal to the boundary layer thickness from the profile.

The target $C_{p_{target}}$ distribution has been obtained by a direct analysis of the NACA0012 profile using linear triangular elements and an adaptive remeshing scheme with a maximum global percentage of error γ fixed to 0.01%. The objective F funcion has been computed as:

$$F = \int (C_p - C_{p_{target}})^2 ds \,. \tag{1}$$

Figure 13 shows the initial shape of the profile corresponding to a NACA63215 and the mesh used for the analysis of this initial design. Linear triangles (P1 elements) have been used for the discretization of the domain. The limit of the allowable global percentage of error γ has been fixed to a 0.1%.

Fifteen design variables have been used for the definition of the shape of each design. This variables are y coordinates of 7 points sited on the upper surface, 7 points sited on the lower surface and the leading edge. The 7 points sited on both the upper and lower surfaces are distributed in a sinusoidal form with a bigger concentration in the leading and trailing edges.

Figure 14 shows the shape and the mesh the final design obtained after 100 iterations.

Figure 15 shows the evolution of the objective function during the iterative process. The process seems to be converged after this number of iterations.

Figure 16 shows the superposition of the C_p distributions corresponding to the initial, final and target profiles. It can be observed how the agreement between the final and the target profiles is almost perfect.

Figure 17 shows the superposition of the geometries corresponding to the initial, final and target profiles. Here also, the agreement between the final and the target profiles is almost perfect.

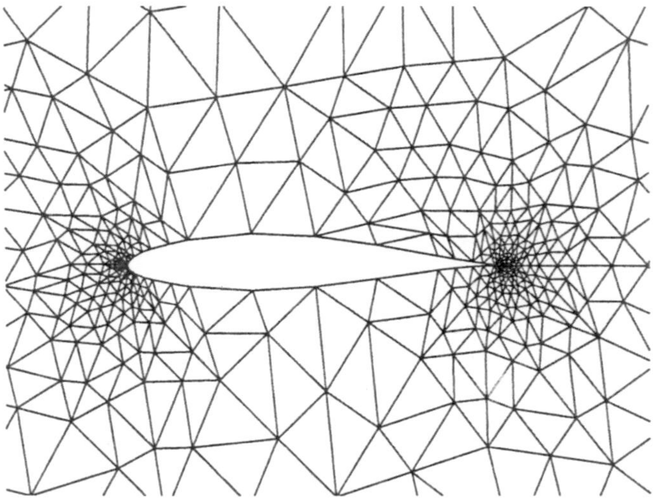

Figure 13 Initial shape and initial mesh

Figures 18, 19 and 20 show the evolution of the lift, drag and pitching moment coefficients. All of them remain almost constant after some big changes at the beginning of the iterative process.

12.10.1 Analysis of the results

The conclusions from this test case are similar to those obtained with test case TE1:

- A very good agreement between the target and the final designs is obtained. Nevertheless, the quality of the convergence process is very influenced by the parametrization. In fact, the number of points used for the definition of the geometries is critical.
- The quality of the meshes used for the analysis of each design is well controlled and the good quality of the results obtained for each design is assured.
- In order to get a good convergence of the iterative process it is necessary to use a very strict limit in the global percentage of error.

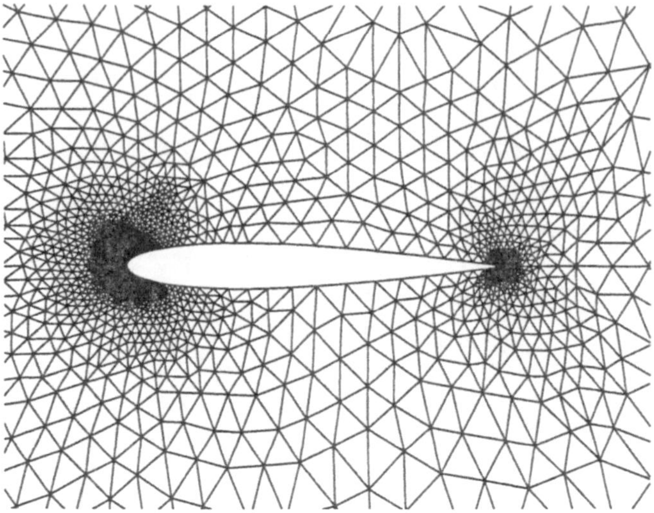

Figure 14 Final shape and final mesh

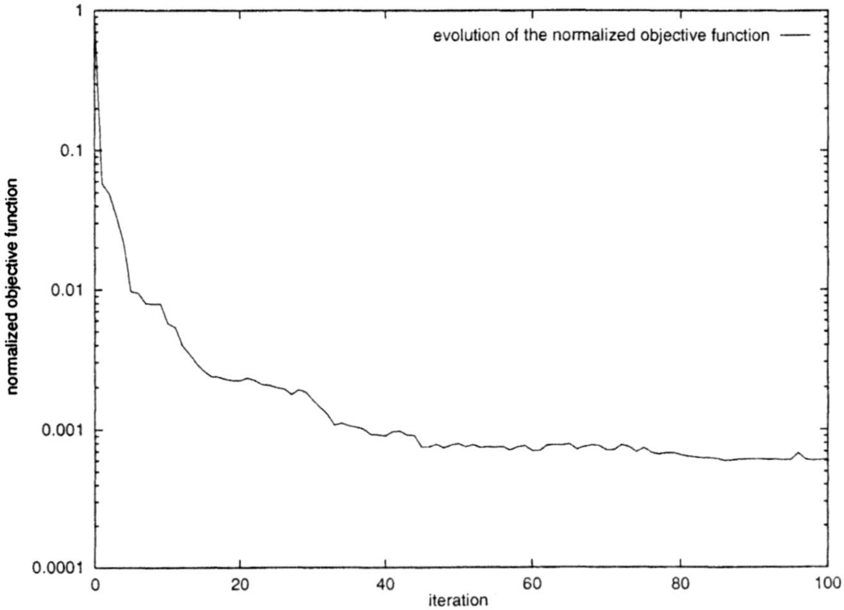

Figure 15 Evolution of the normalized objective function

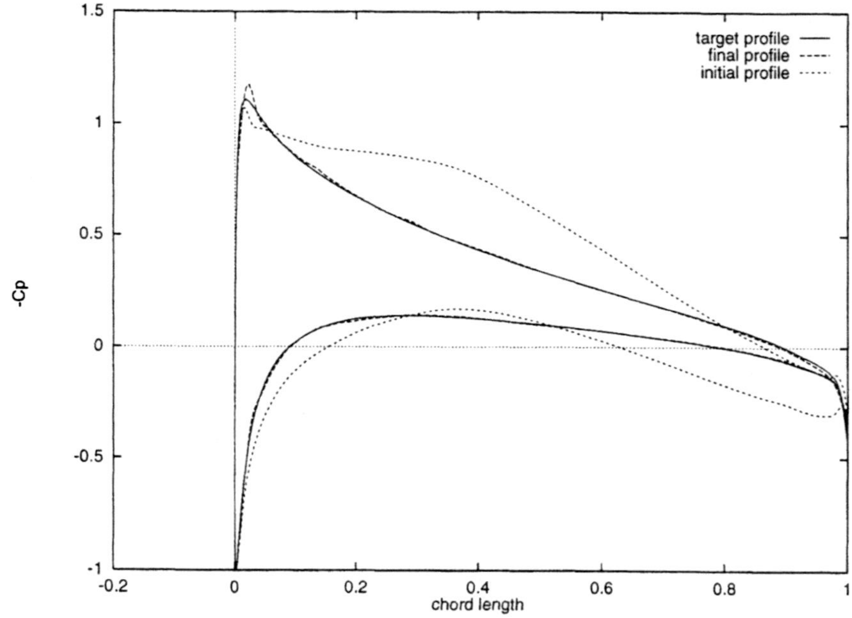

Figure 16 Superposition of the initial, final and target C_p distributions

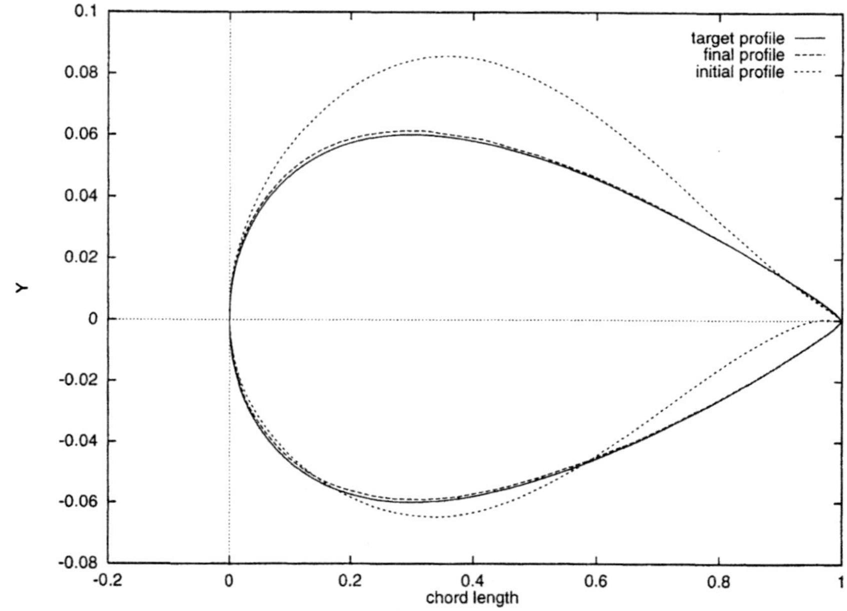

Figure 17 Superposition of the initial, final and target C_p geometries

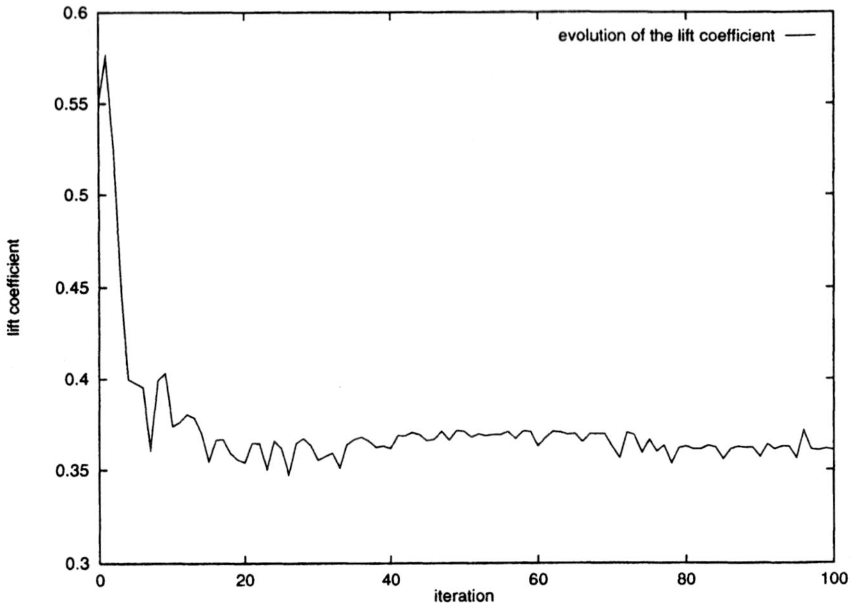

Figure 18 Evolution of the lift coefficient

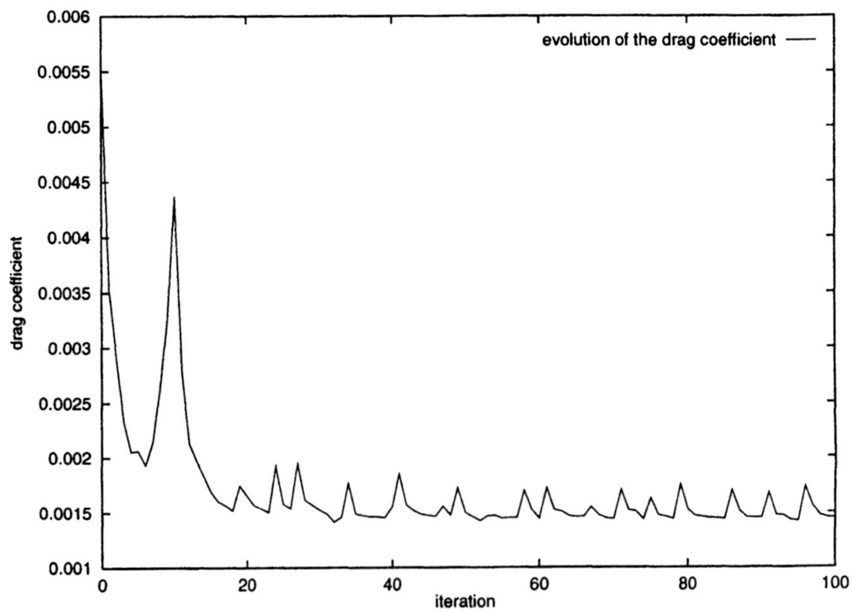

Figure 19 Evolution of the drag coefficient

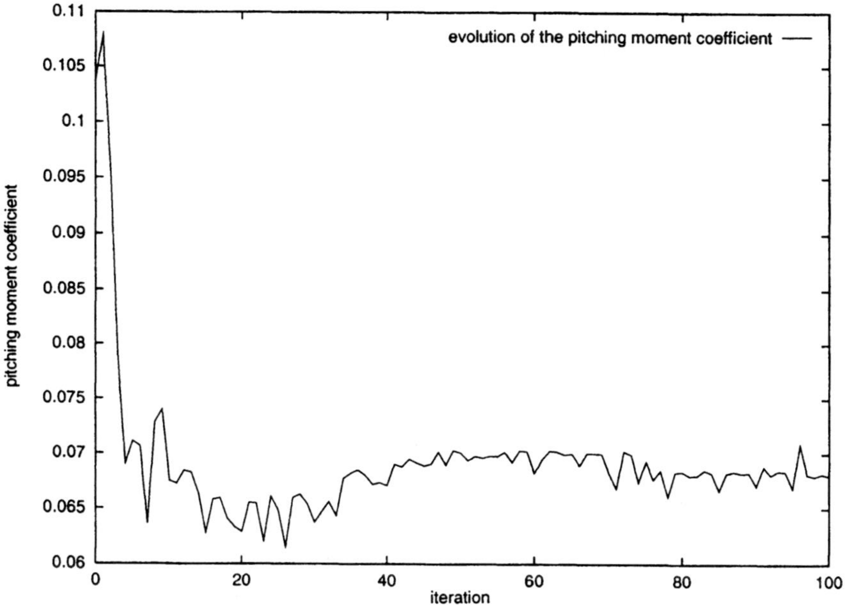

Figure 20 Evolution of the pitch moment coefficient

IV Synthesis of Test Cases

This section contains the synthesis of the different contributions data stored in the workshop Database. The synthesis includes, for each test case, a brief description, the special difficulties for the resolution and a comparison of the results presented by different contributors.

Different approaches (choice of approximation, optimizer and parametrization technique) have been used by the contributors as shown in Tables 1,2,and 3.

Table 4 provides some general information concerning the contributions presented for each test case. It shows the number and the name of the contributors and some key information about the tools used for the solution. (fp=full potential; bl=boundary layer; ct=control theory; ga=genetic algorithms)

Table 1 Flow analysis solvers

Aerospatiale	3-D panels, incompressible
Alenia	3-D full potential, structured grids
CASA	VII strong coupling, Euler + Integral boundary layer, structured grids
Dassault-INRIA	Euler, unstructured grids
Dasa-M	3-D panels, incompressible
Dasa-Airbus	VII, panels + Integral boundary layer
Dornier	VII strong coupling, Euler + Integral boundary layer, structured grids
NLR	Potential, Euler, Navier Stokes, structured grids
NTUA	VII strong coupling, potential + Integral boundary layer, Euler, structured grids
UPC	VII weak coupling, full potential + Integral boundary layer, unstructured grids

A detailed description of each design tool can be found in the corresponding authors contribution section. A brief summary of these contributions is presented here for completeness.

1 Methodologies and contributions from each partner

1.1 Aerospatiale

- methodology: A panel method is coupled to an industrial minimization package for designing engined wings. The wing parametrization is performed within the CAD environment. Designer parameters like twist, thickness, camber and local curvature of the wing sections are selected. Each degree of freedom is simulated by injection distributions to avoid re-meshing and re-computation of the influence matrix.

Table 2 Optimization technique

Aerospatiale	Industrial minimization package, injection techniques for the d. o. f.
Alenia	Commercial package based on the feasible direction approach
CASA	Steepest descent
Dassault-INRIA	Adjoint operator, hierarchical optimization, genetic algorithms
Dasa-M	Augmented Lagrangian, multi-gradient algorithm
Dasa-Airbus	Levenberg-Marquardt. Unification of steepest descent and inverse Hessian
Dornier	Vanderplaats, hill-climber strategy
NLR	Residual correction method
NTUA	One shot inverse approach
UPC	BFGS Quasi-Newton

Table 3 Parametrization technique

Aerospatiale	Designer parameters like twist, thickness, camber, curvature
Alenia	Span-wise linear interpolation, Bezier curve segments and B-splines on the wing sections
CASA	Tchebicheff polynomials
Dassault-INRIA	Piecewise continuous splines, Bezier curves
Dasa-M	Higher order span-wise interpolation, B-splines on the wing sections
Dasa-Airbus	Span-wise linear interpolation, B-splines expansions on the wing sections
Dornier	Bezier weighting point technique
NLR	Surface nodes
NTUA	Surface nodes
UPC	Cubic B-splines

- contribution to the workshop: Definition and computations for the engined wing optimization case.

1.2 ALENIA

- methodology: A non-conservative full potential solver is coupled to an optimization module which employs the method of Zoutendijk, based on the feasible direction algorithm. First order finite difference schemes are used

Table 4 Contributions to the workshop test cases

test case	partner	control variabl.	flow solver	convergence	Obj. funct. Evaluations
TE1	NTUA	3-5	inverse fp+bl	1-5 drag count	one-shot
	UPC	4	fp+bl	2 drag counts	100
TE2	CASA		Euler +bl	-	7
	Dassault INRIA	20	Euler +bl	2 orders	200
	NLR		Navier Stokes	<1 order	10
	NTUA	140	inverse fp+bl		one-shot
	UPC	15	fp+bl	3 orders	100
TE3	ALENIA	24	fp	drag 0.003 to 0.001	150
	CASA		Euler		60
1.	Dassault INRIA	20	Euler+ct	drag 0.0675 to 0.0055	12
2.	Dassault INRIA		Euler+ga	drag 0.0675 to 0.0032	120
1.	Dornier	10	Euler	drag 0.00665 to 0.000024	88
2.	Dornier	10	Euler+bl	drag 0.001 to 0.00089	143
3.	Dornier	10	Euler+bl 2 point		118
TE4	Dassault INRIA	35	Euler+ga	<1 order	30
1.	Dornier	10	Euler	<1 order	284
2.	Dornier	10	Euler+bl	<1 order	296
1.	NLR		panel	1 order	10
2.	NLR		Euler	-	6
TE5	Aerosp.	283	3-D panel	170 to 79	277
1.	Dasa-Airbus	105	3-D panel		4
2.	Dasa-Airbus	105	3-D panel+bl		4
TE6 1.	Alenia	16	Euler	drag 0.0126 to 0.0094	120
2.	Alenia	32	Euler	drag 0.0126 to 0.0095	
TE7	NTUA	160	Inverse Euler		one-shot

to compute the gradients of the objective and constraint functions. Bezier splines are used for the parametrization of the profiles/wing sections.

- contribution to the workshop: Definition and computations for the transonic wing wave drag minimization case. Computations are also performed for the two-dimensional - equivalent case TE3.

1.3 CASA

- methodology: The aerodynamic code is based on the viscous inviscid interaction technique where a conservative streamlined Euler solver is strongly coupled to an integral boundary layer solver. The discrete equations are solved together by means of a Newton solver. The optimization algorithm is based on the steepest descent method. Tchebicheff polynomials are used for the parametrization of the profiles.

- contribution to the workshop: computations are presented for the single airfoil reproduction case TE2 and the 2-D wave drag minimization case TE3.

1.4 DASSAULT-INRIA

- methodology: The flow solver is based on the conservative Euler equations discretized on an unstructured triangular mesh. Two different optimization techniques are adopted. The first employs control theory techniques where the gradient of the cost function is computed by means of the adjoint equation. A conjugate gradient method has been selected as descent method with a possible combination with hierarchical optimization. Parametrization is based on splines. The second method employs a genetic algorithm with real coding of the design variables. In this case the parametrization relies on a Bezier curve.

- contribution to the workshop: definition of TE3 and TE1 (with NTUA). Computations are presented for the test cases TE2 (airfoil reproduction) using a weak VII approach, TE3 and TE4 the multi-point airfoil design.

1.5 Dasa-M

- methodology: A 3-D panel method is coupled to an optimization module based on the augmented Lagrangian, multi-gradient algorithm. Wing sections are parametrized by means of B-splines. Fourth order interpolation schemes are used in the span-wise direction for wing parametrization.

- contribution to the workshop: computations are presented for the engined wing optimization case TE5.

1.6 Dasa-Airbus

- methodology: An inverse design by direct optimization method is used. The flow code employs a panel method based on a potential formulation for the perturbation velocity. Viscous effects are introduced with a transpiration model derived from integral boundary layer calculations along streamlines. The optimization scheme is based on the Levenberg-Marquardt method. B-spline parametrization of profiles is combined to a linear span-wise interpolation scheme for wing analysis.

- contribution to the workshop: computations are presented for the engined wing case TE5, with and without viscous effects.

1.7 Dornier

- methodology: The aerodynamic code is based on the viscous-inviscid interaction technique where a streamlined conservative Euler solver is strongly coupled to an integral boundary layer solver of the semi-inverse type. A fully coupled Newton scheme, which also produces sensitivities at no additional cost, is employed to solve the discrete system of equations. A hill-climber strategy is adopted for the optimization problem. A Bezier weighting point technique is used for profile parametrization.

- contribution to the workshop: Three different computations are presented for case TE3, corresponding to a single point drag reduction problem, with and without viscous effects, and to a two-point viscous drag reduction test-run. Viscous and inviscid results are also presented for the multi-point design case TE4.

1.8 NLR

- methodology: The optimization scheme is based on the residual correction method, where the shape correction results after specifying an equivalent incompressible multi-point problem. The equivalent incompressible perturbation velocities are specified according to the pressure defect splitting technique (to distinguish between the subsonic and the transonic part). The method can be combined with a potential, an Euler and a Navier-Stokes direct solver.

- contribution to the workshop: results are presented for the test case TE2 using a Navier-Stokes flow solver and TE4 using a panel and an Euler direct code. First computations with Navier Stokes solvers or GA optimizers are also reported for TE4

1.9 NTUA

- methodology: The design approach is based on an inverse target pressure one-shot method. The method is valid for both irrotational and rotational single or multi-domain flows. Viscous effects are taken into account by means of a strong viscous-inviscid interaction scheme. The computational cost of the method is similar to the cost of a single direct solution (one shot). A target pressure optimization scheme is also available for duct and airfoil design.

- contribution to the workshop: definition of the test cases TE1 (with Dassault,INRIA),TE2 and TE7 (the axi-symmetric afterbody). Computations are presented for all three cases.

1.10 UPC

- methodology: The flow code is a full potential solver for unstructured triangular meshes, weekly coupled to an integral boundary layer solver. The optimization approach employs shape design using adaptive remeshing, so that both the mesh parameters and the error estimator are projected to the next enhanced design. The direction of change is obtained using the BFGS quasi-Newton or the GMRES scheme. Cubic B-splines are used for parametrization.

- contribution to the workshop: Computations are performed for the viscous drag reduction case TE1 and the airfoil reproduction case TE2.

2 Specific information for each test case

2.1 Test case TE1

Test case TE1 is a viscous drag minimization case for duct flows. Viscous effects are treated through a weak viscous inviscid interaction technique. The drag is computed from the boundary layer end- properties through the semi-empirical formula of Squire and Young.

TE1 was considered by the groups of UPC and NTUA. Both contributors concluded that the adopted boundary optimization scheme is very sensitive to the number of d.o.f.. When a relatively large number of d.o.f. is selected the obtained solutions exhibit a wiggled pressure distribution, leading to a consecutive acceleration and deceleration of the boundary layer. For this reason UPC introduced 4 d.o.f. only, while NTUA introduced 3 d.o.f.. The obtained drag reduction is of the order of 5indicating that the initial shape is a nearly optimum one. The NTUA solution with 5 d.o.f. resulted to a 20UPC. However, it is not apparent that the semi-empirical Squire and Young formula, which estimates the total drag (pressure and skin friction) is valid under such conditions. This last remark addresses the main difficulty of TE1. It would be, therefore, interesting to re-investigate the case using a turbulent Navier-Stokes flow solver instead of the VII scheme.

The obtained best duct shapes in the area of interest (x [0,2]) and the corresponding pressure coefficients are presented in figures TE1.1 and TE1.2 respectively. The computational meshes (which consist part of the solution for both approaches) and field plots of the Mach number for the best design are shown in figures TE1.3 and TE1.4.

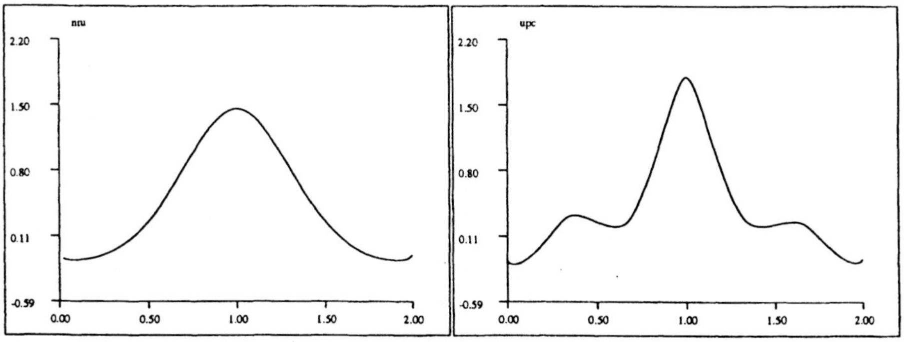

Figure TE1.1 Optimized duct shape

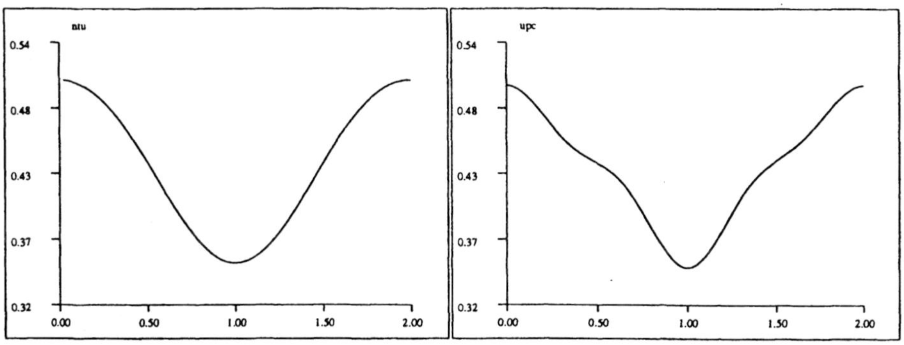

Figure TE1.2 Optimized pressure coefficient

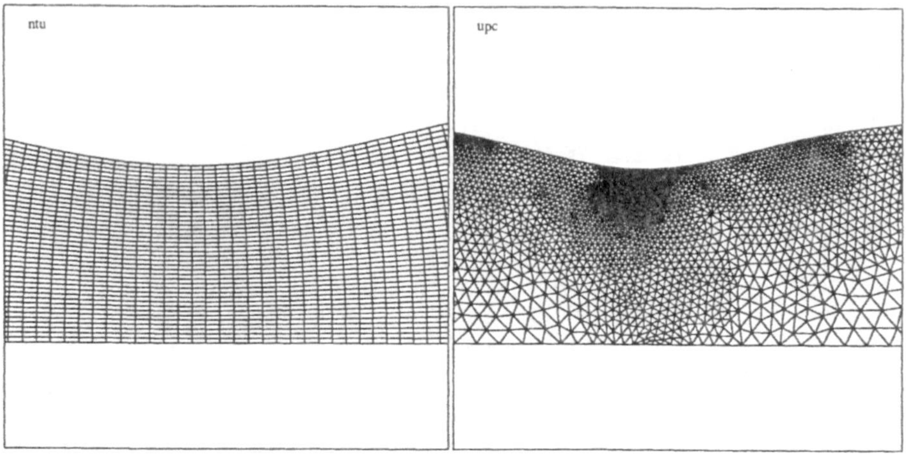

Figure TE1.3 Grid structures for the best design

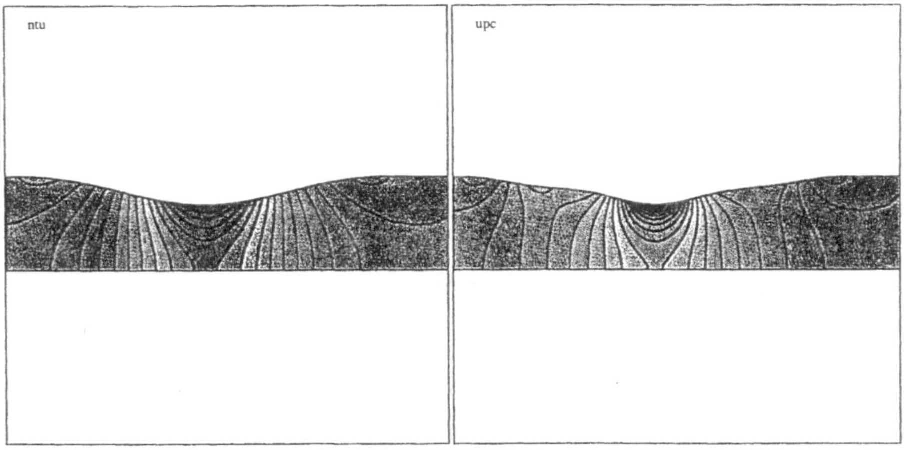

Figure TE1.4 Mach number contours for the best design

2.2 Test case TE2

TE2 is an airfoil reproduction case under subsonic viscous flow conditions. With the exception of NLR, which employed a Navier-Stokes flow solver, all the rest contributors tackled the problem with VII schemes. The target pressure distribution, corresponding to the NACA 0012 profile at 3 degrees incidence, M = 0.3 and Re = 3 millions, is computed by each partner using his own analysis tool. These target pressures along with the reproduced ones are presented in figure TE2.1. The comparison of target pressures shows some differences which are due to the analysis method (location of the transition points, boundary layer solver used, etc). These differences are more pronounced in the NLR and DLT computations and may be either due to convergence problems or to inaccurate grid resolution at the near-leading edge region, where the suction peak occurs.

The reproduced and the target profiles are compared in figure TE2.2. The quality of the reproduction is very good in most of the cases. Convergence problems seem to spoil the accuracy of the DLT and the NLR results. In the NLR method the cost function is computed by means of a Navier-Stokes solver (the only attempt in this direction in the present context), while the gradient is evaluated by an equivalent potential solution. As reported by NLR this inconsistency creates convergence problems in the stagnation point region and, thus, the iterative process is stopped after the first 10 iterations.

The convergence properties of the different optimisation schemes employed are presented in figure TE2.3. The normalized L2 distance norm between the final and the target profiles is plotted against the number of cost function evaluations. The one-shot method of NTUA is, obviously, missing from this plot. It is seen that in all cases the convergence is fast during the first few iterations but stalls as the number of iterations increases. In all cases the adopted parametrization schemes did not allow the error norm to drop more than one order of magnitude.

2.3 Test case TE3

TE3 addresses the wave-drag minimization problem in transonic flows. The problem is posed in two different ways. In the first version a target pressure problem is formed where the drag is added to the cost function through a penalty. In this case the expected solution is an airfoil which satisfies the initial pressure distribution everywhere except the shock wave region, where the shock is replaced by a smooth, drag-free, Mach number drop. In its second variant the problem is posed as a drag minimization one where the lift (and the incidence) are constrained. With the absent of any additional constraints the problem has now more degrees of freedom.

Both full-potential and Euler solvers have been used by the participants for the flow analysis. Two contributors (DOR and DRA) tackled the total drag minimization problem (wave drag plus viscous) using viscous inviscid interaction schemes. DOR faced a two-point minimization variant of the problem, as well. DRA presents results obtained with a full potential, an Euler and a full potential VII scheme. DLT is also presenting four different results using the first

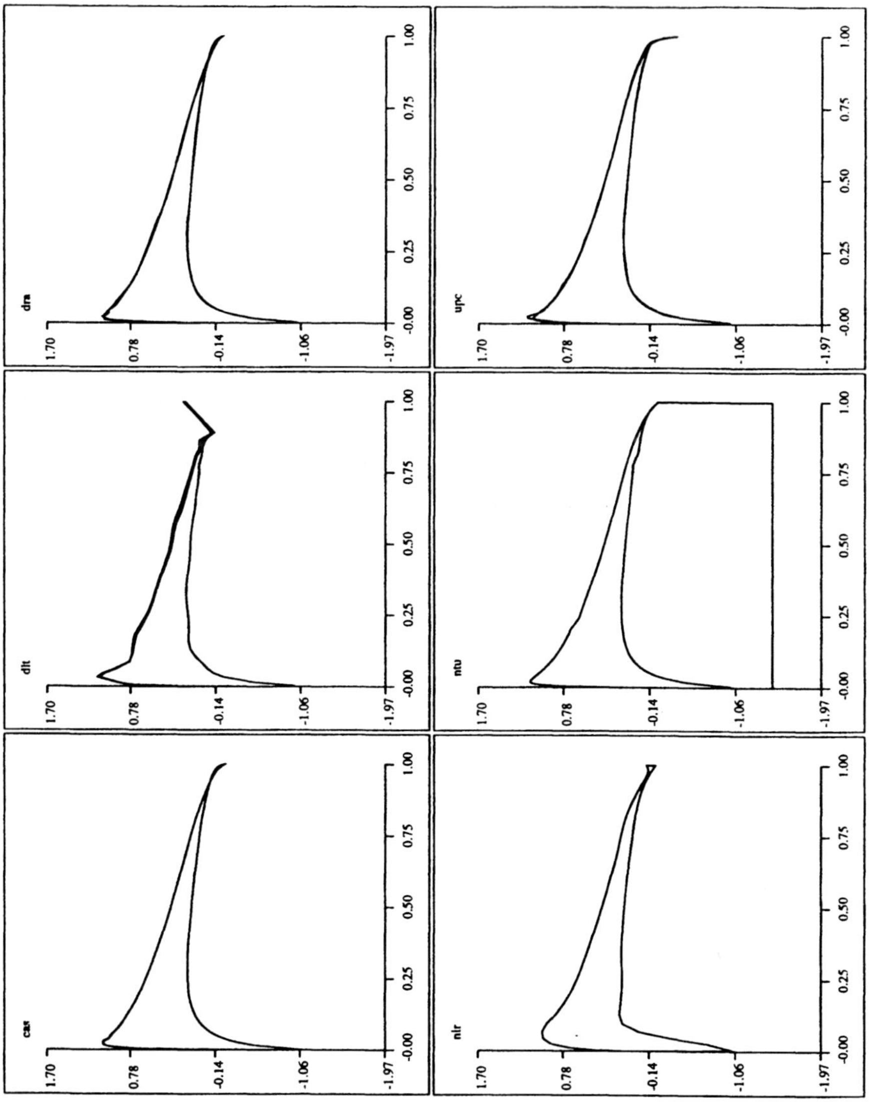

Figure TE2.1 Target and reproduced pressure coefficient distribution

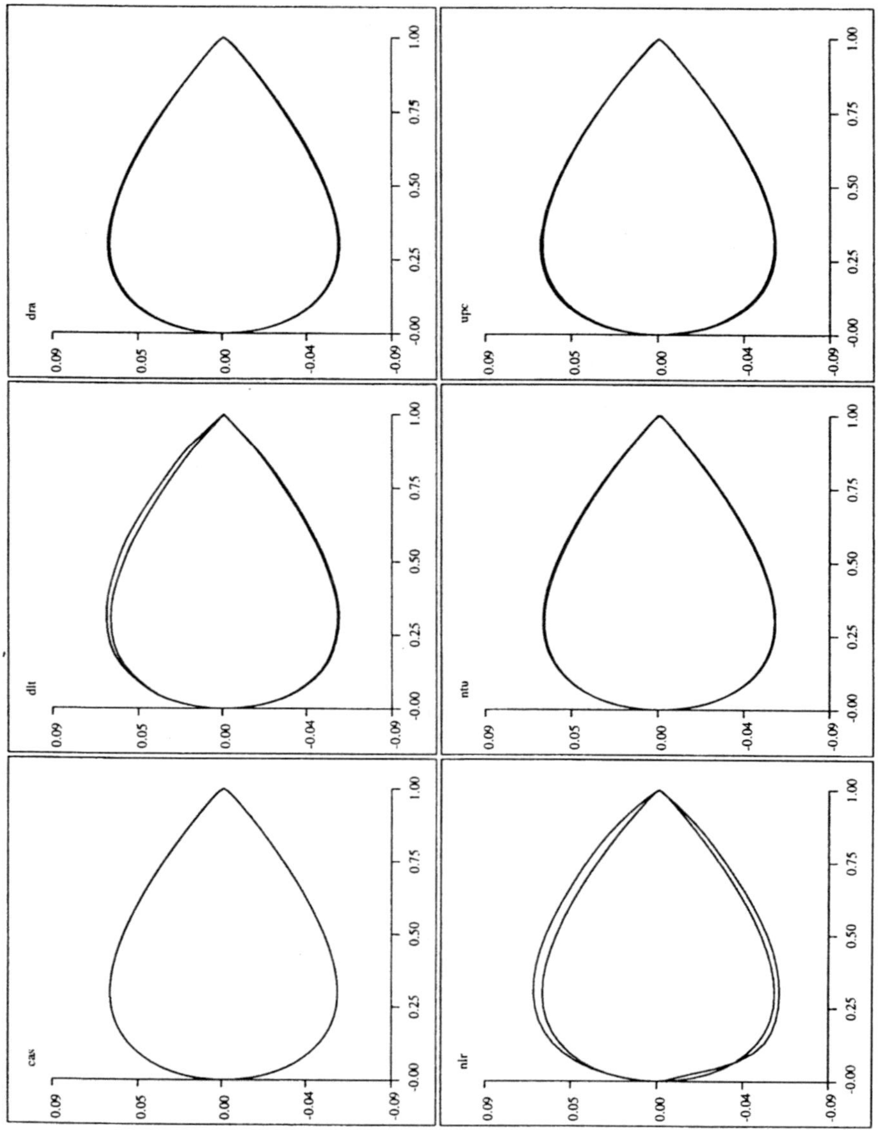

Figure TE2.2 Target and reproduced profiles

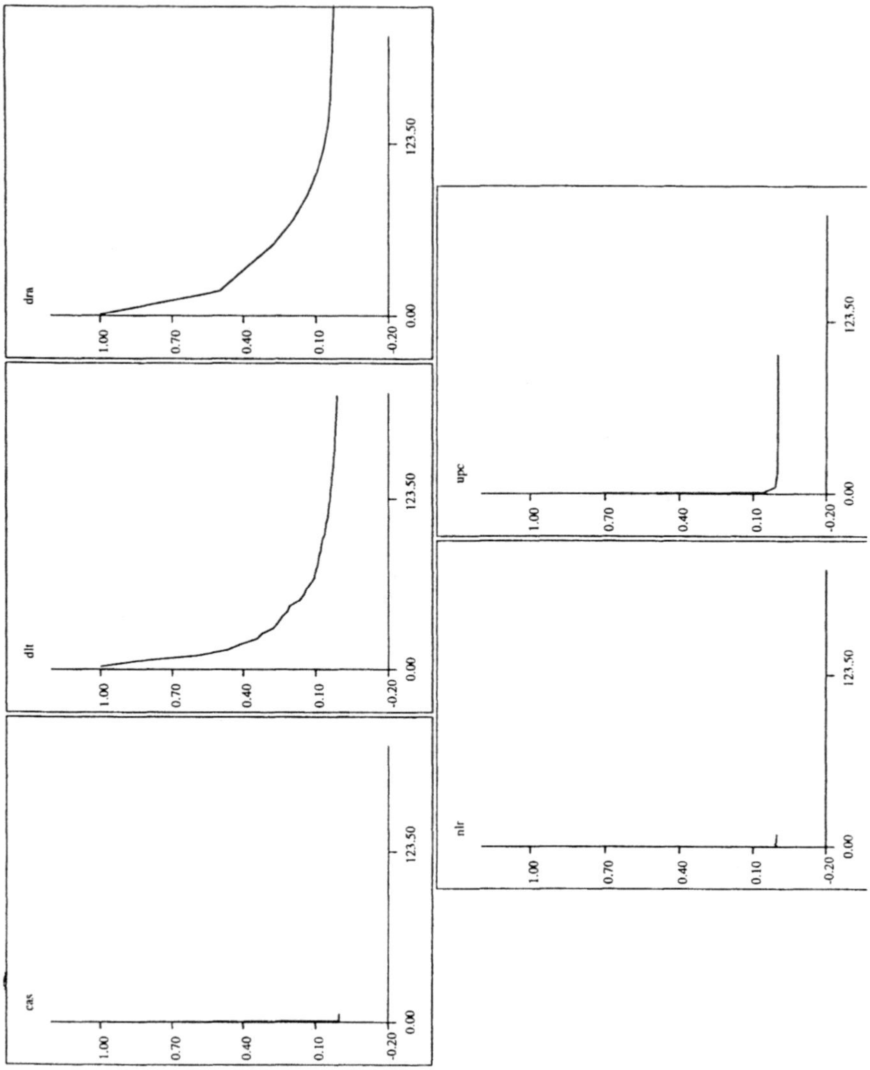

Figure TE2.3 Convergence versus number of cost function evaluation

variant of the problem with different weight factors and two different optimization techniques (a gradient method and a genetic algorithm one).

The initial and optimized pressure coefficient distributions calculated by all the contributors are presented in figure TE3.1. It can be seen that the initial pressure distributions already exhibit quite large differences (the shock strength as well) which are due to the flow solver (conservative or non- conservative full potential or Euler) and the resolution of the mesh (see the DLT cases for instance). As expected, the solutions obtained with different cost function definitions are quite different. With the exception of DLT which adopted the first variant of the problem, all the rest contributors used the second definition and their results are quite comparable. When no viscous effects are taken into account the resulting optimum shape is characterized by a pressure distribution which on the pressure side of the airfoil is quite similar to the initial one. On the suction side, however, the optimized flow seems to decelerate gradually starting from the leading edge region where a stronger suction peak appears compared to the initial profile which is characterized by a long flat pressure regime. Viscous optimization alters the pressure side distribution, as well, in an attempt to minimize its contribution to the viscous drag.

The comparison of the initial and the optimized profiles is shown in figure TE3.2. It is seen from the DLT results that the first variant of the problem produces minor local changes to the profile shape. The larger differences in pressure distribution resulting from the second definition of the problem produce larger differences in shape, as well. The profiles are now becoming thinner in their front part and their maximum thickness is moved backwards. For the inviscid case DOR reported the need of an additional constraint on maximum profile thickness. When viscosity is also taken into account the pressure side of the profile is drastically reshaped.

The evolution of the drag ratio in respect to the number of cost function evaluations of each optimization scheme is shown in figure TE3.3. A one order of magnitude drag reduction is obtained in some cases. However, as DOR comments, this usually corresponds to a local deep minimum which is not desirable in practice. DOR performed full polar-curve computations to show that small incidence variations around the shock-free design point increase the drag drastically. To avoid local minima DOR proposed a two-point optimization scheme.

The complete set of optimized profiles is presented in figure TE3.4.

2.4 Test case TE4

TE4 is a multi-point design case for airfoils. Two design points are defined, a subsonic high lift and a transonic low drag one. The cost function is defined as the best fit of the two individual target pressure distributions properly weighted. Results are presented by DLT, NLR and DOR for the 0.5, 0.5 weights. NLR used three different flow models, an incompressible panel method, an Euler solver and a Navier- Stokes solver. DLT used an Euler solver coupled to a genetic algorithm optimizer. DOR performed inviscid and viscous runs by means of an Euler solver and a viscous inviscid interaction technique. During the workshop the NLR-

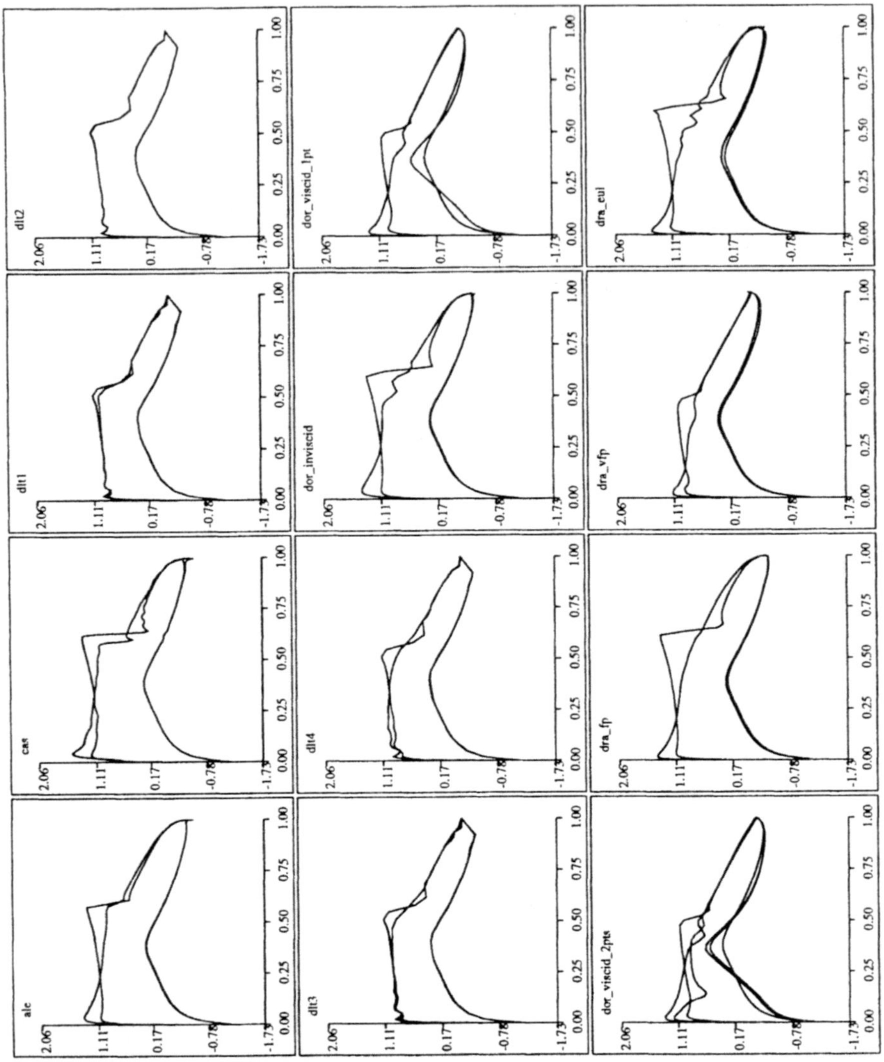

Figure TE3.1 Initial and optimized pressure coefficient distributions

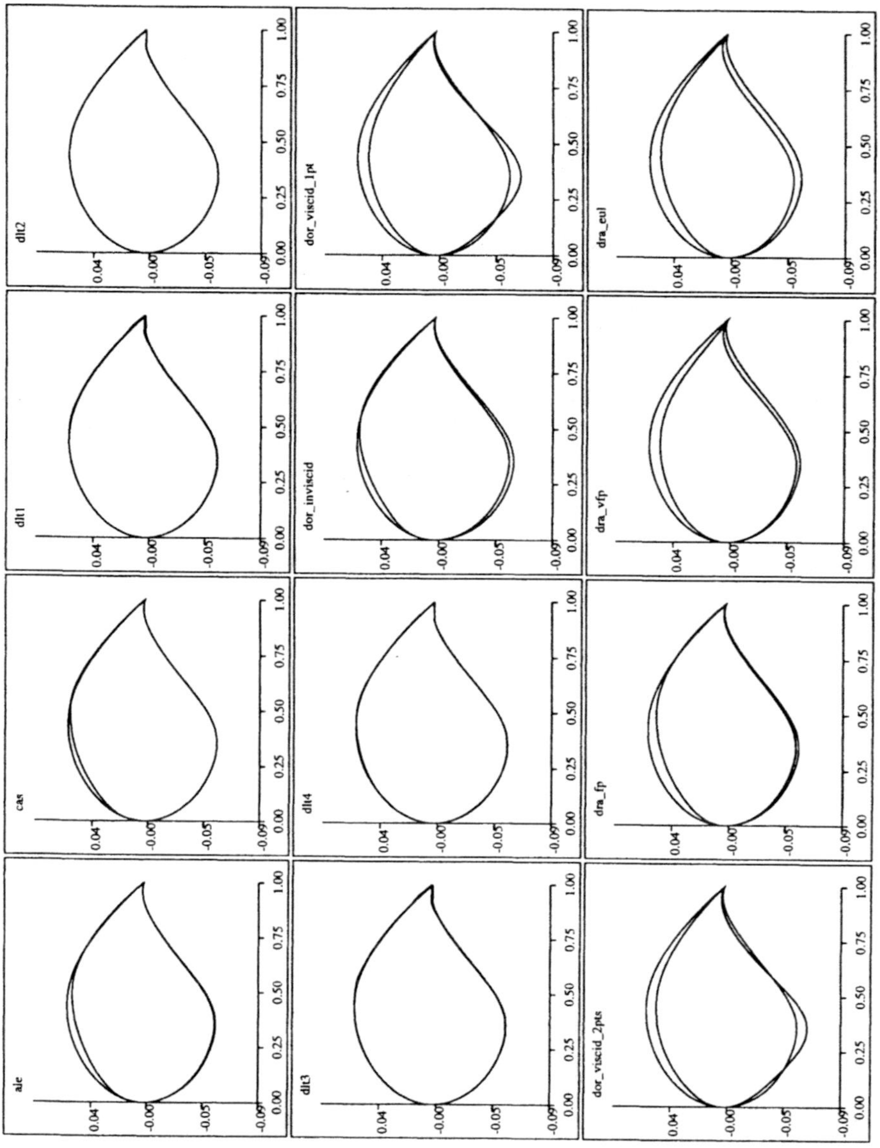

Figure TE3.2 Initial and optimized profiles

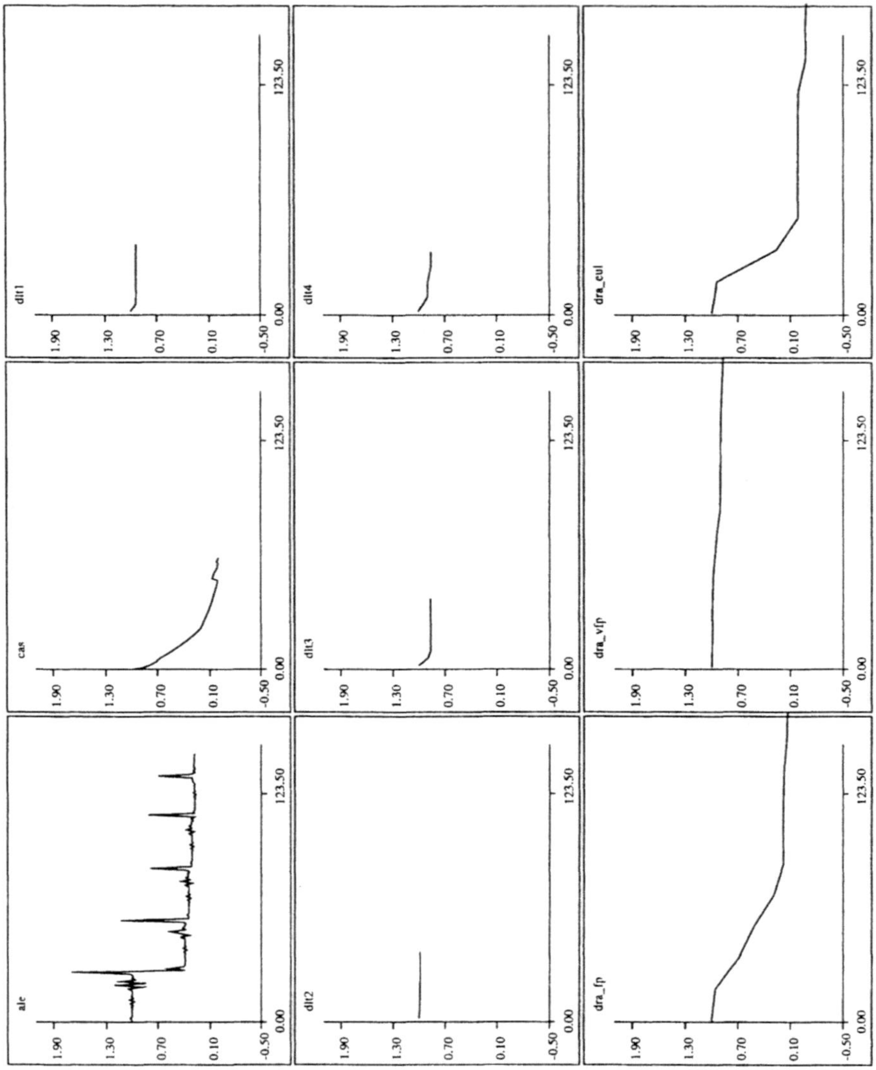

Figure TE3.3 Evolution of the drag ratio versus the number of cost function evaluations

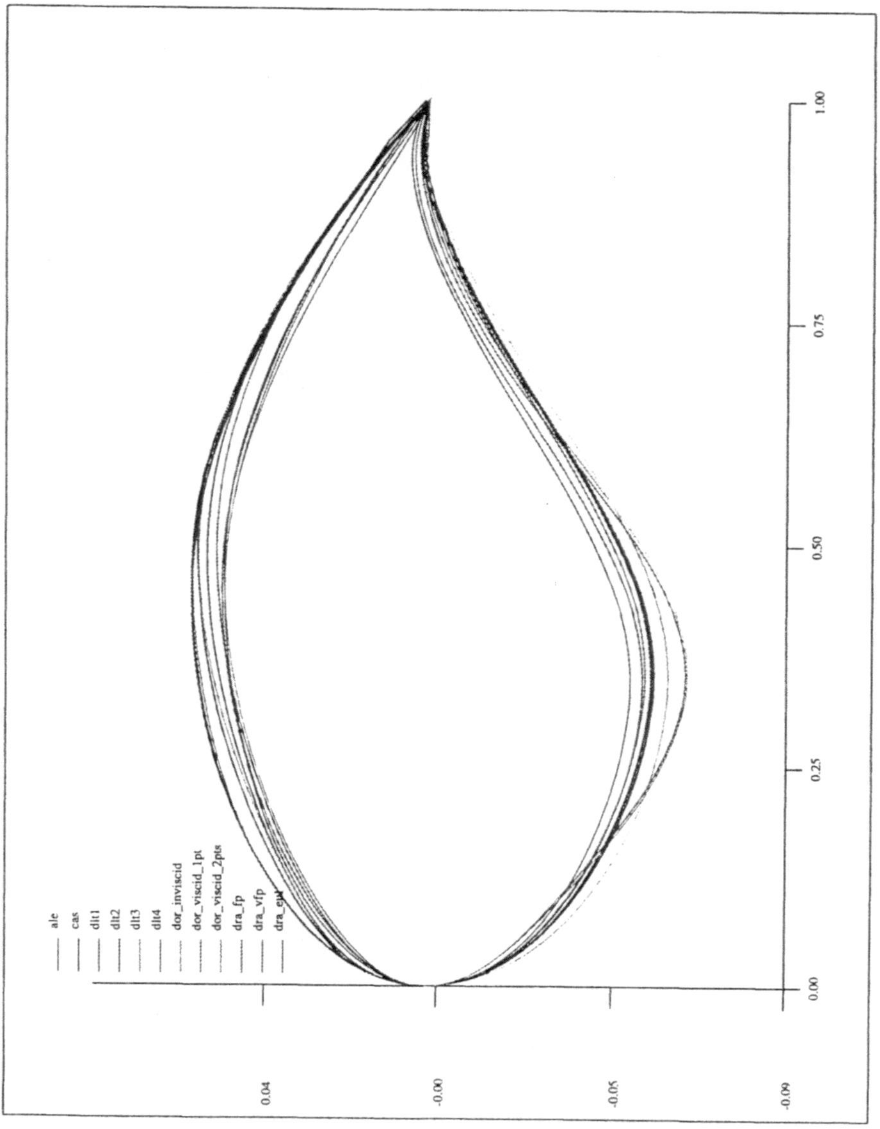

Figure TE3.4 Optimized profiles

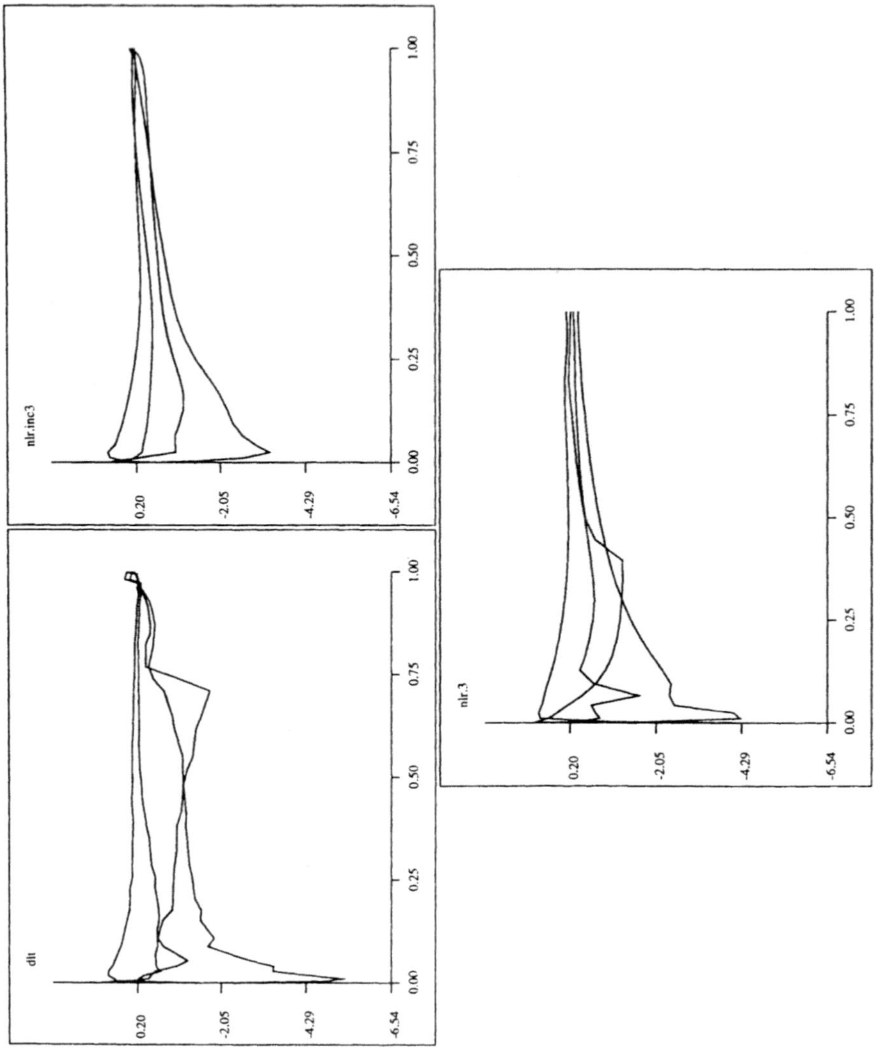

Figure TE4.1 Target C_p distributions for the two design points

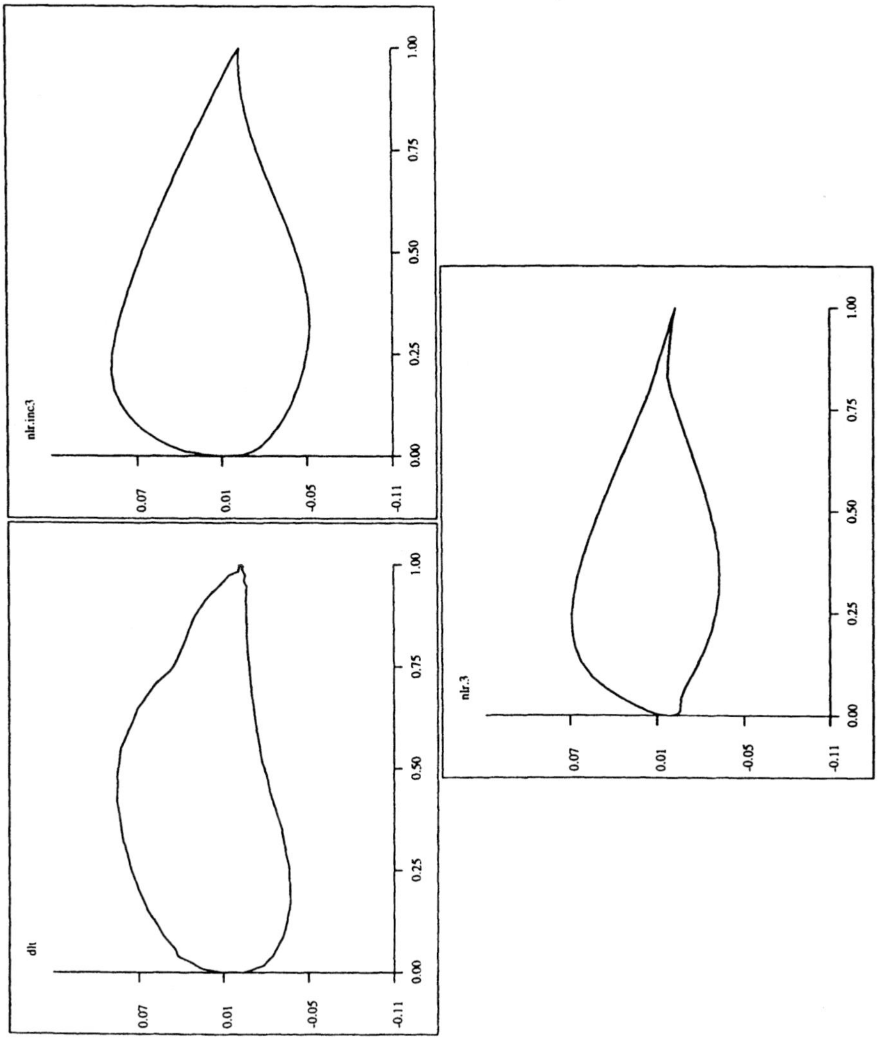

Figure TE4.2 Optimized profiles

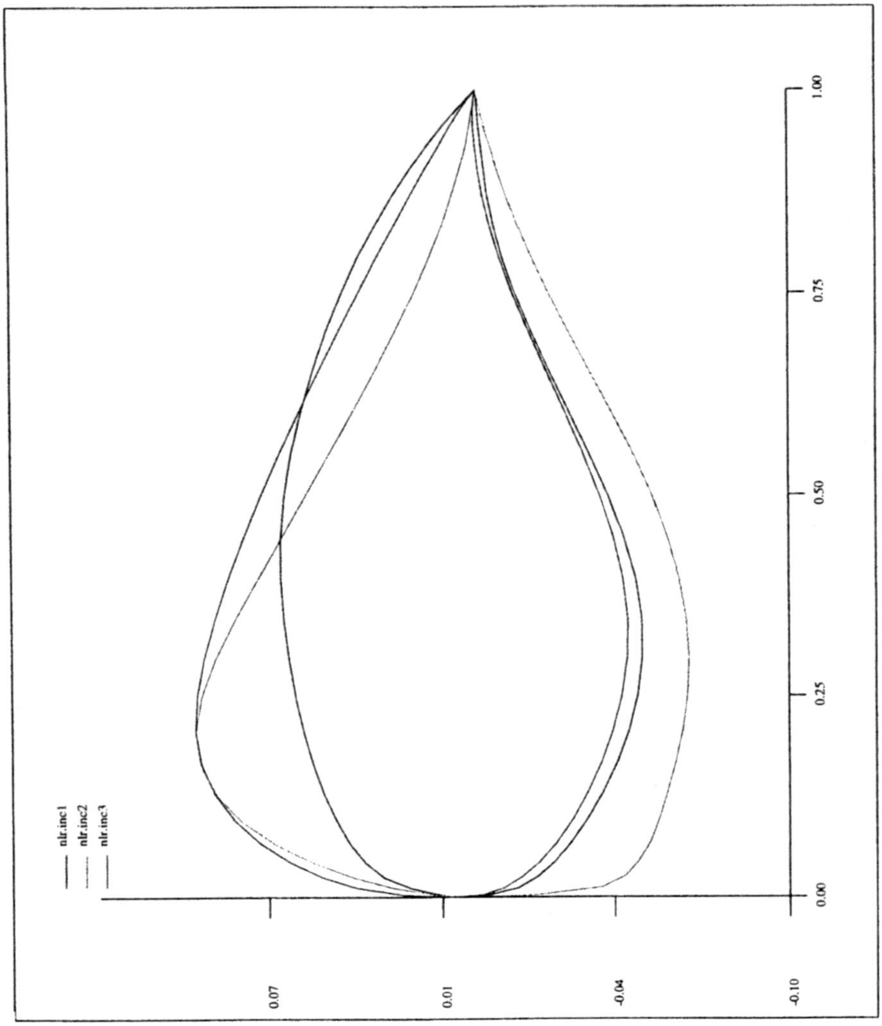

Figure TE4.3 Initial and optimized profiles

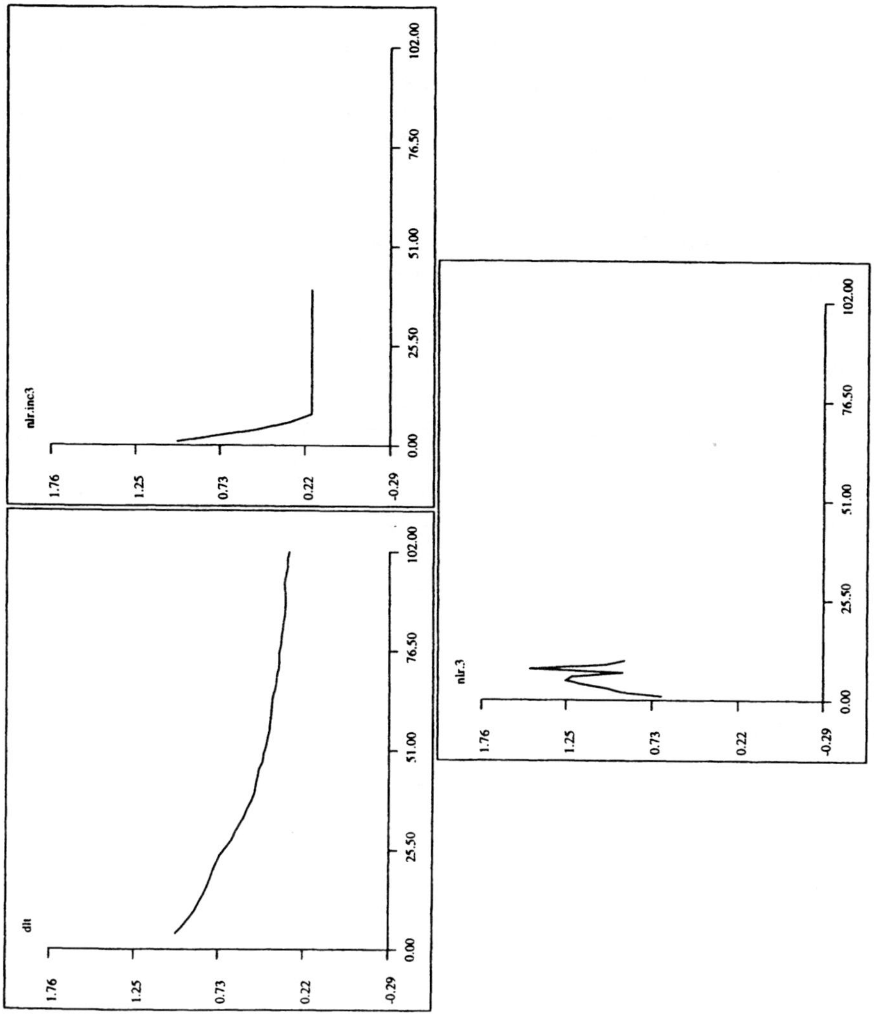

Figure TE4.4 Cost function versus number of cost function evaluations

Euler and DOR results were not available and, thus, they are not included in the following figures. It is suggested to the reader to search for additional information into the individual presentations of these two contributors.

The target pressure coefficient distributions computed by DLT and NLR for the two design points are shown in figure TE4.1. Large differences are observed which are due to the different flow models (DLT-Euler, NLR-potential, Navier-Stokes) and the grid resolution. The corresponding optimum shapes are presented in figure TE4.2. It is seen that the target pressure differences are reflected to the profile shapes. The DLT result, however, suffers from some wiggles on the profile geometry which are due to the lack of some geometry smoothing scheme. On the other hand the NLR.3 (Navier-Stokes) result is not properly converged because of numerical problems during the optimization process. The NLR Euler and the DOR inviscid results are pretty close to the NLR incompressible result. All contributors remarked that the best (inviscid) fit corresponds to a profile shape that locally follows the higher between the two pressure coefficient values. This is clearly seen in figure TE4.3 where the two initial profiles are plotted together along with the optimized one by NLR.

The convergence history of the three runs is presented in figure TE4.4. It is seen that in all cases the cost function drops less than an order of magnitude since the two targets are essentially incompatible.

2.5 Test case TE5

TE5 is a design problem of industrial interest, where a wing-pylon-nacelle configuration is optimized in terms of the wing sections geometry aiming to recover the clean-wing loading. All the involved participants, D.B.A. AIRBUS, Dasa-M and AEROSPATIALE used panel codes for the flow analysis, posing the problem as an inverse target pressure one. The three solutions are characterized by a different number of design parameters. Dasa-M used the largest number of degrees of freedom, while AEROSPATIALE used the least but the most designer oriented ones, including the sections twist, thickness and camber. Some differences on the cost function definition are also observed concerning the weighting of the target pressures, D.B.A. AIRBUS performed an additional calculation where the viscous effects were included by means of a viscous inviscid interaction scheme (incompressible flow plus integral boundary layer). Further details concerning the applied optimization and parametrization techniques can be found in tables 1 to 4.

The addition of the pylon-nacelle configuration to the clean wing produces a local loss of lift on the outboard wing, an overspeed located at the inboard wing section aside the pylon and also influences the pressure recovery of the pylon on the lower surface of the engined wing. Within a potential flow context it can be said that the strength of the induced perturbations follows the law of 1/r**2 and that the near- nacelle wing sections feel this induction as a reduction of the local angle of attach. For this reason the optimized solution tries to recover the clean wing loading by increasing the local twist angle. This is clearly seen in figure TE5.1 where the span-wise distribution of the angle of attack is presented for

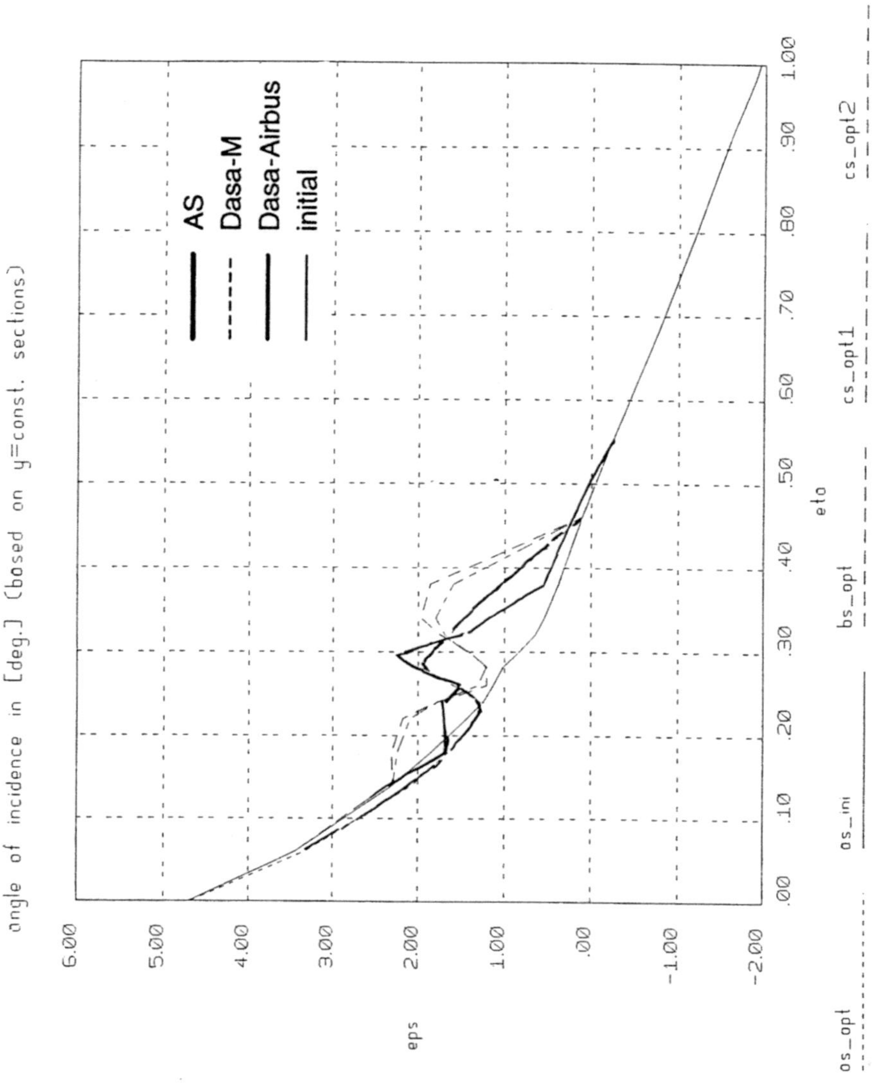

Figure TE5.1 Span-wise distribution of the angle of attack for the initial configuration and the optimized wings

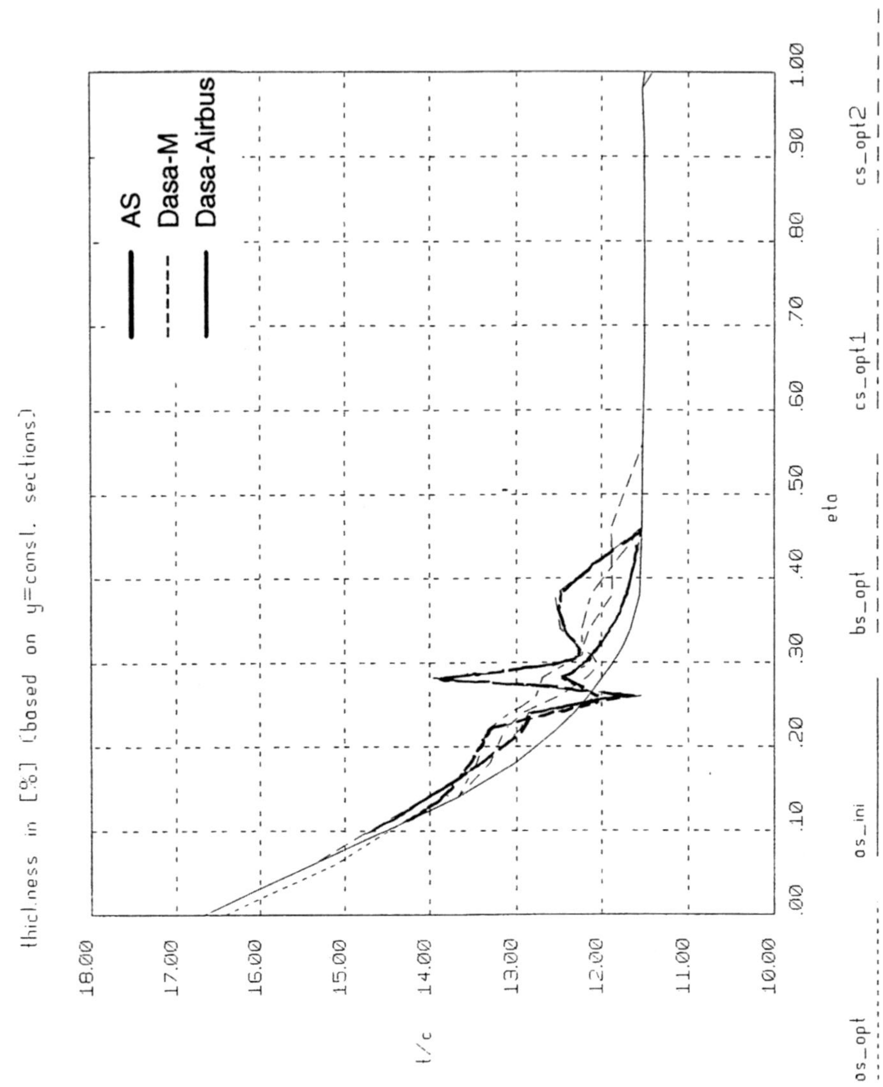

Figure TE5.2 Initial profiles thickness distribution against the optimized ones

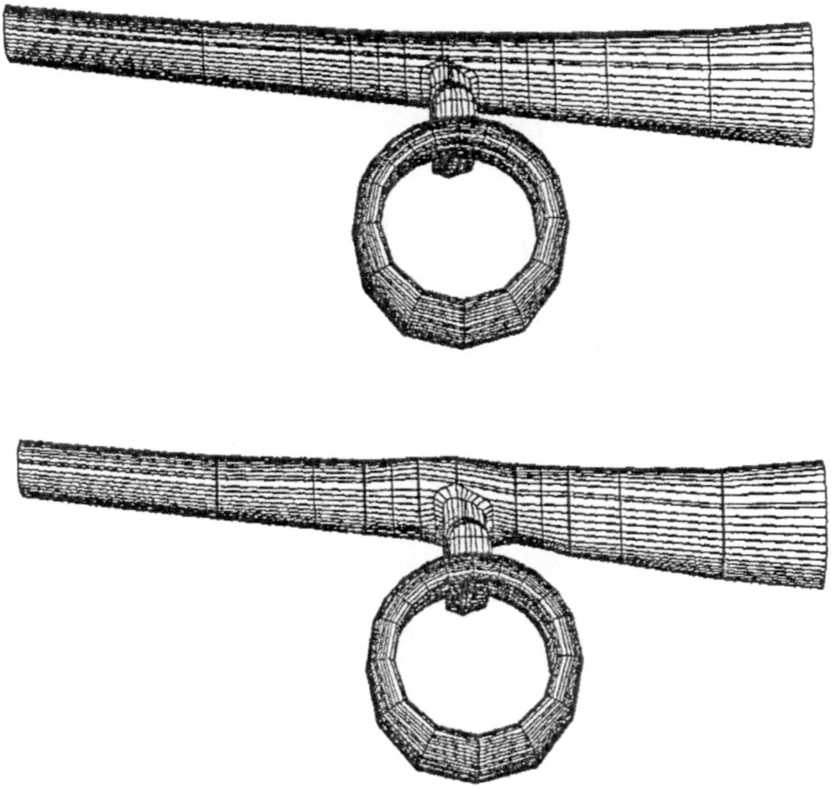

Figure TE5.3 Initial and final optimized solution obtained by Daimler Benz Airbus

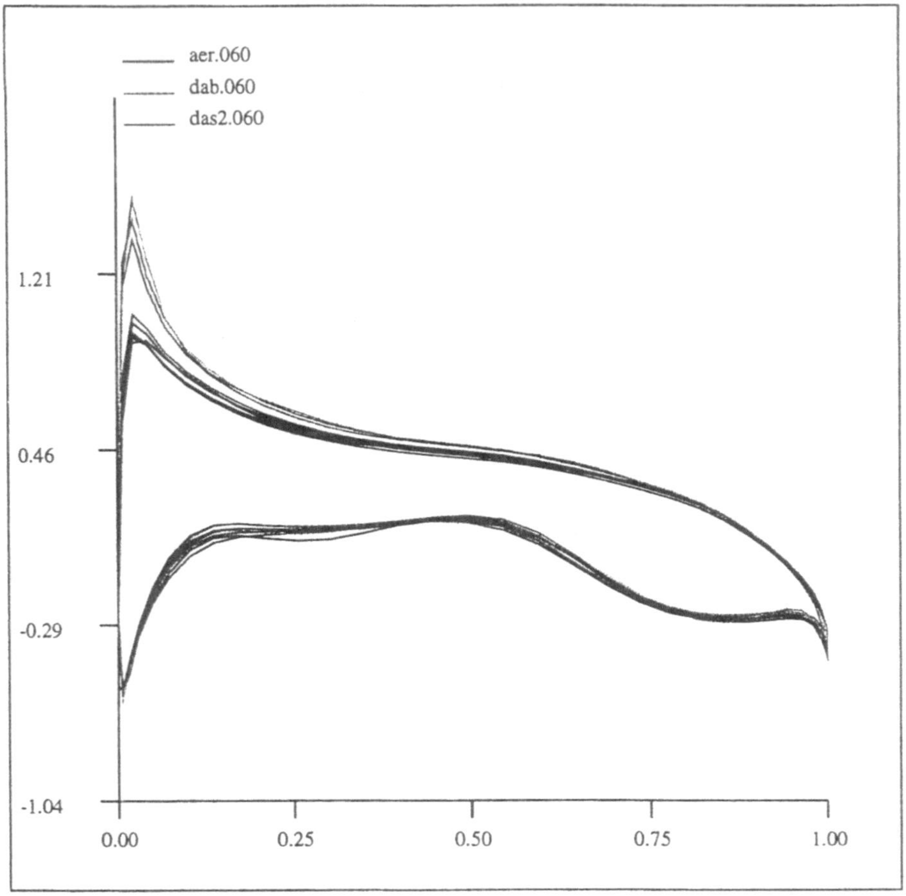

Figure TE5.4 Initial, target and optimized pressure coefficient distribution, section 060

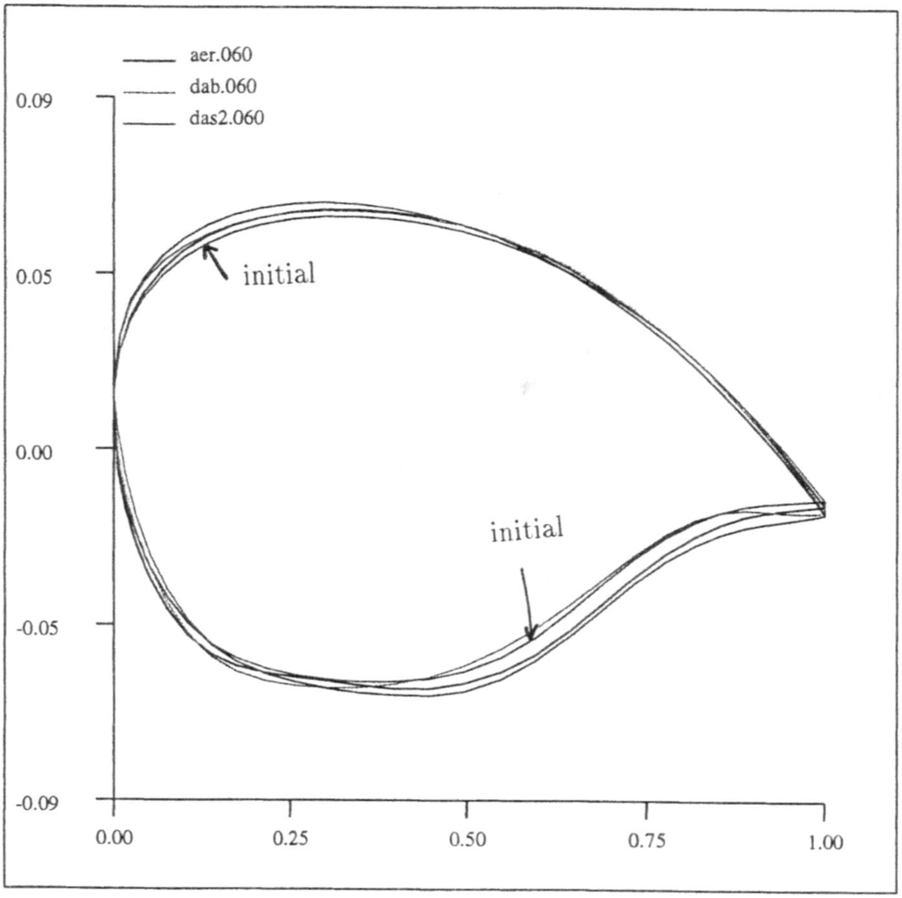

Figure TE5.5 Initial and optimized profiles, section 060

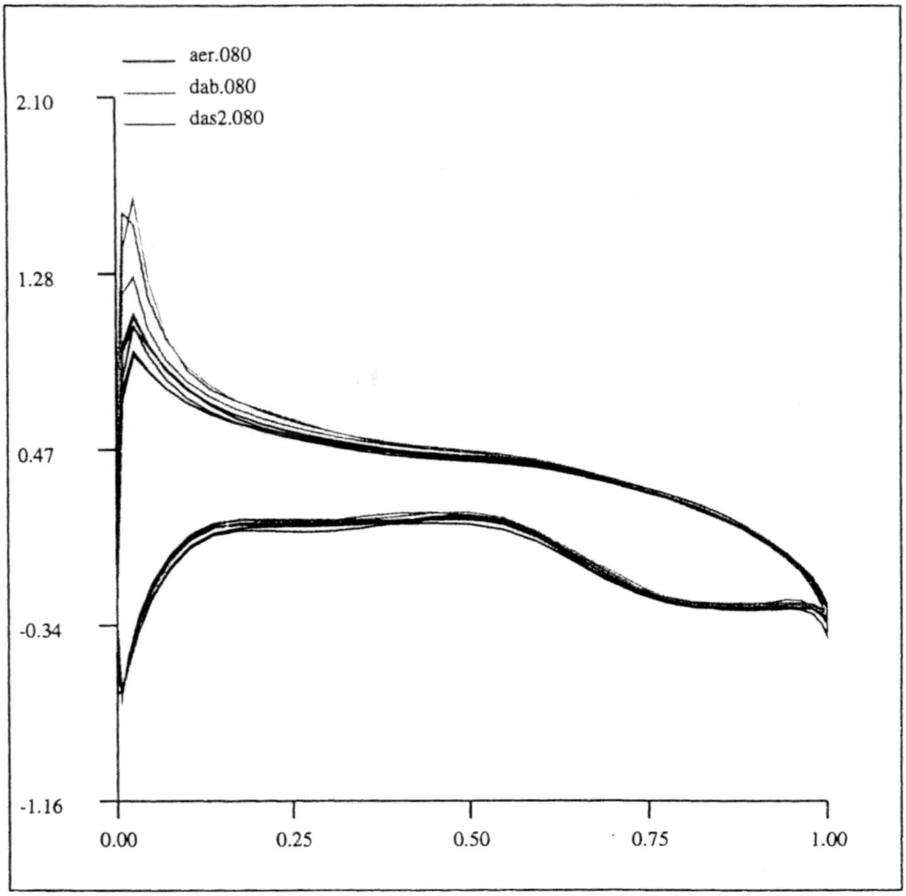

Figure TE5.6 Initial, target and optimized pressure coefficient distribution, section 080

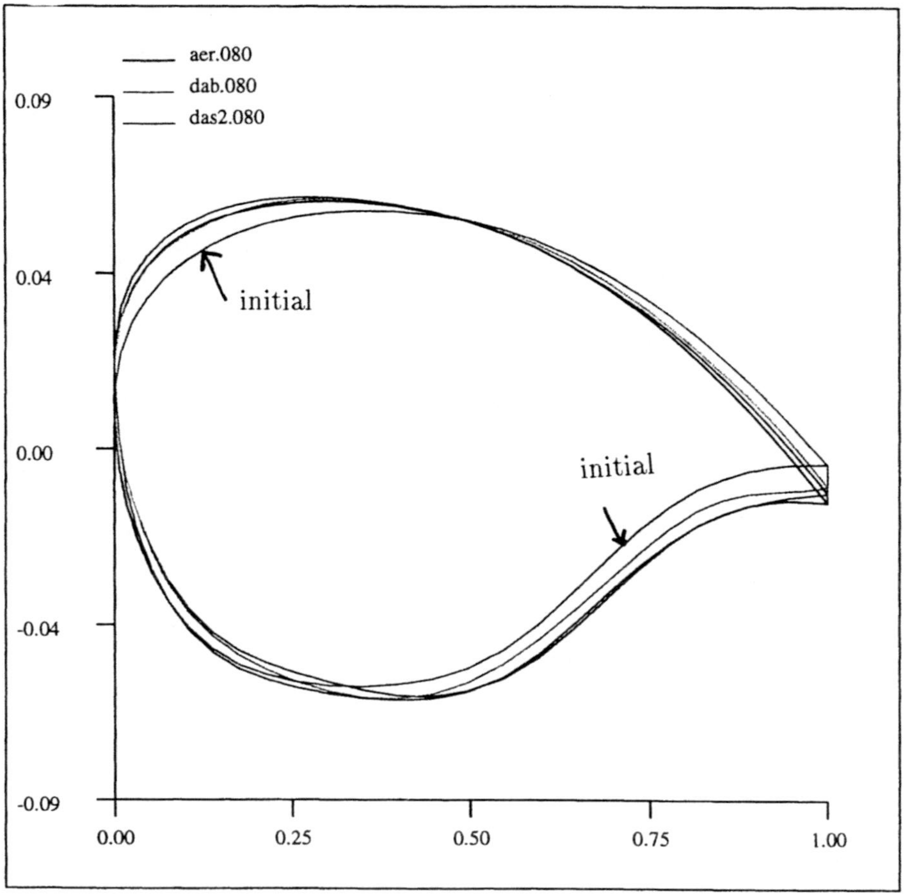

Figure TE5.7 Initial and optimized profiles, section 080

the initial configuration and the optimized wings. The AS and AIRBUS results present a local increase of the twist angle around the pylon position while the Dasa-M results present a local increase of the twist angle around the pylon position while the Dasa-M results present a two-peak distribution, probably because a larger number of design parameters is used and the optimum solution is non-unique. From the AS results it appears that the cost function may be significantly reduced by just adjusting the local twist and thickness distribution. Figure TE5.2 compares the initial profiles thickness distribution against the optimized ones. In all cases the optimum solution demands an increase of profile thickness in the outboard part of the wing. This can be clearly seen in figure TE5.3 where the initial configuration is compared against the optimized solution of D.B.A. AIRBUS.

More detailed results are presented for two wing sections in the following figures. Figure TE5.4 presents the initial, target and optimized pressure coefficient distributions obtained by the three contributors for the wing section 060. It is seen that the AIRBUS panel code has a tendency to overestimate the suction peak compared to the two other codes. This may be due either to differences in handling the compressibility effect or to the accuracy of the different singularity schemes employed. The corresponding optimized section shapes are shown in figure TE5.5. The increase of the optimum profiles thickness is evident. Similar conclusions can be drawn from figures TE5.6 and 7 where pressure distributions and profile shapes are shown for the section 080. According to D.B.A. AIRBUS who was the only contributor that presented viscous results, as well, the interference between clean and engined wing is weakened when viscosity is taken into account.

2.6 Test case TE6

TE6 is a transonic 3-D wing drag minimization case with constrained lift. The initial wing is the ONERA M6, which under the selected inflow conditions experiences a strong lambda shock. Results for TE6 are presented by ALENIA, only. The flow code is a non-conservative full potential solver while the drag minimization is obtained by means of a commercial constrained optimization package. Two runs with different number of control parameters have been performed. At the first run only the root section was parametrized while at the second both the root and the tip sections were parametrized. A linear interpolation scheme along the span-wise direction has been used in both cases.

The pressure distribution of the original and the optimized wing along equidistant span-wise stations is presented in figure TE6.1 for the root section optimization case. It is seen as a general trend of the solution that the strong lambda shock is replaced by a weaker single shock. The corresponding optimized profiles are compared against the initial profiles in figure TE6.2. The convergence history of the optimization procedure is presented in figure TE6.3 as the ratio of the current wave drag value and the initial one versus the number of cost function evaluations. It is seen that a 25in 130 iterations where the convergence stalls. Quite similar results are obtained when both the root and the tip sections are parametrized. The corresponding pressure distributions, profile shapes and convergence history are shown in figures TE6.4 to TE6.6. The same 25now obtained in 364 cost function evaluations. Most probably a further reduction could be obtained if more iterations were performed, since the number of degrees of freedom is now doubled compared to the previous run.

Surface Mach contours for both runs are plotted in figures TE6.7 and TE6.8. The replacement of the initial lambda-shock by a weaker single shock is clearly seen once again.

2.7 Test case TE7

A by-pass axi-symmetric afterbody is reproduced under transonic flow conditions. The difficulties of the test case are associated to the multi domain configuration and the choked flow in the inner nozzle. Some additional difficulties are encountered when treating the block interfaces where vorticity is concentrated because of the mixing of flow stream-tubes with different energy content. NTUA is the only contributor to this test case. The geometry reproduction is obtained by means of a one-shot inverse Euler method which may treat multi-domain configurations.

The original (target) and the reproduced configurations are compared in terms of surface geometry and Mach number contours in figure TE7.1. It is seen that the reproduction with the one-shot method is good considering the difficulty of the case and also that the final shape includes the accumulated error of both the direct (to compute the target pressure) and the inverse (to compute the corresponding geometry) solver, which are of different numerical kind. A detail of flows and geometries in the proximity of the inner nozzle is presented in figure TE7.2. In both target and reproduced solutions the flow in the inner nozzle is shocked because of a strong normal shock wave. The reproduced wave seems to be rather curved compared to the target one. This is because the inner grid of the inverse solution is built on the potential function-stream-function space and the shock follows the equi-potential lines. In this one-shot method the resulting final grid consists part of the solution.

3 Additional comments

From the synthesis achieved with the design data collected in the database different comments on the progress accomplished and future direction of research can be outlined:

- Flow models: design with viscous flows including mainly inviscid flows with viscous boundary layer effects, and also first Navier Stokes flow analysis solvers;
- Three dimensional design: complex configurations with simple models or simple configurations with complex models; the situation of complex geometries under complex flow conditions is still out of range for a lack of

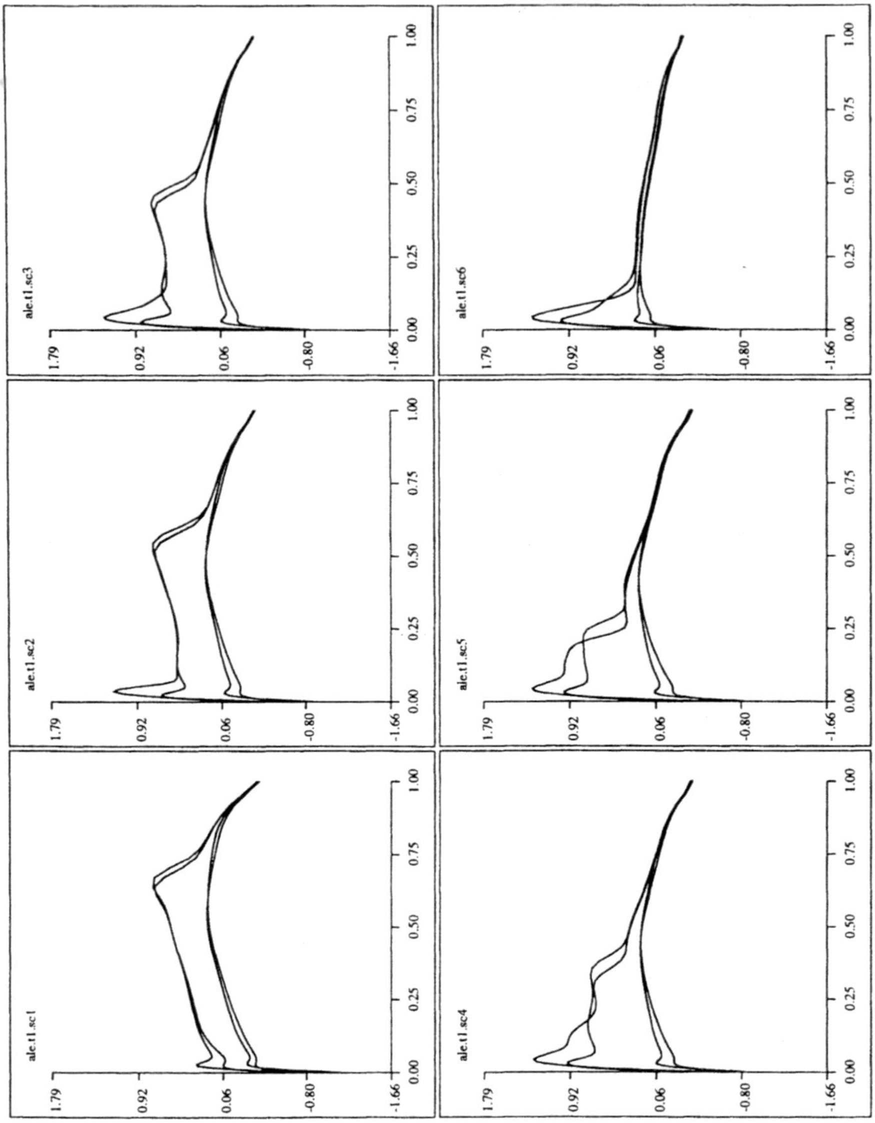

Figure TE6.1 Initial and optimized pressure distribution along equidistant wing sections (root to tip). Parametrization of the root section.

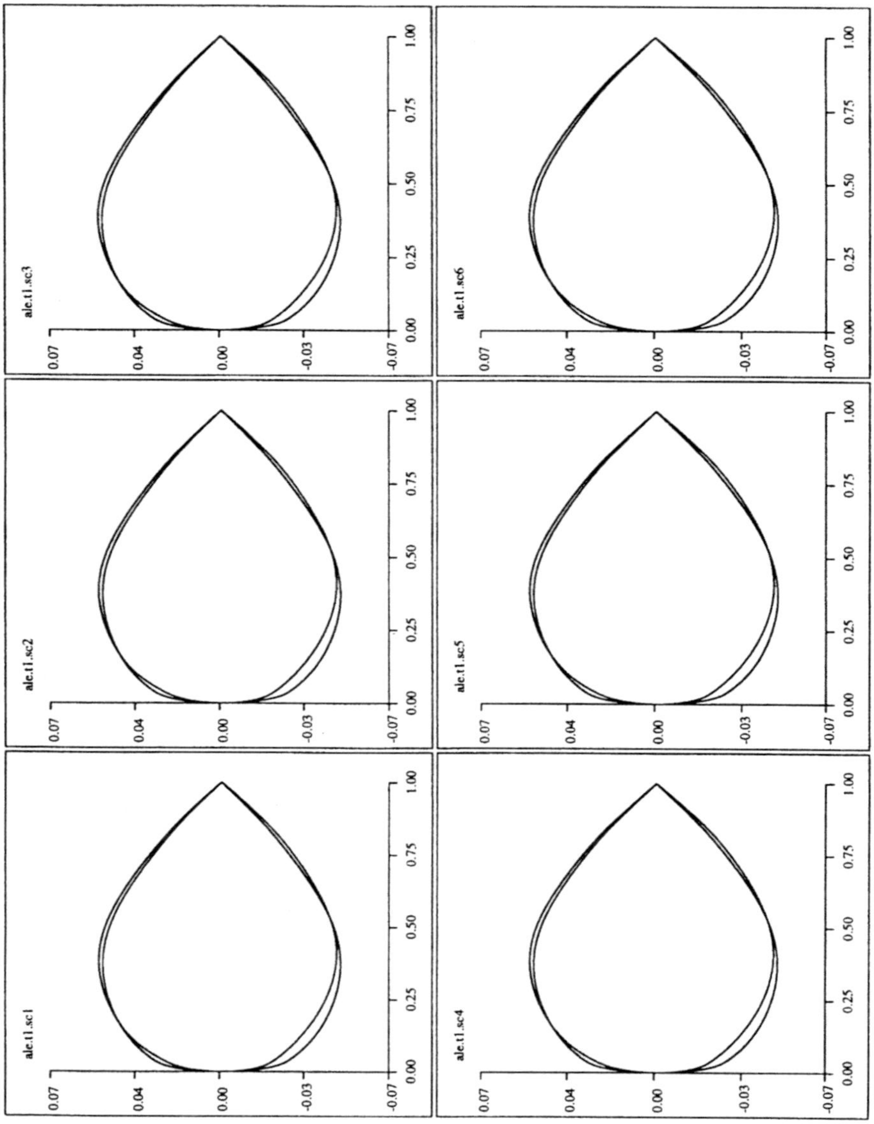

Figure TE6.2 Initial and optimized profiles along equidistant wing sections (root to tip). Parametrization of the root section.

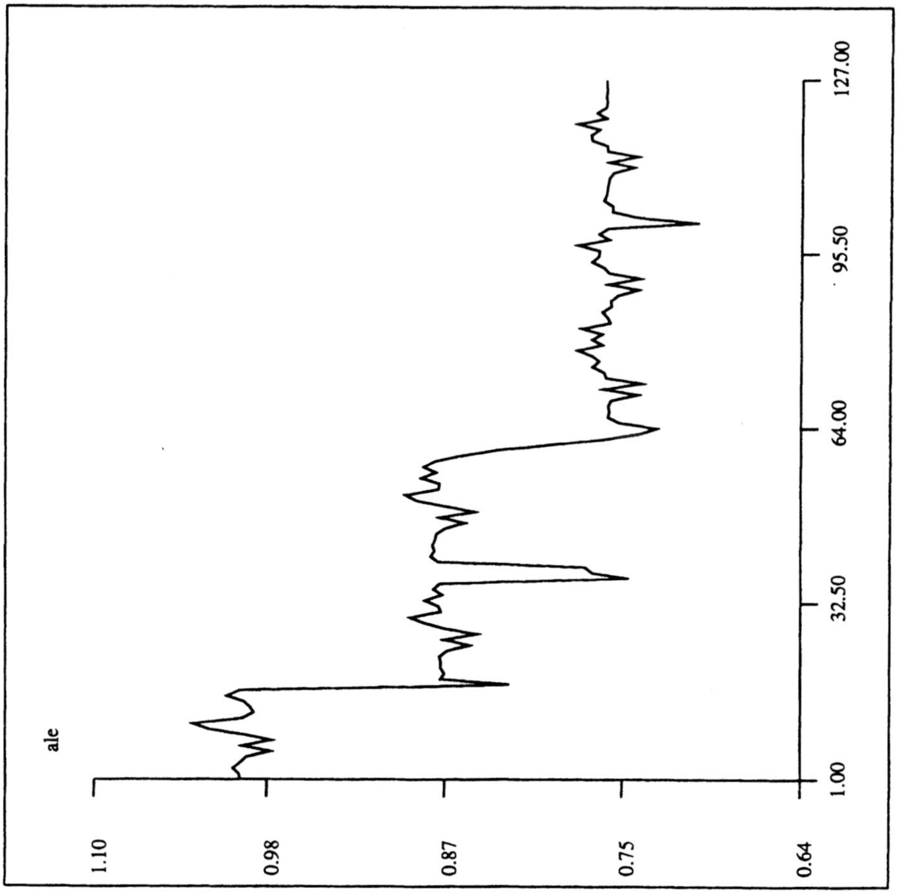

Figure TE6.3 Drag ratio reduction versus number of objective function evaluations. Parametrization of the root section.

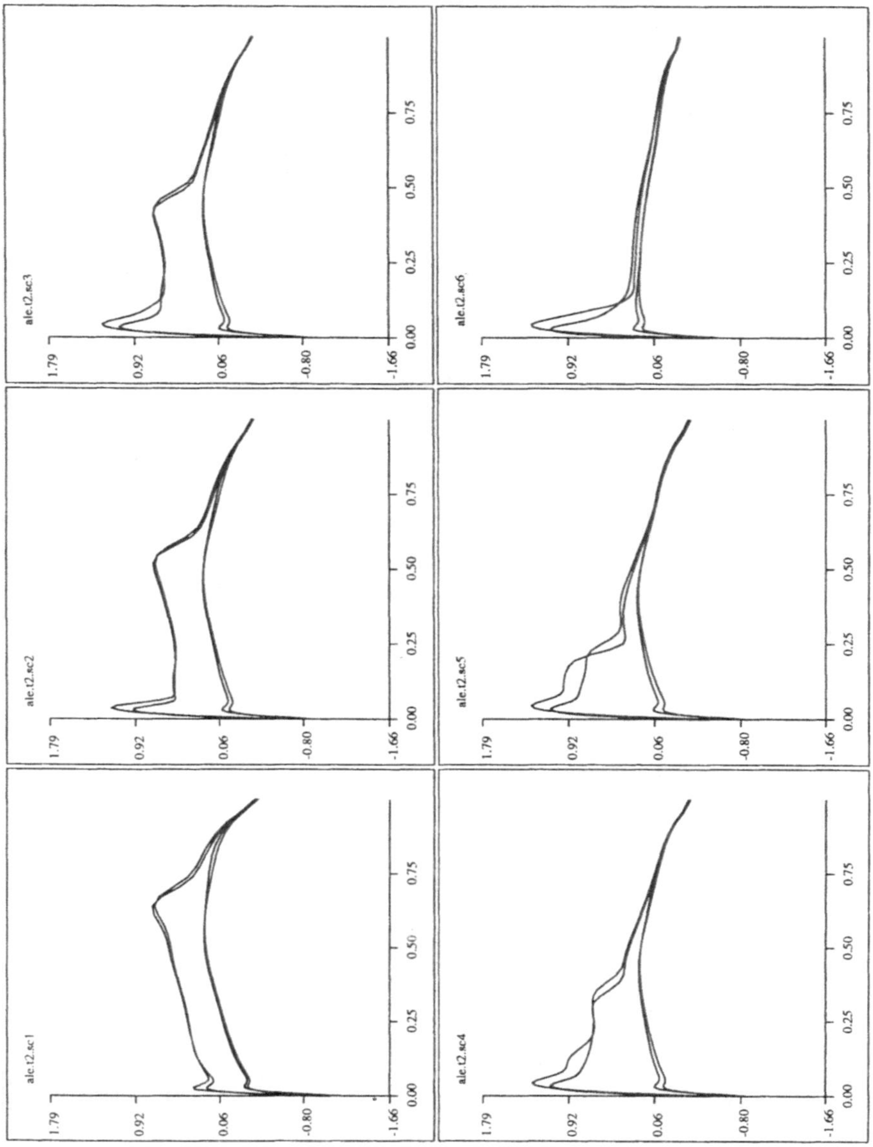

Figure TE6.4 Initial and optimized pressure distribution along equidistant wing sections (root to tip). Parametrization of the root and tip sections.

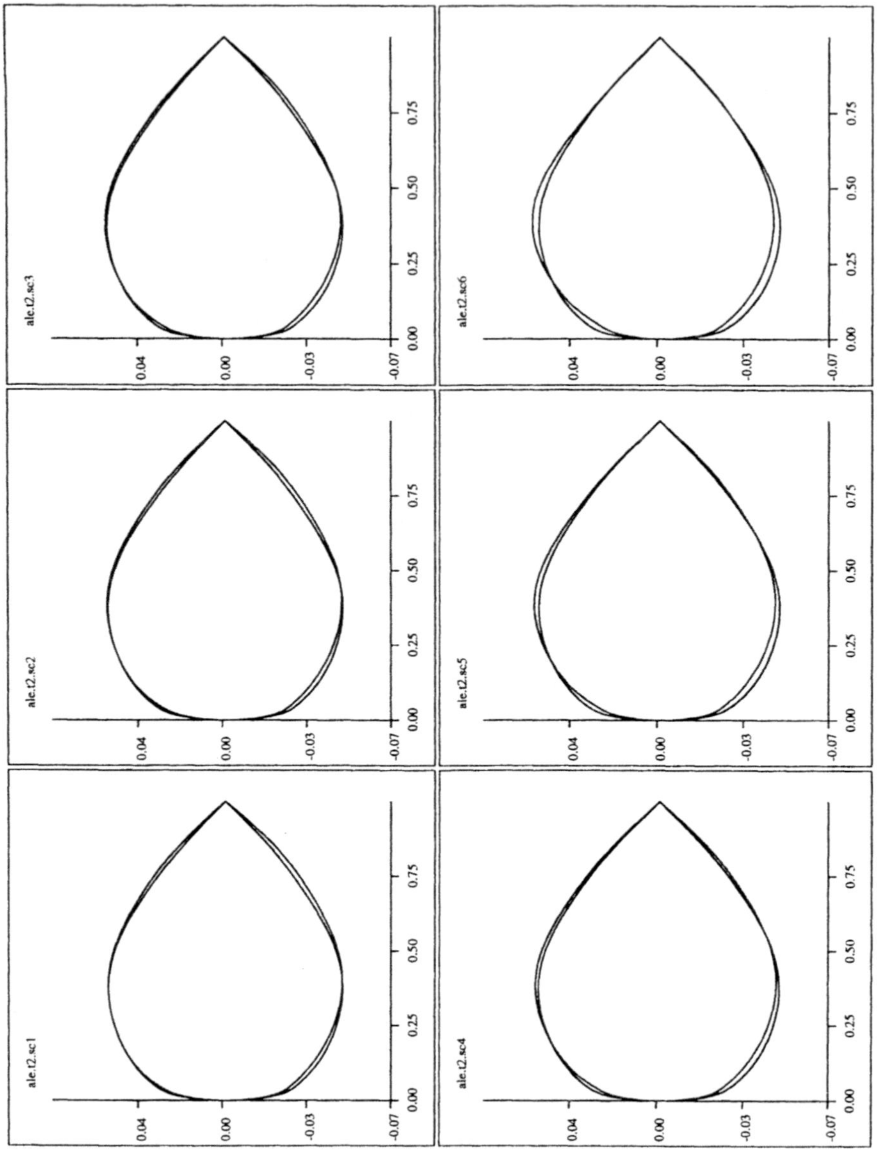

Figure TE6.5 Initial and optimized profiles along equidistant wing sections (root to tip). Parametrization of the root and tip sections.

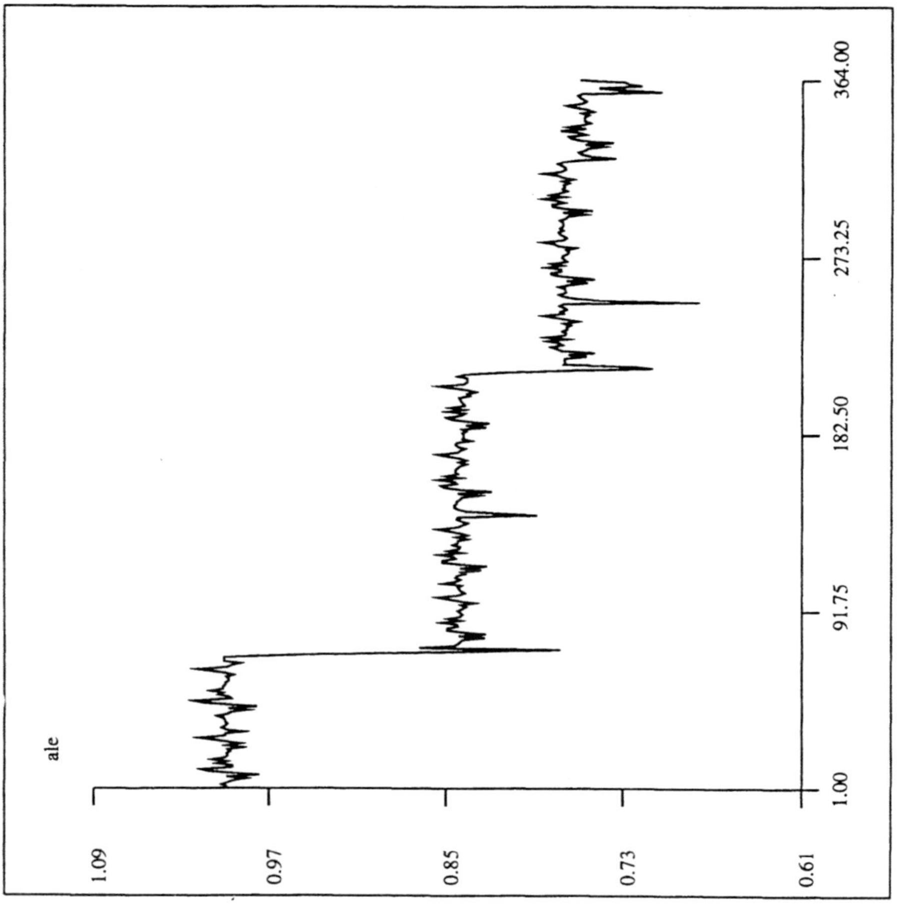

Figure TE6.6 Drag ratio reduction versus number of objective function evaluations. Parametrization of the root and tip sections.

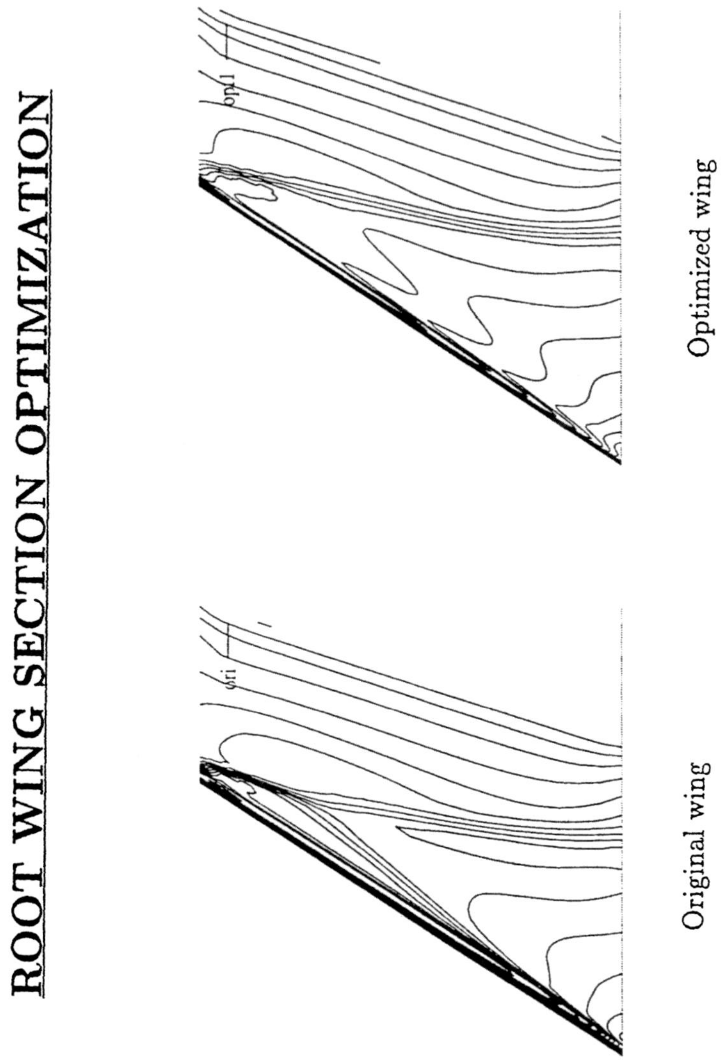

Figure TE6.7 Initial and optimized Mach contours. Parametrization of the root section.

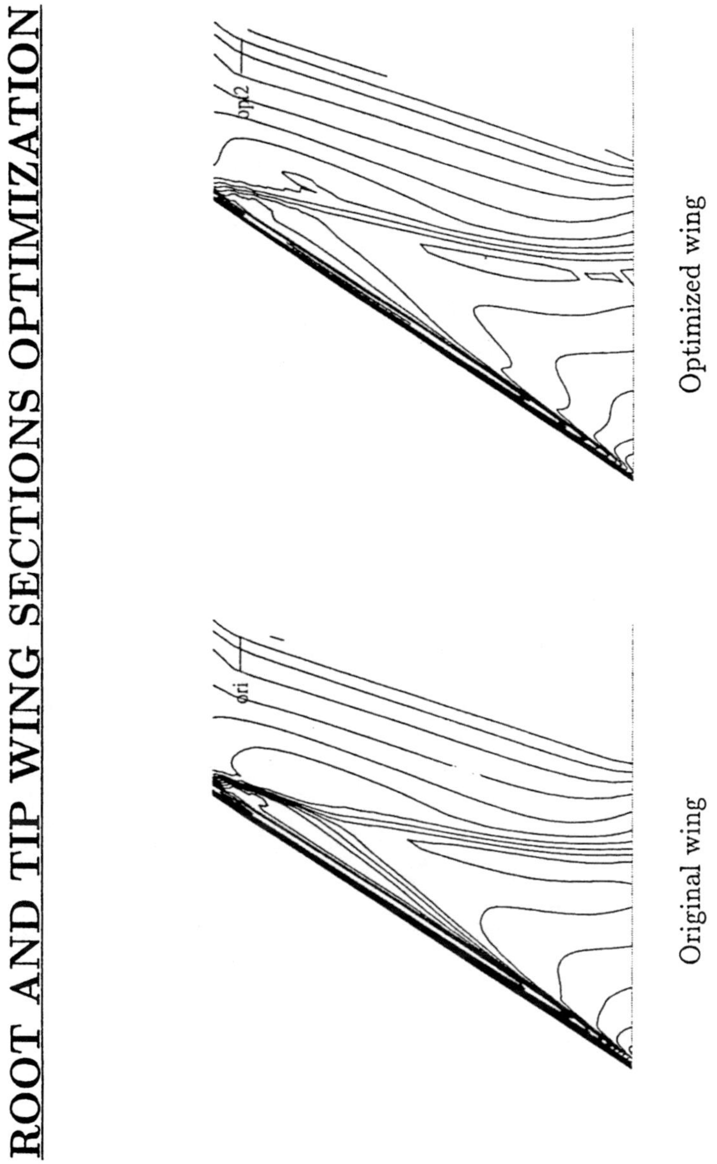

Figure TE6.8 Initial and optimized Mach contours. Parametrization of the root and tip sections.

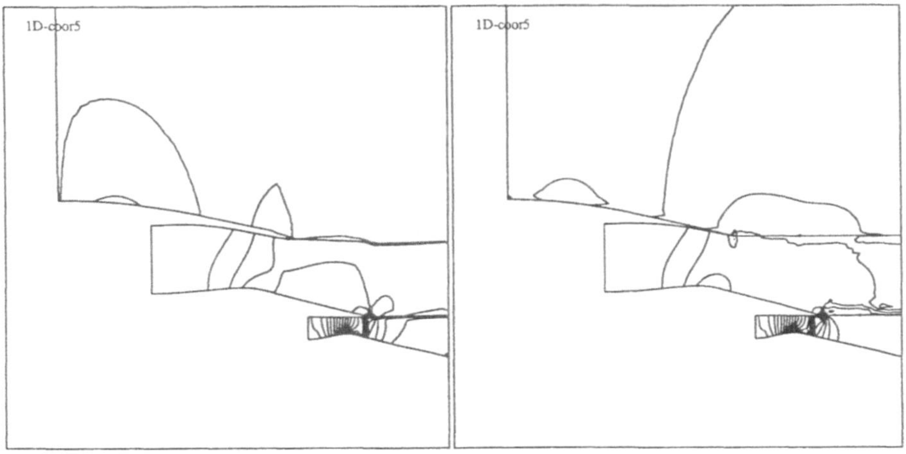

Figure TE7.1 Geometry and Mach number contours. Reproduced (left) and original (right) configurations.

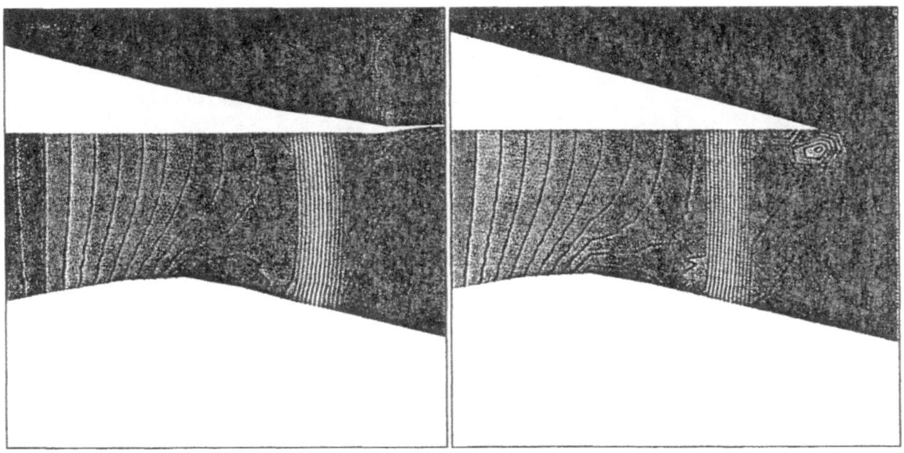

Figure TE7.2 Geometry and Mach contours detail for the inner nozzle. Reproduced (left) and original (right) configurations.

control of the mesh quality and also severe computing costs of the optimization loop on current computer technology;

- Parametrization: an extreme sensitivity of the design to the choice of the design variables ; Bezier splines seem nowadays the favorite compact parametrization tools of design engineers;
- Multi point design: under control with different inviscid design conditions but much more difficult with viscous design conditions. Pareto front strategy is an interesting approach for design of engineers for solving these technical and economical multi-criteria problems;
- The CFD design discipline has been considered in this project; forthcoming concurrent and conflictual multidisciplinary will reduce significantly the design cycle on HPCN environment
- Commercial gradient based optimizers with constraints are currently used by design engineers for their design; they have not been tested in a HPCN environment and automated differentiators like ODYSSEE introduced by INRIA looks a promising approach for differentiable problems. However it is well known that traditional deterministic optimization tools can be trapped in local minima.New evolutionary tools which are stochastic based search space methods able to capture global minima have to be further investigated for complex design problems;
- A good assessment of an optimization process satisfy the three followings conditions simultaneously : quality of the mesh generator, validation of the flow analysis solver and efficiency of the optimizer; a defect of one of them will deteriorate significantly the design; examples from industrial test cases have proved this severe conditions to be enforced (see the different designs and associated pressure distribution - named suction peaks - superposed on Figs. TE5.080.1 and TE5.080.2

4 Directions of research

Data of the contributions presented during the database workshop suggest several new directions of research ; the most important of them are listed thereafter:

- multi objectives: investigation cooperative strategies like Pareto for multi criteria design; non cooperative strategies like Nash are also other optimization candidates according to the problem under consideration;
- HPCN computation: intelligent use of distributed HPCN facilities for optimizers and solvers; these HPCN environment should increase both the convergence speed up and the robustness of the optimization process and not only the scalability factor;

- design variables: parametrization of the design should contain more sensitive features depending of the physical environment and not only geometry depending ;
- do we need criteria in the future?: prescribed criteria remain until now a very sensitive and crucial choice for the designer; they have been recognized as prescriptive design rules for a long time with the property that good criteria should provide good designs but there exists other prescriptive rules to investigate - viability based- to get better design without criteria;
- robustness of stochastic algorithms: designed products of increasing complexity will require tracking of global solution in more and more complex search spaces for which stochastic algorithms hybridized with traditional tools seem the most promising optimization tools.

Some encouraging feasibility studies have emerged from ECARP designs with viscous flows and genetic algorithms and results presented in this workshop already gave a preliminary answer to the above directions of research.

V Conclusion and Perspectives

The results of EUROPT being a starting point a specific aim of ECARP Area 3 was to extend the scope of the existing design procedures to more relevant industrial 3-D applications operating in inviscid flows with viscous interactions and more complex optimisation strategies including multi objectives. Dealing with more realistic design problems the project aimed to develop and assess automated optimization tools implemented on current computer technology for reduction of drag of relevant civil aircraft configurations operating under real flow conditions. The ultimate objective was two folds: (1) reduce the design time cycle and also (2) improve the final quality of aerodynamic design for civil aircraft with the use of the automated design software.

The corresponding optimization software was developed, applied and assessed during three main phases of the project: the development phase with (i) methods, algorithms & parametrization, the real problem phase with (ii) applications and the database design workshop phase (iii) with assessment of the methods on selected test cases.

An important validation step of the developed optimization tools was achieved by means of a database workshop on industrial configurations such as airfoils, wings and multi component systems. Improved results of partners on 8 design test cases of industrial interest, including inverse problems with target pressure distributions, optimization problem with 2-D/3-D wing optimal with minimum drag criteria and transonic multi point airfoil design at different conflictual conditions,were stored in a database allowing immediate comparisons and synthesis on workstations to evaluate the values of the different methodologies in terms of accuracy and efficiency.

Optimum Design (OD) codes developed during the project represent undoubtedly an increasing industrial value for airframe manufacturers. Considerable improvements in methods used and more complex applications treated have brought a more proper and reliable basis for the real use of those methods in design work by aerodynamic project engineers. An increased effort on interaction between the main ECARP areas would have improved (OD) results available today on a database since the discipline itself incorporate the main areas, mesh adaption, validation and optimization.As it is, the database forms today a basis for future (OD) investigations.

Various specific traditional optimization procedures were utilized as fast special purpose deterministic algorithms but necessarily limited when taking into account of the implementation of constraints or replaced by not very fast universal deterministic algorithms with constraints. A new class of robust stochastic algorithms with inherent good parallelization properties, namely Genetic Algorithms (GAs), were exercised during the workshop perfectly suited for problems where it is difficult - if not impossible -to compute exact gradient information.Robustness of evolution algorithms as global optimization search procedures seems very promising for high tech products of increasing complexity.

Major efforts in optimization procedures for 3-D aerodynamic shapes at the edge of the 21st century concern multi-disciplinary aircraft design and optimisation tools: among the main difficulties to be investigated are (1) intelligent parametrization of more complex 3-D configurations with respect to physics and not only geometry , (2) control of mesh quality with more enhanced analysis tools modeling namely turbulent flows, (3) robust hybridization of deterministic and stochastic optimization tools on distributed HPCN environments and last but not least (4) reduced cost of each flow solution.

Optimum Design (OD) in industry critically depends upon several areas to be considered as a whole highly interactive and integrative process, namely mesh generation/adaption, analysis code validation , pure optimization and database post processing tools.Only intensive interaction of engineers involved in the design loop of the above areas will fit the value of OD to the needs of the airframe manufacturers.

Under these conditions the above improvements will provide immediate industrial benefits in concurrent multidisciplinary/multi objective design.

Part B: Navier Stokes Solution on MPP Computers

I Introduction

Evaluating performances and accuracy of CFD codes for scientists and engineers on parallel hardware environment is of major importance in Design Industry. Parallel computing offers a huge potential for CFD applications in Aerospace Industry in particular at the condition that these parallel systems can be utilized as new paths to device innovative algorithms not only in terms of efficiency but also in robustness in a concurrent multidisciplinary production environment. Portability of software including the main issues of scalability and load balancing has become an important step in the HPCN area since new distributed parallel architectures including MPP systems with fast networking capabilities occupy the center the Information Technology to solve larger scale non linear society and industry problems at a lower cost on a daily basis.The modern solutions of complex CFD problems require now various form of parallelism ranging from scalability to intelligent processors interfaces minimizing communication and offering new distributed algorithms which increase the speed of convergence of the global system.

The aim of a concerted action on Parallel Computing with flow algorithms was to determine through a database workshop the impact of parallel architectures on the usability of Navier Stokes software for compressible viscous flows. Previous studies had targeted vector computers but alternative architectures offering high speed at lower cost was rapidly proposed by hardware vendors as distributed parallel processor machines.The price to pay by users - scientists and engineers- to benefit of MPP power was not only to simply divide computational domain into parts but also to construct software carefully to ensure scalability, load balancing and low interprocessor communication by intelligent interfaces between sub-domains.

The target of this concerted action was first consider a wide range of MPP platforms on which to run a particular code; second evaluate the performance of implementation with respect to various parameters including the size, the data structure and the type of problem.

The focus of the comparative database workshop was to get ultimately a clear understanding of the value of parallel architecture machines with an accurate parallel solution of the compressible Navier-Stokes equations.The final synthesis should explain how scalability will translate into significant reduced CPU and communication times for intensive industrial design simulations.

Part B of the volume contains 5 sections organized as follows: Section 2 entitled "ECARP database workshop: MPP for Navier Stokes flows" describes 5 laminar or turbulent problems with their references and their selected parameters of each test case . In this section is also presented the ingredients of a contribution: output formats with o-platform identity, cost evaluation, methodology, quality of the output, short synthesis and also the requested information to be stored on the database for the two mandatory test problems only .

The different results obtained by 8 ECARP test case contributors with their Navier Stokes codes on different parallel architectures are presented in a partner by partner basis in Section 3 as "Contribution to the resolution of the Database Workshop test cases". Each contribution provides computerized data of the test case performed on several parallel machines with the formulation of the method and its parallelization aspects, a conclusion and a bibliography.

Section 4 contains a synthesis of the test cases with a classification of the workshop contributions and the analysis of the accuracy of the numerical predictions and the efficiency of the parallel codes. Comparative information on contributions is provided test case by test case with superposed data on pressure coefficient distribution, friction coefficient distribution and speed up curves extracted from the database.

A conclusion focused on the main issues of large scale problems run on different parallel architectures relevant to aerospace industry ends the second part of this volume in Section 5. Further comments on several directions of research with new computer instruction and data distributed simulations in HPCN environments useful to future aircraft manufacturer's design applications are suggested.

The data stored of the MPP ECARP database will be a useful reference in CFD for the evaluation of other viscous flow solutions in terms of cost and accuracy in the future.

II Definition of the Problems for the Analysis

1 Description

The purpose of this workshop is to determine the potential impact of parallel architectures on the usability of laminar Navier-Stokes solvers for compressible viscous flows. Each participant will carry out calculations in at least one out of the two mandatory "academic" test cases. The first test-case (TC1) is a two-dimensional flow problem while the second (TC2) is a three-dimensional one. Note that five test cases where originally defined; three of them are optional and as such they are not included in the data base. Therefore eventual contributions to these three test cases (TC3 to TC5) will not be included in the synthesis phase even though they will be part of the individual participant presentations during the workshop.

In order to evaluate the implementation on a common footing, a set of common grids, structured and unstructured, are available for both test cases. Having agreed on a set of common performance measures, the contibutors will use available Navier-Stokes solvers for the two test cases.

Due to the ranges of machine types and flow codes considered, a fairly clear understanding will be reached concerning the true value of parallel architecture machines. From the comparative workshop, it is intended not only to establish the likely levels of scalability that can be achieved with these new classes of computer architectures, but also how such scalability will translate into a reduced turn-around time for industrial simulation of complex configuration aerodynamics operating in the transonic regime. Similarly, comparisons will be made between turn-around time on existing computer architectures and parallel computers in order to establish relative performance measures.

2 Test cases definition

The test cases considered have been widely investigated by researchers in the CFD community so that experimental as well as numerical results are available. They are:

TC1 : two-dimensional flow around a NACA 0012 airfoil. Numerical results are available in [Mavriplis89] with free stream Mach number = 0.5 and angle of attach = 0°. The Reynolds number based on free stream conditions and airfoil chord is 5,000. The flow is subsonic and laminar with a tiny separation bubble near the trailing edge. The Reynolds number for this case approaches the upper limit for steady laminar flows prior to the onset of turbulence. Zero heat flux is prescribed across the airfoil surface;

TC2 : three-dimensional flow around a finite wing ONERA M6. Characteristics of the flow are: free stream Mach number 0.8, angle of attach = 0°, Reynolds number based on free stream conditions and the mean aerodynamic chord = 1,000. The wing section is the ONERA-D profile. It is a subsonic laminar flow and numerical results are presented in [Hollanders85].

TC3 : two-dimensional flow around a NACA 0012 airfoil. Experimental results are available in [Harris81] with free stream Mach number = 0.7 and angle of attach = 1.49°. The Reynolds number based on free stream conditions and airfoil chord is 9,000,000. The flow is attached and just slightly supersonic near the leading edge, along the upper surface. Comparison between experimental results and computational results from various numerical methods are included in [Holst87]. This case is optional.

TC4 : two-dimensional flow around a NACA 0012 airfoil. Experimental results are available in [Harris81] with free stream Mach number = 0.55 and angle of attach = 8.34°. The Reynolds number based on free stream conditions and airfoil chord is 9,000,000. The flow is attached and just slightly supersonic near the leading edge, along the upper surface. Comparison between experimental results and computational results from various numerical methods are included in [Holst87]. This case is optional.

TC5 : three-dimensional flow around a finite wing ONERA M6. Characteristics of the flow are: free stream Mach number 0.84, angle of attach = 3.06°, Reynolds number based on free stream conditions and the mean aerodynamic chord = 11,700,000. The wing section is the ONERA-D profile. It is a supersonic laminar attached flow with a weak inviscid/viscous interaction leading to a lambda shoch pattern on the wing surface. Experimental results are available in [Schmitt] while some numerical results are presented in [Vatsa86] . This case is optional.

3 References

[Mavriplis89] Mavriplis, P., Jameson, A., Martinelli, L., (1989) : Multigrid solution of the Navier-Stokes Equations on Triangular Meshes", ICASE Rep. No 89-11.

[Hollanders85] Hollanders, H., Lerat, A., Peyret, R., (1985) : Three-Dimensional Calculation of Transonic Viscous Flows by an Implicit method, AIAA J., **23**, No. 11, 1670-1678.

[Harris81] Harris, C.D., (1981) : Two-Dimensional Aerodynamic Characteristics of tha NACA 0012 Airfoil in the Langley 8-foot Transonic Pressure Tunnel, NASA TM 81927.

[**Holst87**] Holst, T.L., (1987) : Viscous Transonic Airfoil Workshop Compedium of Results, AIAA Paper 87-1460.

[**Schmitt**] Schmitt V., Charpin, F., Pressure Distribution on the ONERA-M6 Wing at Transonic Mach Numbers, AGARD-AR-138, Chapter B-1.

[**Vatsa86**] Vatsa, V.N., (1987) : Accurate Solutions for Transonic Viscous Flow over Finite Wings, AIAA Paper 86-1052 May 1986.

4 Output formats

The focus is on solving the compressible Navier-Stokes equations at the lowest cost for a given accuracy. A contribution would consist in filling the following types of outputs.

4.1 Platform identity

For each configuration (computer, system, compilers) used: location of the site, manufacturer, model, configuration, cycle time, memory, peak rate and how it has been evaluated (possibly "rectified" through some LINPACK benchmarks), communication network type, languages and compilers.

4.2 Cost evaluation

The following items should be given for each computation.

4.2.1 Elapsed and/or CPU time

Elapsed times should be given for :

(a) : computations and communications,

(b) : wait/idle, namely a measure of the unbalanced load (optional),

(c) : the required pre-processing (e.g. mesh partitioning), loading, deloading, post-processing for the particular application (optional).

The analysis of elapsed time should expose how communication time is measured, compare and explain deviations from theoretical machine FLOPS rates. Comments on load balancing techniques used, static (e.g. using a priori partitioned mesh) or dynamic (e.g. dynamically handled by the flow code or the system software), should also be included.

4.2.2 Operation and resources count

The number of FLOPS should be evaluated on a sequential computer. Requested memory should be estimated from the code and measured from trials on the platform (e.g. give the maximum number of nodel points for which computations can be carried out). The total cost of the computation should be given either from existing facturation or from machine price and maintaining cost.

4.3 Methodology

A brief description of the methodology used must be provided, including the type of approximation, the solution method used and the type of meshes.

4.4 Quality of the output

For cross-evaluation of the accuracy of the solutions, we will consider only one-dimensional curves and concentrate on surfacic fields. For two-dimensional flows : pressure coefficient, skin friction coefficient on the airfoil. For three-dimensional flows : pressure coefficient, skin friction coefficient at different span sections on the wing.

For each computation, convergence curves in terms of elapsed time should also be given. More generally, mesh convergence should be evaluated : a rigorous approach could consist in increasing mesh size in the geometric directions. Optionally, influence of far field truncation error could be studied by increasing the size of the flow domain.

5 Synthesis

To have a global view of the results, we could adopt for each test case, the following procedure :

(a) : from accuracy analysis, define a "best" numerical solution,

(b) : plot the solution error with respect to this best solution versus : elapsed time, number of flops and number of unknowns.

6 Deliverables

The following are the required deliverables for the preparation of the synthesis that will be presented during the workshop. Deliverables are specified for the two mandatory test problems only. Please, follow the instructions given in the document **guide.ps** concerning the storage formats associated to the following data.

6.1 Test case TC1

(a) : the one-dimensional distribution of the calculated pressure coefficient C_p, along the solid wall of the airfoil, versus the non-dimensional axial distance (x-coordinate/chord, where x/c=0.0 corresponds to the leading edge and x/c=1.0 to the trailing edge). The pressure coefficient is defined as :

$$C_p = \frac{p - p_\infty}{0.5\rho_\infty u_\infty^2}$$

where quantities subscribed by ∞ correspond to the infinite flow conditions.

(b) : the one-dimensional distribution of the calculated friction coefficient C_f, along the solid wall of the airfoil, versus the non-dimensional axial distance (as defined previously). The friction coefficient is defined as :

$$C_f = \frac{\tau_w}{0.5\rho_\infty u_\infty^2}$$

where τ_w is the wall shear stress.

(c) : the two-dimensional velocity components/Mach/Pressure field, around the airfoil. If a cell-centered discretisation scheme is used, these values must be provided at the grid nodes, using an appropriate 2-D interpolation scheme (that must be provided as well);

(d) : convergence history for the energy equation versus both iterations and CPU time;

(e) : one-dimensional curves related to parallel efficiency :

- the speed-up i.e. the elapsed time versus the number of involved processors for a fixed global problem size;
- the communication time versus the number of involved processors for a fixed global problem size;
- the scalability i.e. the elapsed time versus the number of involved processors for an increasing global problem size (so that the sizes of the local problems remain quasi-constant as the number of processors increases, optional).

6.2 Test case TC2

(a) : the pressure coefficient distribution C_p at a fixed number of spanwise locations located at 0,20,40,60,80 and 100 percents of the span. Coordinates are normalized with the local planform c(y), y being the spanwise direction;

(b) : the surface skin friction vector field, defined as :

$$C_f = \frac{\vec{\tau}_w}{0.5\rho_\infty u_\infty^2}$$

where $\vec{\tau}_w$ is the vector of wall shear stress components.

(c) : the three-dimensional velocity components/Mach/Pressure field around the airfoil. If a cell-centered discretization scheme is used, these values must be provided at the grid nodes, using an appropriate 3D interpolation scheme (that must be provided as well);

(d) : convergence history for the energy equation versus both iterations and CPU time;

(e) : data related to to parallel efficiency as depicted above.

7 Coordination

Coordination by: K.C. Giannakoglou* and S. Lanteri**

*** N.T.U.A., Athens, Greece**
Tel. 30-1-7759584,7798776
Fax. 30-1-7784582
E-mail : kgianna@zeus.central.ntua.gr

**** INRIA, Sophia-Antipolis, France**
Tel. 33-93-65-77-34
Fax. 33-93-65-79-80
E-mail : lanteri@sophia.inria.fr

III Contributions to the Resolution of the Data Workshop Test Cases

The following sections contain the different contributions from the ECARP consortium to the resolution of the workshop test cases defined in section II. Each contribution contains a theoretical description of the optimization technique used by each partner and the results obtained by using this techniques for the resolution of some test cases.

The contributions are ordered according with the following list of partners:

- 1. Dasa-M - University of Stuttgart
- 2. Dassault Aviation
- 3. FFA
- 4. INRIA Sophia Antipolis
- 5. NTUA
- 6. UPC
- 7. University of Rome
- 8. VUB

The above numeration of the partners will also be used in the section corresponding to the synthesis of the workshop results.

1 Flow computation for a NACA0012 with the Parallel Navier-Stokes Solver CGNS

E. H. Hirschel, DASA, München, T. Michl and S. Wagner, IAG, Univ. of Stuttgart

Introduction

During a collaboration (from 1991 to about 1993) with the *Lehrstuhl für Rechnertechnik und Rechnerorganisation* at the *TU München* (Michael Lenke and Arnd Bode) the authors developed and investigated in detail a highly efficient parallel implementation of the sequential nsflex-method on almost any multiprocessor-system (ParNsflex) [1]. Based on these experiences, the authors developed in the last 1 1/2 years an almost new parallel Navier-Stokes solver based on BI-CGSTAB (CGNS). The use of BI-CGSTAB instead of the red-black Gauß-Seidel of ParNsflex for the implicit time-integration rised the numerical efficiency.

CGNS is based on a right-preconditioned BI-CGSTAB algorithm for time-integration. The exchange of boundary-values for a correct treatment of boundary conditions leads to point to point communication, which is efficiently implemented. The driving of BI-CGSTAB includes the computation of some scalar products, which lead to the necessity of several global sums. Global operations in inner loops are critical for massive-parallel systems.

Governing Equations

The basic equations are the time-dependent Reynolds-averaged compressible Navier-Stokes equations. Conservation laws are used with body-fitted arbitrary coordinates ξ, η, ζ building a structured Finite-Volume grid. In cartesian coordinates, using vector notations, the basic equations read:

$$\frac{\partial\rho}{\partial t} + \nabla\left(\rho\vec{V}\right) = 0$$

$$\frac{\partial(\rho u)_i}{\partial t} + \nabla\left((\rho u)_i\vec{V}\right) - \frac{\partial}{\partial x_j}(\tau_{ij} - p\delta_{ij}) = 0$$

$$\frac{\partial(\rho e)}{\partial t} + \nabla\left((\rho e)\vec{V}\right) - \frac{\partial}{\partial x_j}\left(k\frac{\partial T}{\partial x_j} - pu_j + u_i\tau_{ij}\right) = 0$$

where e is the internal energy, τ is the stress-tensor, δ is the Kronecker-symbol. The letters ρ, p, T, k denote density, static pressure, temperature and heat conductivity coefficient, respectively. Turbulent viscosity is computed form the algebraic turbulence-model of Baldwin & Lomax [2].

Rewriting the Navier-Stokes equations with $U^T=(\rho, \rho u, \rho v, \rho w, \rho e)$ and E, F, G representing the fluxes and source terms in ξ, η, ζ and first order discretization in time leads to the backward Euler implicit scheme:

$$D\frac{U^{n+1} - U^n}{\Delta t} + E_\xi^{n+1} + F_\eta^{n+1} + G_\varsigma^{n+1} = 0.$$

D is the cell volume. Newton linearization of the fluxes at the time level n+1,

$$E_\xi^{n+1} = E_\xi^{n} + \left(A^n \Delta U\right)_\xi \quad \text{, with the Jacobian Matrix: } A^n = \frac{\partial E^n}{\partial U^n} .$$

leads to the linearized flow equation:

$$\frac{D}{\Delta t}\Delta U + \left(A^n \Delta U\right)_\xi + \left(B^n \Delta U\right)_\eta + \left(C^n \Delta U\right)_\varsigma = -(E_\xi + F_\eta + G_\varsigma)^n .$$

The inviscid part of the Jacobian on the left-hand-side is discretized with an up-wind scheme, which is based on the split van Leer fluxes [3]. The viscous part of the Jacobian is discretized in central differences. The inviscid fluxes on the right-hand-side are discretized up to third order in space. Shock-capturing capabilities are improved by a modified van Leer flux-vector-splitting method [4].

First order space-discretization on the left-hand-side leads to a linear system of equations,

$$\mathbf{Ac} = \mathbf{b}$$

where A is a sparse, non-symmetric, block-banded and positive-definite matrix. c represents the time difference of the solution vector (ΔU), **b** the right-hand side, which can be relaxed by a factor ω.

Solution Method

The linear system of equations is, for each time-step, solved by a right-preconditioned BI-CGSTAB algorithm (figure 1)[5].

The right-preconditioned version of BI-CGSTAB (figure 1) has proved to be significantly more efficient than the left preconditioned variant [7]. One reason for this is, that the first preconditioning step of the right-preconditioned variant is used to compute a starting vector $\mathbf{c_0}$. Using the result $\mathbf{c_k}^{n-1}$ of the last time-step as $\mathbf{c_0}^n$ in the present time-step makes the first preconditioning operation unnecessary.

With respect to computing time, preconditioning is the critical part of the algorithm. On the one hand it should cluster the Eigen-values, which essentially improves the performance of CG-like schemes. On the other hand it should not spend much computing time. A suitable preconditioner for the Navier-Stokes equations is a Stabilized-ILU method [4]. Sufficient preconditioning is gained within one ILU-step.

An improved stability and convergence of the scheme was gained by smoothing the residual vector $\mathbf{r_0}$.

Parallel Implementation

The parallel implementation follows the concept of domain decomposition (figure 3). The whole computational domain is decomposed in overlapping blocks. To gain great flexibility in domain partitioning and definition of boundaries, boundary conditions are given cellwise. This supports the computation of complex configurations and helps to subdivide the computational domain in almost equally sized blocks. For efficient communication these cellwise given interconnections between blocks have to be collected to large vectors. This is done in the setup-phase of the algorithm (figure 2a). First each block scans its connections and writes the cells from which information is needed, ordered in blocks, in a receive table. These communication requirements are sent to

$$
\begin{aligned}
&[\text{Solve: } \mathbf{M}\mathbf{c}_0 = \mathbf{b} \text{ for } \mathbf{c}_0] \\
&\mathbf{r}_0 = \mathbf{b} - \mathrm{A}\,\mathbf{c}_0 \\
&\text{Smooth } \mathbf{r}_0 \\
&\text{Set: } \mathbf{q}_{-1} = \mathbf{p}_{-1} = 0,\ \rho_{-1} = \omega_{-1} = \alpha_{-1} = 1 \\
&\text{do } k=0,\ 1,\ldots\ldots,\ \text{itmax} \\
&\qquad \rho_k = (\mathbf{r}_0,\mathbf{r}_k) \\
&\qquad \beta_k = \frac{\rho_k \alpha_{k-1}}{\rho_{k-1}\omega_{k-1}} \\
&\qquad \mathbf{p}_k = \mathbf{r}_k + \beta_k\,(\mathbf{p}_{k-1} - \omega_{k-1}\,\mathbf{t}_{k-1}) \\
&\qquad \text{Solve: } \mathrm{M}\,\mathbf{y}_k = \mathbf{p}_k \text{ for } \mathbf{y}_k \\
&\qquad \mathbf{q}_k = \mathrm{A}\,\mathbf{y}_k \\
&\qquad \alpha_k = \frac{\rho_k}{(\mathbf{r}_0,\mathbf{q}_k)} \\
&\qquad \mathbf{s}_k = \mathbf{r}_k - \alpha_k\,\mathbf{q}_k \\
&\qquad \text{Solve: } \mathrm{M}\,\mathbf{z}_k = \mathbf{s}_k \text{ for } \mathbf{z}_k \\
&\qquad \mathbf{t}_k = \mathrm{A}\,\mathbf{z}_k \\
&\qquad \omega_k = \frac{(\mathbf{t}_k,\mathbf{s}_k)}{(\mathbf{t}_k,\mathbf{t}_k)} \\
&\qquad \mathbf{c}_{k+1} = \mathbf{c}_k + \alpha_k\,\mathbf{y}_k + \omega_k\,\mathbf{z}_k \\
&\qquad \text{If } \mathbf{c}_{k+1} \text{ is accurate enough, exit loop} \\
&\qquad \mathbf{r}_{k+1} = \mathbf{s}_k - \omega_k\,\mathbf{t}_k \\
&\text{end do}
\end{aligned}
$$

Figure 1 Right preconditioned BI-CGSTAB.

the processes, which compute the flow on the according blocks. With these requirements each process can build a send table, which contains all cells from which information is needed by another block and the corresponding process ID where the data has to be sent to. During the iterative process (figure 2b) the two tables allow an efficient and secure communication of boundary-values. Communicated are the solution vector U and its time difference ΔU at the beginning of each time-step.
The solution of the linear system of equations with BI-CGSTAB introduces further communication requirements, which concern the scalar products in the computation of the driving parameters α, ρ, ω and the convergence criterion. These scalar variables must be uniform over the massive-parallel system and lead to corresponding global communications, which are critical operations on multiprocessor systems. CGNS is presently implemented on Intel Paragon XP/S-5 and Cray T3D systems. The use of pvm 3.2 for global operations has proved to be inefficient for more than 8 processors, whereas the use of NX on the Intel Paragon XP/S-5 and the virtual shared memory features of the Cray T3D lead to acceptable global communication times.

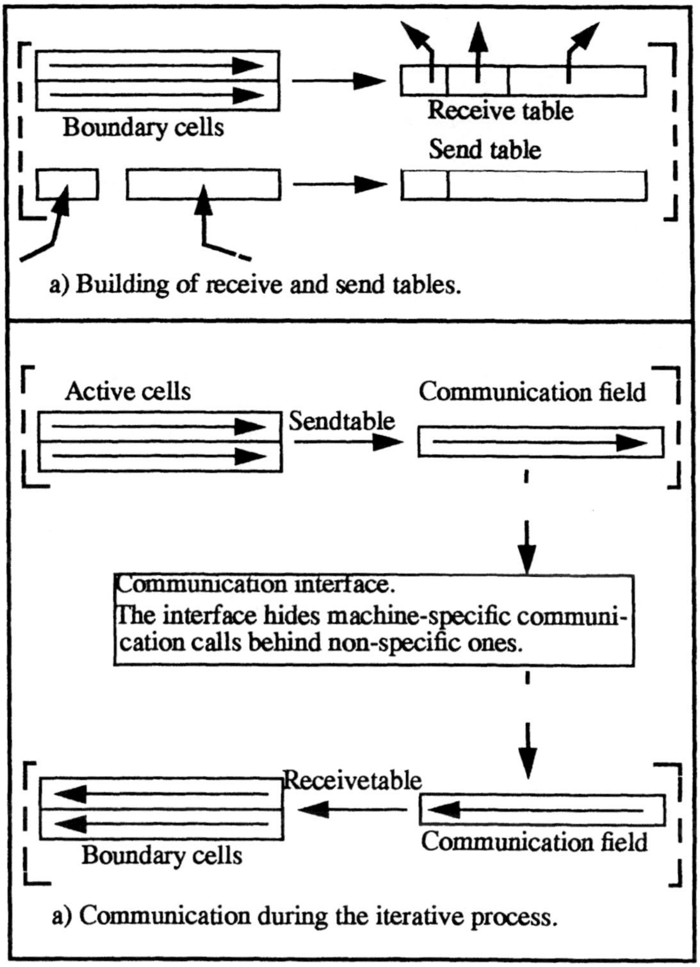

Figure 2 Communication of boundary values.

NACA0012 Flowfield Results

The flow around a NACA0012-profile is computed at a Mach-number of 0.5 and a Reynolds-number of 5 000 (angle of attack $\alpha = 0.0^{o}$). The resulting flowfield leads to small separation bubbles at the trailing edge of the profile.

The computation is done on a C-type grid with 256 x 79 active cells. The basic grid was created by Alex Vinckier and Jens Jacobson (IAG) with an algebraic gridgenerator for non-orthogonal meshes, which are used for the development of multidimensional Riemann-solvers. For this computation the grid has been orthogonalized in the boundary-layer-region, which lead to an acceptable grid-quality. The grid is subdivided in overlapping blocks along lines I=const, orthogonal to the body. The resulting blocks are all of the same size (figure 3), which supports balanced parallel runs.

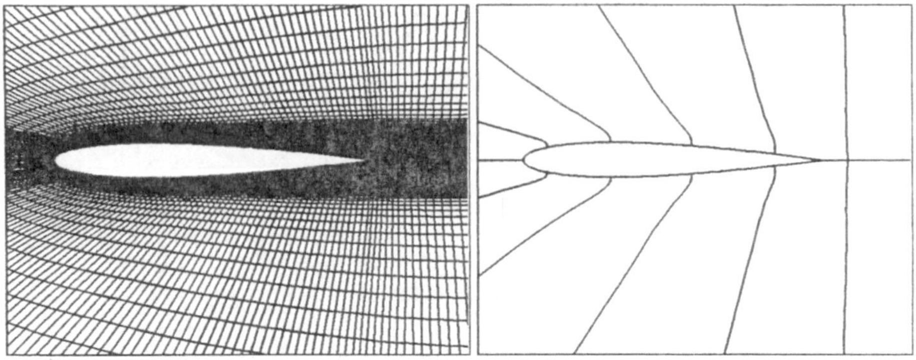

Figure 3 NACA0012 grid (256 x 79 cells) and partitioning in 16 blocks.

The flowcomputation was done with a low CFL-number in the boundary-layer-region ($CFL_{Layer} = 1.0$) and CFL = 10 in the outer region. An optimization of the CFL-number, especially CFL_{Layer} was not done, but would improve the performance significantly. A sketch of the Mach- and C_p-distribution as well as the streamlines at the trailing edge are given in figure 4.

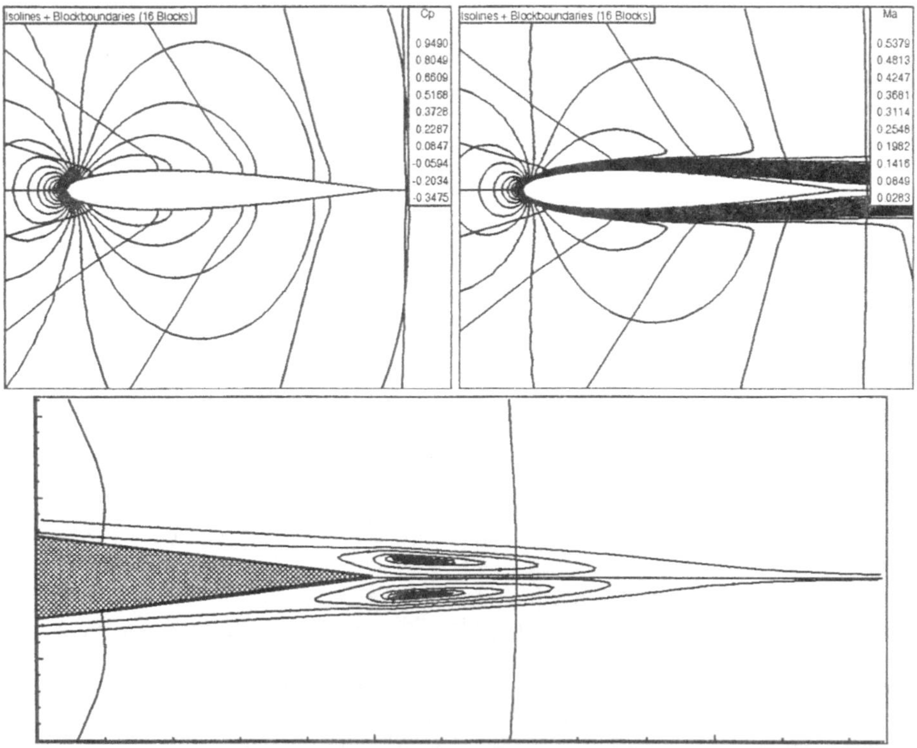

Figure 4 NACA0012 flowfield result (C_p, Ma, streamlines) for a 16 block parallel computation on Intel Paragon XP/S-5 (including block-boundaries, fine grid).

Figure 5 gives C_p and C_f on the profile. C_f was computed from the flowfield (up to third order accurate in space) with first order (solid line) and second order (dashed line) accuracy in space. Both lines are almost identical, which means that at least two cells lie within the linear region of the boundary-layer. The used grid has some 30 cells within the boundary-layer. Flowseparation was computed at x = 0.830 (first order) and x = 0.827 (second order).

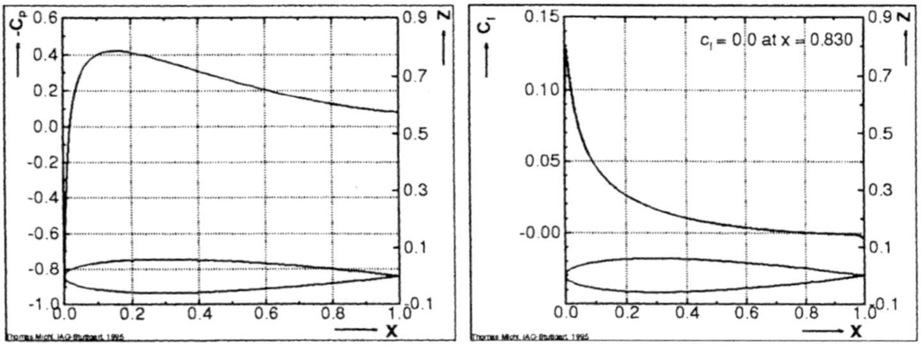

Figure 5 NACA0012 line-plots (C_p, C_f) for fine grid.

An additional computation was done with the grid provided on the ECARP-database. This grid has 280 x 50 cells. The boundary-layer is resolved with about 10 cells at the trailing edge and about 5 cells at the leading edge. The reduced number of cells in the boundary-layer does not allow a first order computation of C_f. Calculating C_f first order in space lead to a separation at about 86% of the chord-length. Calculating C_f second order in space lead to a separation point of x=0.827. Also, the peak in the C_f-distribution rised from about 0.1 to about 0.13 (figure 6).

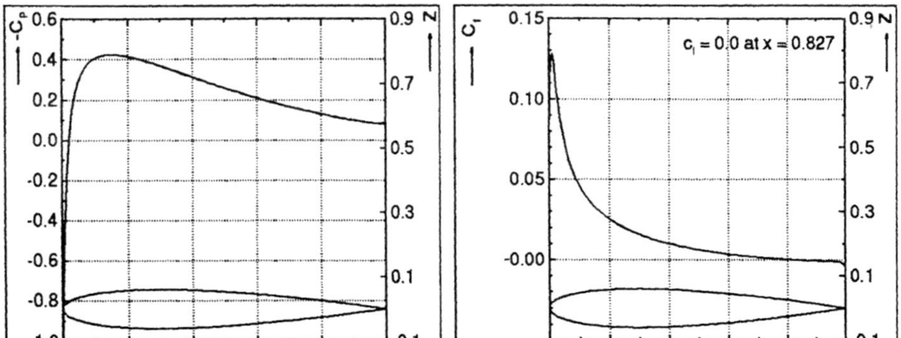

Figure 6 NACA0012 line-plots (C_p, C_f) for ECARP-grid.

Computing the wall-friction second order in space can only improve the results when the first cells are not within the linear region of the boundary-layer. Although the results are almost identical using the ECARP-grid or the fine-grid, we think that the ECARP-grid is not fine enough in the boundary-layer region.

Performance on Massive-Parallel Systems

Presently CGNS runs on Intel Paragon XP/S-5 with the NX-communication library and on Cray T3D with pvm3.2 for the communication of boundary-values and virtual shared memory for global operations.

Computing and communication times as well as speed-ups referred to 500 iterations on the fine grid are given in tables 1 and 2. The computation of the problems took 3 000 iterations. For runtime evaluations we prescribed a convergence-criterion within BI-CGSTAB from 10^{-5} * *actual residual.* Recent tests showed that the convergence criterion must not be so strict, which reduces the runtimes given below significantly.

Table 1 Performance of CGNS on Intel Paragon XP/S-5 (fine grid, 500 iterations).

Processors	4	8	16	32	64
Elapsed time [s]	11 836.0	7 030.0	2 865.0	1 353.0	699.0
Communication of boundary values [s]	11.8	7.1	5.7	4.1	4.9
Global communication [s]	23.7	28.1	37.2	23.0	22.4
Total communication [s]	35.5	35.2	42.9	27.1	27.3
Speed-up [-]	4.0	6.7	16.5	35.0	67.7

The runs on the Intel Paragon XP/S-5 were done with the compiler options *-O4 -Kieee = strict.* A significantly faster run is possible with the option *-O4 -Knoieee -Mvect -Mquad.* Unfortunately the later option lead to some errors in program execution (Meanwhile a new release of the operation system enables that option).

The runs on the Cray T3D were done with the compiler option *-Wf"-o aggress" -Wl" -Drdahead=on"* for Fortran77 and *-O3* for Fortran90. These were the best options found for that system.

Table 2 Performance of CGNS on Cray T3D (fine grid, 500 iterations).

Processors	4	8	16	32
Elapsed time [s]	3 325.0	2 447.0	897.0	417.0
Communication of boundary values [s]	6.7	4.9	5.4	2.5
Global communication [s]	3.2	4.9	9.8	14.6
Total communication [s]	9.9	9.8	15.2	17.1
Speed-up [-]	4.0	5.4	14.8	31.9

CGNS runs on both systems highly parallel efficient. Measured speed-ups are in the linear range. The 32- and 64-processor runs on the Intel Paragon XP/S-5 are slightly superlinear. This is an effect from improved preconditioning within the ILU-scheme, which reduces the required iterations within BI-CGSTAB. The improved numerical efficiency of the ILU-scheme results mainly from a more adequate numbering for these partitionings of the grid.

Both systems spend, as expected, more time on global operations than on the exchange of boundary-values. The reduction of global communication times for

32 and 64 processors on the Intel Paragon XP/S-5 is an effect from faster convergence within BI-CGSTAB for these two cases. Less iterations within BI-CGSTAB lead to less global operations. The time spent for communication of boundary-values decreases with the increase of processors. Communication times include waiting times for other processes. All given times are the maximal times, which all involved processes spend. But the runtime of all processes do not differ significantly.

For the 32-block case the number of floating point operations were counted with a sequential run on a Cray YMP-C90. This lead to 251 MFlops performance on the Cray T3D with 32 processors. But this number must really be treated with care. It's not at all clear if the flop-count and its transfer to the Cray T3D is valid.

Convergence History

The convergence-history for the 32-block-case (fine grid) on the Cray T3D is given in figure 7. The pressure term is converged within 1 500 iterations. The development of

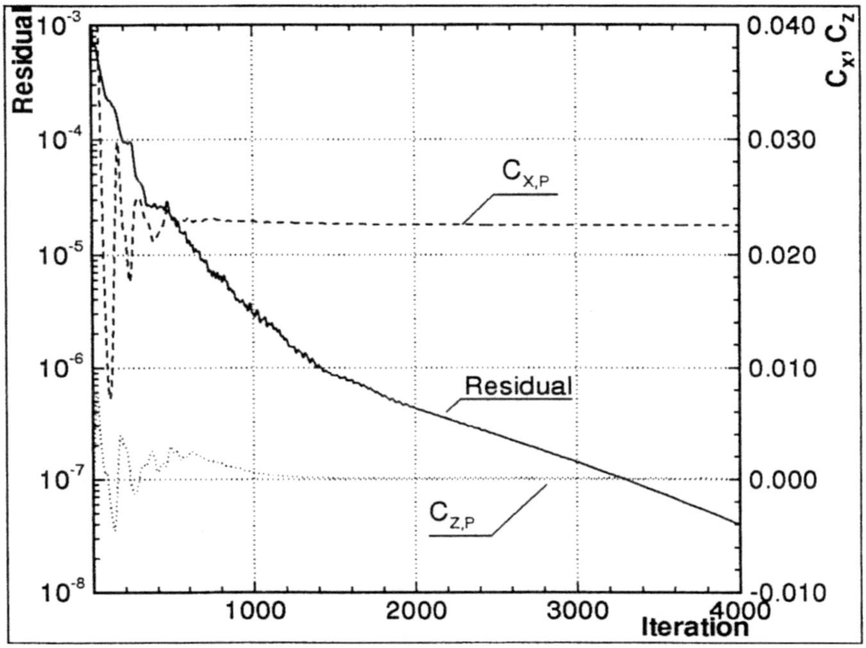

Figure 7 Convergence history (fine grid).

the separation bubbles takes more time and is converged for about 3 000 iterations. A faster convergence would be possible with an increased CFL-number. The convergence history is given for a convergence-criterion within BI-CGSTAB from 10^{-3} * *actual residual*. The runtime for 4 000 iterations is 2 560 [s], which is 77% of the runtime given in table 2. A first test with BI-CGSTAB and preconditioning in Fortran90 (CF90 1.0.2.0, compiler option: *-O3*) lead to a runtime of 2 854 [s] for the same problem. Especially the recursive ILU-preconditioning looses performance.

A convergence analysis of the solver on the ECARP-grid lead to a significantly reduced runtime. The computation was done with CFL = 10 and CFL_{Layer} = 4. Con-

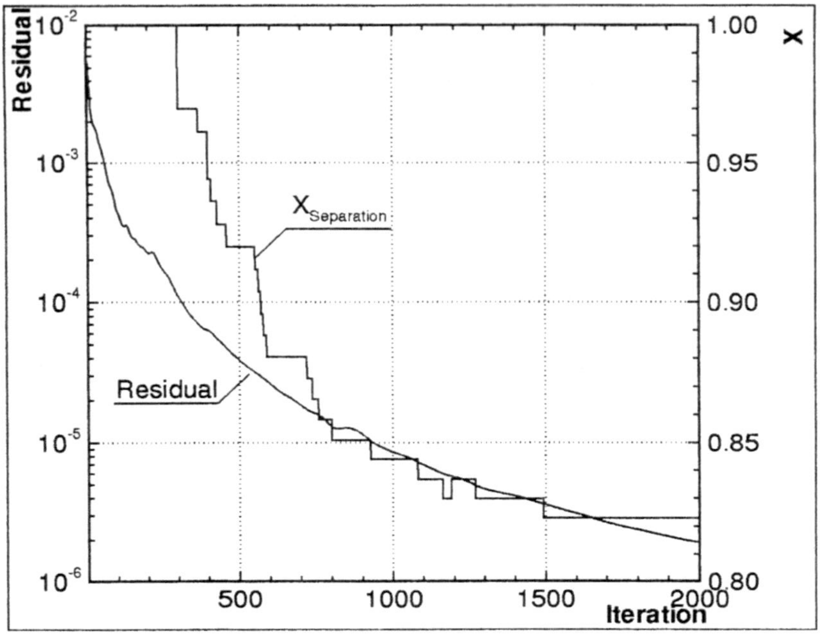

Figure 8 Convergence history (ECARP-grid).

vergence is gained within 1 600 iterations (figure 8) which takes only about 720 [s] on 32 processors of the Cray T3D (whole implicit part in Fortran90). The improved convergence is, besides a rised CFL_{Layer}, probably a result from the reduced number of cells in the boundary-layer, which means that there are also less stiff cells to compute on. Table 3 gives times and speed-ups for the parallel runs on that grid. Due to the

Table 3 Times until convergence of CGNS on Cray T3D (ECARP-grid).

Processors	4	8	16	32
Iterations [-]	1 600	1 600	1 600	1 600
Elapsed time [s]	6 087	3 148	1 577	720.0
Communication of boundary values [s]	12.2	15.7	15.8	10.1
Global communication [s]	6.1	9.4	20.5	33.8
Total communication [s]	18.3	25.1	36.3	43.9
Speed-up [-]	4	7.7	15.4	33.8

slightly unbalanced computation - there are some blocks which have one line I = const more than others - the times for global communications increased a bit. The measured parallel efficiency of CGNS on the ECARP-grid is better than that on the fine-grid. Again, speed-up is almost in the linear range. The 32-block run is superlinear due to improved numerical efficiency within the ILU-preconditioning, which reduces the required iterations for BI-CGSTAB.

Additional Times for Pre- / Postprocessing

The parallel computation requires the partitioning of the grid in overlapping blocks. The corresponding tool can handle any grid of C-, O-, and H-type. The splitting is done in such a way, that the best possible load-balance is achieved. The tool also determines the connectivity along the inner boundaries. The run-time is only a few seconds and is negligeable.

As each parallel process reads its own input-data, the grid-file must be split in several files on the multiprocessor. The corresponding tool takes also only a few seconds.

After the parallel run all parallel written result-files are collected. This program runs only a few seconds.

A bit more time is spend by postprocessing for graphical output. The postprocessor needs about ten seconds to provide all graphic files.

Computing Costs

The computing cost for that problem can only be given for the Intel Paragon XP/S-5. The computing centre at the University of Stuttgart charges 0.2 DM per cpu and hour for institutes of the university, and 2.00 DM for others. This leads to computing costs for the fine grid problem given in table 4 (corresponding to the times given in table 1). The computations on the Cray T3D are presently not accounted.

Table 4 Computing costs on Intel Paragon XP/S-5.

Processors	4	8	16	32	64
	fine grid (until convergence)				
Costs (University) [DM]	15.8	18.7	15.3	14.4	14.9
Costs (Externals) [DM]	158.	187.	153.	144.	149.
	ECARP-grid (estimated costs, until convergence)				
Costs (University) [DM]	5.9	7.0	5.7	5.4	5.6

The computation on the ECARP-grid is about 2.67 times faster. That way the computing costs on the ECARP-grid are by a factor of 2.67 cheaper than on the fine grid.

Scaled Computation

A small scaled computation was done with a 128 x 79 cells grid on 16 processors and 256 x 79 cells on 32 processors of the Cray T3D. That way each processor has for each case the same amount of cells (632) to work on. Computed are 4 000 iterations with the same specification. The computing times are given in table 4.

As the solver reacts dynamically on instabilities and converges within BI-CGSTAB to a convergence criterion, these two cases are not easily comparable. Additionally the cellsize of both grids naturally differ, which means that the numerical diffusion (which is related to cellshape, -size and flowdirection) might also differ.

The computation was done with a different driving than the runtime-evaluations above.

The run on 32 processors took about 9.3% longer than the run on 16 processors. The time spend for the communication of boundary-values increased slightly by 6.2%. Significantly increased (147.7%) is the time spent for global sums within BI-CGSTAB.

Table 5 Times for a scaled computation on Cray T3D.

Processors	Total time [s]	BI-CGSTAB [s]	Preconditioning [s]	Global sums [s]	Communication [s]
16	2246	1166	501	34	155
32	2455	1328	506	83	164

The time spent for preconditioning is about the same for both cases. With the increase of time for global operations the time spent in BI-CGSTAB is also increased (13.8%). Additionally there might have been some more BI-CGSTAB-iterations to converge. All in all the increase in computing time is not considerable. And one has to keep in mind that scaled computations work on different grids and with that not exactly on the same problem.

Conclusion

The parallel computations were done on two different grids, a fine grid with a high resolution of the boundary-layer and a grid provided on the ECARP data-base. The computed results on both grids are almost identical, although we think that the ECARP-grid does not resolve the boundary-layer properly.

The computation on the ECARP-grid is a lot faster than the computation on the fine grid (0.051 [s/cell] versus 0.095 [s/cell] for 32 processors on Cray T3D). One reason for that is the increased CFL-number in the boundary-layer region, another reason might be the reduced number of cells in the boundary-layer. The reduction for the specific computing time would be even more, if the computation on the ECARP-grid would be done with Fortran77 instead of Fortran90 as it was done with the fine-grid case.

The parallel code runs with a good parallel efficiency on the investigated massive-parallel systems. Although global operations are costly on a multiprocessor, the time spent for global operations is moderate. The more modern system Cray T3D is more than 3 times faster than the Intel Paragon XP/S-5, although the processor is only two times faster. Also communication times are a lot faster.

CGNS is a research code. The code is designed for a flexible implementation of new algorithmic parts. In order to save memory (which is still necessary for the computation of 3-D problems on multiprocessors), CGNS does not store all constans, as metric terms, but computes it when they are needed. Additionally many parts of the solver compute a pseudo-3-D-problem for 2-D. The code is not machine-optimized. Presently the code gets only some 5% of the theoretical peak-performance of the Cray T3D. Thus, the code could be a lot faster with a proper algorithmic and machine-optimization.

References

[1] T. Michl and S. Wagner: Parallel Computing in Fluid Mechanics, Part I. ERCOFTAC Course on Parallel Computational Fluid Dynamics, Universität Stuttgart, 1994.

[2] B. S. Baldwin and H. Lomax: Thin Layer Approximation and Algebraic Model for Separated Turbulent Flow. Report AIAA 78-0257, (1978).

[3] R. Wirtz: Eine Methodik für die areodynamische Vorentwurfsrechnung im Hyperschall. Dissertation am Institut für Aerodynamik und Gasdynamik der Universität Stuttgart, 1993.

[4] D. Hänel and R. Schwane: An Implicit Flux Vector Splitting Scheme for the Computation of Viscous Hypersonic Fluxes. Report AIAA 89-0274, (1989).

[5] Henk A. van der Vorst: BI-CGSTAB: A Fast and Smoothly Converging Variant of BI-CG for the Solution of Nonsymmetric Linear Systems. SIAM J. Sci. Stat. Comput., 13(2): 631-644, (1992).

[6] Henk A. van der Vorst and Kees Dekker: Conjugate Gradient Type Methods and Preconditioning. Journal of Computational and Applied Mathematics, 24:73-87, 1988.

[7] T. Michl: Effiziente Euler- und Navier-Stokes-Löser für den Einsatz auf Vektor-Hochleistungsrechnern und massiv-parallelen Systemen. Dissertation am Institut für Aerodynamik und Gasdynamik der Universität Stuttgart, 1995

2 Massively Parallel Computers for Navier-Stokes Solutions on Unstructured Meshes

Q. V. Dinh, Dassault Aviation, St. Cloud, France,
P. Leca and F. X. Roux, Onera, Châtillon, France

Abstract

This report describes the work carried out by the authors within the scope of the ECARP project for the Concerted Action 4.2 : MPP for Navier-Stokes. It includes contributions to databases of the workshop held on July 11, 1995 at Inria, Sophia-Antipolis, France, where comparison on accuracy and efficiency of numerical solutions were made interactively.

During the course of the project, a two/three-dimensional implicit Navier-Stokes solver using unstructured meshes has been tested on two M(ultiple)-I(nstruction)-M(ultiple)-D(ata) platforms, namely Intel IPSC 860 and IBM SP2, and benchmark results are available. The Top-Up phase of the project has allowed us to produce valid results for contributions to the above workshop.

I Introduction

For this joint work, the contributions of the two partners were :

- for Dassault-Aviation, a proven two/three-dimensional implicit Navier-Stokes solver based on unstructured meshes which has been validated for subsonic, transonic and hypersonic flows.

- for Onera, an expertise on parallel computing in a distributed environment and access to a 128-processors Intel IPSC 860 located at Onera, Châtillon.

The three-dimensional solver has first been ported on the Intel IPSC 860 and in order to measure scalability, benchmark results (i.e. results for a fixed, usually low, number of iterations) concerning test cases TP1 and TP2 of the workshop were produced.

In May 1995, the Intel IPSC 860 has been removed from Onera. In the meantime, a 22-processors IBM SP2 has been installed at Dassault-Aviation, Saint-Cloud, allowing us to rerun the two test cases TP1 and TP2 for the purpose of validating the numerical results. They were the final contributions to the workshop.

II Methodology

1 Overview of the solver

We refer to [Chalot92] for a detailed description of this solver. It suffices to know that it uses unstructured meshes made up of elements (triangles in two-dimensions and tetrahedra in three-dimensions), and a Galerkin/least-squares formulation which leads to a compact numerical stencil. Convergence to steady-state is achieved through an implicit iterative time-marching algorithm. At each time step, linearized problems are solved using the iterative G(eneralized)-M(inimal)-RES(idual) (c.f. [Saad86]) algorithm.

Its finite element implementation is based on the so-called *matrix-free* procedure : matrices are never assembled and matrix-vector products are done by loops through elements. In this case, the work load is directly proportional to the number of elements and the solver, although numerically implicit, behaves from a implementational point of view much like an explicit method.

2 Parallel implementation

To run our solver on a distributed environment, we rely on the domain decomposition paradigm and, making use of message-passing techniques, we proceed in two steps :

1. Data distribution step : we decompose our meshes into non-overlapping geometric blocks which will be stored in the local memory of each processor
2. Code adaptation step : working in the S(ingle)-P(rogram)-M(ultiple)-D(ata) mode, we add to the original Fortran code of our solver, several procedures based on message-passing primitives (at most 10), to insure the global coherence of our numerical solution

We now give some details on these two steps.

2.1 Mesh decomposer

To decompose a general unstructured mesh into non-overlapping blocks, several algorithms are available (c.f. [Farhat93]) in the parallel computing litterature. They all aim to satisfy more or less the following criteria :

C1. Balancing the work-load on the different processors. In our case, it means that we are looking for equal size blocks in terms of number of elements. A coefficient to measure this would be :

$$LBF = \frac{\mathcal{T}_{moy}}{\mathcal{T}_{max}}$$

where $\mathcal{T}_{moy}$ (resp. $\mathcal{T}_{max}$) designates the mean (resp. maximum) number of block elements. The nearer LBF is to unity, the better is the load balancing.

C2. Minimizing communication between processors. In practice, this requires smaller interfaces and a smaller number of interfaces per block. We can define a coefficient to measure this :

$$COM = \frac{\Sigma_b N_b - N}{N}$$

where N_b (resp. N) designates the number of nodes in block b (resp. the total number of nodes of the original mesh). We can see that COM represents the proportion of ***redundant*** nodes created on the interfaces, and is directly related to the communication cost.

We have tried several algorithms to decompose our meshes. Due to the *explicit* nature of our solver and because of our *matrix-free* approach, we have privileged criterium C1 over criterium C2.

2.2 Porting on one processor of the IPSC 860

Due to limited processor-memory size (8 Mega-bytes in this case), this work consisted mainly in optimizing the memory requirements of our solver and making sure that the message-passing procedures we have added remained inactive. Thus, we were able to run on :

- meshes with at most 8000 nodes for two-dimensional cases
- meshes with at most 4000 nodes for three-dimensional cases.

2.3 Porting on several processors of the IPSC 860

Once optimization on one processor was accomplished, we proceeded to test the message-passing procedures that we have added to insure the global coherence of our solver. They were of two types :

1. Residual computations and matrix-vector products : they require local processor-to-processor communications and for this, use inputs from the mesh partitioner. We noticed that their work loads are directly proportionnal to the number of elements in each block and thus, are well balanced.
2. The GMRES algorithm requires a certain number of global communications like : inner-products, norms, factorization of Hessenberg matrices ... This work load and the corresponding communication cost are proportionnal to the number of nodes and the dimension of Krylov spaces used. The required communications have been grouped together so as to minimize communication overhead.

2.4 Porting on several processors of the IBM SP2

The domain decomposition paradigm has allowed us to have a very portable solver. Starting with the Fortran code adapted to IPSC 860, to port it on IBM SP2 we only need to change several (at most 10) subroutines requiring local processsor-to-processor communication or global communication.

III Discussion of results

We have run our solver on two test-cases : TP1 and TP2.

1 Test-case TP1

It is a two-dimensional flow around a NACA0012 profile with the following characteristics :

- Laminar flow at Reynolds number 5000.

- Free stream Mach number = 0.5

- Angle of attack = 0.

The two-dimensional unstructured mesh used contains :

- 8008 nodes

- 15794 triangles.

For our benchmark results on the IPSC 860, we have decomposed this mesh into 4, 8, 16 and 32 blocks and the run was performed for only 100 non-linear iterations.

Table (1) shows the relatively good scalability of our code. The run on one processor was possible and speed-up was measured based on it.

We validated our numerical results by re-running the test-case on one node of the IBM SP2 and the appropriate numerical data is available in the Workshop data base. This computation lasted 500 iterations and 65 minutes.

2 Test-case TP2

This three-dimensional flow around the Onera M6 wing is defined by the following parameters :

- Laminar flow at Reynolds number 1000.

- Free-stream Mach number = 0.8

- Angle of attack = 0.

The three-dimensional unstructured mesh used contains :

- 62967 nodes

- 349808 tetrahedra.

For our benchmark results on the IPSC 860, we have decomposed this mesh into 32, 64 and 128 blocks and the run was performed for only 5 non-linear iterations.

Table (2) shows the relatively good scalability of our code. Since the run on one processor was not possible, speed-up is measured based on the run with 32 processors.

To validate our numerical results, we have rerun the test-case on a 4 processors IBM SP2. The appropriate data is available in the Workshop data base. This computation lasted 500 iterations and 50 hours. It should be noted that we were not able to run in a dedicated mode, so that computing time as well as communication time are not reliable. Nevertheless, this is the real industrial environment where the solver must work.

Table 1 Test case TP1 : Benchmark results

Number of processors	Load balance LBF	Communication COM	Elapsed time (seconds)	Speed-up
4	0.9999	0.0348	669.6	3.76
8	0.9996	0.0572	341.6	7.37
16	0.9991	0.0978	187.7	13.42
32	0.9991	0.1445	100.4	25.10

Table 2 Test case TP2 : Benchmark results

Number of processors	Load balance LBF	Communication COM	Elapsed time (seconds)	Speed-up
32	0.9999	0.2729	77.6	1.
64	0.9999	0.3782	41.4	1.9
128	0.9999	0.4993	22.3	3.5

IV Conclusion and Future work

Using the domain decomposition paradigm, we have implemented a two-three dimensional Navier-Stokes solver based on unstructured meshes on two MIMD platforms : Intel IPSC 860 and IBM SP2. During the MPP workshop at Inria, Sophia-Antipolis, our results compared favorably with results from solvers based on structured meshes.

For the future, we feel that for parallel computers of MIMD type, with a medium to large (i.e. *10 to 100*) number of processors with large local memory (i.e. more than *32 Mega-bytes*), and for stiff Computational Fluid Dynamics problems (i.e. *Unsteady Navier-Stokes solutions with turbulence*), research and development effort could be spent on :

i) improving existing parallel iterative linear solvers.

ii) parallelizing multilevel methods in a domain decomposition context.

Bibliography

[Chalot92] *Chalot F., Johan Z., Mallet M., Ravachol M., Roge G. (1992) :* Development of a finite element Navier-Stokes solver with applications to turbulent and hypersonic flows, AIAA Paper **92-0670**.

[Farhat93] *Farhat C., Lesoinne M. (1993) :* Automatic partitioning of unstructured meshes for the parallel solution of problems in computational mechanics, Int. J. Num. Meth. Eng. **36**, 745-764.

[Saad86] *Saad Y., Schultz M.H., (1986) :* GMRES, a generalized minimal residual algorithm for solving nonsymmetric linear equations, SIAM J. Sci. Stat. Comput. **7**, 856-869.

3 Parallel computations on an IBM SP2 and SGI cluster

P. Eliasson, FFA, The Aeronautical Research Institute of Sweden

3.1 Parallel computations on an IBM SP2 and an SGI cluster for the NACA0012 airfoil

Abstract

The finite-volume Navier-Stokes code Euranus has been parallelized. The parallelization is based on distributing blocks to different processors with message passing between the processors. The message passing is done under PVM. A special treatment of the boundary conditions ensures that the convergence is independent of how the blocks are distributed and that the convergence is the same for a sequential and a parallel computation. The parallel code is based on explicit Runge-Kutta time marching accelerated with implicit residual smoothing and full multigrid. The message passing is done in all Runge-Kutta stages at all grid levels.

The NACA0012 airfoil was computed at Reynolds number $Re = 5000$ and $\alpha = 0°$ angle of attack. The parallel facilities were an IBM SP2 and a cluster of SGI workstations. 16 processors were used at most.

Introduction

As a result of the physical limitation on the speed up of monoprocessor computers, computer constructers developed multiprocessor systems. The computer systems addressed here are MIMD distributed memory systems in which the processors handle different computers connected in a cluster.

Message passing environments have been developed to make the communication machine independent and, as a result, the codes more transportable. Different such environmentsare available, PVM (parallel Virtual Machine) being one of the most poular.

In the present paper the sequential multiblock/multigrid Navier-Stokes solver Euranus, Rizzi (1993), is ported to different MIMD architectures by distributing blocks to the processors and using PVM for communication. The code is characterized by a software engineering-based coding approach to produce a structured and flexible program. It has a dynamic memory handling to facilitate the storage and maintenance of multiple blocks of different sizes. The data structure supports both computer architecture's with all data in the internal memory and architecture's with limited internal memory with disk read and write.

The parallel code is based on explicit Runge-Kutta time marching accelerated with implicit residual smoothing and multigrid. As each processor has its own local memory and performs its own separate program instructions the different blocks need not to be uniform or of exactly the same size. On the block level the CFD code looks much the same for a MIMD computer as for the sequential computer. The difference lies in the method of exchange of boundary data between the blocks, where the data has to be sent explicitly to neighbouring processors on the MIMD computer by message passing. A special treatment of the boundary conditions ensures that the convergence is independent of the order in which the blocks are treated and that the same convergence is obtained for the parallel multiblock and the serial monoblock case.

Multiblock/multigrid implementation

Since Euranus uses structured meshes, a multiblock approach is adopted to handle complex geometries. Patched blocks are for the moment used. For flexibility, load balancing and overall efficiency Euranus allows any possible number of boundary conditions on a block face. In addition, Euranus imposes no restriction on the number of blocks, allowing any block to be artificially split up into more blocks.

Euranus uses so-called *ghost cells* or *dummy cells* to set up the boundary conditions. Two rows of dummy cells are required on all boundaries for second order accuracy. In addition to the 'physical' dummy cells, an extra copy of all dummy cells is used for independency of the order in which the blocks are treated and to obtain the same convergence in the parallel and sequential cases, Eliasson (1992).

Figure 1 shows two blocks that share a common interface. Each block writes updated boundary information in its own extra copy of dummy cells if it concerns a non-connected boundary and in the extra copy of the connected block if it is a connected boundary. The latter operation requires a communication between the connected blocks, indicated with thicker arrows in Figure 1.

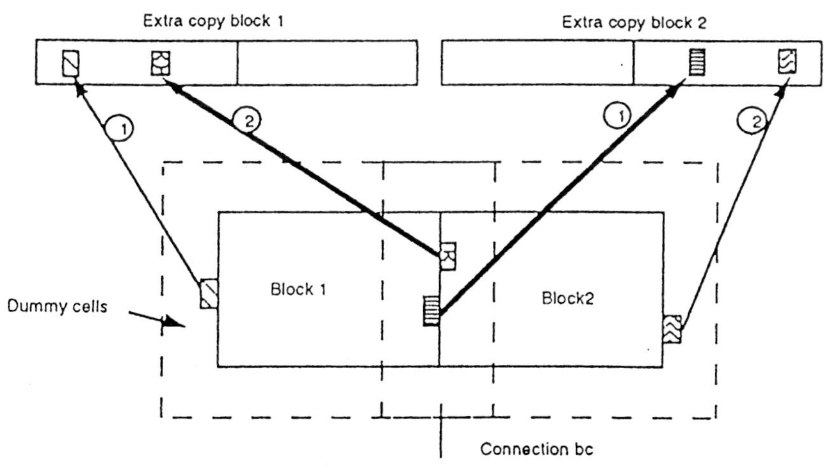

Figure 1 Use of extra dummy cells for multiblock on multiprocessors.

To prevent dead locks, all processors loop through their assigned block(s) and update the boundary conditions where non-blocking send of boundary data (properly labeled) are sent to connecting blocks/processors. When all data has been sent, blocking receives collect the proper messages and the data is put into the extra dummy cells. Having received all messages, the data in the extra dummy cells can be copied into the 'physical' dummy cells. The message passing is done under PVM but the implementation is general and can easily be substituted for some other package.

The FORTRAN loop over the blocks is inside the loop over the grids. The block loop is also inside the loop over the different Runge-Kutta stages, Figure 2. This means that the blocks communicate on every multigrid level and at each Runge-Kutta stage.

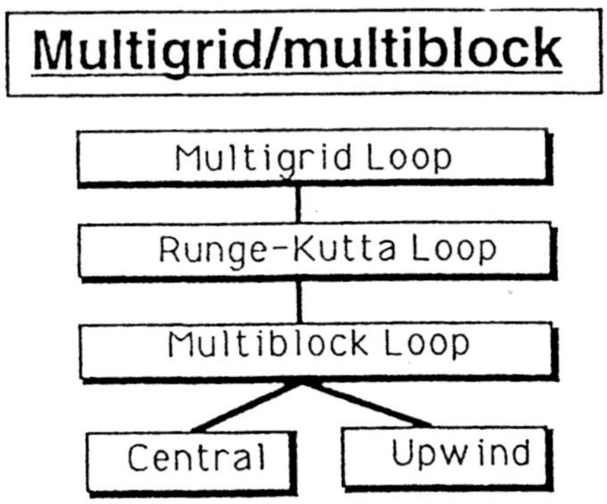

Figure 2 Order of FORTRAN loops within Euranus.

The multigrid cycles in Euranus are sawtooth cycles, V-cycles or W-cycles. For efficiency, full multigrid is also available. The prolongation from a coarser to a finer grid is based on tri linear interpolation (in 3D). The restriction used is the transpose of the prolongation resulting in a higher degree accurate restriction, Figure 3. With the higher order restriction, smoothing of the corrections in the prolongation is not needed.

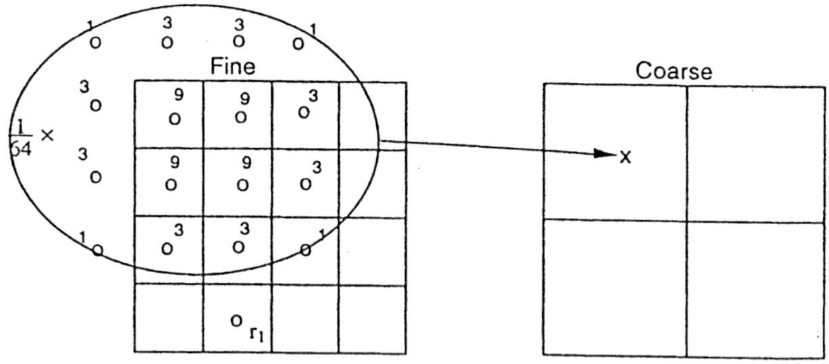

Figure 3 Higher order restriction in 2D.

Spatial discretization

In Euranus the numerical flux can generally be expressed as

$$(\vec{F}\cdot\vec{n})^{*}_{i+\frac{1}{2}} = \frac{1}{2}\{(\vec{F}\cdot\vec{n})_i + (\vec{F}\cdot\vec{n})_{i+1}\} - d_{i+\frac{1}{2}} \tag{1}$$

where the first term represents a central discretization and $d_{i+\frac{1}{2}}$ represents a numerical dissipation. The numerical dissipation is either an artificial dissipation for the central scheme or of upwind type.

The central scheme results from a Jameson type of dissipation using a blend of second and fourth order derivatives of conservative variables

$$d_{i+\frac{1}{2}} = \epsilon^{(2)}_{i+\frac{1}{2}}\delta U_{i+\frac{1}{2}} + \epsilon^{(4)}_{i+\frac{1}{2}}\delta^3 U_{i+\frac{1}{2}} \tag{2}$$

where δ is the difference operator. The scalar coefficients ϵ are given by

$$\begin{aligned}\epsilon^{(2)}_{i+\frac{1}{2}} &= \kappa^{(2)}\rho_{i+\frac{1}{2}}\max(\nu_{i-1},\nu_i,\nu_{i+1},\nu_{i+2})\\ \epsilon^{(4)}_{i+\frac{1}{2}} &= \max(0,\kappa^{(4)}\rho_{i+\frac{1}{2}} - \epsilon^{(2)}_{i+\frac{1}{2}}).\end{aligned} \tag{3}$$

The variables $\kappa^{(2)}$ and $\kappa^{(4)}$ are constants given by the user to control the amount of dissipation. $\rho_{i+\frac{1}{2}}$ is the cell face spectral radius. The variables ν_i are sensors to determine strong shocks. They measure pressure variations and are defined as:

$$\nu_i = \left|\frac{p_{i+1} - 2p_i + p_{i-1}}{p_{i+1} + 2p_i + p_{i-1}}\right|. \tag{4}$$

The dissipation for upwind schemes in Euranus is expressed as

$$d_{i+\frac{1}{2}} = \frac{1}{2}R_{i+\frac{1}{2}}diag(\alpha^l_{i+\frac{1}{2}})R^{-1}_{i+\frac{1}{2}}(U_{i+1} - U_i). \tag{5}$$

The matrices R and R^{-1} contain the right and left eigenvectors of the Jacobian matrix. $diag(\alpha^l_{i+\frac{1}{2}})$ represents a diagonal matrix with $\alpha^l_{i+\frac{1}{2}}$ the element in row and column l.

Different upwind schemes are obtained depending on the expression of α. In Euranus a TVD version of the Flux Difference Splitting scheme (FDS TVD) and symmetric TVD (STVD) are available.

For both schemes α can be expressed as:

$$\alpha_{i+\frac{1}{2}} = |\lambda_{i+\frac{1}{2}}|(1 - Q_{i+\frac{1}{2}}) \tag{6}$$

where $\lambda_{i+\frac{1}{2}}$ represents the eigenvalue of the Jacobian matrix and $Q_{i+\frac{1}{2}}$ is a limiter function ensuring monotonicity and second order spatial accuracy.

The limiter $Q_{i+\frac{1}{2}}$ acts on ratios $r_{i+\frac{1}{2}}$ of variations of the characteristic variables, defined as:

$$r^-_{i+\frac{1}{2}} = \frac{w_i - w_{i-1}}{w_{i+1} - w_i}, \quad r^+_{i+\frac{1}{2}} = \frac{w_{i+2} - w_{i+1}}{w_{i+1} - w_i}. \tag{7}$$

For the FDS TVD scheme Q is a function of only one of these ratios; r^- if $\lambda_{i+\frac{1}{2}}$ is positive and r^+ if $\lambda_{i+\frac{1}{2}}$ is negative. The 'classical' limiters like the minmod, van Albada, van Leer and the superbee can be used as they are all defined as functions of one ratio.

In the STVD scheme a novel approach is used, Lacor (1993). The limiter function Q is chosen as:

$$Q = Q(r^*) \tag{8}$$

with r^* a so-called effective ratio obtained by averaging r^- and r^+. It can be shown that any of the limiters mentioned above applied to r^* leads to a monotonic scheme provided $r^* \geq 0$. Some possible definitions of r^* are:

$$\min(r^+, r^-) \leq \frac{r^+ {r^-}^2 + r^- {r^+}^2}{{r^-}^2 + {r^+}^2} \leq \underline{\frac{2r^+r^-}{r^- + r^+}} \leq \sqrt{r^+r^-} \leq \frac{1}{2}(r^+ + r^-) \leq \max(r^+, r^-) . \tag{9}$$

Definitions with a higher value leads to a more compressive limiter. The effective ratio used for the computations in this paper is the underlined harmonic mean.

Results

The mandatory test case (TP1), the 2D flow over a NACA0012 airfoil at Reynolds number $Re = 5000$ and $\alpha = 0°$ angle of attack, was computed.

The computations were carried out in parallel on an IBM SP2 computer located at KTH in Stockholm using up to 16 processors. The computer has altogether 55 processors with 128 or 512 Mbyte of memory each. For the NACA0012 airfoil computations were also made on a cluster of SGI work stations at FFA with R4000 processors connected via Ethernet.

The common C-mesh of 289 × 57 was used which allowed 4 multigrid levels. A coarse mesh was also computed using every two points from the fine mesh, 145 × 29 points.

The parallelization was done by splitting the supplied mesh into equally sized blocks. 2,4,8 and up to 16 blocks were used. The partitioning can be seen in Figure 4.

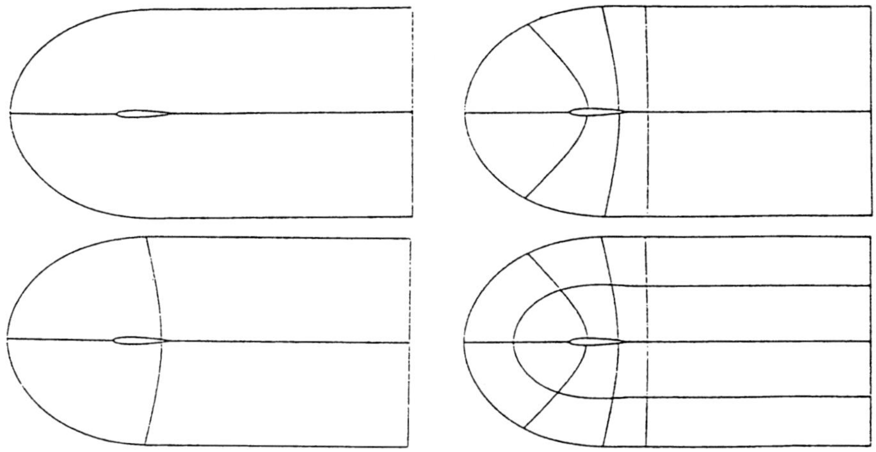

Figure 4 Partitioning of the supplied C-mesh.

Full multigrid was used with 4 level W-cycles on the fine mesh and 3 levels on the coarse grid. One Runge-Kutta smoothing is done both going down and up in the multigrid cycle.

Three different discretizations were used: the symmetric TVD with van Leer effective ratio and van Leer limiters, the flux difference splitting with minmod and van Leer limiter and the central scheme with no second order dissipation ($\kappa^{(2)} = 0$) and with different amount of fourth order dissipation.

A five stage Runge-Kutta scheme was used, the coefficients being

$$\alpha_1 = 0.0814\ ,\ \alpha_2 = 0.1906\ ,\ \alpha_3 = 0.342\ ,\ \alpha_4 = 0.574\ ,\ \alpha_5 = 1.0 \qquad (10)$$

which has good high frequency smoothing for both central and upwind schemes, Eliasson (1993). Implicit residual smoothing was used with coefficient by Radespiel& Rossov (1989), allowing in theory the CFL number to be doubled. $CFL = 5$ was used in all computation.

The rate of convergence for the central scheme ($\kappa^{(4)} = 1/32$) can be seen in Figure 5. The RMS residual of the energy is plotted against number of multigrid iterations and elapsed CPU time on one SGI R4000 processor in a sequential computation. About 80 multigrid iterations are required to reduce the residual 5 orders of magnitude. The convergence is similar for the upwind schemes.

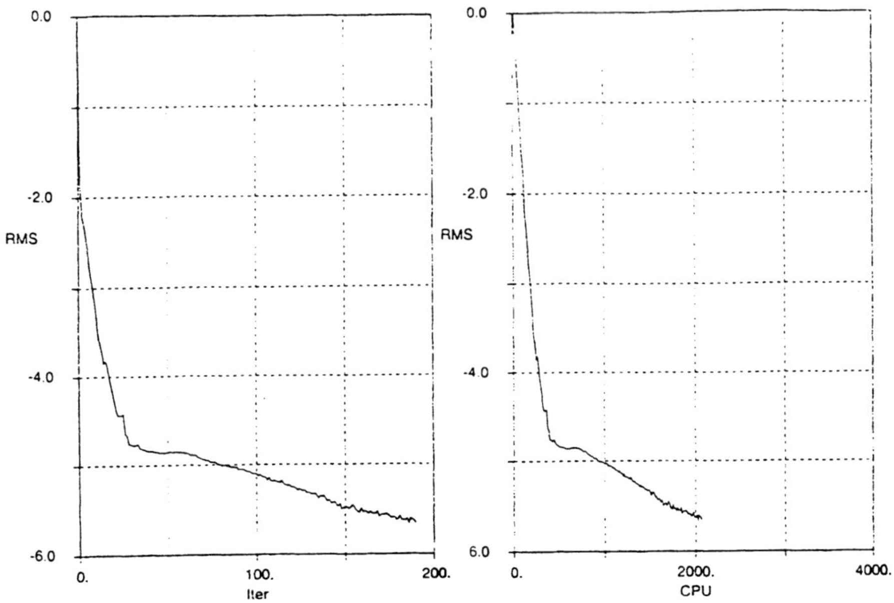

Figure 5 Rate of convergence for the central scheme ($\kappa^{(4)} = 1/32$) on the fine mesh

The timings, speedups, for the fine mesh can be seen in Table 1. The times are given per multigrid iteration on a central scheme, the speedup is defined as the sequential time for a given number of processors divided with the parallel elapsed time. The speedup for the upwind schemes is about the same. The Mflop rate is estimated ignoring grid transfer operations. Results for the coarse mesh can be seen in Table 2, the speedup is also plotted in Figure 6.

Table 1 Timings/iteration and speedup on the fine grid

Computer	# proc.	mesh size	CPU sequ.	elapsed	speedup	Mflop
IBM SP2	1	289×57	8.84 s	8.84 s	1	75
IBM SP2	2	$2 \times 145 \times 57$	8.98 s	4.74 s	1.89	142
IBM SP2	4	$4 \times 73 \times 57$	9.32 s	2.61 s	3.57	268
IBM SP2	8	$8 \times 37 \times 57$	10.15 s	1.55 s	6.54	491
IBM SP2	16	$16 \times 37 \times 29$	10.73 s	0.97 s	11.06	830
SGI R4000	1	289×57	25.61 s	25.61 s	1	26
SGI R4000	2	$2 \times 145 \times 57$	25.33 s	16.5 s	1.54	40
SGI R4000	4	$4 \times 73 \times 57$	26.62 s	11.3 s	2.36	61
SGI R4000	8	$8 \times 37 \times 57$	27.68 s	8.1 s	3.41	88

Table 2 Timings/iteration and speedup on the coarse grid

Computer	# proc.	mesh size	CPU sequ.	elapsed	speedup	Mflop
IBM SP2	1	145×29	2.34 s	2.34 s	1	70
IBM SP2	2	$2 \times 73 \times 29$	2.48 s	1.41	1.76	123
IBM SP2	4	$4 \times 37 \times 29$	2.75 s	0.85 s	3.23	226
IBM SP2	8	$8 \times 19 \times 29$	3.23 s	0.64 s	5.05	353
IBM SP2	16	$16 \times 19 \times 15$	3.95 s	0.59 s	6.69	468
SGI R4000	1	289×29	6.48 s	6.48 s	1	25
SGI R4000	2	$2 \times 73 \times 29$	6.77 s	3.71 s	1.82	45
SGI R4000	4	$4 \times 37 \times 29$	7.63 s	4.3 s	1.77	44
SGI R4000	8	$8 \times 19 \times 29$	8.83 s	3.9	2.26	56.5

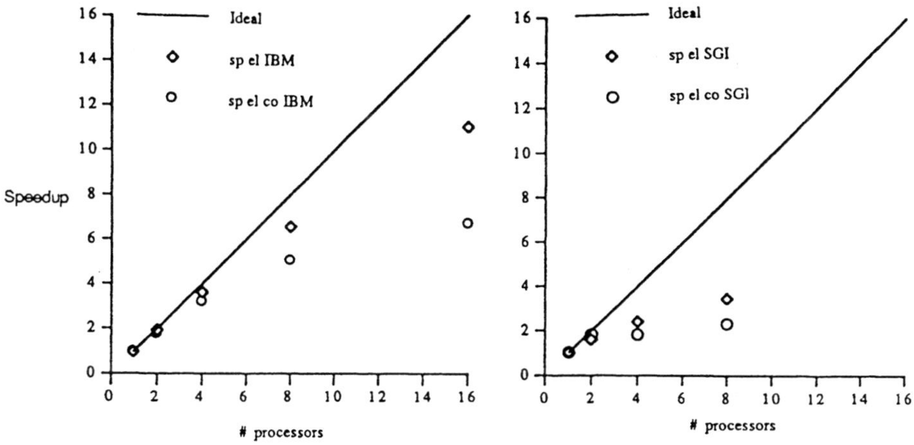

Figure 6 Speedup on the IBM SP2 and the SGI cluster on the fine and coarse grids.

As can bee seen the parallelization on the coarse mesh on the SGI cluster is not efficient, the problem is too small. On the fine mesh the paralellization is good. Note that over 800 Mflop was obtained on the SP2. With 16 processors on the SP2 and 80 full multigrid iterations a fully converged solution is obtained in about 90 seconds.

The quality of the results is measured in terms of pressure and viscous drag and point of separation. This is given in Table 3 where results from Radespiel (1989), Mavripilis (1989) on a 512×128 mesh is included. As can be seen, the agreement with those results is fine from the central scheme with a low amount of dissipation. The pressure and skinfriction on the airfoil is plotted in Figure 7 for the central scheme ($\kappa^{(4)} = 1/64$) and on the coarse grid (dashed line).

Table 3 Comparison of pressure and viscous drag and point of separation

Type of scheme	mesh size	CD_p	CD_v	Separation point
Symm. TVD	289 × 57	0.0235	0.0325	82.9%
FDS TVD	289 × 57	0.0245	0.0321	84.4%
Central $\kappa_4 = \frac{1}{32}$	289 × 57	0.0220	0.0338	82.8%
Central $\kappa_4 = \frac{1}{64}$	289 × 57	0.0224	0.0332	81.9%
Central $\kappa_4 = \frac{1}{128}$	289 × 57	0.0226	0.0329	81.3%
Central $\kappa_4 = \frac{1}{64}$	145 × 29	0.0224	0.0349	85.0%
Central	512 × 128	0.02235	0.03299	81.4%

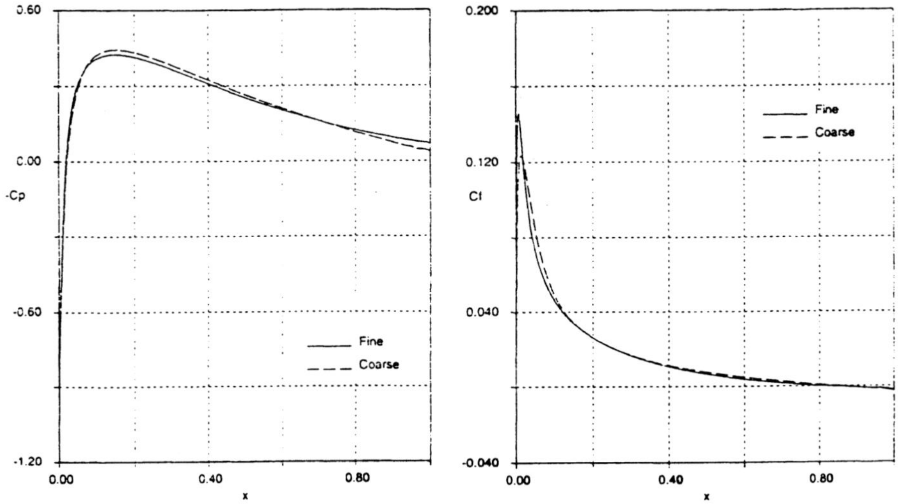

Figure 7 Pressure and skinfriction on the airfoil on a fine and coarse grid ($\kappa^{(4)} = 1/64$).

3.2 Parallel computations on an IBM SP2 for the M6 wing

Abstract

The finite-volume Navier-Stokes code Euranus has been parallelized. The parallelization is based on distributing blocks to different processors with message passing between the processors. The message passing is done under PVM. A special treatment of the boundary conditions ensures that the convergence is independent of how the blocks are distributed and that the convergence is the same for a sequential and a parallel computation. The parallel code is based on explicit Runge-Kutta time marching accelerated with implicit residual smoothing and full multigrid. The message passing is done in all Runge-Kutta stages at all grid levels.

The technical aspects of the parallelization of the Euranus code on an IBM SP2 computer are identical to those described in point 3.1.

Results

The optional test cases (TP5) was computed, the 3D flow over the M6 wing at $Re = 11.7\,10^6$ and $\alpha = 3.06°$ angle of attack.

The computations were carried out in parallel on an IBM SP2 computer located at KTH in Stockholm using up to 16 processors. The computer has altogether 55 processors with 128 or 512 Mbyte of memory each.

An O-O mesh consisting of $257 \times 97 \times 49$ points was created at FFA with the external boundary 15 chords from the wing. 257 points in the stream wise direction, 97 points normal to the wing and 49 points span wise. The distribution normal to the wing was made to obtain $y^+ \simeq 5$ in first point outside the boundary. Results were also obtained on a coarser mesh by removing every two points, totally $129 \times 49 \times 25$ points.

The Baldwin-Lomax turbulence model was used, the central scheme with $\kappa^{(2)} = 1.0$ and $\kappa^{(4)} = 1/20$ was used for the spatial discretization. A five stage Runge-Kutta scheme was used, the coefficients being

$$\alpha_1 = 0.0814\,,\ \alpha_2 = 0.1906\,,\ \alpha_3 = 0.342\,,\ \alpha_4 = 0.574\,,\ \alpha_5 = 1.0 \qquad (10)$$

which has good high frequency smoothing for both central and upwind schemes, Eliasson (1993). Implicit residual smoothing was used with coefficient by Radespiel& Rossov (1989). 5 levels of full multigrid W-cycles with $CFL = 3$ was used

This case was only run on the IBM SP2 with up to 16 processors. The mesh was split up into equally sized blocks.

Figure 8 shows the rate of convergence, 3.5 orders of reduction is obtained in 300 multigrid iterations which is sufficient for the pressure to converge.

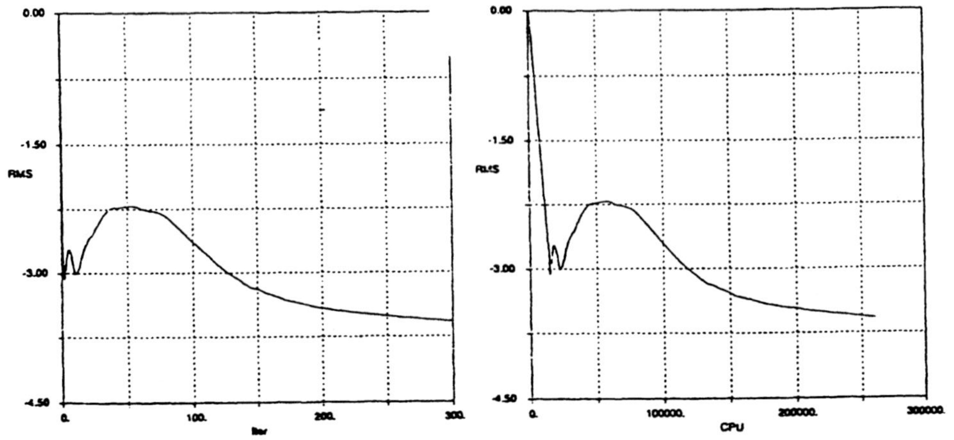

Figure 8 Convergence for the M6.

The timings and speedups per multigrid iteration can be seen in Table 4. Note that over 1 Gflop was obtained with 16 processors.

Table 4 Timings/iteration and speedup on the fine grid

Computer	# proc.	mesh size	CPU sequ.	elapsed	speedup	Mflop
IBM SP2	1	$257 \times 97 \times 49$	180 s	180 s	1	75
IBM SP2	2	$2 \times 129 \times 97 \times 49$	181 s	92 s	1.97	147
IBM SP2	4	$4 \times 65 \times 97 \times 49$	181 s	48 s	3.78	283
IBM SP2	8	$8 \times 65 \times 49 \times 49$	183 s	24.7 s	7.40	555
IBM SP2	16	$16 \times 33 \times 49 \times 49$	185 s	13.1 s	14.10	1060

Finally the pressure distributions on the fine and coarse mesh are shown in Figure 9 compared to experiments. The solution is not yet mesh converged even though the mesh is very fine. The agreement is rather good though, somewhat better than for Vatsa (1986).

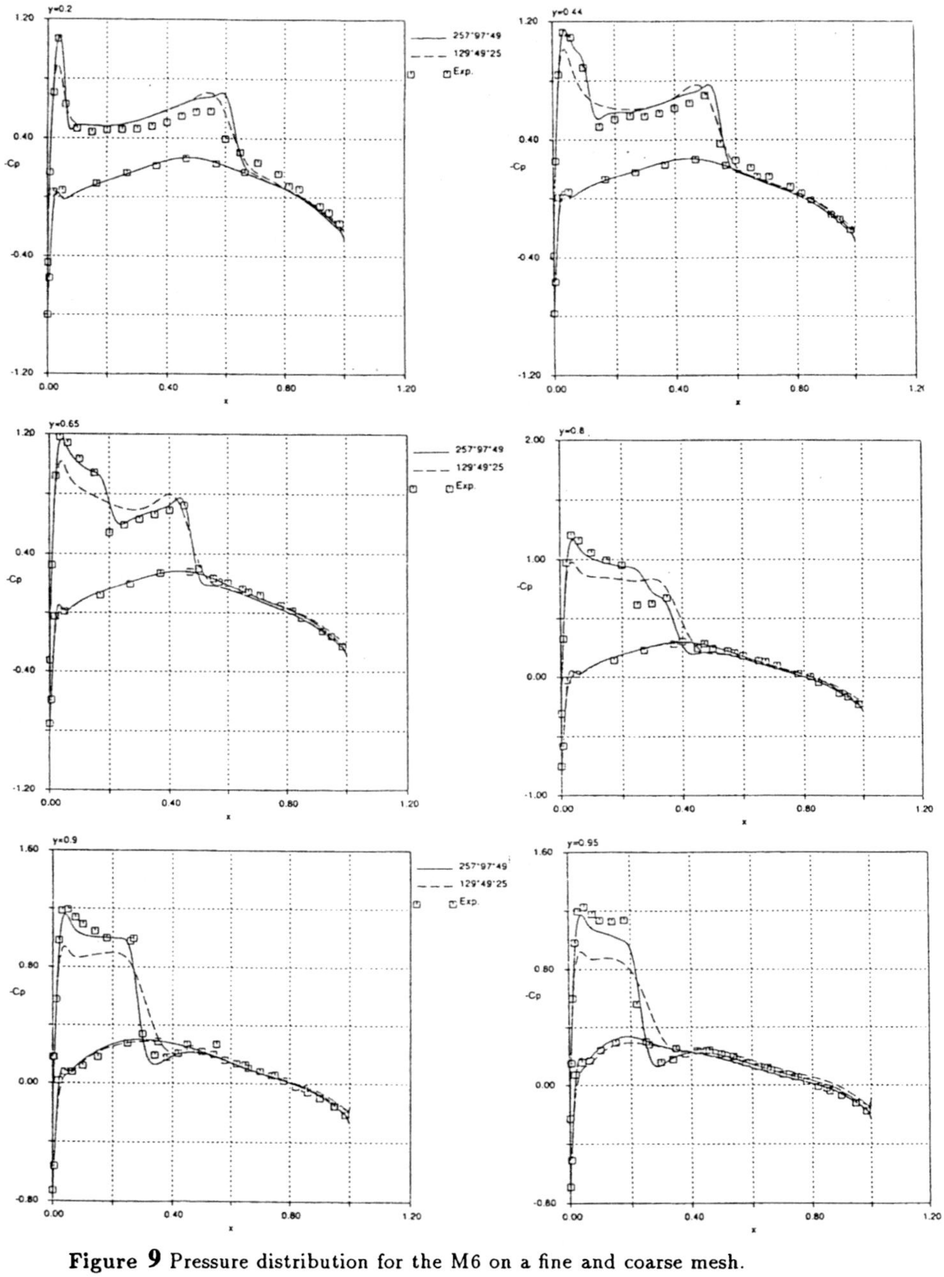

Figure 9 Pressure distribution for the M6 on a fine and coarse mesh.

Bibliography

Rizzi, A., Eliasson, P., Lindblad, I., Hirsch, C., Lacor, C. and Haeuser, J. (1993) The Engineering of Multiblock Multigrid Software for Navier-Stokes Flows on Structured Meshes, Computers and Fluids, **Vol. 22,** , pp. 341-367.

Eliasson, P., Lindblad, I., Wahlund, P. (1992) Implementation of a Multidomain Navier-Stokes Code on the Intel iPSC2 Hypercube, FFA TN 1992-37.

Lacor, C., Zhu, Z.W. and Hirsch, Ch. (1993) A New Family of Limiters Within the Multigrid/Multiblock Navier-Stokes Code Euranus, AIAA/DGLR 5th Int. Aerospace Planes and Hypersonics Technologies Conf., Munich.

Eliasson, P. (1993) Multigrid Solutions in a Multiblock Solver for Compressible Flow, Doctoral Thesis, KTH, TRITA-NA-R9314, ISSN-0348-2952

Radespiel, R., Rossow, C. and Swanson, R.C. (1989) An Efficient Cell-Vertex Multigrid Scheme for the Three-Domensional Navier-Stokes Equations, AIAA Paper 89-1953-CP.

Vatsa, V.N. (1986) Accurate Solutions for Transonic Viscous Flow Over Finite Wings, AIAA-86-1052, Atlanta.

Radespiel, R. and Swanson, R.C. (1989) An Investigation of Cell Centered and Cell Vertex Multigrid Schemes for the Navier-Stokes Equations, AIAA Paper 89-0548.

Mavripilis, D.J., Jameson, A. and Martinellim L. (1989) Multigrid Solutions of the Navier-Stokes Equations on Triangular Meshes, ICASE Report No. 89-11.

4 Contributions from INRIA to the test cases

S. Lanteri, INRIA Sophia Antipolis, Sophia-Antipolis Cedex, France

4.1 TP1 test case

1 Discrete Equations and Methodology

Here we report on results obtained in simulating two-dimensional compressible laminar viscous flows on the KENDALL SQUARE RESEARCH KSR-1 MPP (Massively Parallel Processors) system using a second-order accurate Monotonic Upwind Scheme for Conservation Laws (MUSCL) finite volume/finite element method on fully unstructured triangular meshes.

1.1 Governing equations

Let $\Omega \subset \mathbb{R}^2$ be the flow domain of interest and Γ be its boundary. The conservative law form of the equations describing two-dimensional Navier-Stokes flows is given by:

$$\frac{\partial W}{\partial t} + \vec{\nabla}.\vec{\mathcal{F}}(W) = \frac{1}{Re}\vec{\nabla}.\vec{\mathcal{R}}(W) \tag{1}$$

where $W = W(\vec{x}, t)$; $\vec{x}$ and t denote the spatial and temporal variables, and

$$W = \left(\rho \ , \ \rho\vec{U} \ , \ E\right)^T \quad , \quad \vec{\nabla} = \left(\frac{\partial}{\partial x} \ , \ \frac{\partial}{\partial y}\right)^T$$

and where $\vec{\mathcal{F}}(W)$ and $\vec{\mathcal{R}}(W)$ respectively denote the convective and diffusive fluxes (see [Fezoui89] for a detailed description of the mathematical model). In the above equations, $Re = \dfrac{\rho_0 U_0 L_0}{\mu_0}$ is the Reynolds number.

1.2 Spatial approximation method

The flow domain Ω is assumed to be a polygonal bounded region of $\mathbb{R}^2$. Let $\mathcal{T}_h$ be a standard triangulation of Ω, and h the maximal length of the edges of $\mathcal{T}_h$. A vertex of a triangle T is denoted by S_i, and the set of its neighboring vertices by $K(i)$. At each vertex S_i, a cell C_i is constructed as the union of the subtriangles resulting from the subdivision by means of the medians of each triangle of $\mathcal{T}_h$ that is connected to S_i (see Fig. 1). The boundary of C_i is denoted by ∂C_i, and the unit vector of the outward normal to ∂C_i by $\vec{\nu}_i = (\nu_{ix}, \nu_{iy})$. The union of all these control volumes constitutes a discretization of domain Ω :

$$\Omega_h = \bigcup_{i=1}^{N_V} C_i \quad , \quad N_V \text{ : number of vertices of } \mathcal{T}_h \, .$$

Integrating Eq. (1) over C_i yields the following discrete equation:

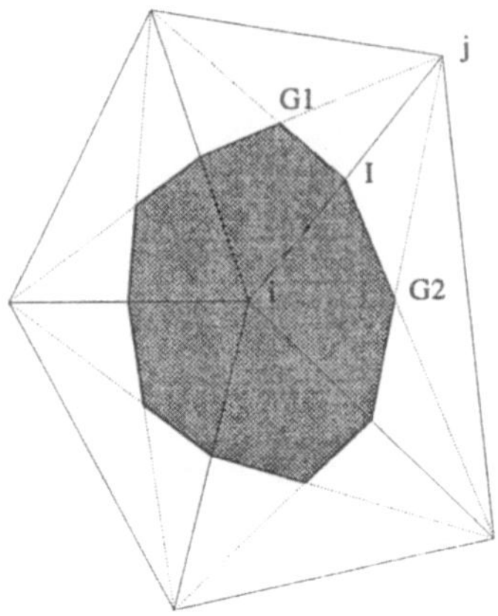

Figure 1 Control volume in an unstructured grid

$$\begin{aligned}\int\!\!\int_{C_i} \frac{\partial W}{\partial t} d\vec{x} \;&+\; \sum_{j\in K(i)} \int_{\partial C_{ij}} \vec{\mathcal{F}}(W).\vec{\nu}_i d\sigma \\ &+ \int_{\partial C_i \cap \Gamma_w} \vec{\mathcal{F}}(W).\vec{n}_i d\sigma \\ &+ \int_{\partial C_i \cap \Gamma_\infty} \vec{\mathcal{F}}(W).\vec{n}_i d\sigma \\ &= -\frac{1}{Re} \sum_{T, S_i \in T} \int\!\!\int_T \vec{\mathcal{R}}(W).\vec{\nabla} N_i^T d\vec{x}\end{aligned} \tag{2}$$

where $\partial C_{ij} = \partial C_i \cap \partial C_j$, and $N_i^T = N_i^T(x,y)$ is the P1 shape function defined at the vertex S_i and associated with the triangle T.

The spatial discretization method adopted here combines the following features (see [Fezoui89] for details):

- a finite volume upwind approximation method for the convective fluxes. Second order spatial accuracy is achieved using an extension of Van Leer's MUSCL technique to unstructured meshes;
- a classical Galerkin finite element centered approximation for the diffusive fluxes.

The computation of the different integral terms of equation (2) will not be detailed here as there are several common points with the approach used for the solution of the three-dimensional Navier-Stokes equations. We therefore refer to our corresponding paper which is also included in this volume.

1.3 Time integration

Assuming that $W(\vec{x},t)$ is constant over the control volume C_i (in other words, a mass lumping technique is applied to the temporal term of Eq. (2)), we obtain the following semi-discrete fluid flow equations:

$$area(C_i)\frac{dW_i^n}{dt} + \Psi(W_i^n) = 0 \quad , \quad i = 1,\cdots,N_V \tag{3}$$

where $W_i^n = W(\vec{x}_i, t^n)$, $t^n = n\Delta t^n$ and

$$\begin{aligned}\Psi(W_i^n) &= \sum_{j\in K(i)} \Phi_{\mathcal{F}}(W_{ij}, W_{ji}, \vec{\nu}_{ij}) + \int_{\partial C_i \cap \Gamma_\infty} \vec{\mathcal{F}}(W).\vec{n}_i d\sigma \\ &+ \frac{1}{Re}\sum_{T, S_i\in T} area(T)\left(R_T\frac{\partial N_i^T}{\partial x} + S_T\frac{\partial N_i^T}{\partial y}\right).\end{aligned} \tag{4}$$

In this study we will use and compare explicit and linearized implicit time integration procedures. The explicit scheme is a 4-step low storage Runge-Kutta algorithm :

$$\begin{cases} W_i^{(0)} &= W_i^n \\ W_i^{(k)} &= W_i^{(0)} + \dfrac{\alpha_k \Delta t^n}{area(C_i)}\Psi(W_i^{(k-1)}) \ , \ k = 1,\cdots,4 \\ W_i^{n+1} &= W_i^{(4)} \end{cases} \tag{5}$$

where $\alpha_1 = 0.11$, $\alpha_2 = 0.2766$, $\alpha_3 = 0.5$ and $\alpha_4 = 1.0$.

Explicit time integration procedures are subjected to a stability condition expressed in terms of a CFL (Courant-Friedrichs-Lewy) number. When one is interested in looking for steady solutions of the Navier-Stokes equations, an efficient time advancing strategy can be obtained by means of an implicit linearised formulation. Here, we briefly recall the main ingredients of this approach which is described in details in [Fezoui89]. A Newton-like implicit linearised version of Eq. (3) writes as :

$$\frac{area(C_i)}{\Delta t^n}\delta W_i^{n+1} + \Psi(W_i^{n+1}) = 0 \quad , \quad i = 1,\cdots,N_V \tag{6}$$

where $\delta W_i^{n+1} = W_i^{n+1} - W_i^n$. The resulting Euler implicit time integration scheme is given in matrix form by :

$$P(W^n)\delta W^{n+1} = \left(\frac{I}{\Delta t^n} + J(W^n)\right)\delta W^{n+1} = \delta\hat{W}^n \tag{7}$$

where $J(W^n)$ denotes the approximate Jacobian matrix and $\delta\hat{W}^n$ is the explicit part of the linearisation of $\Psi(W^{n+1})$. The matrix $P(W^n)$ is sparse and has the suitable properties (diagonaly dominant in the scalar case) allowing the use of a relaxation procedure (Jacobi or Gauss-Seidel) in order to solve the linear system of Eq. (7). Moreover, an efficient way to get second order accurate steady solutions while keeping the interesting properties of the first order upwind matrix is to use the second order elementary convective fluxes in the right-hand side of Eq. (7).

1.4 Parallelisation Strategy

In addition to efficiency and parallel scalability, portability should be a major concern. With the proliferation of computer architectures, it is essential to adopt a programming model that does not require rewriting thousands of lines of code — or even worse, altering the architectural foundations of a code — every time a new parallel processor emerges. We believe that the above goals can be achieved by using mesh partitioning techniques and by programming in a message-passing model. Essentially, the underlying mesh is assumed to be partitioned into several submeshes, each defining a subdomain. The same "old" serial code can be executed within every subdomain. The assembly of the subdomain results can be implemented in a separate software module and optimized for a given machine. This approach enforces data locality, and therefore is suitable for all parallel hardware architectures. For example, we have shown in [Lanteri93] that for unstructured meshes, this approach produces substantially better performance results on the KSR-1 than the acclaimed virtual shared memory programming model. Note that in this context, message-passing refers to the assembly phase of the subdomain results. However, it does not imply that messages have to be explicitly exchanged between the subdomains. For example, message-passing can be implemented on a shared memory multiprocessor as a simple access to a shared buffer, or as a duplication of one buffer into another one.

The reader can verify that for the computations described herein, mesh partitions with overlapping simplify the programming of the subdomain interfacing module. However, mesh partitions with overlapping also have a drawback: they incur redundant floating-point operations. On the other hand, non-overlapping mesh partitions incur little redundant floating-point operations but induce one additional communication step. While nodal variables are exchanged between the subdomains in overlapping mesh partitions, partially gathered nodal gradients and partially gathered fluxes are exchanged between subdomains in non-overlapping ones. In addition, special care must be taken in the treatment of the convective fluxes in the case of non-overlapping mesh partitions (because of the possible differences in the orientation of the interface edges which are not part of the original mesh but are instead constructed during a preprocessing phase of the parallel algorithm). In other words, the programming effort is maximized when considering non-overlapping mesh partitions. We refer to Farhat and Lanteri [Farhat94] for a comparison of these two approaches in the context of two-dimensional simulations. In the present study we will consider one element wide overlapping mesh partitions for second order accurate computations. We will therefore exchange physical states and nodal gradients.

2 Numerical results and analysis

2.1 Platform identity

The KENDALL SQUARE RESEARCH KSR-1 is a massively parallel processors system with shared address space and distributed physical memory modules. The shared address space provides the ease of use of the shared memory programming model, without worrying about allocating storage, managing sharable data or passing messages from one processor to another, while the distributed memory design provides the ease of scalability. These two features are provided by the unique design of the KSR-1's distributed memory scheme, the ALLCACHE system. The KSR-1 is built as a group of ALLCACHE engines, connected in a fat tree hierarchy of rings; 34 rings can be connected by a single second-level ring for a maximum configuration of 1088 processors (computational cells). Each computational cell consists of a 64-bit RISC superscalar processor operating at 20 Mhz and capable of a theoretical floating point rate of 40 Mflop/s, a 0.25 Mbytes instruction sub-cache, a 0.25 Mbytes data sub-cache and a 32 Mbytes local cache. The KSR-1 is a UNIX compatible system based on OSF/1.

2.2 Physical solutions and performance results

Here we focus on test case **TC1** of the **ECARP** MPP for Navier-Stokes flows workshop which deals with the laminar flow around a NACA 0012 airfoil at a Mach number equal to 0.5 and a Reynolds number equal to 5000. The angle of attack is set to 0°.

We first present results obtained with fully unstructured meshes that were not available on the **ECARP** data base. Three meshes have been generated; their characteristics are summarized in Tab. 1 below where N_v and N_t respectively denote the number of vertices and triangles. A partial view of the coarsest mesh is shown on Fig. 2. Computations have been performed on the KSR-1; in that case the exchange phase at artificial submesh boundaries is realised through write/read operations in a shared array. In the following tables, timing measures concern the main parallel loop. All performance results reported herein are for 64-bit arithmetic.

Parallel scalability is evaluated for problems where the subdomain size is fixed, and the total size is increased with the number of processors. Note that because we are dealing with unstructured meshes, some slight deviations are inevitable. One triangle wide overlapping mesh partitions were generated using a recursive inertial bisection algorithm [Loriot92]. Tab. 2 and 3 summarizes the performance results obtained on the KSR-1 parallel system using the explicit (5) and implicit (6)-(7) time marching strategies. The number of processors is denoted as N_p.

Table 1 Characteristics of fully unstructured meshes for the NACA0012 airfoil

N_v	N_t
3114	6056
12284	24224
48792	96896

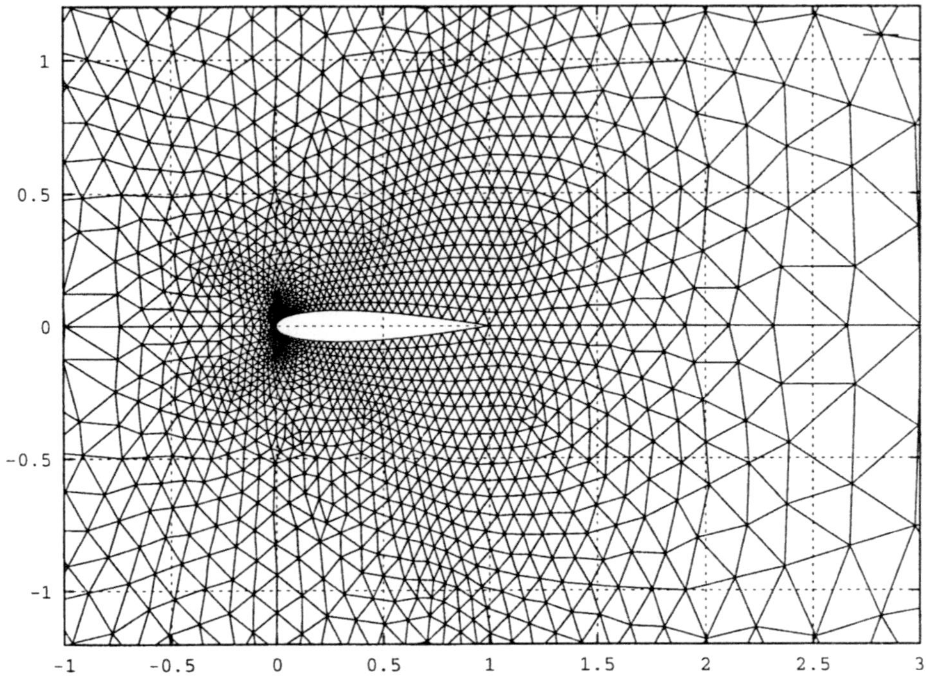

Figure 2 Unstructured mesh around a NACA0012 airfoil : $N_v = 3114$

On the other part, "Loc Comm" measures the total time spent in the exchange phase. Timing measures are given for 10 iterations (pseudo time steps); in the case of the implicit time marching strategy the number of Jacobi relaxations has been respectively fixed to 36 and 72. Here we expect the total CPU time to remain almost constant when switching from $N_p = 1$ to $N_p = 16$. For the implicit time marching strategy, we observe a 13% degradation of the scalability when using 36 Jacobi relaxations; this figure increases to 17% for $N_{relx} = 72$ due to additional communication costs in the linear system solution phase. These degradations in scalability are mainly attributed to the overlapping strategy that is to the necessity to perform redundant arithmetic operations.

We are now interested in computing the steady state solution for the selected test case. For this purpose we have used the implicit time marching procedure (6)-(7) with a local time step computed from the CFL law CFL=$MAX(100, it)$ where *it* denotes the non-linear iteration. Tab. 4 summarizes the obtained results;

Table 2 Parallel scalability of the explicit time marching procedure

N_v	N_p	CPU	Loc Comm
3114	1	13.0 s	0.0 s
12284	4	14.5 s	0.5 s
48792	16	15.5 s	1.0 s

Table 3 Parallel scalability of the implicit time marching procedure

N_v	N_p	N_{relx}	CPU	Loc Comm
3114	1	36	68.0 s	0.0 s
-	-	72	109.0 s	0.0 s
12284	4	36	73.0 s	2.0 s
-	-	72	119.0 s	3.5 s
48792	16	36	77.0 s	3.0 s
-	-	72	128.0 s	5.5 s

N_{relx} denotes the number of Jacobi relaxations while "# Iter" refers to the effective number of non-linear iterations necessary to reach the steady state. The convergence tolerance for the non-linear iteration (main time-stepping loop) has been fixed to 10^{-12} (normalized $l2$-energy residual) while the convergence tolerance for the linear iteration (linear system solution) has been fixed to 10^{-2}. For the case $N_v = 48792$ using $N_{relx} = 36$ Jacobi relaxations the final energy residual after a maximum of 300 non-linear iterations is equal to 2×10^{-10}. We therefore increase the maximal number of Jacobi relaxations to 72 and this time we obtain the targeted value of the energy residual after 222 non-linear iterations. The non-linear convergence curves are depicted on Fig. 3. The obtained steady Mach lines on the coarsest and finest meshes are visualised on Fig. 4 and 5.

Table 4 Steady state solutions around a NACA0012 airfoil : $M_\infty = 0.5$, $Re = 5000$

N_v	N_p	N_{relx}	# Iter	CPU	Loc Comm
3114	1	36	118	771.0 s	3.0 s
12284	4	36	184	1330.0 s	32.0 s
48792	16	36	300	2324.0 s	87.0 s
-	-	72	222	2915.0 s	146.0 s

We consider now solutions of the same test case using the NACA0012 structured triangular mesh provided on the **ECARP**. This mesh (see Fig. 6) consists of 14106 vertices, 27636 triangles and 41742 edges. In order to obtain the steady state solution of this flow we have used and compare the two following time advancing procedures :

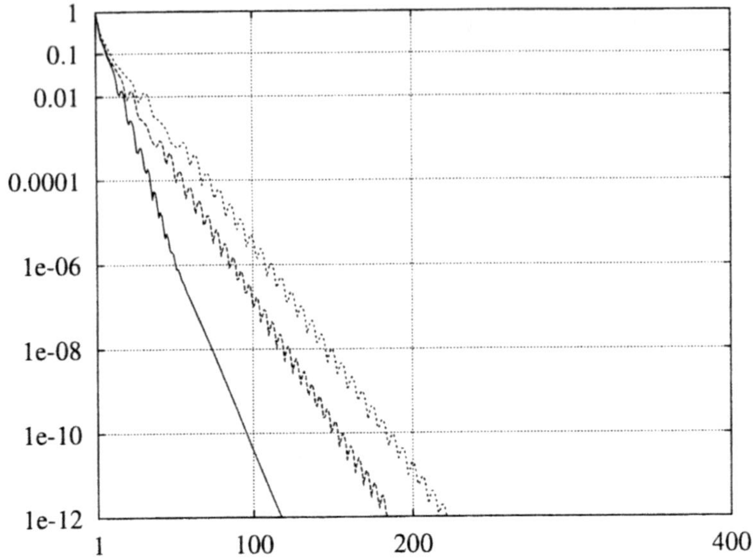

Figure 3 Non-linear convergence (energy residual) - implicit time integration
Leftmost curve : $N_v = 3114$ and $N_{relx} = 36$ - Rightmost curve : $N_v = 48792$ and $N_{relx} = 72$

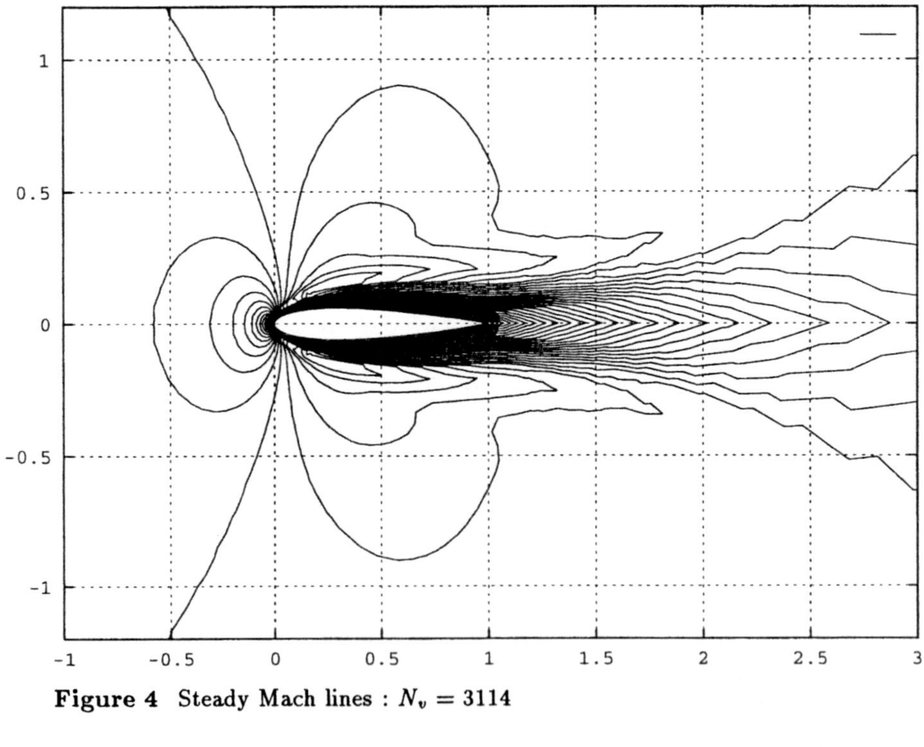

Figure 4 Steady Mach lines : $N_v = 3114$

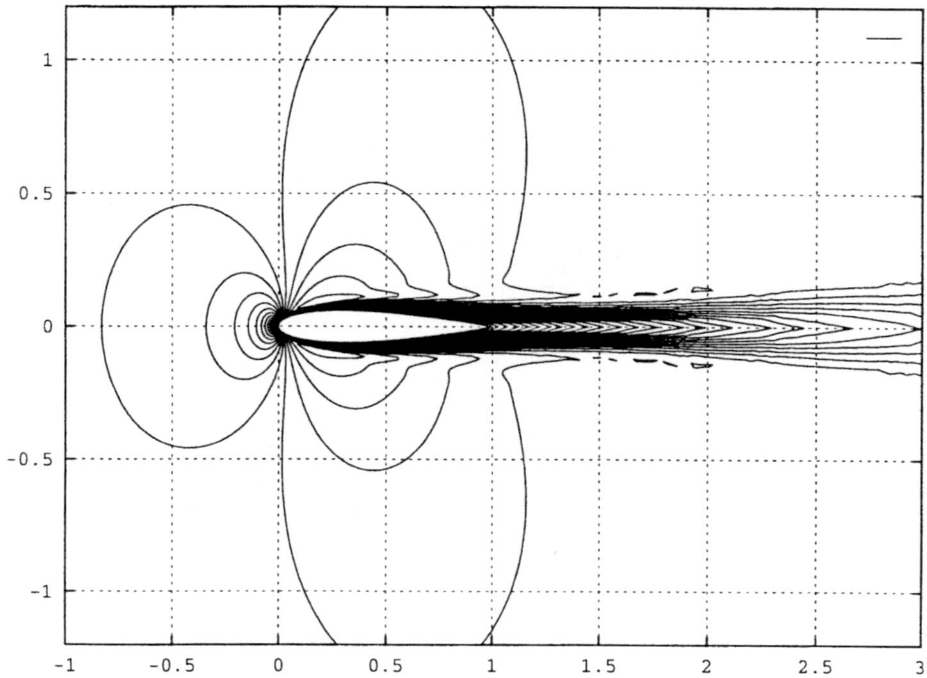

Figure 5 Steady Mach lines : $N_v = 48792$

- the explicit Runge-Kutta scheme (5) with a CFL fixed to 1.9;
- the implicit linearised scheme (6) with a CFL given by the law CFL=*it* where *it* denotes the non-linear iteration. The Jacobi relaxation method is used to solve the linear systems (7) arising at each time steps; the residual tolerance for the linear system solution is set to 10^{-2} while the maximal number of Jacobi relaxations is fixed to 128.

The obtained steady Mach lines are visualised on Fig. 7. The one-dimensional distributions of the pressure and friction coefficients are shown on Fig. 8 and 9. Fig. 10 visualises the non-linear convergence behaviors corresponding to the two time advancing procedures while Fig. 11 gives the effective number of Jacobi relaxations necessary to reach the imposed residual tolerance for the first 100 non-linear iterations.

Tab. 5 gives the CPU times for 1000 non-linear iterations of the explicit Runge-Kutta time marching procedure using 8 processors of the KSR-1. Tab. 6 compares the measured total CPU times for 100 non-linear iterations of the implicit time advancing procedure (6)-(7). In this table "Matrix Inver" gives the total time spent in the linear system solutions using the Jacobi relaxation method which is clearly the dominant part the total CPU time; $S(N_p)$ denotes the parallel speed-up computed from the calculation realised on one processor of the KSR-1. These experiments show that 1000 non-linear explicit iterations cost as musch as 100 non-linear implicit iterations (using 8 processors). However there is a two-order of magnitude difference in the non-linear convergence of these two schemes.

Table 5 Steady state solutions around a NACA0012 airfoil : $M_\infty = 0.5$, $Re = 5000$
Structured triangular mesh : $Nv = 14106$
Computations on the KSR-1 : explicit time marching procedure

N_p	CPU	Loc Comm
8	879.0 s	99.0 s

Table 6 Steady state solutions around a NACA0012 airfoil : $M_\infty = 0.5$, $Re = 5000$
Structured triangular mesh : $Nv = 14106$
Computations on the KSR-1 : implicit time marching procedure

N_p	CPU	Loc Comm	Matrix Inver	$S(N_p)$
1	5636.0 s	0.0 s	4714.0 s	1
4	1506.0 s	54.0 s	1222.0 s	3.8
8	907.0 s	122.0 s	665.0 s	6.2
16	541.0 s	108.0 s	369.0 s	10.4

3 References

[Fezoui89] Fezoui, L., Lanteri, S., Larrouturou, B., Olivier, C., (1989) : Résolution numérique des équations de Navier-Stokes pour un fluide compressible en maillage triangulaire, Rapport INRIA, No. 1033.

[Lanteri93] Lanteri, S., Farhat, C., (1993) : Viscous Flow Computations on M.P.P. Systems : Implementational Issues and Performance Results for Unstructured Grids, Proceedings of the Sixth SIAM Conference on Parallel Processing for Scientific Computing, Norfolk, Virginia, 65-70.

[Farhat94] Farhat, C., Lanteri, S., (1994) : Simulation of Compressible Viscous Flows on a Variety of MPPs: Computational Algorithms for Unstructured Dynamic Meshes and Performance Results, Comp. Meth. in Appl. Mech. and Eng. **119**, 35-60.

[Loriot92] Loriot, M., (1992) : MS3D : Mesh Splitter for 3D Applications, User's Manual, Simulog.

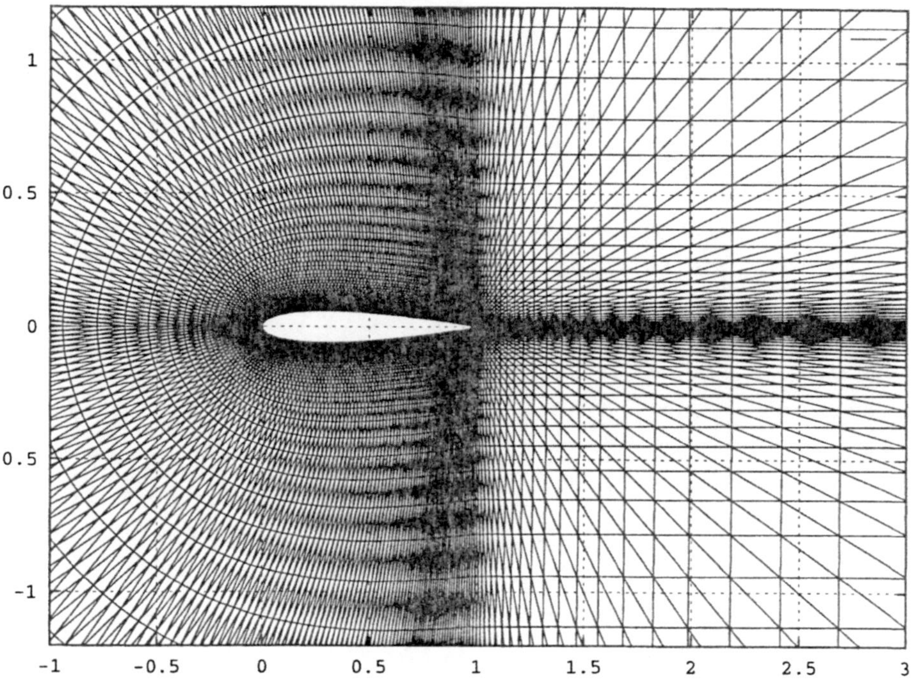

Figure 6 Structured triangular mesh around a NACA0012 airfoil : $N_v = 14106$

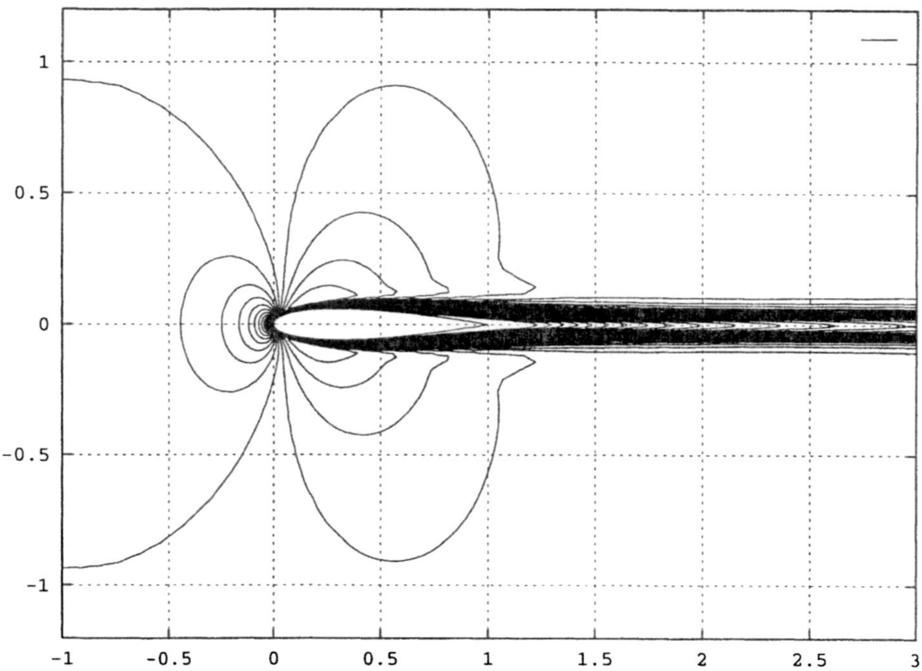

Figure 7 Steady Mach lines : $N_v = 14106$

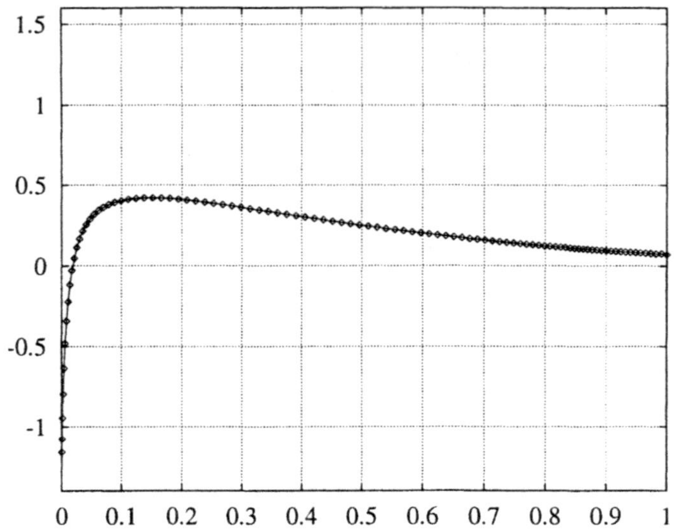

Figure 8 Distribution of the pressure coefficient on the airfoil

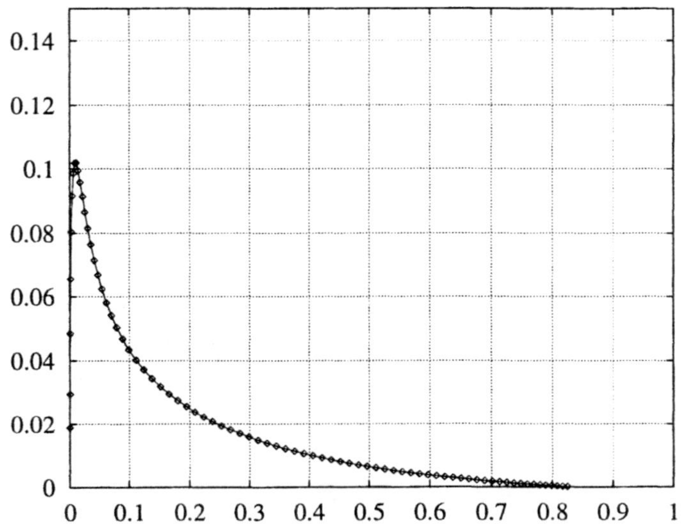

Figure 9 Distribution of the skin-friction coefficient on the airfoil

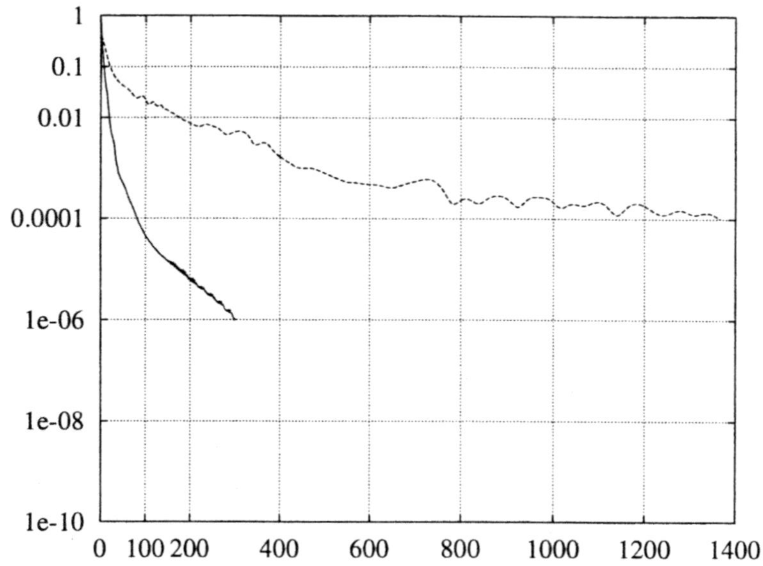

Figure 10 Non-linear convergence curves (energy residual)
Leftmost curve : implicit time integration - Rightmost curve : explicit time integration

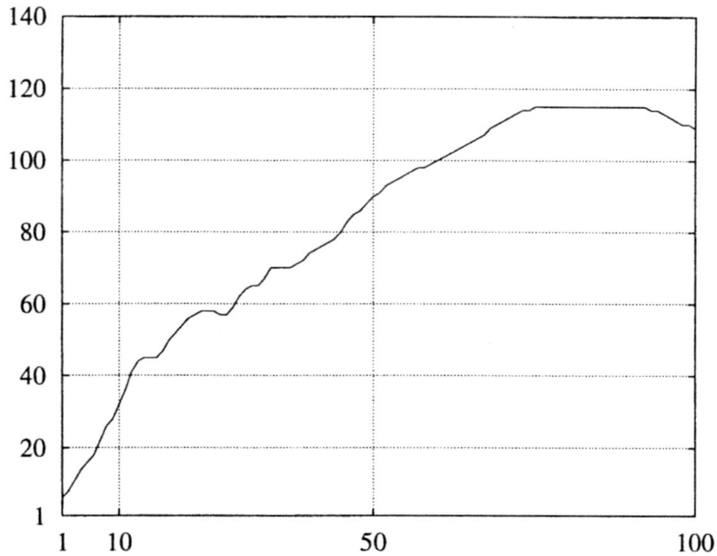

Figure 11 Effective number of Jacobi relaxations (implicit time integration)

4.1 TP2 test case

Abstract

Here we report on results obtained in simulating three-dimensional compressible laminar viscous flows on the INTEL PARAGON MPP (Massively Parallel Processors) system using a second-order accurate Monotonic Upwind Scheme for Conservation Laws (MUSCL) finite volume/finite element method on fully unstructured tetrahedral meshes. We focus on test case **TC2** of the **ECARP** MPP for Navier-Stokes flows workshop which deals with the laminar flow around an `ONERA M6` wing at a Mach number equal to 0.8 and a Reynolds number equal to 1000. The angle of attack is set to 0°.

1 Discrete equations and methodology

Governing Equations

Let $\Omega \subset IR^3$ be the flow domain of interest and Γ be its boundary. The conservative law form of the equations describing three-dimensional Navier-Stokes flows is given by:

$$\frac{\partial W}{\partial t} + \vec{\nabla}.\vec{\mathcal{F}}(W) = \frac{1}{Re}\vec{\nabla}.\vec{\mathcal{R}}(W) \tag{1}$$

where $W = W(\vec{x}, t)$; $\vec{x}$ and t denote the spatial and temporal variables, and

$$W = \left(\rho\ ,\ \rho\vec{U}\ ,\ E\right)^T \quad , \quad \vec{\nabla} = \left(\frac{\partial}{\partial x}\ ,\ \frac{\partial}{\partial y}\ ,\ \frac{\partial}{\partial z}\right)^T$$

and:

$$\vec{\mathcal{F}}(W) = \begin{pmatrix} F_x(W) \\ F_y(W) \\ F_z(W) \end{pmatrix} \quad , \quad \vec{\mathcal{R}}(W) = \begin{pmatrix} R_x(W) \\ R_y(W) \\ R_z(W) \end{pmatrix}.$$

$F_x(W)$, $F_y(W)$ and $F_z(W)$ denote the convective fluxes and are given by:

$$F_x(W) = \begin{pmatrix} \rho u \\ \rho u^2 + p \\ \rho uv \\ \rho uw \\ u(E+p) \end{pmatrix} \quad , \quad F_y(W) = \begin{pmatrix} \rho v \\ \rho uv \\ \rho v^2 + p \\ \rho vw \\ v(E+p) \end{pmatrix} \quad , \quad F_z(W) = \begin{pmatrix} \rho w \\ \rho uw \\ \rho vw \\ \rho w^2 + p \\ w(E+p) \end{pmatrix}$$

while $R_x(W)$, $R_y(W)$ and $R_z(W)$ denote the diffusive fluxes and are given by:

$$R_x(W) = \begin{pmatrix} 0 \\ \tau_{xx} \\ \tau_{xy} \\ \tau_{xz} \\ u\tau_{xx} + v\tau_{xy} + w\tau_{xz} + \dfrac{\gamma k}{Pr}\dfrac{\partial \varepsilon}{\partial x} \end{pmatrix}$$

$$R_y(W) = \begin{pmatrix} 0 \\ \tau_{xy} \\ \tau_{yy} \\ \tau_{yz} \\ u\tau_{xy} + v\tau_{yy} + w\tau_{yz} + \dfrac{\gamma k}{Pr}\dfrac{\partial \varepsilon}{\partial y} \end{pmatrix}$$

$$R_z(W) = \begin{pmatrix} 0 \\ \tau_{xz} \\ \tau_{yz} \\ \tau_{zz} \\ u\tau_{xz} + v\tau_{yz} + w\tau_{zz} + \dfrac{\gamma k}{Pr}\dfrac{\partial \varepsilon}{\partial z} \end{pmatrix}.$$

In the above expressions, ρ is the density, $\vec{U} = (u, v, w)^T$ is the velocity vector, E is the total energy per unit of volume, p is the pressure, ε is the specific internal energy, k is the normalised thermal conductivity, $Re = \dfrac{\rho_0 U_0 L_0}{\mu_0}$ where ρ_0, U_0, L_0 and μ_0 denote the characteristic density, velocity, length, and diffusivity is the Reynolds number, and $Pr = \dfrac{\mu_0 C_p}{k_0}$ is the Prandtl number; τ_{xx}, τ_{xy}, τ_{xz}, τ_{yz}, τ_{yy} and τ_{zz} are the components of the three-dimensional Cauchy stress tensor. The velocity, energy, and pressure are related by the equation of state for a perfect gas:

$$p = (\gamma - 1)(E - \frac{1}{2}\rho \parallel \vec{U} \parallel^2)$$

where γ is the ratio of specific heats ($\gamma = 1.4$ for air), and the specific internal energy is related to the temperature via:

$$\varepsilon = C_v T = \frac{E}{\rho} - \frac{1}{2} \parallel \vec{U} \parallel^2.$$

The boundary Γ of the flow domain is partitioned into a wall boundary Γ_w and an infinity boundary Γ_∞: $\Gamma = \Gamma_w \cup \Gamma_\infty$. Let $\vec{n}$ denote the outward unit normal at any point of Γ, and T_w denote the wall temperature. On the wall boundary Γ_w, a no-slip condition and a Dirichlet condition on the temperature are imposed:

$$\vec{U} = \vec{0} \quad , \quad T = T_w \, . \tag{2}$$

No boundary condition is specified for the density. Hence, the total energy per unit of volume and the pressure on the wall are given by:

$$p = (\gamma - 1)\rho C_v T_w \quad , \quad E = \rho C_v T_w \, . \tag{3}$$

The viscous effects are assumed to be negligible at infinity, so that a uniform free-stream state vector W_∞ is adopted as a representation of the solution on Γ_∞:

$$\rho_\infty = 1 \quad , \quad \vec{U}_\infty = (\, u_\infty \, , \, v_\infty \, , \, w_\infty \,) \quad \text{with} \quad \parallel \vec{U}_\infty \parallel = 1 \quad , \quad p_\infty = \frac{1}{\gamma M_\infty^2} \tag{4}$$

where α is the angle of attack, and M_∞ is the free-stream Mach number.

Spatial approximation method

The flow domain Ω is assumed to be a polyhedral bounded region of IR^3. Let $\mathcal{T}_h$ be a standard tetrahedrisation of Ω, and h the maximal length of the edges of $\mathcal{T}_h$. A vertex of a tetrahedron T is denoted by S_i, and the set of its neighboring vertices by $K(i)$. At each vertex S_i, a control volume C_i is constructed as the union of the subtetrahedra resulting from the subdivision by means of the medians of each tetrahedron of $\mathcal{T}_h$ that is connected to S_i (see Fig. 1). The boundary of C_i is denoted by ∂C_i, and the unit vector of the outward normal to ∂C_i by $\vec{\nu}_i = (\nu_{ix}, \nu_{iy}, \nu_{iz})$. The union of all these control volumes constitutes a discretisation of domain Ω:

$$\Omega_h = \bigcup_{i=1}^{N_V} C_i \quad , \quad N_V : \text{number of vertices of } \mathcal{T}_h \, .$$

The spatial discretisation method adopted here combines the following features (see Chargy [Chargy93] for more details):

- a finite volume upwind approximation method for the convective fluxes. Second order spatial accuracy is achieved using an extension of van Leer's [VanLeer79] MUSCL technique to unstructured meshes;

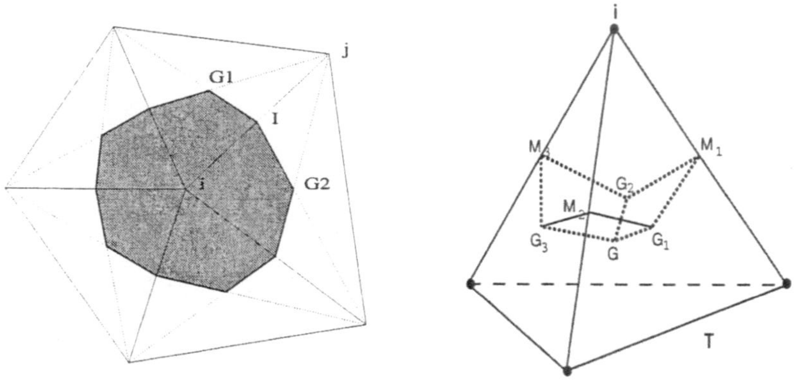

Figure 1 2D control surface and contribution to a 3D control volume

- a classical Galerkin finite element centered approximation for the diffusive fluxes.

Integrating Eq. (1) over C_i yields:

$$\iiint_{C_i} \frac{\partial W}{\partial t} d\vec{x} + \iiint_{C_i} \vec{\nabla}.\vec{\mathcal{F}}(W) d\vec{x} = \iiint_{C_i} \frac{1}{Re} \vec{\nabla}.\vec{\mathcal{R}}(W) d\vec{x} . \tag{5}$$

Finally, integrating Eq. (5) by parts leads to:

$$\begin{aligned} \iiint_{C_i} \frac{\partial W}{\partial t} d\vec{x} \; &+ \sum_{j \in K(i)} \int_{\partial C_{ij}} \vec{\mathcal{F}}(W).\vec{\nu}_i d\sigma && <1> \\ &+ \int_{\partial C_i \cap \Gamma_w} \vec{\mathcal{F}}(W).\vec{n}_i d\sigma && <2> \\ &+ \int_{\partial C_i \cap \Gamma_\infty} \vec{\mathcal{F}}(W).\vec{n}_i d\sigma && <3> \\ &= -\frac{1}{Re} \sum_{T, S_i \in T} \iiint_T \vec{\mathcal{R}}(W).\vec{\nabla} N_i^T d\vec{x} && <4> \end{aligned} \tag{6}$$

where $\partial C_{ij} = \partial C_i \cap \partial C_j$, and $N_i^T = N_i^T(x, y, z)$ is the P1 shape function defined at the vertex S_i and associated with the tetrahedron T.

Convective Fluxes Computation

A first order finite volume discretisation of $< 1 >$ goes as follows:

$$< 1 > = W_i^{n+1} - W_i^n + \Delta t \sum_{j \in K(i)} \Phi_{\mathcal{F}}(W_i^n, W_j^n, \vec{\nu}_{ij}) \tag{7}$$

where $\Phi_{\mathcal{F}}$ denotes a numerical flux function such that:

$$\Phi_{\mathcal{F}}(W_i, W_j, \vec{\nu}_{ij}) \approx \int_{\partial C_{ij}} \vec{\mathcal{F}}(W).\vec{\nu}_i d\sigma \, . \tag{8}$$

Upwinding can be introduced in the computation of Eq. (8) by using Roe's [Roe81] approximate Riemann solver thus computing $\Phi_{\mathcal{F}}$ as follows:

$$\Phi_{\mathcal{F}}(W_i, W_j, \vec{\nu}_{ij}) = \frac{\vec{\mathcal{F}}(W_i) + \vec{\mathcal{F}}(W_j)}{2}.\vec{\nu}_{ij} - \mid \mathcal{A}_R(W_i, W_j, \vec{\nu}_{ij}) \mid \frac{(W_j - W_i)}{2} \tag{9}$$

where $\mathcal{A}_R$ is Roe's mean value of the flux Jacobian matrix $\frac{\partial \vec{\mathcal{F}}(W)}{\partial W}.\vec{\nu}$.

Following the MUSCL technique, second order accuracy is achieved in Eq. (8) via a piecewise linear interpolation of the states W_{ij} and W_{ji} at the interface between control volumes C_i and C_j. This requires the evaluation of the gradient of the solution at each vertex as follows:

$$\tilde{W}_{ij} = \tilde{W}_i + \frac{1}{2}(\vec{\nabla}\tilde{W})_i.\vec{S_iS_j} \quad , \quad \tilde{W}_{ji} = \tilde{W}_j - \frac{1}{2}(\vec{\nabla}\tilde{W})_j.\vec{S_iS_j} \tag{10}$$

where $\tilde{W} = \left(\rho \, , \, \vec{U} \, , \, p\right)^T$ — in other words, the interpolation is performed on the physical variables instead of the conservative variables. The approximate nodal gradients $(\vec{\nabla}\tilde{W})_i$ are obtained by means of a linear interpolation of the Galerkin gradients computed on each tetrahedron of C_i:

$$\left(\vec{\nabla}\tilde{W}\right)_i = \frac{\displaystyle\iiint_{C_i} \vec{\nabla}\tilde{W}|_T d\vec{x}}{\displaystyle\iiint_{C_i} d\vec{x}} = \frac{1}{\mathrm{vol}(C_i)} \sum_{T \epsilon C_i} \frac{\mathrm{vol}(T)}{4} \sum_{k=1, k \epsilon T}^{4} \tilde{W}_k \vec{\nabla} N_k^T \, . \tag{11}$$

Diffusive Fluxes Computation

The viscous integral $< 4 >$ is evaluated using a classical Galerkin finite element P1 method. The components of the stress tensor and those of ∇N_i^T are constant in each tetrahedron. The velocity vector in a tetrahedron is computed as follows:

$$\vec{U}_T = \frac{1}{4} \sum_{k=1,k\epsilon T}^{4} \vec{U}^k$$

and the viscous fluxes are approximated as follows:

$$\vec{\mathcal{R}}_i(T) = \iiint_T \vec{\mathcal{R}}(W).\vec{\nabla} N_i^T d\vec{x} = \text{vol}(T)\left(R_x(T)\frac{\partial N_i^T}{\partial x} + R_y(T)\frac{\partial N_i^T}{\partial y} + R_z(T)\frac{\partial N_i^T}{\partial z}\right)$$

where $R_x(T)$, $R_y(T)$ and $R_z(T)$ are constant values on the tetrahedron T.

Time integration

Assuming that $W(\vec{x},t)$ is constant over the control volume C_i (in other words, a mass lumping technique is applied to the temporal term of Eq. (6)), we obtain the following semi-discrete fluid flow equations:

$$\text{vol}(C_i)\frac{dW_i^n}{dt} + \Psi(W_i^n) = 0 \quad , \quad i = 1, \cdots, N_V \tag{12}$$

where $W_i^n = W(\vec{x}_i, t^n)$, $t^n = n\Delta t^n$ and:

$$\Psi(W_i^n) = \sum_{j \in K(i)} \Phi_{\mathcal{F}}(W_{ij}, W_{ji}, \vec{\nu}_{ij}) + \int_{\partial C_i \cap \Gamma_\infty} \vec{\mathcal{F}}(W).\vec{n}_i d\sigma + \frac{1}{Re} \sum_{T, S_i \in T} \vec{\mathcal{R}}_i(T). \tag{13}$$

Explicit time advancing procedure

A predictor-corrector can be selected for time integrating the semi-discrete equations Eq. (13). This explicit scheme which is of second-order accuracy and cheap in terms of CPU costs (this is often a requirement of industrial codes). First we predict a state $\tilde{W}^{n+\frac{1}{2}}$ using the Euler equations :

$$\tilde{W}_i^{n+\frac{1}{2}} = \tilde{W}_i^n - \frac{\Delta t^n}{2}\tilde{A}(W_i^n).(\vec{\nabla}\tilde{W})_i \ . \tag{14}$$

In the second phase (correction), the fluxes are evaluated using the predicted state:

$$W_i^{n+1} = W_i^n - \Delta t^n \left[\sum_{j \in K(i)} \Phi(\tilde{W}_{ij}^{n+\frac{1}{2}}, \tilde{W}_{ji}^{n+\frac{1}{2}}, \vec{\nu}_{ij}) + \frac{1}{Re} \sum_{T, S_i \in T} \vec{\mathcal{R}}_i(T) \right]. \tag{15}$$

Implicit time advancing procedure
Explicit time integration procedures are subjected to a stability condition expressed in terms of a CFL (Courant-Friedrichs-Lewy) number. When one is interested in looking for steady solutions of the Euler or Navier-Stokes equations, an efficient time advancing strategy can be obtained by means of an implicit linearised formulation. Here, we briefly recall the main ingredients of this approach which is described in details in Fezoui and Stoufflet [Fezoui89]. A Newton-like implicit linearised version of Eq. (12) writes as :

$$\frac{\mathrm{vol}(C_i)}{\Delta t^n}\delta W_i^{n+1} + \Psi(W_i^{n+1}) = 0 \quad , \quad i = 1,\cdots,N_V \tag{16}$$

where $\delta W_i^{n+1} = W_i^{n+1} - W_i^n$. The resulting Euler implicit time integration scheme is given in matrix form by :

$$P(W^n)\delta W^{n+1} = \left(\frac{I}{\Delta t^n} + J(W^n)\right)\delta W^{n+1} = \delta \hat{W}^n \tag{17}$$

where $J(W^n)$ denotes the approximate Jacobian matrix and $\delta\hat{W}^n$ is the explicit part of the linearisation of $\Psi(W^{n+1})$. The matrix $P(W^n)$ is sparse and has the suitable properties (diagonaly dominant in the scalar case) allowing the use of a relaxation procedure (Jacobi or Gauss-Seidel) in order to solve the linear system of Eq. (17). Moreover, an efficient way to get second order accurate steady solutions while keeping the interesting properties of the first order upwind matrix is to use the second order elementary convective fluxes in the right-hand side of Eq. (17).

Parallelisation strategy
The parallelisation strategy adopted in this study has been already successfully applied in the two-dimensional case (see Farhat and Lanteri [Farhat94]). It combines mesh partitioning techniques and a message-passing programming model. The underlying mesh is assumed to be partitioned into several submeshes, each defining a subdomain. Basically the same "old" serial code is going to be executed within every subdomain. Modifications occured in the main time-stepping loop in order to take into account one or several assembly phases of the subdomain results, depending on the order of the spatial approximation and on the nature of the time advancing procedure (explicit/implicit). The assembly of the subdomain results can be implemented in a separate software module and optimized for a given machine. This approach enforces data locality, and therefore is suitable for all parallel hardware architectures.

The reader can verify that for the computations described herein, mesh partitions with overlapping simplify the programming of the subdomain interfacing module. However, mesh partitions with overlapping also have a drawback: they incur redundant floating-point operations. On the other hand, non-overlapping mesh partitions incur little redundant floating-point operations but induce one additional communication step. While nodal variables are exchanged between the subdomains in overlapping mesh partitions, partially gathered nodal gradients and partially gathered fluxes are exchanged between subdomains in non-overlapping ones. In addition, special care must be taken in the treatment of the convective fluxes in the case of non-overlapping mesh partitions (because of the possible differences in the orientation of the interface edges which are not part of the original mesh but are instead constructed during a preprocessing phase of the parallel algorithm). In other words, the programming effort is maximized when considering non-overlapping mesh partitions. We refer to Farhat and Lanteri [Farhat94] for a comparison of these two approaches in the context of two-dimensional simulations. In the present study we will consider one element wide overlapping mesh partitions for second order accurate computations. We will therefore exchange physical states and nodal gradients.

2 Numerical results and analysis

Platform identity

Like its predecessors, the prototypical TOUCHSTONE DELTA system and the iPSC-860, the INTEL PARAGON XP/S is a distributed memory computer. Its architecture supports Multiple Instruction Multiple Data stream (MIMD) and most notably Single Program Multiple Data (SPMD) styled applications. The INTEL PARAGON processing nodes are arranged in a two-dimensional rectangular grid. The system contains compute nodes, service nodes and I/O nodes. Compute nodes are used for execution of parallel programs; service nodes offer the capabilities of a UNIX system including compilers and program development tools thus making a traditional front-end computer unnecessary; I/O nodes are interfaces to mass storage and Local Area Networks (LANs). The data network is constructued on the basis of Mesh Routing Chips (MRCs) which are connected by high-speed channels. The network has a theoretical bidirectional bandwith of 175 Mb/s. Each processing node consists of two i860XP RISC processors. One of them is working as the *application processor*, the other one as the *message processor*. They share a common memory. The purpose of the *message processor* is to relieve the *application processor* from the overhead work related to message-passing. The *message processor* sends and receives messages from the other nodes via a network interface. The i860XP processor runs at 50 Mhz. It has two pipes for floating-point operations, an adder and a multiplier, each able to provide a result (64 bit IEEE arithmetic) at every cycle. This adds up to 75 Mflop/s peak performance for a single i860XP processor.

Physical solutions

Five meshes with increasing sizes have been generated. Their characteristics are summarized in Tab. 1 where N_V denotes the number of vertices, N_T the number of tetrahedra.

In order to compute the steady state corresponding to the **ECARP** test case **TC2** the implicit time advancing procedure (16) is used with 36 Jacobi relax-

Table 1 Five meshes and their characteristics for an **ONERA M6** wing

MESH	N_V	N_T
M1	2203	10053
M2	15460	80424
M3	31513	161830
M4	63917	337604
M5	115351	643392

ations for the approximate solution of the linear system resulting from (17). The pseudo-time step is computed according to the law CFL=*it* where *it* denotes the current non-linear iteration. The steady state solution is considered as obtained when the initial normalized non-linear energy residual has been divided by 10^8. The non-linear convergence behaviors are shown on Fig. 2. A partial view of mesh *M1* is shown on Fig. 3; Fig. 4 visualises the obtained steady pressure lines for mesh *M5* while Fig. 5 to 7 give the corresponding pressure coefficients at several spanwise locations on the wing.

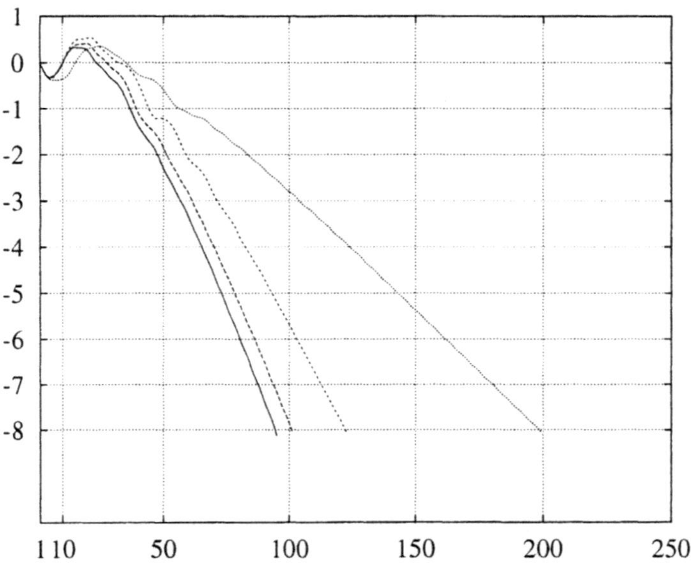

Figure 2 Non-linear convergence curves (energy residual)
Leftmost curve : mesh *M2* - Rightmost curve : mesh *M5*

Figure 3 Mesh on the skin of on an **ONERA M6** wing (mesh *M1*)

Performance results

Timing measures concern the main parallel loop. All performance results reported herein are for 64-bit arithmetic; the redundant floating-point operations are not accounted for when evaluating the Mflop rate. Unless stated otherwise, the reported CPU times always refer to the maximum of the individual processor measures. In the following tables N_p is the number of involved processors (submeshes) while "Loc Comm" and "Glb Comm" respectively denote the local

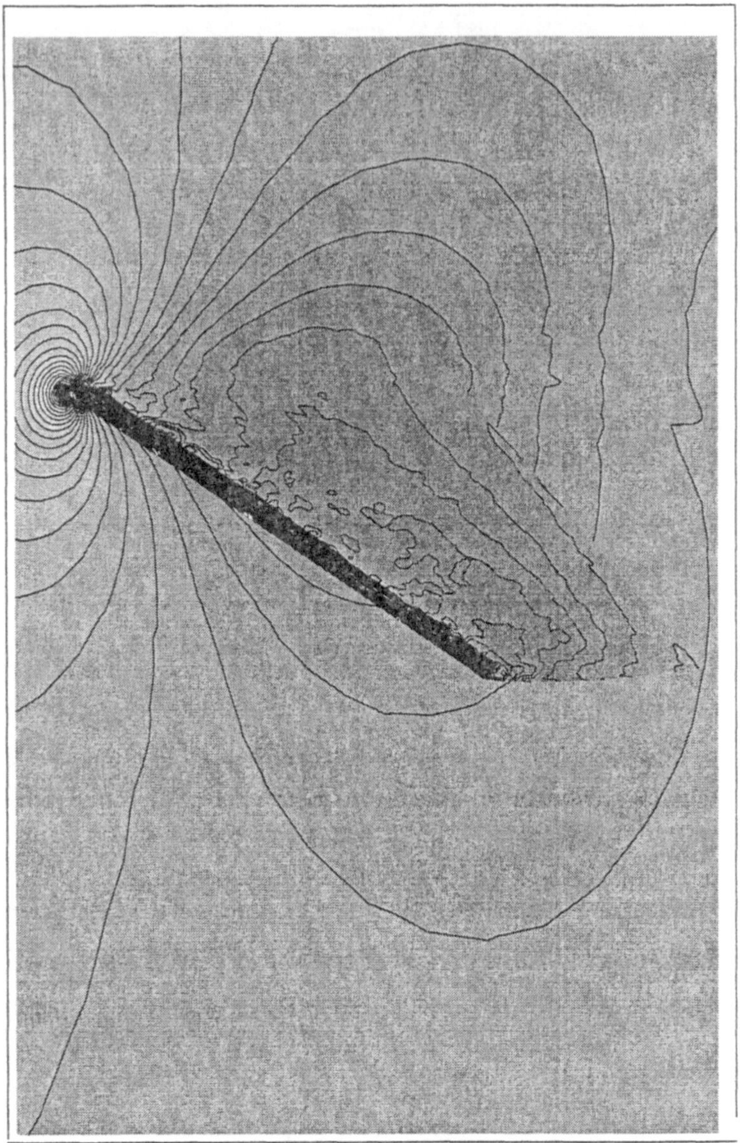

Figure 4 Steady pressure lines on an `ONERA M6` wing (mesh *M5*)

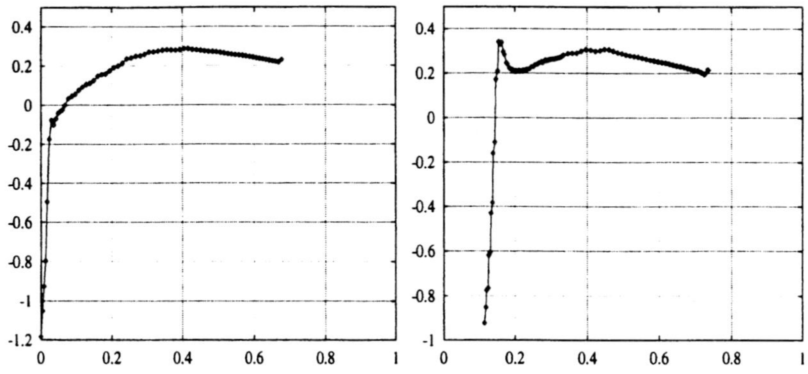

Figure 5 One dimensional pressure coefficient distributions
Left figure : 0% spanwise location - Right figure : 20% spanwise location

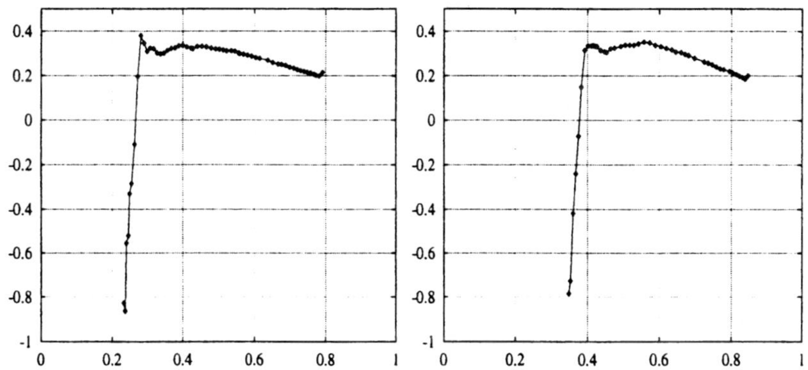

Figure 6 One dimensional pressure coefficient distributions
Left figure : 40% spanwise location - Right figure : 60% spanwise location

(send/receive at artificial submesh boundaries) and global communication times. In the case of local communication operations, the corresponding measures include the time spent in packing and unpacking message buffers. On the other part, explicit synchronisation points have been inserted prior to entering the update phases at the artificial submesh boundaries; this allow a precise timing of pure local communication operations without taking into account the idle times due to computational load imbalance.

Combined computational/numerical scalability assessment
Here, we make a simple assessment of the combined computational/numerical scalability properties of the parallel algorithm under consideration. In order to do so, we compare the required numbers of non-linear iterations and total CPU times for steady state computations using meshes *M2* to *M5* on the `Intel`

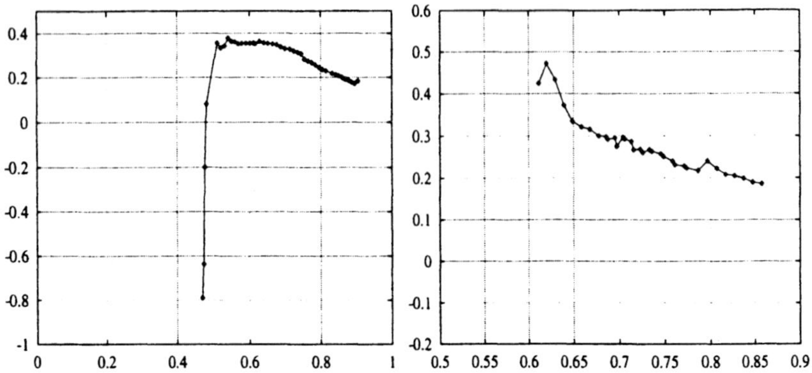

Figure 7 One dimensional pressure coefficient distributions
Left figure : 80% spanwise location - Right figure : 100% spanwise location

Paragon system. Tab. 2 summarizes the obtained results; "# *it*" denotes the number of non-linear iterations while "CPU/*it*" gives the corresponding CPU cost per non-linear iteration. We also report in this table the minimum and maximum values of the local communication times.

Table 2 Implicit Navier-Stokes computations on the ONERA M6 wing
Computations on the Intel Paragon : NX communication library
Combined computational/numerical scalability assessment

MESH	N_p	# *it*	CPU	Loc Comm		CPU/*it*
				Min	Max	
M2	8	95	2112.0 s	16.0 s	26.0 s	22.2 s
M3	16	101	2511.0 s	25.0 s	53.0 s	24.9 s
M4	32	123	3303.0 s	19.0 s	50.0 s	26.8 s
M5	64	199	5662.0 s	46.0 s	96.0 s	28.4 s

Switching from mesh *M2* to mesh *M5* one can observe a 28% degradation in the measured CPU per non-linear iteration times. There are two main reasons to this behavior. The introduced redundant floating-point operations at the artificial submesh boundaries cause a degradation of the computational scalability. On the other part, the Jacobi solver is well known for its poor numerical scalability properties; here, the maximum number of relaxations is kept constant when increasing the global problem size, which means that the linear system solution is better on the coarsest mesh that it is on the finest one for large time steps as it can be seen on Fig. 8. This in turn as a noticeable influence on the non-linear convergence as far as the linear convergence is low which is the case here.

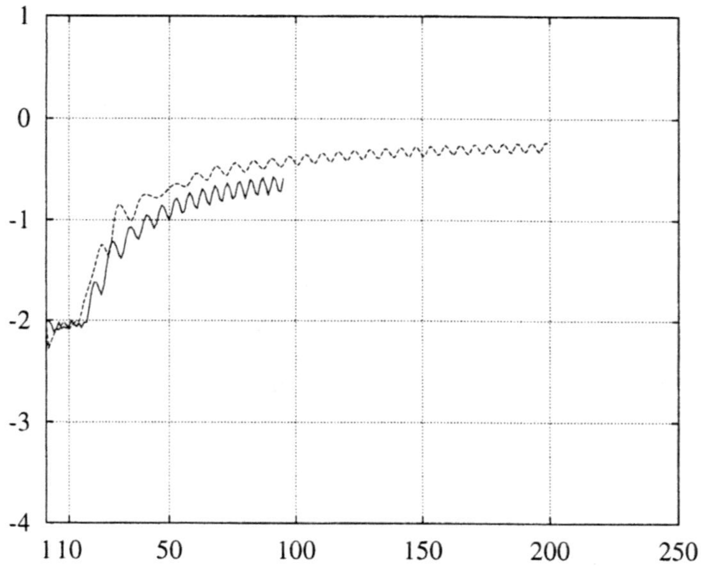

Figure 8 Linear system solution convergence curves (Jacobi solver)
Straight-line curve : mesh *M2* - Dashed-line curve : mesh *M5*

Evaluation of the parallel speed-up
We can write the total run time T_1^s of a serial program running on a single processor node as composed of two parts :

$$T_1^s = T_{serial}^s + T_{parallel}^s$$

where T_{serial}^s and $T_{parallel}^s$ respectively denote the serial execution time of the non-parallelisable and the parallelisable part of the application. If we assume an ideal situation where the parallel part can be parallelised by distributing equal shares among the processor nodes, we can expect the total execution time $T_{N_p}^p$ for the parallel program running on N_p processors to be :

$$T_{N_p}^p = T_{serial}^p + T_{parallel}^p = T_{serial}^s + \frac{T_{parallel}^s}{N_p} .$$

The parallel speed-up $S(N_p)$ can be expressed as :

$$S(N_p) = \frac{T_1^s}{T_{N_p}^p} .$$

The execution time of the parallel part $T_{parallel}^p$ can be split into two terms :

$$T^p_{parallel} = T^p_{comp} + T^p_{comm} = \frac{T^s_{parallel}}{N_p} + T^p_{comm}$$

where T^p_{comp} and T^p_{comm} respectively denote the computation time and the communication time of the parallel part of the application. We obtain the following expression for the parallel speed-up :

$$S(N_p) = \frac{T^s_1}{T^s_{serial} + \frac{T^s_{parallel}}{N_p} + T^p_{comm}} = \frac{T^s_1}{(1-p)T^s_1 + \frac{pT^s_1}{N_p} + T^p_{comm}} .$$

However, for large size problems, the total run time T^s_1 of the serial program running on a single processor node is generally not available because of memory limits. It is therefore necessary to devise another expression of the parallel speed-up which doesn't make use of T^s_1. While doing so, we assume that the application under consideration is such that $p = 1$ so that $T^s_{serial} = 0$ (which is a realistic assumption for the SPMD programming model we are using here) and $T^s_{parallel} = T^s_1$. We will write the total execution time T^s_1 as :

$$T^s_1 = N_p \times T^p_{comp}$$

and the parallel speed-up takes the following form :

$$S(N_p) = \frac{T^s_1}{T^p_{N_p}} = \frac{N_p \times T^p_{comp}}{T^p_{comp} + T^p_{comm}} = N_p \times (\frac{1}{1 + \frac{T^p_{comm}}{T^p_{comp}}}) .$$

This expression will be used in the sequel in order to evaluate the speed-up of the parallel algorithms based on overlapping and non-overlapping mesh partitions.

The parallel speed-up has been evaluated for computations performed using mesh *M5*. Tab. 3 summarizes the obtained results. Even though the speed-up measures are very good, the main point to be noted here concerns the associated relatively low Mflop rates; based on the figure for $N_p = 128$ we obtain 4 Mflop/s per processor which roughly corresponds to 5% of the theoretical performance of the I860XP processor for double precision floating-point operations.

3 References

[Chargy93] Chargy, D., (1993) : NSTC3D : a 3D Compressible Navier-Stokes Solver, User's Manual, Simulog.

[VanLeer79] Van Leer, B., (1979) : Towards the Ultimate Conservative Difference Scheme V : a Second-Order Sequel to Godunov's Method, J. of Comp. Phys., **32**, 361-370.

[Fezoui89] Fezoui, L., Stoufflet, B., (1989) : A Class of Implicit Upwind Schemes for Euler Simulations with Unstructured Meshes, J. of Comp. Phys., **84**, 174-206.

Table 3 Implicit Navier-Stokes computations on the **ONERA M6** wing
Computations on the **Intel Paragon** : **NX** communication library
Parallel speed-up evaluation using mesh *M5*

N_p	CPU	Loc Comm		Glb Comm		$S(N_p)$	Mflop/s
		Min	Max	Min	Max		
64	5662.0 s	46.0 s	96.0 s	8.0 s	9.5 s	63	242
80	5439.0 s	32.0 s	86.0 s	8.0 s	10.5 s	78	252
96	4151.0 s	23.0 s	75.0 s	8.0 s	11.0 s	94	330
128	2922.0 s	24.0 s	65.0 s	9.0 s	11.5 s	124	468

[**Farhat94**] Farhat, C., Lanteri, S., (1994) : Simulation of Compressible Viscous Flows on a Variety of MPPs: Computational Algorithms for Unstructured Dynamic Meshes and Performance Results, Comp. Meth. in Appl. Mech. and Eng., **119**, 35-60.

[**Roe81**] Roe, P.L., (1981) : Approximate Riemann Solvers, Parameters Vectors and Difference Schemes, J. of Comp. Phys., **43**, 357-371.

5 Viscous flow computations using structured and unstructured grids on the INTEL-PARAGON

D. Koubogiannis, K. C. Giannakoglou, K. D. Papailiou, NTUA, Athens, Greece

Abstract

A computational technique for the numerical solution of the Navier-Stokes equations is ported on the distributed memory *Intel-Paragon* computing system. A key element in the present work is that the solution method possesses the capability of using either structured or unstructured grids, through a common finite-volume discretization technique and an explicit time integration scheme. The parallelization of the method is based on the multi-domain concept, where each subdomain is assigned to a different processor. Different discretization algorithms for the control volumes along or close to the interfaces and consequently different communication techniques are employed, depending on the type of the grid, in order to minimize the inter-processor communication cost. The laminar flow around an isolated NACA0012 profile, at zero incidence, infinite Mach number equal to 0.5 and a Reynolds number equal to 5,000 is analyzed, by making use of up to 40 compute nodes.

I Method Formulation

In a Cartesian coordinate system, the mass, momentum and energy conservation equations are written in a vector form, as follows ([Hirsch90])

$$\frac{\partial \vec{W}}{\partial t} + \frac{\partial \overrightarrow{F^{inv}}}{\partial x} + \frac{\partial \overrightarrow{G^{inv}}}{\partial y} - \frac{\partial \overrightarrow{F^{vis}}}{\partial x} - \frac{\partial \overrightarrow{G^{vis}}}{\partial y} = \vec{0} \tag{1}$$

by separately introducing inviscid (superscripted by *inv*) and viscous (superscripted by *vis*) terms. The solution variable array is denoted by $\vec{W}$, while $\vec{F}$ and $\vec{G}$ stand for the flux vectors in the x and y directions, respectively.

Equation 1 is integrated over control volumes which are formed around any grid node. For both structured and unstructured grids, the finite volumes are defined in a similar way, by successively connecting the midpoints of the segments incident upon the node at hand, with the barycenters of the (triangular or quadrangular) grid elements surrounding this node. In figure 1, the finite volumes for the structured and the unstructured meshes are denoted by the dashed areas.

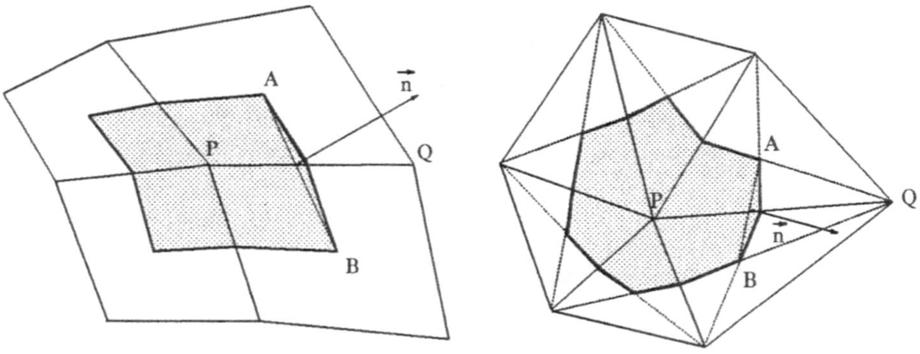

Figure 1 Grid nodes and grid cells, as well as finite volumes for structured and unstructured meshes

For the integration of equation 1 over each control volume C_P the Green's theorem is used, so

$$\iint_{C_P} \frac{\partial \overrightarrow{W}}{\partial t} dxdy + \sum_{Q \in K(P)} \int_{\partial C_{PQ}} (n_x \overrightarrow{F^{inv}} + n_y \overrightarrow{G^{inv}} - n_x \overrightarrow{F^{vis}} - n_y \overrightarrow{G^{vis}}) d\gamma = \overrightarrow{0} \quad (2)$$

where K(P) stands for the set of grid nodes which are directly linked to node P. The summation $\sum$ applies over the segments AB (figure 1) which correspond to the ∂C_{PQ} part of the control volume boundary ∂C_P. Associated with each part ∂C_{PQ}, which constitutes the interface between cells C_P and C_Q, is the normal outward vector $\overrightarrow{n} = (n_x, n_y)$, its length being equal to the length of the straight segment AB.

For the steady flows considered herein, the discretization of the above equation leads to

$$\left(\frac{\partial \overrightarrow{W}}{\partial t}\right)_P \iint_{C_P} dxdy + \sum_{Q \in K(P)} (\overrightarrow{\Phi_{PQ}}^{inv} - \overrightarrow{\Phi_{PQ}}^{vis}) = \overrightarrow{0} \quad (3)$$

where the surface integral is equal to the area of the volume C_p, while $\overrightarrow{\Phi_{PQ}}$ is the numerical approximation of the flux directed from P to Q. The approximated flux vectors, either inviscid or viscous, are written in a compact form as

$$\overrightarrow{\Phi_{PQ}} = \frac{1}{2}(\overrightarrow{\Phi_P} + \overrightarrow{\Phi_Q}) + \overrightarrow{T_{PQ}} \, . \quad (4)$$

The term $\overrightarrow{T_{PQ}}$ is zero for the viscous fluxes (this corresponds to the standard central differencing scheme, for the structured grids). For the inviscid fluxes, the Roe flux difference splitting scheme is used ([Roe81]), according to which the extra term $\overrightarrow{T_{PQ}}$ is calculated as follows

$$\overrightarrow{T_{PQ}} = \frac{1}{2}\,|A^*|\,(\overrightarrow{W_P} - \overrightarrow{W_Q})\,. \tag{5}$$

In equation 5, A^* denotes the flux jacobian which is built using the Roe-averaged quantities and is extended to second order accuracy through the standard MUSCL extrapolation technique ([vanLeer81]) in structured grids or an appropriate implementation of the latter for unstructured grids ([Fezoui89]).

In the structured grids, all nodes are swept and inviscid and viscous fluxes are computed along the right and the upper edges of the corresponding control cells. In the unstructured grids, the inviscid fluxes are computed via a loop over segments, during which contributions at their two nodal edges are collected. For the same grids, viscous fluxes are calculated by sweeping over triangular elements and collecting contributions to their three nodes. All quantities are stored at grid nodes.

The time derivative term in equation 3 is discretized by using the forward Euler scheme. Consequently, the solution is updated at each node in an explicit manner. Local time-stepping is used. Boundary conditions are employed through the adherence condition along the solid walls, while the fluxes leaving or entering the domain along the farfield boundaries are computed through a proper implementation of the Steger-Warming scheme ([Steger81],[Fezoui89]).

II Parallelization Aspects

1 The *Intel-Paragon* Computing System

The *Intel-Paragon* is a scalable distributed-memory multi-processing system. Its network topology is planar and supports Multiple Instruction/ Multiple Data (MIMD) applications. It is based on *Intel*'s i860XP/S RISC processors with a high-speed inter-connection network. In the present application, the *Intel-Paragon* machine is used in a Single Program/Multiple Data (SPMD) mode.

This computer has been installed in the Supercomputing Center of the National Technical University of Athens (NTUA) and is equipped with 51 nodes having a theoretical peak performance of 48x75 Mflops, in double precision computations. In the current configuration there are 48 compute (lying on a two-dimensional 4x12 mesh backplane) and 3 service nodes. Each node consists of two i860XP/S RISC processors, the first of them being the application processor while the other stands for the message processor. Both processors share the same memory, which is equal to 32 Mb per node with a cycle time of 50 MHz. The message processor is responsible for the message-passing operations, its role being to send to and receive messages from the other nodes via the network interface. The application processor is carrying out the primary work. Two RAID disks of 4.8 Gb each are connected to the two I/O nodes with SCSI interfaces.

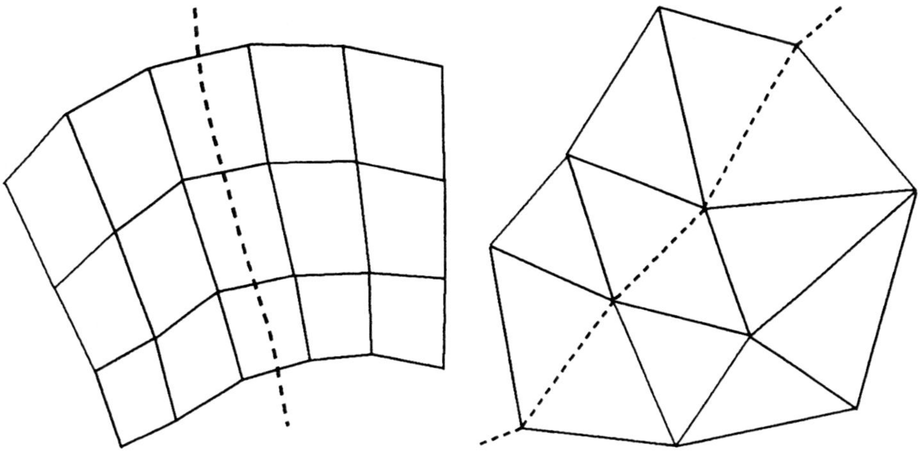

Figure 2 Grid partitioning for structured and unstructured meshes

In the configuration used in the present study, the 100 x 100 Linpack benchmark, executed on 36 processors, resulted to the following figures: $R_{max} = 1.558$ $Gflops$, $N_{max} = 10500$, $N_{1/2} = 2600$ and for the 1000 x 1000 Linpack: $R_{1000} = 0.237 Gflops$.

The interprocessor communication is carried out through a row-column routing. Since the disk I/O takes place via the same communication network ,it burdens the communication cost. Nevertheless, the communication time is only slightly affected by the number of nodes interfering between two communicating compute nodes. The communication cost is estimated as the cost of sending and receiving information between compute nodes. For the exchange of a double-precision variable message, this communication cost consists of a latency (being equal to about 65 μsec, regardless of the message length) in addition to a linear part which is proportional to the size of the message (the proportionality factor being equal to about 0.12 μsec per double-precision variable) ([Arbenz94]). The type of communication to be used is defined by the user. In the present work the sunchronous communication mode has been used by exploiting the existing NX library.

2 Grid Partitioning and Parallel Implementation

The approach adopted herein for the parallelization of the solution algorithm is based on the domain-decomposition concept. The global grid is first partitioned into a number of subdomains each of which is assigned to a different processor.

For the structured grids, the global mesh is partitioned by defining 'equidistant' (measured in terms of the number of grid lines in between them) grid lines, as interfacing boundaries. The subdomains are defined as in figure 2, where the dashed line plays the role of the interfacing boundary between the two adjacent

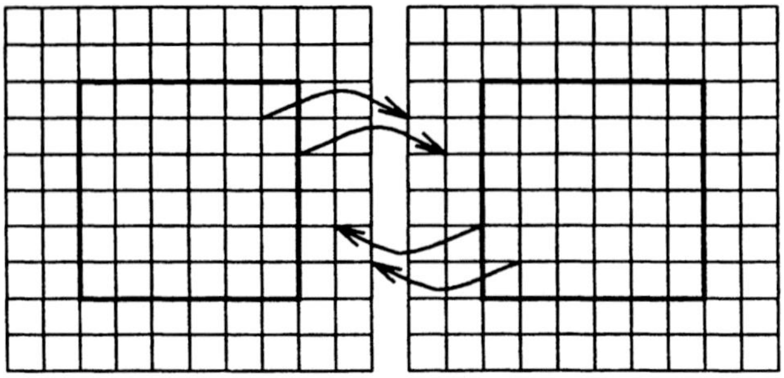

Figure 3 Communication patterns for two adjacent subdomains (and consequently processors). Structured mesh, second order accuracy

subdomains. As shown in figure 2, adjacent subdomains do not share common nodes, while each control volume strictly belongs to a single subdomain. The grid of each subdomain is perimetrically extended using two rows of fake nodes; the nodes along the first row coincide with the boundary nodes of the adjacent domains. The second row allows the application of second order accurate schemes. The communication task is associated with these two rows of nodes in order to properly assign them the dependent variables calculated at the boundary and near-boundary grid lines of the adjacent subdomains. Thus, at the end of each iteration, the dependent variables at the fake nodes of each subdomain are updated by receiving information from the adjacent domains. In this way, the numerical fluxes for all the nodes within each subdomain are computed using values already loaded to the corresponding processor. By retaining a well ordered communication schedule, any internal subdomain exchanges information only with four adjacent subdomains (east, west, north and south), while communication with four additional subdomains (in the cross directions, line north-west etc.) is avoided. Figure 3 shows the information exchange between two adjacent subdomains, in an illustrative manner.

For the purposes of the present study, the unstructured grid was generated by subdividing each quadrangular element of the structured grid into two triangles. The partitioning of the unstructured grid was carried out on the basis of the structured grid. Consequently, all internal subdomains communicate with four adjacent ones (by means of a common set of edges) and with four other subdomains (by means of a single common node). The control volumes corresponding to the interface nodes split among two or more adjacent subdomains (figure 2).

For the unstructured grids, the communication takes place in two phases. In the first phase, the gradients of the four primitive variables (to be used within the second order accurate scheme) and as well as time-steps are exchanged over each interface node. At the end of the first phase, the fluxes computation is taking place. However, the so calculated fluxes over the interface nodes which belong to more than one subdomains, are incomplete. For this reason, in the second phase, the four fluxes per node are exchanged. By adding these fluxes, interface nodes are separately updated, as members of the subdomains where they belong to.

Due to the way the subdomains have been defined, the connectivity of either structured or unstructured subdomains retains the exact topology of the interprocessor communication network. Since C-type grids are in use, the aforementioned rule is perturbed only along the split line springing from the trailing edge, but this is of minor importance.

The above partitioning methods fulfill the standard requirements of equiloading the processors while minimizing the interprocessor communication.

III Assesment of the Parallel Method

The case used to assess the numerical tool ported on the *Intel-Paragon* is concerned with the two-dimensional flow around a NACA0012 profile. The flow is considered to be laminar, with a free stream Mach number of M=0.5 and zero incidence. The Reynolds number based on the free stream conditions and the airfoil chord, is 5000. The flow is subsonic and it is characterized by a tiny separation bubble near the trailing edge. The Reynolds number for this case approaches the upper limit for steady laminar flows prior to the onset of turbulence. Zero heat flux is prescribed along the airfoil surface. Numerical results for this case are available in [Mavriplis89].

As discussed in a previous section, the starting point for all grids used in the calculations is the same structured grid. This is a C-type mesh, generated using a standard hyperbolic grid generation method ([Rizzi81]). The grid dimension is 283x50 and the leading edge is located at the 44th node. The distance of the first grid node off the wall is equal to 0.2 percent of the chord, while the outer boundary is placed at a distance of 75 chords from the airfoil.

Some numerical aspects will be elaborated first. The solution obtained on the basis of the unstructured grid and a second order accurate scheme has been proved to be stable. On the contrary, the second order calculation on the structured grid appeared to be unstable and the instability occured after some thousands of iterations. The resulting unsteady flow, was attributed to the very high Reynolds number of the case examined. From a numerical point of view this is a consequence of the insufficient amount of artificial dissipation added by the upwind scheme itself, in this case. As expected, the first order solution scheme was always stable. So, in order to overcome the instability problems, a hybrid scheme was finally implemented. Thus, for the computation of the inviscid fluxes over the midnodes, using the equation

$$\overrightarrow{\Phi_{PQ}}^{inv} = \overrightarrow{\Phi}^{inv}(\overrightarrow{W_P}, \overrightarrow{W_Q}) \tag{6}$$

a blend of first and second order accurate schemes was applied. This was realized, in practice, by expressing the solution variables array at the nodal points as

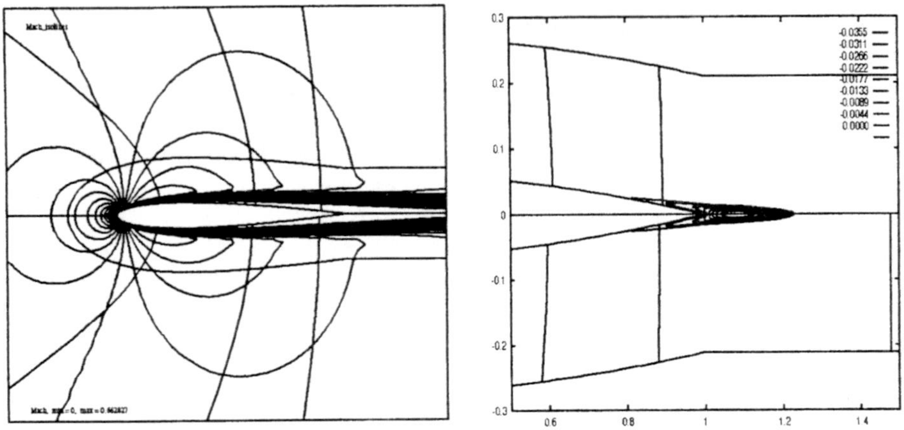

Figure 4 Mach number isolines in the vicinity of the airfoil and non-dimensional u-velocity contours close to the trailing-edge

$$\overrightarrow{W_P} = (1-\omega)\overrightarrow{W_P^{(1)}} + \omega\overrightarrow{W_P^{(2)}} \tag{7}$$

where $\overrightarrow{W_P^{(1)}}$ and $\overrightarrow{W_P^{(2)}}$ stand for the first and second order accurate approximations of the corresponding variable vectors. A similar expression is used for $\overrightarrow{W_Q}$ as well. The value $\omega = 0.8$ was found to be the maximum one which allowed a stable solution, while providing a very satisfactory comparison with some reference results.

Results to be used for the validation of the approach adopted herein are then presented. These results have been obtained by running on 30 processors for both the structured and the unstructured grids.

Figure 4 illustrates the iso-Mach contours in the vicinity of the airfoil and the negative non-dimensional u-velocity contours, in order to make clear the separation region that appears close to the trailing edge of the airfoil. Its origin is located at the 80 percent of the airfoil, which is in accordance with the reference calculations [Mavriplis89].

In figure 5, the predicted pressure C_p and friction C_f coefficient distributions along the airfoil wall are shown. For the C_p distribution, results obtained through the structured and the unstructured grids are illustrated. For the sake of convenience, the C_f distribution was also obtained using only the structured grid. The predicted distributions are compared with the reference results [Mavriplis89].

Figure 6 shows the convergence history of the structured grid algorithm for the flow case under consideration, running on 30 processors. Since the method is explicit, its convergence history is identical to that of the sequential algorithm running on a single processor.

Finally, the speedup S_p and the efficiency E versus the number of processors N_p are tabulated in Table 1 and plotted in figure 7, for both structured

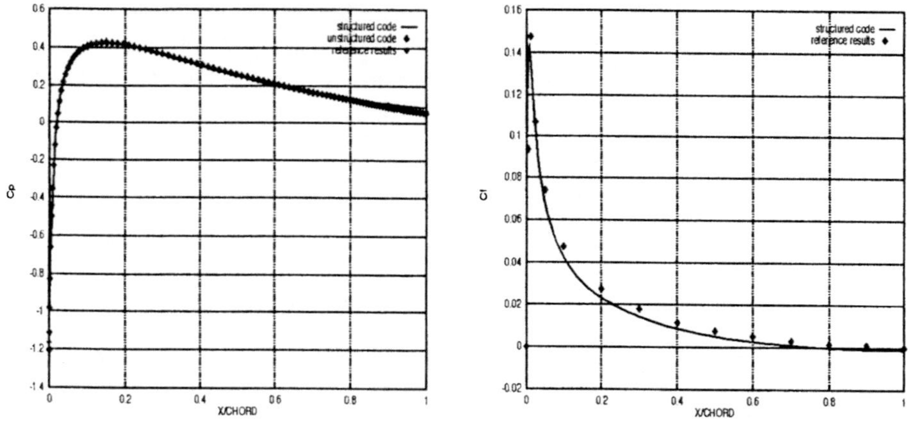

Figure 5 Pressure and friction coefficient distributions along the airfoil wall

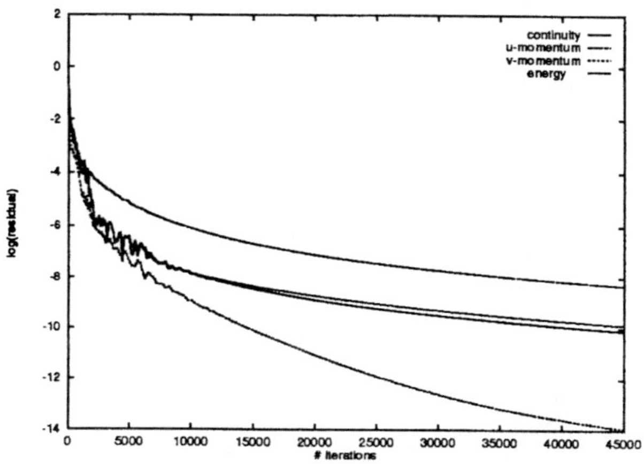

Figure 6 Convergence history for the structured grid on 30 processors

and unstructured grid computations. Table 1 was filled in after performing 5000 iterations, which is a typical number of iterations required for a convergence of about five orders of magnitude. The structured grid computation presents a slightly better speed-up. This can be attributed to (a) the fact that the code for structured grids is 25 % slower than the unstructured one when the same number of nodal points are used (this is due to the different ways used for the calculation of the viscous terms) and (b) the increased communication required by the unstructured grid concerning neighbouring subdomains in the cross direction. In this table, T_{ca} is the CPU time required for the calculation, T_{el} is the elapsed time and T_{co} is the communication time spent for the communication task. All these times have been measured in seconds.

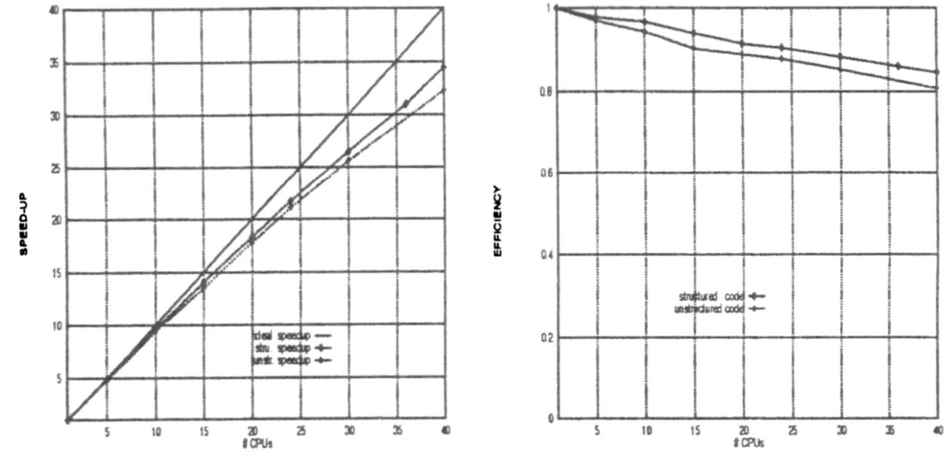

Figure 7 Speed-up and the efficiency curves for both structured and unstructured grid computations

Table 1 Calculation and communication time (sec) for 5000 iterations

	Structured grid					*Unstructured grid*				
N_p	E	S_p	T_{ca}	T_{el}	T_{co}	E	S_p	T_{ca}	T_{el}	T_{co}
1	1	1	40302	40302	0	1	1	30313	30313	0
5	0.98	4.89	41208	9000	822	0.97	4.85	31244	6680	940
10	0.97	9.67	41680	4800	1246	0.94	9.43	32128	3480	1819
15	0.94	14.10	42870	3550	2572	0.90	13.55	33569	2410	3243
20	0.91	18.30	44040	3000	3986	0.89	17.79	34084	1860	3789
24	0.90	21.72	44520	2650	4478	0.88	21.06	34536	1620	4213
30	0.88	26.48	45660	2400	5496	0.85	25.60	35520	1330	5214
40	0.85	34.46	46782	2233	7112	0.81	32.32	37519	1160	7245

IV Conclusions

An explicit, finite-volume solution algorithm for the solution of the compressible Navier-Stokes equations has been migrated to a distributed memory *Intel-Paragon* parallel computer. Aiming at a better understanding and assessment of the parallelizing techniques used, the laminar flow around an isolated profile was examined using either structured or unstructured grids. The unstructured grid calculations were performed by generating a pseudo-unstructured grid, resulting from the splitting of the C-type structured grid elements into triangles. Depending on the type of the grid, the solution and the communication algorithms are different, aiming at an optimum efficiency per case. For the structured grids, each domain is perimetrically extended using two rows of fake nodes, which resulted to increased computational loading per subdomain and minimum communication. The upwind scheme used in conjuction with a structured grid was proved to be insufficient for a stable solution in the examined high Reynolds number case and a hybrid scheme, formulated by merging first and second order accurate schemes, was used instead. In the case of unstructured grids with second order accurate schemes, better efficiency was obtained through a two-phase communication scheme.

Acknowledgment
This work constitutes the contribution of the Laboratory of Thermal Turbomachines of NTUA in the Workshop entitled "MPP for Navier-Stokes Flows" held during the last phase of the ECARP Project (Concerted Action 4.2 "Cost Effective Solutions of the Compressible Navier-Stokes Equations on Massively Parallel Computers").

Bibliography

[Hirsch90] *Hirsch, C., (1990) :* , Numerical Computation of Internal and External Flows, John Willey & Sons.

[Roe81] *Roe, P., (1981) :* , Approximate Riemann Solvers, Parameter Vectors, and Difference Schemes Journal of Comput. Physics **43** , 357-371.

[vanLeer81] *van Leer, B., (1972) :* , Towards the Ultimate Conservative Difference Scheme I : The Quest of Monotonicity Lecture Notes in Physics, **18** , 163.

[Fezoui89] *Fezoui, L., et al., (1989) :* , Resolution Numerique des Equations de Navier-Stokes pour un Fluide Compressible en Maillage Triangulaire INRIA Report, No 1033, Programme 7.

[Steger81] *Steger, P., Warming, R., F., (1981) :* , Flux Vector Splitting of the Inviscid Gasdynamic Equations with Application to the Finite-Difference Methods Journal of Comput. Physics **40** , 263-293.

[Arbenz94] *Arbenz, P., (1994) :* , First experiences with the Intel Paragon Speedup Journal **8** , No 2 , 53-58.

[Mavriplis89] *Mavriplis, P., Jameson, A. and Martinelli, L., (1989) :* , Multigrid Solution of the Navier-Stokes Equations on Triangular Meshes ICASE Rep **89-11**.

[Rizzi81] *Rizzi, A., (1981) :* , Computational Mesh for Transonic Airfoils, GAMM Workshop, Numerical Methods for the Computation of Inviscid Transonic Flows with Shock Waves, Notes on Numerical Fluid Dynamics **3** , 222-263 .

6 Development of finite element algorithms for compressible viscous flows for parallel SIMD machines

T. Fischer and G. Bugeda, UPC, Barcelona, Spain

Abstract

This final report describes the work carried out by the authors within the scope of the ECARP project, subtask 4.4.5 of MPP for Navier-Stokes. It includes a contribution to the workshop using a database tool to compare the accuracy and efficiency of numerical and experimental solutions interactively.

During the course of the project, the proven explicit Taylor-Galerkin method [1, 2, 3] using unstructured meshes for solving viscous compressible flows, has been extended to a parallel SIMD environment on the Connection Machine CM-200. For this, practical problems such as the implementation of new data structures, communication and processor management routines, are also addressed.

This work should be seen as a contribution to comparison of different massively parallel computing environments and solvers of other partners. Hence, performance measures have been made on the basis of a common test case using common structured and unstructured meshes. These calculations were performed during the Top-Up phase of the project.

I INTRODUCTION

The finite element and finite volume methods are well established numerical techniques whose main advantage is their ability to deal with complicated domains in a simple manner while maintaining a local character in the approximation. Thus, even the more complex problems in CFD, such as some 3D solutions of the Navier-Stokes equations, can be accurately tackled nowadays providing a good 3D mesh is available. However, future demands in CFD, especially aerodynamic design, require a huge amount of computational power that can only be attended by parallel systems in the next years. For this, parallel solvers need to be developed which provide the same degree of accuracy and robustness as their serial versions.

In this contribution, we present a parallel extension of the explicit Taylor-Galerkin finite element solver using unstructured meshes. The Taylor-Galerkin approach has proved to be an efficient and accurate numerical procedure capable of simulating laminar viscous flows around complex geometries in two dimensions [1, 2, 3]. The spatial discretization is based on linear triangular finite elements.

The advantage of the two-step algorithm with a lumped mass matrix formulation versus its one-step form is a reduction of stored memory [1] resulting in a lean algorithm for parallelization. Due to the nature of the high order scheme, artificial dissipation of Lapidus [4] or Jameson [5] is added to account for discontinuities such as shocks or flow near the boundary. The implementation involves the use of local timestepping to accelerate convergence to the steady state solution. However, other convergence acceleration techniques such as multigrid or implicit residual averaging have not been added during the course of this project.

The following list summarizes the main features of the flow solver that was used for this project:

- 2D Taylor-Galerkin finite element formulation
- Two-step explicit scheme
- Addition of artificial dissipation (Lapidus and Jameson)
- Local time steps
- Adaptive remeshing and error estimation (not used with the reference mesh)
- Parallel implementation

II METHODOLOGY

1 Navier-Stokes equations

The Navier-Stokes equations for a compressible fluid in conservative (or divergence) form for laminar flux within a two dimensional domain can be written, in the absence of source terms:

$$\frac{\partial \mathbf{u}}{\partial t} + \frac{\partial \mathbf{f}_i}{\partial x_i} = \frac{\partial \mathbf{g}_i}{\partial x_i} \qquad i = 1, 2 \tag{1}$$

where the nodal unknowns $\mathbf{u}$, the advective fluxes $\mathbf{f}_i$ and the viscous fluxes $\mathbf{g}_i$ are:

$$\mathbf{u} = \left\{ \begin{array}{c} \rho \\ \rho u_1 \\ \rho u_2 \\ \rho\epsilon \end{array} \right\} \quad \mathbf{f}_i = \left\{ \begin{array}{c} \rho u_i \\ \rho u_i u_1 + p\delta_{1i} \\ \rho u_i u_2 + p\delta_{2i} \\ (\rho\epsilon + p) u_i \end{array} \right\} \quad \mathbf{g}_i = \left\{ \begin{array}{c} 0 \\ \sigma_{1i} \\ \sigma_{2i} \\ k \frac{\partial T}{\partial x_i} + u_j \sigma_{ji} \end{array} \right\}. \tag{2}$$

Assuming a bulk viscosity of $(3\lambda + 2\mu) = 0$, the law of Navier-Poisson for a Newtonian fluid leads to the components of the viscous stress tensor:

$$\sigma_{ij} = \mu\left(\frac{\partial u_i}{\partial x_j} + \frac{\partial u_j}{\partial x_i}\right) - \frac{2}{3}\mu\frac{\partial u_k}{\partial x_k}\delta_{ij} . \tag{3}$$

In above: ρ is the density, p the pressure, T the temperature, u_i the Cartesian components of the velocity, ϵ the total energy per unit mass, δ_{ij} the Kronecker delta, k the material heat conductivity, and $\mu = \mu(T)$ is the dynamic coefficient of viscosity which is calculated from Sutherland's equation.

The pressure is obtained from the equation of state for a perfect gas:

$$p = (\gamma - 1)\rho(\epsilon - 0.5u_j u_j) \tag{4}$$

where $\gamma = C_p/C_v$ is the ratio of specific heats . In the present computations we have taken $\gamma = 1.4$ for the simulation of air.

2 Taylor-Galerkin algorithm

The numerical algorithm for solving the partial differential equation (1) is the well known Taylor-Galerkin method, successfully developed and used by different authors [1, 2, 3]. Non structured meshes of linear triangular finite elements are used.

The solution vector $\mathbf{u}$ is written in Taylor series up to second order as:

$$\mathbf{\Delta u} = \Delta t\frac{\partial \mathbf{u}^n}{\partial t} + \frac{\Delta t^2}{2}\frac{\partial^2 \mathbf{u}^n}{\partial t^2} + O(\Delta t^3) . \tag{5}$$

The time derivative of eq. (5) can be replaced by eq. (1) in one step and discretized using standard Galerkin finite elements:

$$\mathbf{u}^n = \sum_i N_i \mathbf{a}_i^n = \mathbf{N}\mathbf{a}^n . \tag{6}$$

Performing an integration by parts leads to the Taylor-Galerkin one-step scheme. However, it would be necessary to compute matrices of vector gradients, $\partial \mathbf{f}_i/\partial \mathbf{u}$ and $\partial \mathbf{g}_i/\partial \mathbf{u}$ which is time and memory consuming. Instead, a two-step scheme is implemented which is equivalent to the one-step scheme, but avoids the computation of these jacobians.

2.1 Two-step algorithm

In the first step, a Taylor expansion at time $n + 1/2$ is performed and using eq. (1) (neglecting the viscous terms):

$$\mathbf{u}^{n+1/2} \simeq \mathbf{u}^n + \frac{\Delta t}{2}\frac{\partial \mathbf{u}^n}{\partial t} = \mathbf{u}^n - \frac{\Delta t}{2}\frac{\partial \mathbf{f}_i^n}{\partial x_i} . \tag{7}$$

From this step $\mathbf{u}^{n+1/2}$ is obtained. For the second step, again the use of Taylor expansion for $\mathbf{f}_i^{n+1/2}$ and equation (1) without third order derivatives, provides the tools to replace the jacobians by values of fluxes at time $n + 1/2$ [1]. The complete system of equations now becomes:

$$\frac{1}{\Delta t}\left(\int_{\Omega}\mathbf{N}^T\mathbf{N}d\Omega\right)(\mathbf{u}^{n+1}-\mathbf{u}^n) = \frac{\mathbf{M}\Delta\mathbf{u}}{\Delta t} =$$

$$\int_{\Omega}\frac{\partial\mathbf{N}^T}{\partial x_k}\left(\mathbf{f}_i^{n+1/2}-\mathbf{g}_j^n\right)d\Omega - \int_{\Gamma}\mathbf{N}^T\left(\mathbf{f}_i^{n+1/2}-\mathbf{g}_j^n\right)_n d\Gamma. \tag{8}$$

Subscript i and j indicate mean nodal values and elemental values respectively; subscript n refers to boundary normal values and $\mathbf{M}$ is evaluated using a lumped form to preserve the explicitness of the scheme. For time accurate calculations the consistent matrix can be recovered by performing two Jacobi iterations [1].

3 Artificial dissipation

To account for discontinuities in the solution of fluid flow using a high order scheme (e.g. Taylor expansion) a certain kind of artificial viscosity or diffusion must be introduced. First an algorithm of the Lapidus kind [4] and then an approach similar to Jameson [5] are described.

3.1 Lapidus Algorithm

It can be shown that for the ith node of an unidimensional solution, a second derivative of the unknown can be obtained as:

$$C\left[\frac{\partial}{\partial x}(h^2\frac{\partial\phi}{\partial x})\right]_i = [\mathbf{M}_l^{-1}(\mathbf{M}_c-\mathbf{M}_l)\phi]_i\,. \tag{9}$$

$\mathbf{M}_l$ is the diagonalized or lumped mass matrix, $\mathbf{M}_c$ is the consistent mass matrix, C is a constant, and ϕ is the unknowns variable.

An extension of this idea to a multidimensional problem gives:

$$\Delta\mathbf{u}_s = \mathbf{u}_s^{n+1} - \mathbf{u}^{n+1} = \Delta t\frac{\partial}{\partial\mathbf{l}}\left(k^\star\frac{\partial(\mathbf{v}^{n+1})}{\partial\mathbf{l}}\right). \tag{10}$$

$k^\star$ is variable to determine the amount of viscosity necessary, $\mathbf{l}$ is the normalized vector of velocity gradients, $\Delta\mathbf{u}_s$ are incremental smoothed values, $\mathbf{u}_s^{n+1}$ is the smoothed solution at time n+1, and $\mathbf{v}$ is the vector of velocity.

$k^\star$ and vector $\mathbf{l}$ are obtained as follows:

$$k^\star = C_k(h^{el})^2\left|\frac{\partial(\mathbf{v}\mathbf{l})}{\partial\mathbf{l}}\right| \quad ; \quad C_k : \textit{Lapidus constant} \tag{11}$$

$$\mathbf{l} = \begin{Bmatrix} l_1 \\ l_2 \\ l_3 \end{Bmatrix} = \frac{1}{|\text{grad }\mathbf{v}^2|}\begin{Bmatrix} \text{grad}_1\mathbf{v}^2 \\ \text{grad}_2\mathbf{v}^2 \\ \text{grad}_3\mathbf{v}^2 \end{Bmatrix}. \tag{12}$$

3.2 2nd and 4th order artificial dissipation

Instead of the artificial viscosity of Lapidus, an artificial dissipation of a blend of second and fourth order can be chosen [5]. This formulation is well suited for both supersonic and subsonic flow. The second order dissipation accounts for strong discontinuities such as shocks. The fourth order terms introduce a backgorund dissipation to avoid oscillations where $|u| \to 0$, ie. stagnation zone. Since no shocks occur, best results were obtained using only the fourth order background dissipation, just like in the reference test case by Mavripilis and Jameson [6].

The general expression for the dissipation flux (which is subtractred from the other fluxes) is as follows in a finite element fashion:

$$f_d = \mathbf{M}_l^{-1} \left[\alpha^{(2)}(\mathbf{M}_c - \mathbf{M}_l)\mathbf{u} - \alpha^{(4)}(\mathbf{M}_c - \mathbf{M}_l)\left(\mathbf{M}_l^{-1}(\mathbf{M}_c - \mathbf{M}_l)\mathbf{u}\right)\right]. \quad (13)$$

The summation over the elements using elemental mass matrices and spreading the values to the nodes is exactly equivalent to performing a summation over the sides in the mesh, ie. [6]. Eq. (13) is the additional term to be added to all the equations (1) with the modification of the energy equation where the enthalpy is constant at steady state. Hence, the last component of $\mathbf{u}$, $\rho\epsilon$, is replaced by the expression ρH.

The two coefficients, $\alpha^{(2)}$ and $\alpha^{(4)}$, are obtained from the expressions

$$\alpha^{(2)} = c^{(2)}(|\mathbf{v}| + c)S_e \quad (14)$$

$$\alpha^{(4)} = \max\left[0, c^{(4)} - c^{(2)}S_e\right] \quad (15)$$

and a pressure switched diffusion coefficient for each node which is calculated according to the following term:

$$S_e = \max_{i \epsilon el} \frac{|(\mathbf{M}_c - \mathbf{M}_l)\, p|}{\mathbf{M}_c\, p}. \quad (16)$$

The terms $c^{(2)}$ and $c^{(4)}$ are user defined constants that can be tuned according to the desired amount of diffusion to be added. As mentioned earler $c^{(2)}$, was set to zero.

4 Adaptivity

4.1 Error estimator

The solution program has the possibility to increase the efficiency of the unstructured solution by adapting the element sizes in the mesh. This reduces the amount of work without sacrificing accuracy. For instance, strong variations may appear in very localized regions of the domain, whereas the rest of the solution remains smooth. A way to perform remeshing is by estimating *a posteriori* the error [7].

The element mean quadratic error for unidimensional elliptic problems is:

$$E^{el} = C(h^{el})^2 \left|\frac{\partial^2 \phi}{\partial x^2}\right|_{el} . \tag{17}$$

h^{el} is the 1D element length, ϕ is the variable chosen, and C is a constant($C =$ 1/11 for linear elements).

If the criterion of equidistribution of the error among the elements is adopted, then:

$$(h^{el})^2 \left|\frac{\partial^2 \mathbf{v}}{\partial x^2}\right|_{el} = \text{constant} . \tag{18}$$

For 2D problems eq. (18) may be extended as follows:

$$(\delta_i)^2 |\lambda_i| = \text{constant} \qquad i = 1,2 \tag{19}$$

where λ_i are the eigenvalues of the Hessian matrix $C_{ij} = \frac{\partial^2 \phi}{\partial x_i \partial x_j}$ and δ_i is the corresponding element size.

This has the advantage of providing directionality to the error estimator which can produce stretched elements [1]. It is a desirable property, since some discontinuities present a strong variation in one direction and smooth behavior in the direction normal to the first.

Due to the vector character of $\mathbf{u}$, the variable $\mathbf{u}$ is not well suited for the error estimation. A scalar variable should be chosen like the Mach number, density etc. The new element sizes δ_1 and δ_2 in the principal directions are defined according to:

$$(\delta_1)^2 |\lambda_1| = (\delta_2)^2 |\lambda_2| = \delta_r \lambda_{\max} . \tag{20}$$

λ_1 and λ_2 being the eigenvalues of matrix $\mathbf{C}$, δ_r is a user specified constant (size) and λ_{max} is the maximum eigenvalue over the whole mesh. The calculations which were performed initially (not using the reference mesh) were done with the adaptive feature.

4.2 Mesh generation

A finite element mesh of linear triangles of 3 nodes is generated using the advancing front technique [1, 2]. The element sizes are decided in accordance with the sizes specified by a background grid. The initial background grid may be a very simple mesh, and the subsequent grids are those used for the latest or seemingly best computation, where the sizes are decided as specified in the previous paragraph. During the Top-Up phase, the common unstructured grid was used and no refinement was performed.

III PARALLEL IMPLEMENTATION

The main parallel computing environment and tools consisted of:

- Computer: Connection-Machine CM-200 with 2048 processors and 1 Mbit memory.
- Communication: Router software libraries (method FASTGRAPH).
- Peak performance: 640 MFlops
- Programming language: Fortran 90 and CMF Fortran

Once the algorithm is determined, some thought must be given to the data structure and interprocessor communication to be able to exploit the speed of the massive parallel environment of the CM-200.

In the present form of the program, no attempt has been made to decompose the domain into partitions, according to the number of processors. This practice is common for MIMD platforms, but seems to be impractical for SIMD machines with more than 2000 processors, especially when using completely unstructured meshes. However, it must be mentioned that Farhat *et al* [8] have shown that it is possible to partition the mesh into structured submeshes of 16 elements each which coincide with the structure of the processor connectivity. This is to reduce the communication bandwidth, and the promising results presented by Farhat *et al* [9] are based on calculations of irregular grids.

1 Data Structure

It is important to find the adequate assignments of the vectors involved in order maximize performance. Three different parameters are involved here: vectors with a size of multiples of number of points, elements and boundary sides. Hence, the optimal layout is assigning these parameters in parallel to the processors and keep the other dimension in serial. Consider, for instance, the conservative variables $\mathbf{u}$ with the following dimensionalization:

$$\textit{dimension}\ \ \mathrm{u}(4, \textit{npoin}).$$

By means of the following compiler directive, we are able to control the layout and performance of that vector:

$$\textit{cmf\$ layout}\ \ \mathrm{u}(:\textit{serial}, :\textit{news}).$$

We obtain 4 virtually parallel assigned vectors of the following form:

$$\mathbf{u} = \begin{Bmatrix} \rho \\ \rho u_1 \\ \rho u_2 \\ \rho\epsilon \end{Bmatrix} \quad => \quad \begin{Bmatrix} \textit{parallel} \\ \textit{parallel} \\ \textit{parallel} \\ \textit{parallel} \end{Bmatrix}.$$

It is interesting to note that the serial version of the code has exactly the opposite structure in order to optimize the cache alignment:

$$\textit{dimension}\ \ \mathrm{u}(\textit{npoin}, \textit{4}) \,.$$

The advantage of inverting the structure is that this vector can be combined with a vector of the same size, but with a different dimension, for example the timestep increment with a dimension and layout of

$$\textit{dimension}\ \ \mathrm{dt}(\textit{npoin})$$

$$\textit{cmf\$}\ \textit{layout}\ \ \mathrm{dt}(:\textit{news}) \,.$$

Any other layout distribution would provide additional difficulties in combining the two vectors.

2 Interprocessor Communication

Once the definition of the layout has been optimized, some data exchange between vectors of different layouts remains which requires that different processors exchange their contents with others. Therefore, interprocessor communication is an essential part in parallel computations, both on MIMD architectures as well as SIMD systems such as the Connection Machine CM-200. In the current implementation, between 35 and 60 percent of the cpu time (depending on the mesh) is devoted to shift data from elemental values to nodal values and vice versa. In particular these are gather and scatter routines [10].

Although the refined structure of the program has avoided the exchange of elemental and nodal values as far as possible, a minimum data transfer still remains which is the bottleneck of the parallel codification of eq. (9).

Basically, two possibilities exist for establishing interprocessor links on the CM-200. One is the North East South West (NEWS) system and the other is the Router system.

2.1 NEWS System

The NEWS system allows a structured data exchange between the neighboring processors making it attractive for flow calculations using structured meshes. However, communication based on the NEWS system for unstructured meshes becomes less evident and was shown to be slower than the Router system in an attempt to optimize interprocessor communication [10].

2.2 Router System

For the current version of the program the router system has been preferred for simplicity even though parts of the mesh could be generated in a structured way using the NEWS system. In order to invoke the Router communication system it is necessary to use the automatic communication routines provided by the CM system library routines. It allows for arbitrary data exchange between processors and the source and target arrays do not have to have the same rank.

We have considered three available gather and scatter routines in this context. The first one is provided by the Connection Machine Scientific Software Library (CMSSL), *sparse_util_gather* or *sparse_util_scatter* modules. The other two possibilities would be the *cmf_send_add* module with option FASTGRAPH or NOP. FASTGRAPH has a speed advantage of about factor 3 for a reasonable size mesh, but with the limitation of long setup times and large memory requirements. The setup for option NOP is fast and less memory intensive but performance is slower. A comparison of these routines can be found in [11]. In the context of this project option FASTGRAPH was used.

IV NUMERICAL EXAMPLES

Numerical experiments have been performed during this project resolving the mandatory test case 1: laminar Navier-Stokes in two dimensions (NACA0012; M=0.5; Re=5000; α=0°) as referenced in [6]. In addition, the results were prepared in digital form to comply with the proposed database format by INRIA (Top-Up phase) and presented at the MPP-workshop (11.07.1995 at INRIA Sophia-Antipolis) for comparison with partners.

The contribution to the database includes two dimensional results (Mach number, pressure, density etc.) as well as some 1D curves (pressure coefficient, skin friction coefficient, convergence history, etc).

During the TOP UP phase, we have obtained a reference unstructured mesh for test case TC1 from the INRIA database containing 14106 nodes and 27636 elements which was used to obtain the following results.

1 Accuracy

The aim of this test case was to compare the obtained solution with a very well documented test case provided by Jameson *et al* [6]. Since our artificial diffusion model is very similar to the one of Jameson, it was very easy to establish a basis for comparison of the results, because we used the same constants.

Our seemingly best computation used $\kappa_4 = \frac{1}{120}$: the calculations show that the overall features of the flow are very similar, if not identical, to the results obtained from the reference solution [6]. In the two dimensional plots like Mach number contours (Fig. 1) around the airfoil or velocity vectors along the trailing edge (Fig. 2) no differences can be visualized. Also, the pressure coefficient (Fig. 3) and the separation point of 82.7% chord agrees extremely well with the reference solution (82.4%) given roughly the same values for the fourth order diffusion coefficient. Fig. 4 shows the location of the separation point as a function of

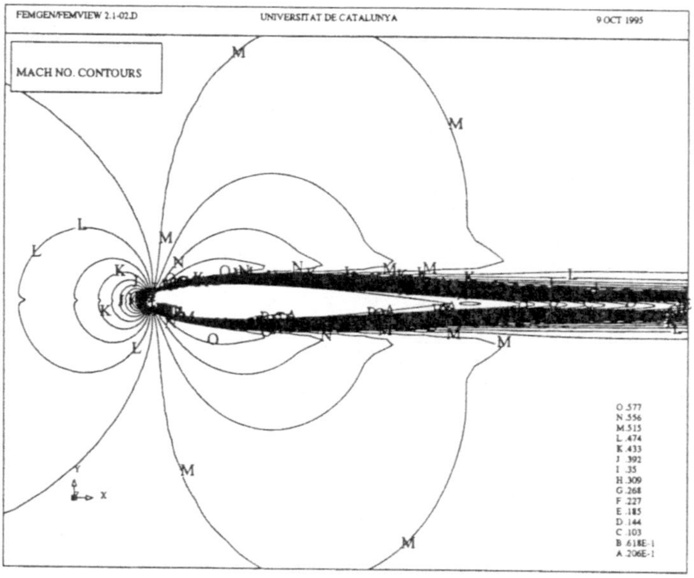

Figure 1 Mach number contours for the solution of viscous flow around a NACA0012 airfoil (Re=5000, M_∞=0.5, α=0°) on the common unstructured mesh with 14106 nodes.

the iteration timestep for two different coefficients κ_4: $\frac{1}{60}$ and $\frac{1}{120}$. Thus, as κ_4 is increased the location of the separation moves toward the rear of the airfoil. Also the pressure drag coefficient shows a very nice agreement with 0.02286 compared to 0.0228 of [6].

From the 2D analysis during the workshop, parallel pressure isolines and a clear definition of the recirculation area were observed which underline the quality of the results. However, the main difference that we observed at the workshop was comparing the skin friction coefficient with the reference results and other partners. Especially near the leading edge the peak value of c_f = 0.15 has not been obtained using this solver. Other partners had similar results. A reason for this is could be of the different meshes that have been used. The normal mesh spacing in Jameson's mesh is about 0.0002 chords, whereas in the mesh used for this calculation it is roughly 8 times higher (0.0017 chords). However, if we incorporate stretching directions of the mesh for the calculation of the artificial diffusion terms [6], the c_f curve was improved compared to the workshop results (Fig. 5). Note also the parallel pressure lines in the wake (Fig. 6) Any other assessment of accuracy is left for the evaluation of the digital results of the database in comparison with other contributors.

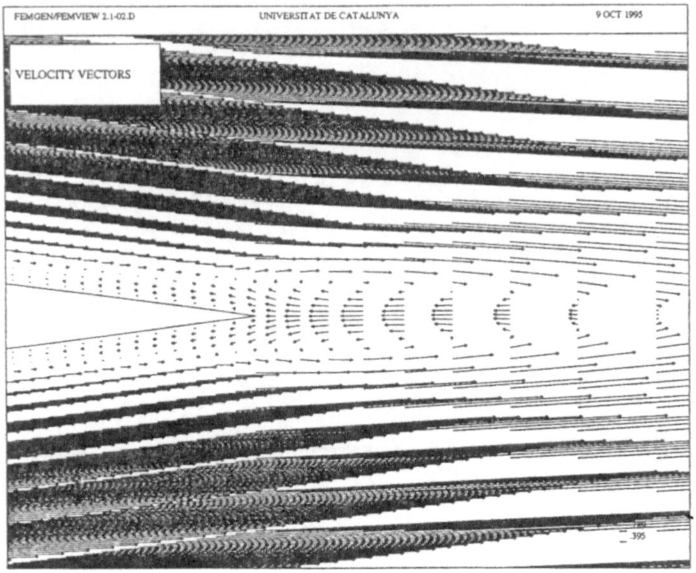

Figure 2 Velocity vectors at the trailing edge. Note the recirculation at tip of the trailing edge.

2 Parallel performance

Fig. 7 shows the convergence history of the energy-residual versus cpu time where convergence to about 5 orders of magnitude can be observed. Total time to reach a steady state where no more significant change is appreciable can be reached for this mesh in about 30 minutes for $\kappa_4 = \frac{1}{60}$ and 90 minutes $\kappa_4 = \frac{1}{120}$ both without multigrid acceleration.

Due to the nature of the problem: unstructured meshes and a finite element method, a large amount of gather, scatter and scatter-add operations can be expected. Each time one of these operations occurs a communication link must be established between processors. Thus we can expect a strong amount of interprocessor communication for this type of problems.

The CM-200 has a performance peak performance of about 640 Mflops using 2048 processors. During the present calculation the communication required between processors consumes nearly 40% of the total computation, which has been drastically reduced during the course of this project. So for the reference mesh, we obtain 0.4515 sec. per time iteration.

Since the comparison of the performance of this type of architecture to MIMD type configurations of the other contributers is very difficult, we will present a possible comparison to the claimed peak performance values of a serial processor: MIPS R8000, running at 300 Mflops peak. A similar version of the code which was optimized for the R8000 yielded for the reference mesh a value of 0.685 sec per time iteration.

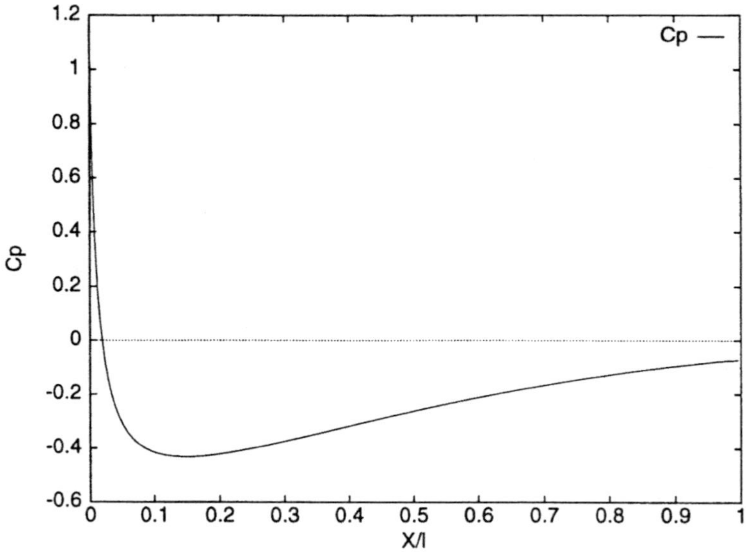

Figure 3 Pressure coefficient along the airfoil.

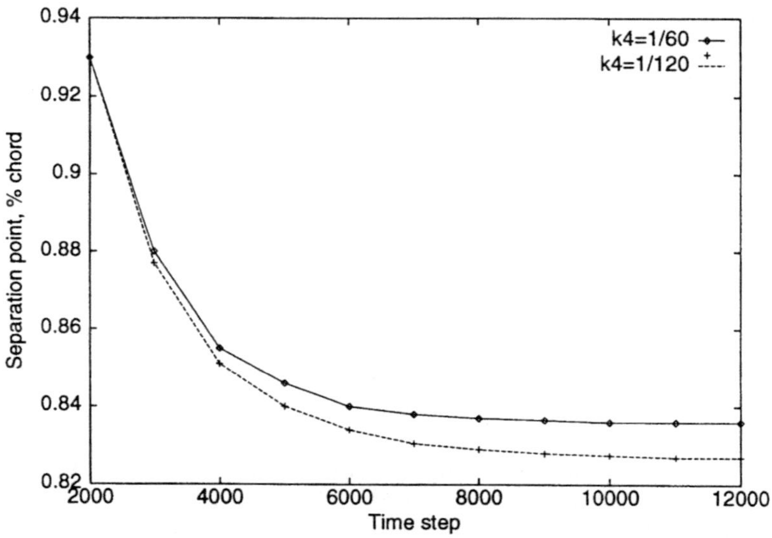

Figure 4 Influence of κ_4 on the evolution of the separation point.

Comparing the results, the loss due to communication becomes obvious: the ratio of peak performances is 2.13, the real gain is 1.51. Even though the comparison of peak flop rates is problem and system dependent we can still draw the following conclusion. The loss of roughly 30% due to parallelization seems very low, considering that we must communicate information between 2048 pro-

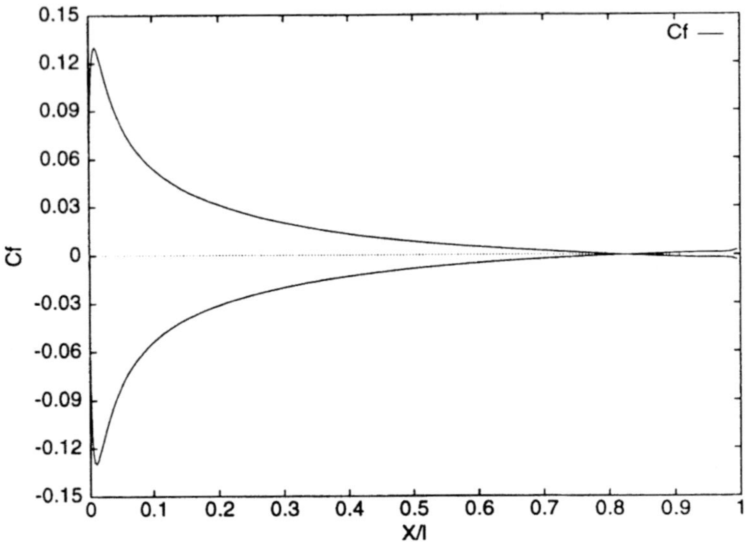

Figure 5 Skin friction coefficient along the airfoil.

cessors, which accounts for 40% of total time. Hence, it seems like the parallel version without communication runs closer to the claimed peak performance than the serial version. As mentioned earlier, the overall performance of the program can be improved by elaborating a better strategy for communication [8] reducing that cost from roughly 40% to below 10%.

V CONCLUSIONS

The authors have presented a highly accurate and robust laminar Navier-Stokes flow solver in 2D on unstructured meshes. Accuracy is demontrated by comparing the presented results to the reference solution [6], by the parallel pressure lines in the wake (Fig. 6)and the steady separation point of about 82.7% (Fig. 4).

The algorithm has been successfully implemented in a parallel SIMD environment (CM-200) yielding identical results to the serial version. The speed increase is very promising, especially because of the increase in MFlops, even though only 2048 processors were used. In addition, the efficiency of the whole approach could be improved by adapting the mesh locally, and saving precious cpu time. However, this potential could not be demonstrated because of accuracy comparisons of the method.

The disdvantages that were encountered are mainly due to hardware limitations such as availibility, service and scalability. Three years ago, when this work was formulated, the contribution of our group to the MPP workshop was to identify the possibilities of exploiting SIMD machines within CFD calculations on

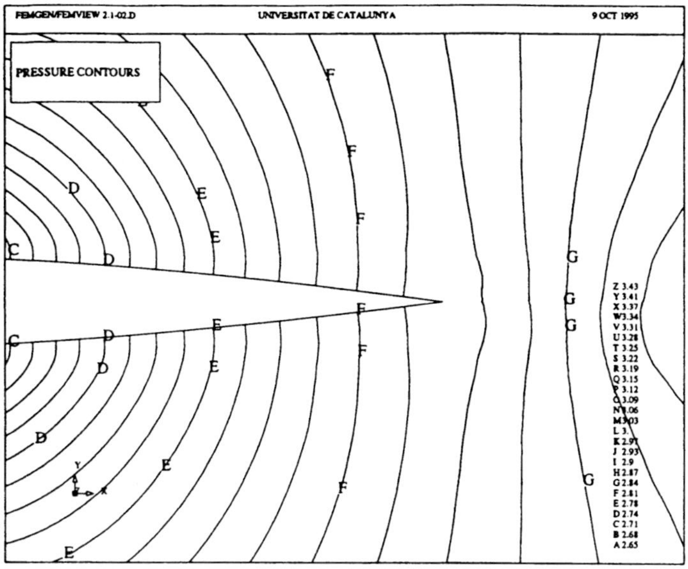

Figure 6 Pressure lines in the wake. Note their parallel position to each other in the wake.

unstructured meshes. The main message from our contribution is that the use of SIMD machines is feasible and that the speed up compared to serial machines is impressive, but the scalability of this kind of architecture has its limitations both in cpu performance and problem size, especially when using irregular grids. The communication overhead remains large and convergence acceleration techniques such as multigrid will be difficult to implement on such a system. The complexity of the mulltigrid strategy will destroy the current data structure and add complication through renumeration.

We also find that the availibility and service of CM-200 can not be guaranteed in the future because of problems by the hardware manufacturer. Thus, the authors' recommendation must go towards strategies in MIMD environments using related ideas like domain decomposition methods to divide the load among different processors if a significant performance increase is to be achieved on unstructured meshes in the future.

Bibliography

[1] Peraire, J. 'A finite element method for convective dominated flows'. PhD Thesis at University College of Swansea, 1986.

[2] Zienkiewicz O.C., Taylor 'The Finite Element Method' 4th Edition, Volume 2,Mc Graw Hill, 1991.

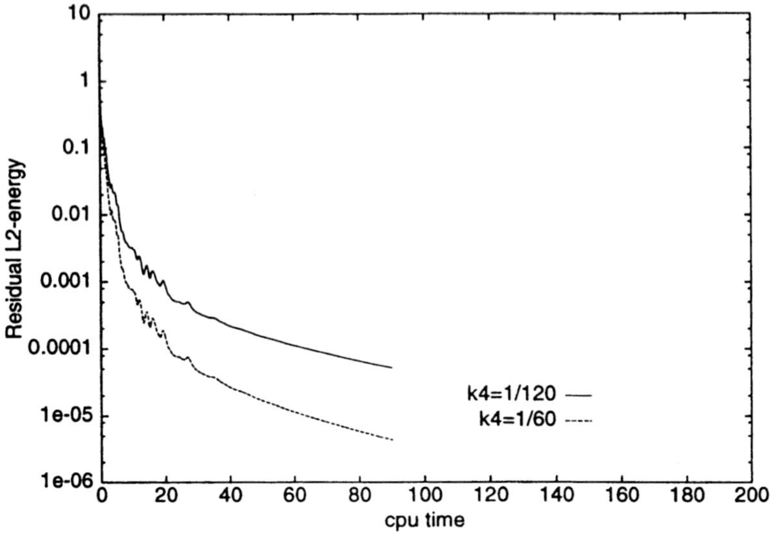

Figure 7 Convergence history of the energy residual for two different values of κ_4 (without multigrid).

[3] Fischer, T.,Codina R., Miquel, J. and Oñate E. "Adaptive finite element computations for viscous high speed flows", Proceeding of "Finite Elements in Fluids", Eds. K. Morgan, E. Oñate, J. Périaux, J. Peraire and O.C. Zienkiewicz, 1993.

[4] Lapidus A. 'A Detached Shock Calculation by Second Order Finite Differences'. J. of Computational Physics 2, 154-177, 1967.

[5] Jameson A. 'Transonic aerofoil calculations using the Euler equations', in Numerical Methods in Aeronautical Fluid Dynamics, 1982.

[6] Mavripilis P., Jameson A., Martinelli L., "Multigrid solution of the Navier-Stokes equations on triangular meshes", ICASE Rep. No 89-11, February 1989.

[7] Oñate, E. 'Error estimations and adaptive refinement techniques for structural and fluid flow problems' in Mecánica Computational Vol. 11, S. Idelsohn, Asociación Argentina de Mecánica Computational, 1991.

[8] Farhat, C., Sobh, N., and Park, K.C. "Transient finite element computations on 65536 Processors: The Connection Machine", Int. J. for Num. Meth. in Eng., Vol. 30, 27-55, 1990.

[9] Farhat, C., Fouzi, L. and Lanteri, S. "Two-dimensional viscous flow computations on the Connection Machine: Unstructured meshes, upwind schemes and massively parallel computations", Comp. Meth. in Appl. Mech. and Eng., 102, 61-88, 1993.

[10] Cante J.C., Joannas D., Oliver J., Oller S. "Experiences in massive parallel computations in a finite element context", IT-95 CIMNE, May 1993.

[11] Fischer, Oñate E. and Miquel, J. "Finite element analysis of high speed viscous flows in a massive parallel computer", Proceedings of "Computational Fluid dynamics '94, Stuttgart", Eds. S. Wagner, E.H. Hirschel, J. Périaux, J. Peraire and E. Piva, 1994.

7 Implementation of Navier Stokes solvers on parallel computers

F. Grasso and C. Pettinelli, Università di Roma, Italy

Abstract

In the present work a multidomain technique is developed for the solution of viscous two-dimensional flows by means of a "patched" subdomain decomposition with continuous grids, that ensures global conservation and it yields no distortion of discontinuities (such as shocks and slip surfaces) that cross (grid/subdomain) interfaces. The technique allows operator adaptation by introducing viscous and inviscid subdomains, and has been implemented on parallel architectures by means of *message passing paradigm*. Results of the flow over the NACA 0012 airfoil are discussed to demonstrate the properties of the method mainly for operator adaptation and parallel applications on the IBM SP2 and on the CRAY T3D.

I Introduction

One of the critical aspects of computational fluid dynamics is related to the disparity of characteristic scales in the different regions of the flow: regions where the behavior is essentially inviscid, boundary layers, shock waves, geometric singularities, etc. Moreover, the solution of engineering problems on modern computers with parallel architecture needs a partitioning of computer instruction and data on the available processors.

Domain decomposition is a "natural" methodology to resolve the intrinsic difficulties of computational fluid dynamics and to exploit parallel architectures ([Gropp91], [Hauser92]). With the use of such a technique, some of the main problems, which CFD encounters in simulating flows over complex configurations, can be overcome. In particular: in each subdomain grids can be independently generated (this is very useful especially for three-dimensional complex geometries); different numerical schemes can in principle be employed in the different regions of the flow (for example accurate schemes, which are computationally expensive, can be used only in the subdomains where high order schemes are needed); different governing equations can be used in the different regions (for example the Navier-Stokes equations can be solved only in the regions of the flow where the viscous effects are important, whereas the Euler equations, or even the potential ones, can be solved in the other zones, with an obvious gain in CPU time); the use of multidomain decomposition allows the execution in parallel of a complex aerodynamic problem, whereby the solution on all domains is obtained concurrently by reducing a large scale problem to several (depending on the number of available processors) small-scale partitions of the same problem.

In general, there are two alternatives for subdividing the computational domain: the first one introduces domains which can "overlay" with an arbitrary orientation; the second one subdivides the computational domain into "patched" zones that share common boundaries without overlapping.

The main advantage of "overlaid" subdomain decomposition is the flexibility in grid generation. However, difficulties may arise in ensuring conservation; moreover, the effects of the size of overlaid zones on the accuracy and convergence rate are difficult to predict. Berger has derived a conservative flux scheme [Berger87] for 2D problems, which, however, cannot be easily extended to 3-D flows. Moon and Liou [Moon89] have introduced a constraint on the conservative quantities such as mass, momentum and total energy.

"Patched" subdomain decomposition introduces some constraints on the grid generation. However, it makes it simple to ensure a conservative treatment of block interfaces. Rai ([Rai86]) has ensured global conservation by enforcing surface fluxes and the continuity of the dependent variables at the interfaces. Yadlin and Caughey ([Yad91a]) have developed a patched block approach for the solution of the Euler equations by implementing a diagonal implicit multigrid approach. At the (subdomain) interfaces they have imposed the continuity of the dependent variables. More recently Yadlin et al. ([Yad91b]) have reported an extension of the algorithm to the solution of the Navier Stokes equations. In particular, they have exploited the advantage of operator adaptation by solving the Navier Stokes equations in the near wall domains and in the wake region, while solving the Euler equations in the far field.

In the present work a multidomain technique is developed for the solution of viscous two-dimensional flows by means of a "patched" subdomain decomposition with continuous grids, that ensures global conservation and it yields no distortion of discontinuities (such as shocks and slip surfaces) that cross (grid/subdomain) interfaces. Results of the flow over the NACA 0012 airfoil are discussed to demonstrate the properties of the method mainly for operator adaptation and parallel applications.

II Numerical Algorithm

The governing equations solved are the two-dimensional compressible Navier-Stokes equations in conservation form.

$$\frac{\partial}{\partial t}\int_S \mathbf{W}\, dS = -\oint_{\partial S} (\mathbf{F}_E - \mathbf{F}_V)\cdot \underline{n}\, ds \qquad (1)$$

where $\mathbf{W}$, $\mathbf{F}_E$ and $\mathbf{F}_V$ are respectively the vector unknown, the inviscid and viscous fluxes, defined as:

$$\mathbf{W} = [\rho\, , \rho\, \underline{u}\, , \rho\, E]^T$$

$$\mathbf{F}_E = \left[\rho\, \underline{u}\, , \rho\, \underline{u}\, \underline{u} + p\, \underline{\underline{U}}\, , \rho\, \underline{u} \left(E + \frac{p}{\rho}\right)\right]^T$$

$$\mathbf{F}_V = \left[0\, , \underline{\underline{\sigma}}\, , -\left(\underline{q} - \underline{u} \cdot \underline{\underline{\sigma}}\right)\right]^T$$

and

$$p = (\gamma - 1)\, \rho\, (E - \underline{u} \cdot \underline{u}/2)$$

$$\underline{\underline{\sigma}} = \mu\left(\nabla\, \underline{u} + \nabla\, \underline{u}^T\right) - \frac{2}{3}\, \mu\, \nabla \cdot \underline{u}\, \underline{\underline{U}}$$

$$\underline{q} = -\lambda\, \nabla T.$$

The basic algorithm for solving the system of governing equations is based on a finite volume approach with a cell centered formulation. A system of ordinary differential equations is obtained by using the method of lines, which allows separation of space and time discretization.

Approximating surface and boundary integrals by means of the mean value theorem and mid-point rule, the governing equations are cast in the following discretized form

$$\frac{d\mathbf{W}_{i,j}}{dt}\, S_{i,j} + \sum_{\beta=1,4} \left(\mathbf{F}_{E,num} - \mathbf{F}_{V,num}\right) \cdot \underline{n}\, ds = 0 \tag{2}$$

where β indicates the cell face, and the subscript num stands for numerical.

The numerical inviscid flux ($\mathbf{F}_{E,num}$) is based on an upwind biased discretization that yields

$$(\mathbf{F}_{E,num} \cdot \underline{n})_{i+1/2,j} = \frac{1}{2}\left(\mathbf{F}_{E_{i,j}} + \mathbf{F}_{E_{i+1,j}}\right) \cdot \underline{n}_{i+1/2,j} + D_{i+1/2,j} \tag{3}$$

where $D_{i+1/2,j}$ represents the numerical antidiffusive flux that makes the scheme upwind biased and ensures the TVD property. Following Harten and Yee [[Yee85]) its expression is obtained by characteristic decomposition along the normal direction to cell face as a function of the differences of the characteristic variables and the right eigenvector matrix (defined in terms of the covariant and contravariant velocity components), with the use of minmod limiter for the antidiffusive flux.

The viscous flux depends on the gradient of the primitive variables (velocity and temperature) that are numerically evaluated by applying Gauss theorem to a computational cell whose vertices are the two grid nodes (I, J) and $(I, J-1)$ and the centers of the two adjacent cell (i, j) and $(i+1, j)$. The time integration is performed by a three-stage Runge Kutta algorithm.

1 Domain Decomposition and Data Structure

Domain decomposition is a natural methodology for an efficient use of parallel computers [Gropp91], [Quart89]), allowing the partitioning of the instruction set and the data on the available processors and memories. Moreover, the technique allows for operator adaptation and mesh refinement, and it is suitable for complex geometries. In the present work the technique has been implemented to exploit its properties for parallel applications and operator adaptation. The computational domain is partitioned in patched-type subdomains, called blocks in logical space. However, at the boundaries, two layers of fictitious cells are introduced for an efficient treatment of the boundary conditions, thus making all blocks overlaid.

For each block (B) the data structure requires:

1. informations for neighboring blocks (i.e. to identify adjacent blocks).
2. the dimensions (in logical space) in x- and y- directions (NX_B and NY_B);
3. the metric variables X_B (coordinates) and S_B (surfaces);
4. the field variables $\mathbf{W}_B$, p_B, T_B etc.;
5. the parameters that characterize the geometrical (wall, wake etc.) and physical topology (viscous or inviscid) of the block.

Global data such as residuals, problem input data, etc. are collected and distributed by a master process, which controls the input/output process as well. However, such an approach does not alter the Single Process Multiple Data nature of the algorithm.

2 Boundary Conditions and Interface Treatment

The partitioning of the computational domain in subdomains introduces internal boundaries corresponding to the interfaces between adjacent blocks. Therefore, the use of a multidomain technique requires boundary conditions along the true boundaries as well as at block interfaces. In order to impose boundary conditions at interfaces, block adjacency relationships must be defined . The continuity of the variables is enforced along the boundaries shared by two adjacent blocks [Quart89]. The introduction of a fictitious layer of cells surrounding the blocks allows to enforce the interface conditions by injecting the solution of the underlying block onto the overlaying one. However, with high order schemes only one fictitious layer of cells is not sufficient to evaluate the flux at the interface . Therefore, to maintain second order accuracy at the interfaces we use a second fictitious layer of cells. For an efficient exploitation of the parallelism, we have also treated all interfaces in a loosely synchronous manner: nodes are constrained to intermittently communicate with each other when a syncronization point is reached.

III Parallel Implementation

The present approach has been implemented on the IBM-SP2 and CRAY-T3D machines whose characteristics are:

- *Large number of processors*

  The maximum number of processors of the SP2 (available at CASPUR-University of Rome "la Sapienza") is 8; the maximum number of processors of the T3D (available at CINECA-Bologna) is 64.

- *Local memory*

  Each processor has it own local memory (respectively 128 and 64 Mb) and communicates over a network

- *MIMD*

  Each node can execute a separate program or instruction stream. In our implementation each node runs an identical program ; however, the nodes perform different instruction, due to data-dependent branching that occurs mostly at the boundaries.

It must be pointed out that in implementing an algorithm (or computational strategy) on a distributed parallel machine some critical issues must be addressed:

- *Selection of the numerical scheme*

  In CFD applications, most of the available and well known schemes are data dependent (due to boundary condition treatment, time and space discretization) as is the case, for example, of implicit schemes and nonlinear high order schemes. In the present work we have selected an explicit method, which can be easily parallelized. In particular, with the use of a multistage Runge-Kutta algorithm, the solution at each grid node can be advanced in time independently from the other grid points. The spatial discretization stencil introduces data dependencies as well. As previously described, we have implemented a TVD scheme, which has a 5 points stencil. Thus, at each grid point values from the previous time step at more than one neighbouring grid point are needed.

- *Load Balancing*

  The ideal parallel program is a program in which either there is no need for communications between nodes or the communications are all asyncronous. However , in most applications of computational fluid dynamics constraints are generally necessary for a correct evolution of the solution. In such a case load balancing plays a crucial role since the computation between two synchronization points is driven from the slower processor (which is the one with the largest workload). Hence, to achieve high parallel computational efficiency, algorithms with the least number of synchronization points have to be devised. In the present work we have introduced syncronization points at the first stage of the Runge-Kutta algorithm to exchange data at subdomain interfaces (by data communication) , and a syncronization at the final stage for convergence analysis (by performing global data summation).

- *Problem Size*

  An optimal problem size must be addressed on each node to maximize the ratio between computation and communication times. Indeed, for a given problem size, the CPU and communication times are roughly proportional, respectively, to N^2 and N (where N is the characteristic size of each subdomain). Consequently, increasing the number of processors reduces the characteristic size, and the ratio of CPU and communication times decrease.

- *Operator Adaptation*

  If the same physical and numerical operators are used in all subdomains, load balancing is naturally achieved by partitioning in subdomains that have the same number of grid points. However, an efficient use of operator adaptation requires an evaluation of the performance of the numerical schemes. For example, the computational time of TVD schemes is roughly twice the computational time of Adaptive Dissipation schemes. Therefore, for load balancing, 50% fewer grid points should be used on subdomains where total variation diminishing schemes are implemented.

1 Speed-up

The standard speed-up is defined as the ratio of single processor to n-processors execution times ($S = T_1/T_n$).
For the purpose of determinig the parallel efficiency we have evaluated S by partitioning the single block grid provided for the test case in blocks having the same number of cells. Results for the flow around the NACA 0012 airfoil are given in Table 1, which reports the speed-up obtained on the IBM-SP2 machine available at CASPUR in Rome, and on the CRAY-T3D available at CINECA in Bologna. In Fig. 1 the speed-up is plotted *vs* the number of processors (only up to 8 for the limited number of processors available on the SP2); the figure shows a nearly linear speed-up of the CRAY-T3D. The differences between the theoretical and true speed-ups can be explained if one accounts for the loss of efficiency due to communication. To understand the differences between the CRAY-T3D and the IBM-SP2 one must consider the ratio (R) between communication and CPU times: we have estimated that $R_{T3D} \simeq R_{SP2}/4$, where the factor 1/4 accounts for the differences in communication and CPU times between the two machines.

IV Results

The viscous flow around the NACA 0012 airfoil of the Workshop has been simulated ([Mavri90]). The test case corresponds to the flow at a Mach number $M_\infty = .5$, an angle of attack $\alpha = 0°$, a Reynolds number $Re = 5\ 10^3$, and the conditions are such that the flow is laminar and subsonic throughout, with separation occurring in the proximity of the trailing edge. For this test case several computations have been performed (see Table 2).

The results of the computations are reported in Figs. 2 and 3, where the iso-Mach and iso-pressure contour lines are reported; the distributions of the pressure and skin friction coefficients vs. x/c are shown in Figs. 4 and 5, while the cunvergence history (in terms of the rms of the energy) is plotted in Fig. 6. The figures clearly show the small recirculation bubble in the near-trailing edge and wake regions; the separation point is predicted at approximately 97% of the chord if the parameter for entropy correction of Harten is taken $\epsilon = .125$. A reduction of ϵ makes the scheme more accurate: with $\epsilon = .0125, .00625$ the flow is found to separate at approximately 87% of the chord independently of ϵ. From Table 2 note that the effect of the entropy correction parameter is mainly on the friction drag: a 10-fold reduction of ϵ yields a 10% reduction in CD_f.

V Conclusions

A multidomain technique has been developed for the solution of viscous transonic (and supersonic) flows around airfoils. The technique uses a patched domain decomposition with continuous grids, it ensures global conservation and it does not produce distortions of discontinuities (such as shocks and slip surfaces) that cross interfaces.

Domain decomposition has been here exploited for operator adaptation and for an efficient exploitation of parallel computers. In general, operator adaptation amounts to either numerical or physical operator adaptation. The former one amounts to use different schemes in different blocks depending upon the degree of accuracy required. The latter one amounts to adapt the governing equations in the different blocks according to the controlling local physical phenomena. The methodology has been developed for physical operator adaptation by introducing "viscous subdomains", in which the full Navier Stokes equations have been solved, and "inviscid subdomains" where the Euler equations are solved. The computed solution is found to be as accurate as the single domain one, indicating the validity of the technique, particularly as far as the interface treatment is concerned. The technique uses an upwind biased total variation diminishing discretization (a central one with adaptive dissipation is also possible). The technique has been applied to compute the flow around the NACA 0012 airfoil of the workshop. The results show that the flow remains subsonic throughout, and it separates in the near-trailing edge region at approximately 87% of the chord.

Acknowledgements

The authors wish to acknowledge the support of the Computer Centers CASPUR at University of Rome "la Sapienza" and CINECA of Bologna for providing, respectively, computer time on the IBM-SP2 and CRAY-T3D machines.

Bibliography

[Gropp91] *Gropp, W.D., and Keyes, D.E. (1991) :, Domain Decomposition Methods in Computational Fluid Dynamics, ICASE Technical Report 91-20.*

[Hauser92] *Hauser, J. and Sion, H.D., (1992) :, Aerodynamic Simulation on Massively Parallel Systems, Parallel CFD '91 Conference Proceedings, Elsevier Science Publishers.*

[Berger87] *Berger, M., (1987) :, On Conservation at Grid Interfaces, SIAM J. Numer. Anal., Vol. 24.*

[Moon89] *Moon, Y.J., and Liou, M.S.,(1989) :, Conservative Treatment of Boundary Interfaces for Overlaid Grids and Multi-Level Grid Adaptations, AIAA Paper 89-1990.*

[Rai86] *Rai, M.A., (1986) :, A Conservative Treatment of Zonal Boundaries for Euler Equation Calculations, Jour. of Comp. Phys., Vol. 62.*

[Yad91a] *Yadlin, Y., and Caughey, D.A., (1991) :, Block Multigrid Implicit Solution of the Euler Equations of Compressible Fluid Flow, AIAA Journal, Vol. 29, pp. 712–719.*

[Yad91b] *Yadlin, Y., Tysinger, T.L., and Caughey, D.A., (1991) :, Parallel Block Multigrid Solution of the Compressible Navier Stokes Equations, AIAA 10th CFD, Conference Proceedings, pp. 965–966, 1991.*

[Yee85] *Yee, H.C., Warming, R.F., and Harten, A., (1985) :, Implicit TVD Schemes for Steady State Calculations, Jour. of Comp. Phys., Vol. 57, pp. 327–360.*

[Quart89] *Quarteroni, A., (1989) :, Domain Decomposition Methods for Systems of Conservation Laws: Spectral Collocation Approximation, ICASE Report 89–5.*

[Mavri90] *Mavripilis, D.J. and Jameson, A., (1990) :, Multigrid Solution of The Navier Stokes Equations on Triangular Meshes, AIAA Journal, Vol. 28, pp. 1415-1425.*

TABLE 1
Speed - up (S)

N_{block}	N_{bx}	N_{by}	S_{SP2}	S_{T3D}
1	1	1	1	1
2	2	1	1.90	1.99
3	3	1	2.82	2.96
4	4	1	3.7	3.93
5	5	1	4.5	4.83
6	6	1	5.4	5.79
7	7	1	5.9	6.69
8	8	1	6.35	7.45

TABLE 2
NACA 0012, $M_\infty = .5$, Re $= 5\ 10^3$, $\alpha = 0°$

$N_{bx} \times N_{by}$	Method	ϵ	CD_W	CD_f	X_{sep}
3 × 2	TVD (NS/E)	.125	0.02512	0.03788	0.97
3 × 2	TVD (NS)	.125	0.02512	0.03788	0.97
6 × 1	TVD (NS)	.125	0.02430	0.03711	0.97
6 × 1	TVD (NS)	.0125	0.02483	0.0336	0.874
6 × 1	TVD (NS)	.00625	0.02482	0.0337	0.867

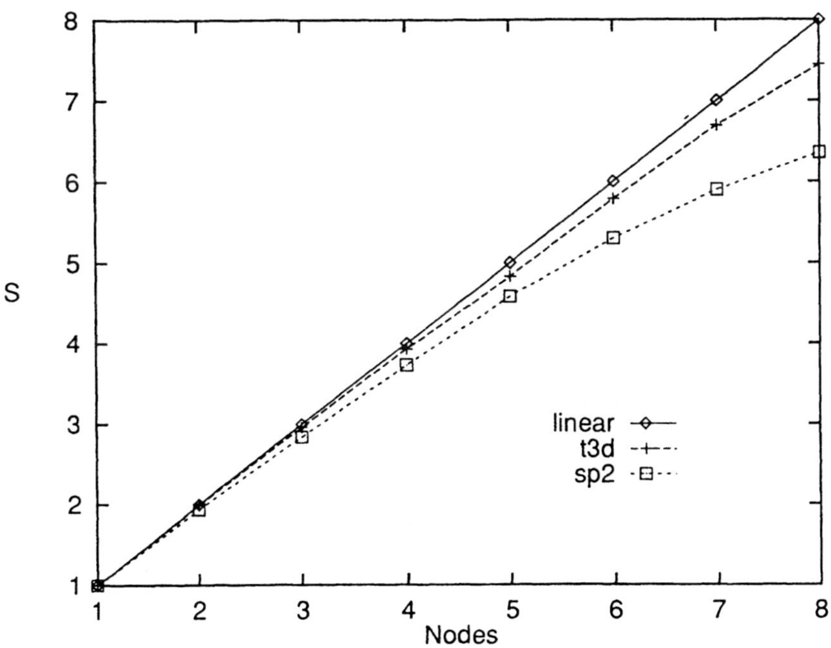

Fig. 1 – Speed-up vs nodes

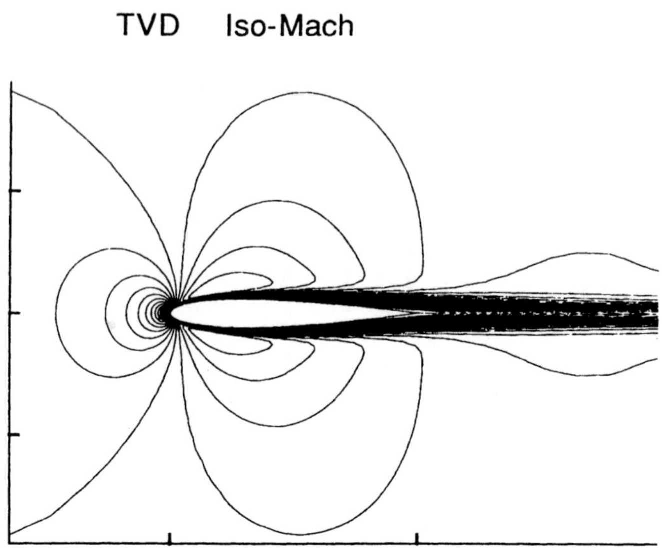

Fig. 2 – NACA 0012: iso-Mach contour lines.

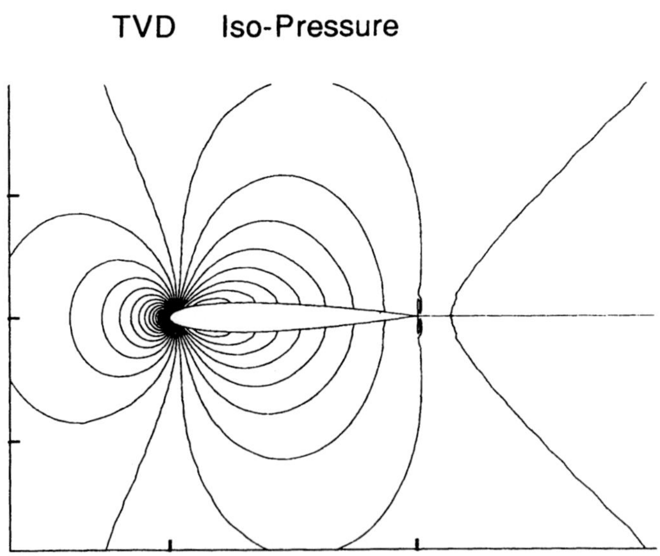

Fig. 3 – NACA 0012: iso-pressure contour lines.

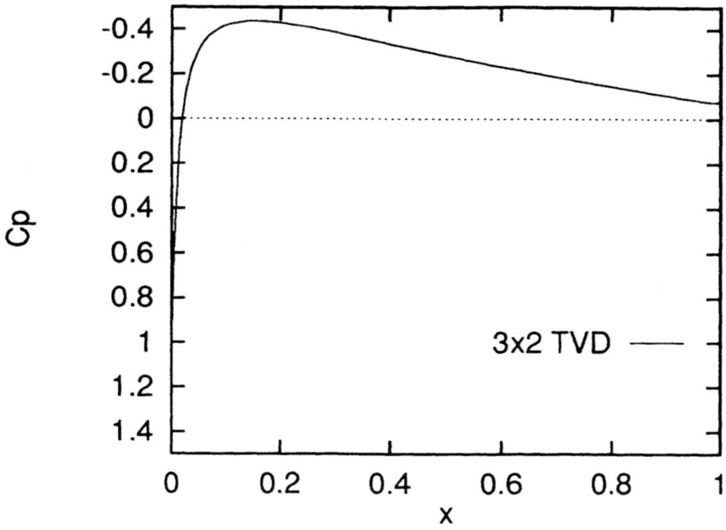

Fig. 4 – NACA 0012: pressure coefficient vs x/c.

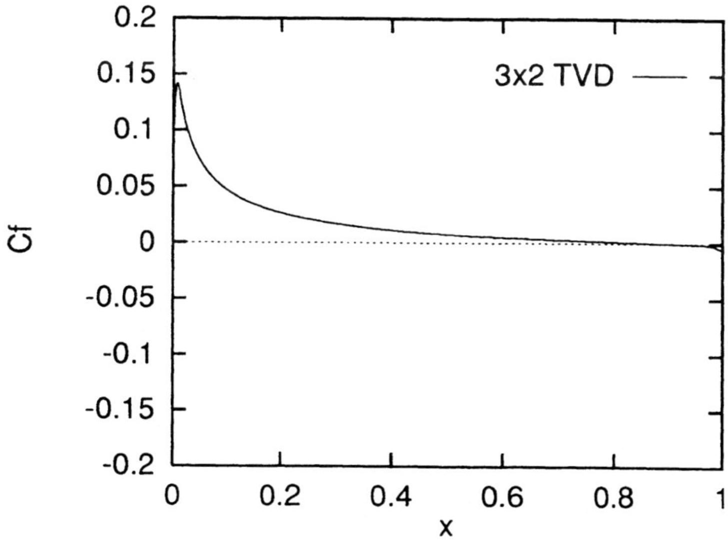

Fig. 5 – NACA 0012: skin friction coefficient vs x/c.

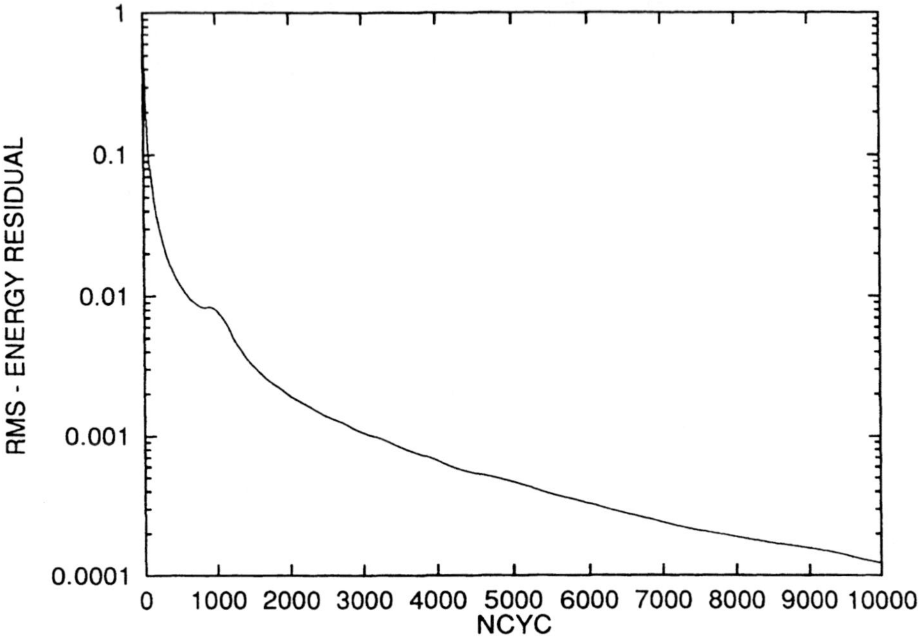

Fig. 6 – NACA 0012: convergence rate vs cycles.

8 Navier-Stokes Simulations on MIMD computers using the EURANUS code

C. Lacor and Ch. Hirsch, VUB, Brussels, Belgium

Abstract

Within this work the 3D Multigrid/Multiblock Navier-Stokes code EURANUS was parallelized for MIMD type parallel computers, using PVM (Parallel Virtual Machine).

The code was applied to both Euler and laminar Navier-Stokes flow around a NACA0012 airfoil. In the Euler calculation both clusters of workstations and parallel computers (SP2 and Parsytec transputer) were used. The Navier-Stokes case which corresponds to the mandatory test case TC1 was calculated on a cluster of workstations.

Good parallel efficiencies were obtained on the MIMD computers. On the workstation clusters the overhead of communication is more pronounced, especially for the Navier-Stokes case.

I Basic Features of the Method

1 The EURANUS code

EURANUS is a 3D Multiblock/Multigrid Navier-Stokes solver developed for ESA/ESTEC, in a joint work between VUB and FFA, [Lac92],[Riz93],[Ala95]. It solves the time-dependent Reynolds Averaged Navier-Stokes equations on structured meshes. Real gas effects are included both for chemical equilibrium and chemical and thermal non-equilibrium. Turbulence modeling is based on algebraic (Baldwin-Lomax) or on two-equation models (k-ϵ and k-τ).

EURANUS uses a cell-centered control volume approach for the discretization in space. It contains both a central scheme with Jameson type dissipation, and TVD type upwind schemes (Flux Difference Splitting TVD and Symmetric TVD).

Time integration is based on multistage Runge-Kutta schemes or on implicit solvers with relaxation methods.

A multigrid procedure ensures fast convergence towards steady state and can be combined with any of the solvers.

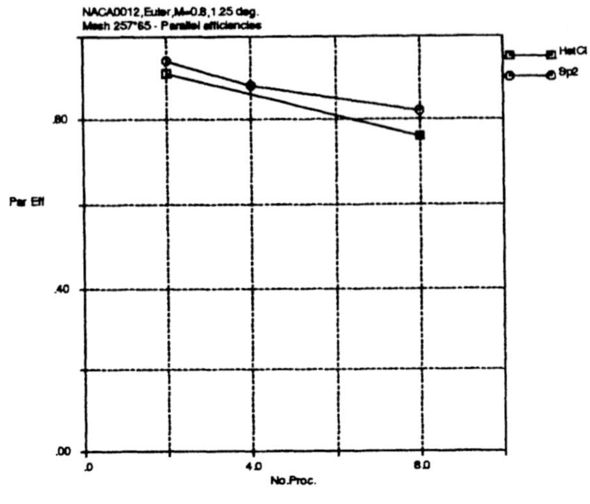

Figure 1 Parallel Efficiencies for the heterogeneous cluster and SP2 on the 257*65 mesh for inviscid NACA0012 test case

2 Parallelization approach

The parallelization is based on the multiblock features of EURANUS. The computational mesh is artificially split into blocks. The different blocks are then assigned to different processors. In order to always be able to ensure a good load balancing, there is no restriction on the total number of blocks, nor on the number of blocks assigned to the same processor.

The block loop is within the multigrid loop. This means that the blocks communicate on each multigrid level such that any grid level is treated in phase on all blocks.

Combined with the Runge-Kutta solver and a careful treatment of boundary conditions, this approach ensures that the convergence (in terms of number of cycles) is exactly the same for a mono- and a multiprocessor case, [Lac94a].

3 Communication

PVM is used as communication library. Load balancing is achieved by putting blocks of similar size on the different processors.

The code was also implemented on a Parsytec GC/32 transputer system. Here PVM was not available and the communication routines provided by the PARIX Operating System were used instead, [Lac94b].

The Parsytec machine however was too limited in memory to run the mandatory test case TC1.

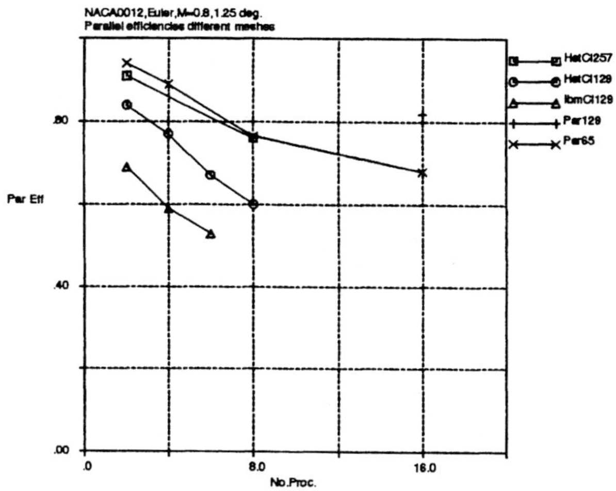

Figure 2 Parallel Efficiencies for the heterogeneous cluster and SP2 on the 257*65 mesh for inviscid NACA0012 test case

II Discussion of results

1 Euler flow around NACA0012

As a first test case the Euler flow around a NACA0012 airfoil at Mach 0.8 and 1.25 degrees angle of attack was calculated. Three meshes were used : a fine (257*65 points), a medium (129*33) and a coarse mesh (65*17). The latter was only used on the Parsytec computer, where the finest mesh did not fit into the relatively small memory.

Multiblock meshes for the parallel calculations were generated with up to 16 blocks. These blocks were connected in 'a serial way', i.e. connecting blocks share only one face.

The following platforms were used : 2 Ethernet clusters of workstations, one heterogeneous consisting of HP9000, DECα and SGI Indigo, the other homogeneous consisting of 6 IBM RS6000; an IBM SP2 MIMD computer; a Parsytec GC/32 transputersystem.

Details of the calculations can be found in [Lac94b], here only the main results are shown.

On all meshes fast convergence to computer accuracy was achieved using multigrid (about 80 cycles on the finest mesh). Figure 1 depicts the parallel efficiencies obtained on the finest mesh for the SP2 and the heteregoneous cluster. The parallel efficiency is defined as the ratio of actual and theoretical speed up. The actual speed up is the CPU time per iteration on 1 processor, divided by the CPU time per iteration on multiprocessors, on the same multiblock mesh.

Good parallel efficiencies are obtained. They are nearly the same on both platforms. As the communication on the cluster (over Ethernet network) is slower

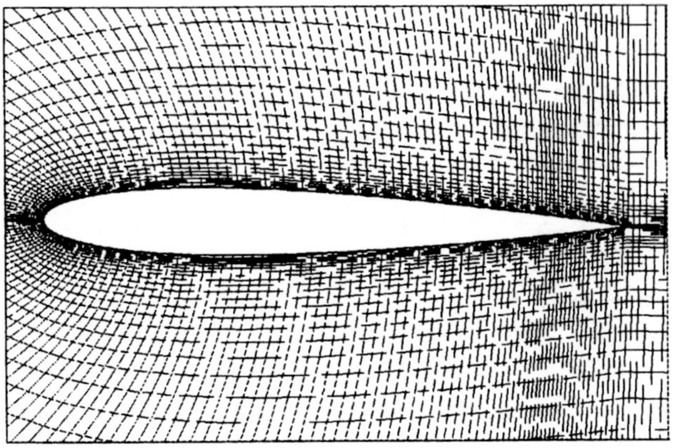

Figure 3 Zoom of the 289*57 mesh of TC1 near the airfoil

than on the SP2, this indicates that the communication overhead for this test case is small. Figure 2 shows the results for the medium mesh. Here no SP2 results are available. For the heterogeneous cluster, the results on the finest mesh are also given for comparison. For the Parsytec, results on the medium mesh could not be obtained with less than 16 processors because of insufficient memory. For this machine, results are also given for the coarse (65*17) mesh. Comparing results on the medium mesh, the best efficiency is obtained with Parsytec (only 1 point on fig. 2 for 16 processors). Note that the processors of this machine are slower than those of the two clusters, such that the relative influence of communication is less.

It is not clear however why the heterogeneous cluster performs better than the homogeneous IBM cluster.

Comparison of the results on different meshes show that the efficiency increases on the finer grid, which is obvious as the processors will have more computing work to do.

2 The mandatory test case TC1

The test case TC1 is a laminar Navier-Stokes calculation around the NACA0012 airfoil. The Mach number is 0.5, the angle of attack 0 degrees and the Reynolds number (based on chord length) equals 5000. A mandatory mesh was used. It is a C-type mesh, consisting of 289*57 points and with the outer boundary at 75 chord lengths. Figure 3 shows a zoom of the mesh near the airfoil.

Calculations were performed using three different schemes : the central scheme with Jameson type dissipation and two upwind schemes, TVD Flux Difference Splitting (FDS) and Symmetric TVD (STVD).

In the central scheme artificial viscosity coefficients of 0 and .016 for resp.

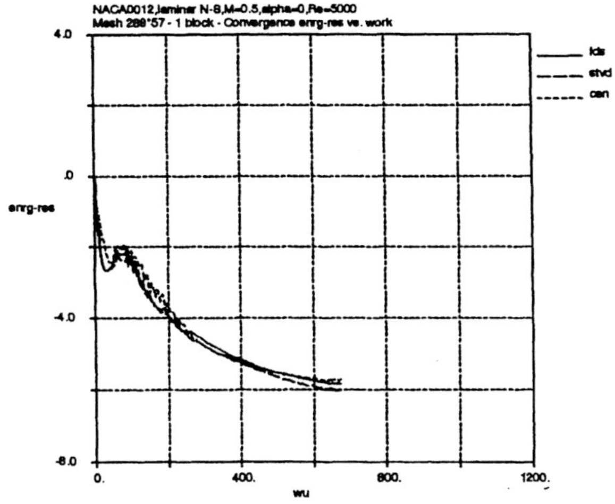

Figure 4 Convergence history for TC1 of energy residual vs. work units for the 3 schemes

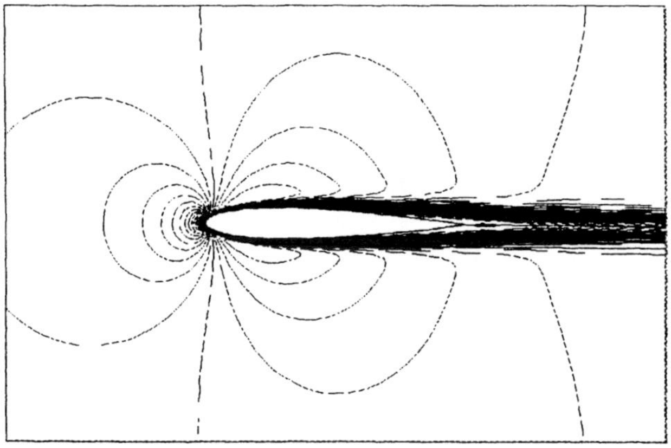

Figure 5 Isomach contours near the airfoil (TC1, STVD solution)

second- and fourth-order dissipation were used.

In the FDS scheme the Van Leer limiter was used. For the STVD scheme a new family of robust and accurate limiters was recently developed, [Lac93], where limiters are applied to so-called effective ratios, i.e. averages of the + and - ratios of FDS schemes. In the present calculation the Van Leer limiter was applied to an effective ratio based on the harmonic average.

All calculations used 4 level FAS multigrid with W-cycle to accelerate convergence. A 4-stage Runge-Kutta scheme is used as smoother in combination with residual smoothing. The Navier-Stokes terms are only calculated in the

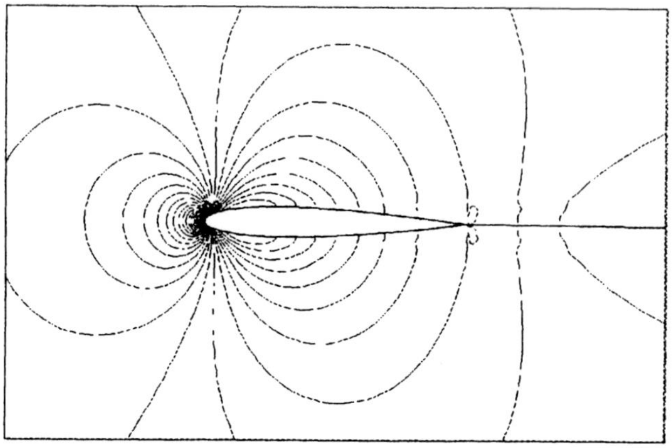

Figure 6 Isopressure contours near the airfoil (TC1, STVD solution)

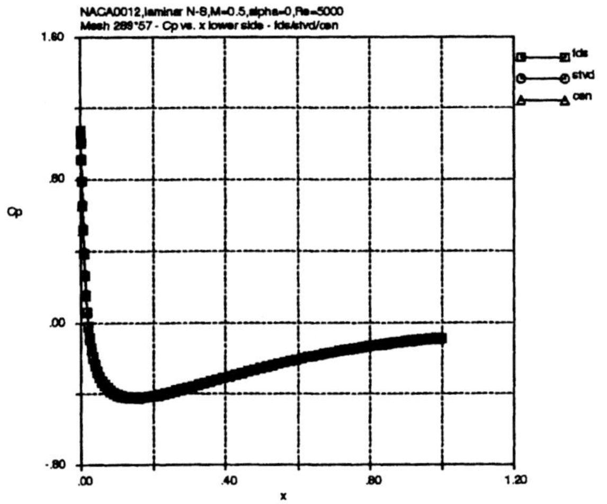

Figure 7 Cp distribution for TC1 along the lower side for the central, FDS and STVD schemes

first stage. On coarse grid levels, the first-order upwind scheme is used (for FDS and STVD) or the central scheme with increased dissipation on solid walls (for central scheme). The CFL number is 4.

Figure 4 shows the convergence in terms of Work Units for the 3 calculations. Convergence to computer accuracy is achieved in about 600 WUs, which corresponds to 275 cycles.

It should be noted that by using Full multigrid the convergence could further be improved.

However, the Full multigrid uses a switch based on residual drop and would

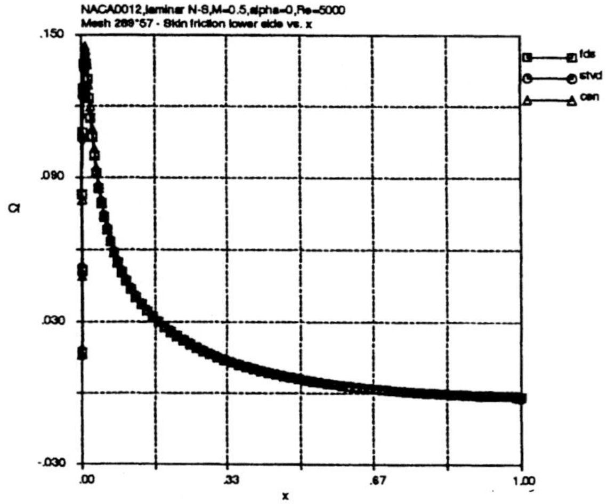

Figure 8 Cf distribution for TC1 along the lower side for the central, FDS and STVD schemes

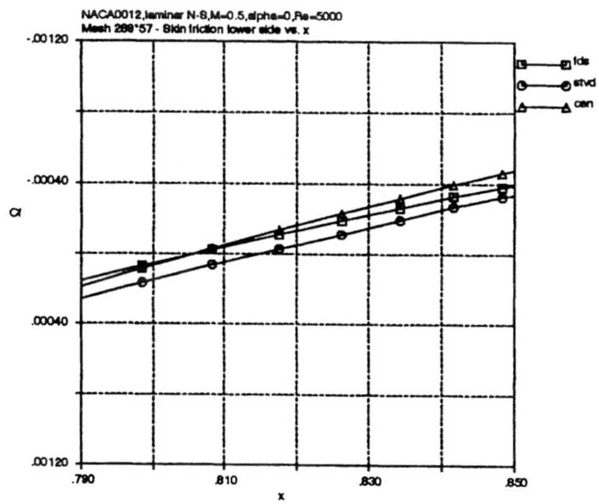

Figure 9 Zoom of Cf near the separation point for TC1 on the lower side for the central, FDS and STVD schemes

therefore behave differently for a different number of mesh blocks. This would not allow a fair comparison of the timings for mono- and multiprocessor calculations. For this reason, Full multigrid was not used.

Figures 5 and 6 show isolines of Mach number and static pressure around the airfoil, as obtained with the STVD scheme. The pressure distribution along the lower side of the airfoil, as obtained with the 3 schemes is shown in figure 7. There is no visible difference between the three solutions. The solution is also

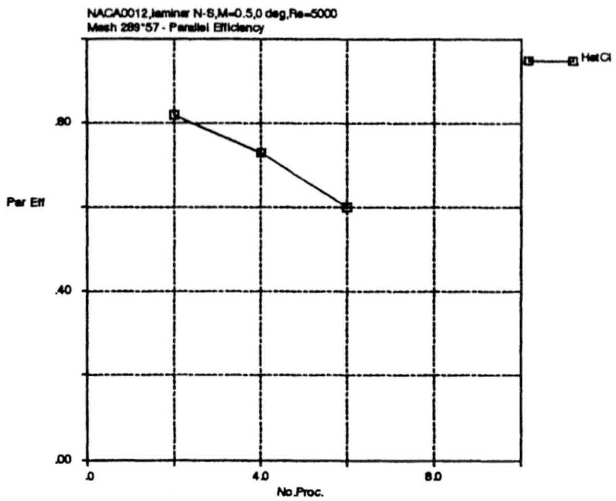

Figure 10 Speed up obtained for 2,4, 6 processors of a cluster of workstations

symmetric, in that the distribution on the upper side (not shown) is identical to that on the lower side. Figure 8 shows the skin friction distribution along the lower side obtained with the 3 schemes. Here some minor differences are found, with respect to the predicted peak (which is slightly higher for the central scheme) and with respect to the separation point. Figure 9 shows a zoom of the skin friction distribution near the separation point on the lower side; the central and FDS scheme predict separation at about 80.5% of the chord and STVD at approximately 81.5%. With respect to the skin friction a slight asymmetry was found (not shown here, cf. [Lac95]). The reason is that the mesh is also not exactly symmetrical.

For the parallel calculations, multiblock meshes with resp. 2, 4, 6 blocks were generated with blocks connected in 'a serial way'.

The calculations were performed on a cluster of HP-workstations using an Ethernet network. Figure 10 depicts the parallel efficiencies obtained. It can be seen that there is an important influence of the communication, which can be explained by the relative slow network in combination with the relative small size of the problem.

III Conclusions

The Navier-Stokes code EURANUS was parallelized using the multiblock approach and PVM as message passing library.

EURANUS was ported to different parallel platforms, including the MIMD computers Parsytec and SP2, as well as clusters of workstations.

Two test cases were calculated : the Euler (M=0.8, α=1.25) and laminar

Navier-Stokes (mandatory TC1) flow around a NACA0012 airfoil.

The use of multigrid allowed fast convergence to computer accuracy in both test cases.

Good parallel efficiencies were obtained on all platforms for the Euler case. The parallel efficiency of the Navier-Stokes calculation - which was only done on a cluster of workstations - indicates an important influence of communication. This can be explained by the relatively slow Ethernet network, in combination with the relative small size of the problem.

Acknowledgements
The authors would like to thank Dr. Giannakoglou of NTUA for kindly providing the mesh for the mandatory test case TC1.

Bibliography

[Lac92] *Lacor C., Alavilli P., Hirsch Ch., Eliasson P., Lindblad I., Rizzi A., (1992) :* Hypersonic Navier-Stokes computations about complex configurations, Proceedings 1st European CFD Conference, 1089-1097.

[Riz93] *Rizzi A., Eliasson P, Lindblad I., Hirsch Ch., Lacor C., Häuser J., (1993) :* The Engineering of Multiblock/Multigrid Software for Navier-Stokes Flows on Structured Meshes, Computers Fluids **22** , 341-367.

[Ala95] *Alavilli P., Lacor C., Hirsch Ch., (1995) :* Nonequilibrium Flow Computations with General Thermochemistry in a Multigrid-Multiblock Framework, AIAA Paper **95-2011**, 30th AIAA Thermophysics Conf., San Diego.

[Lac94a] *Lacor C., Hirsch Ch., Eliasson P., Lindblad I., (1994) :* Study of the Efficiency of a Parallelized Multigrid/Multiblock Navier-Stokes Solver on Different MIMD Platforms, 2nd European CFD Conf., Stuttgart.

[Lac94b] *Lacor C. (1994) :* 3D Navier-Stokes Simulations on MIMD Computers, VUB contribution to Task 4.2, ECARP Mid-Term Report.

[Lac93] *Lacor C., Zhu Z.W., Hirsch Ch., (1993) :* A new family of limiters within the Multigrid/Multiblock Navier-Stokes Code EURANUS, AIAA Paper **93-5023**, AIAA/DGLR 5th Int. Aerospace Planes and Hypersonics Technology Conf., Nov. 30-Dec. 3, München.

[Lac95] *Lacor C. (1995) :* 3D Navier-Stokes Simulations on MIMD Computers, VUB contribution to Task 4.2, ECARP Report July 95.

IV Synthesis of Test Cases

Abstract

The scope of the MPP for Navier-Stokes "Workshop" was to investigate the impact of parallel architectures on the usability of laminar Navier-Stokes solvers for compressible flows. Turbulent flows are also included as optional test-cases. In this text, a brief discussion as well as some main conclusions from the Workshop that took place in Sophia-Antipolis are presented. Emphasis will be given to the 2-D flow problem that has been examined by all contributors. Structured or unstructured grids, explicit or implicit methods and upwind or centered schemes have been used.

1 Classification of the Workshop Contributions

Eight contributors provided numerical predictions for the mandatory flow problems along with the analysis of the relevant computing cost on parallel platforms. They are listed in Table 1, where the parallel system(s) they have used are also tabulated. For the sake of convenience, abbreviations are introduced for all of them.

Apart for the parallelization aspects, the various methods utilized in this workshop present basic differences which are related to the type of grid they are handling (structured or unstructured), the type of the numerical solver they are using (explicit, implicit or point-wise implicit schemes; the latter will be refered to as explicit-like methods), the way the flow domain splits into subdomains for the concurrent execution (patched or overlapping domains) and the scheme used for the analysis of the convective terms (upwind schemes of different formal accuracy or central schemes with a "controllable" amount of artificial dissipation).

The contributions are classified in Table 2, depending on the type of the grid used for the analysis of the standard two-dimensional flow problem *TC1*. For this case, only NTUA presented results using both kinds of grid, but, even in this case, the unstructured grid used has been derived by splitting up each structured grid cell into two triangles. Other contributors (for instance, TUS) have used more than one structured grids to assess their effect on the quality of the predicted flowfield. The type of numerical solvers used by the contributors are given in Table 3 and the approximation for the convective terms in Table 4. There are contributors (for instance, FFA and VUB) that have used more than one discretization scheme (upwind and central differencing).

Table 1 Contributions to the Workshop

Institution	Contributors	Parallel Computer(s)	Abbrev.
DASA & University of Stuttgart	T.Michl, S.Wagner	Intel-Paragon $XP/S-5$, CRAY $T3D$	TUS
Universita di Roma "La Sapienza"	F.Grasso, C.Pettinelli	$IBM - SP2$, CRAY $T3D$	URO
Dassault-Aviation	Q.V.Dinh, P.Leca, F.X.Roux	$IBM - SP2$, Intel $IPSC860$	DLT
FFA, Sweden	P.Eliasson	$IBM - SP2$, Cluster SGI $R4000$	FFA
Vrije Universiteit, Brussel	C.Lacor, Ch.Hirsch	Cluster of Workstations	VUB
UPC de Barcelona	T.Fischer, G.Bugeda	$CM - 200$	UPC
INRIA, Sophia-Antipolis	S.Lanteri	$KSR1$	INR
N.T. University of Athens	D.Koubogiannis, K.Giannakoglou, K.Papailiou	Intel-Paragon $i860XP/S$	NTUA

2 *TC1*: On the Accuracy of the Numerical Predictions

Concerning *TC1*, a common way to measure the suitability of the numerical approximations is through comparisons of the location of the separation point close to the trailing edge, which has been differently predicted by the various methods. Hereafter, this location will be denoted by X_s and will be given as percentage of the airfoil chord C. The X_s/C ratio depends on the grid used (how stretched the grid is close to the solid boundaries) and the dissipation of the numerical scheme. Details on the test case are available in [Mavr89].

The grid distributed for this Workshop had 280 × 50 cells. According to the study carried out in TUS, a much finer grid (256 × 79 cells), with about 30 cells in the boundary layer, is capable of capturing the correct separation point regardless the method accuracy; both first- and second-order methods predict the separation point equally well ($X_s/C = 0.83$ and $X_s/C = 0.827$, respectively). Using the ECARP grid, the same group calculated the separation point at $X_s/C = 0.827$, with a second-order scheme. The excessively diffusive first-order upwind scheme moved the separation point farther downstream, ($X_s/C = 0.86$). This is considered as of minor importance in the present study, since first-order schemes are not in use anymore.

Table 2 Type of Grids Used for TC1

Structured Grid	Unstructured Grid
TUS	INR
URO	DLT
NTUA	NTUA
FFA	UPC
VUB	

Table 3 Type of Numerical Method

Implicit	Explicit or Explicit-like
TUS	INR
DLT	URO
	NTUA
	FFA
	VUB
	UPC

Table 4 Discretization Scheme for the Convective Terms

Contributor	Discretization Scheme for Convective Terms
TUS	Upwind 3rd order, Modified van-Leer FVS
URO	Upwind, Harten-Yee, minmod limiter
DLT	Upwind
FFA	Upwind (TVD-FDS, STVD) and Central with Artificial Dissipation
VUB	Upwind (TVD-FDS, STVD) and Central with Artificial Dissipation
UPC	Central with Artificial Dissipation
INR	Upwind 2nd order, van-Leer's MUSCL
NTUA	Upwind 2nd order, van-Leer's MUSCL

In the Harten-Yee scheme (with the minmod limiter) used by URO, high values of the ε parameter in the entropy correction scheme ($\varepsilon = 0.125$) give a very small separation bubble (separation occurs at $X_s/C = 0.97$), but by lowering the ε value the separation point moves upstream. Values of ε close to 0.00625 stabilize the separation point at $X_s/C = 0.87$. This is the typical behaviour of all upwind schemes where an external parameter may control the scheme diffusivity. This also the case of the central schemes where the artificial dissipation is also controllable through coefficients that adjust the fourth-order

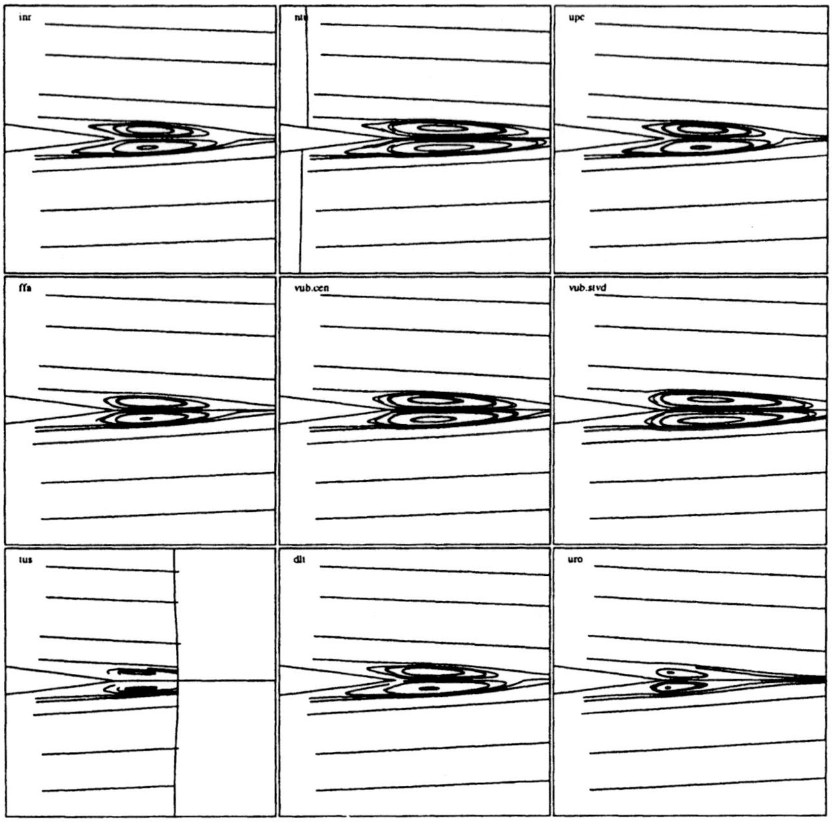

Figure 1 The flowfield in the vicinity of the trailing edge, for TC1

derivatives added.

A similar analysis was performed by FFA. With the same grid and a central finite-difference scheme with variable artificial dissipation, the separation point moves upstream as the artificial dissipation coefficient (of the fourth-order dissipation term, k_4) decreases. For instance, the value $k_4 = 1/32$ gives $X_s/C = 0.828$, while the value $k_4 = 1/128$ gives $X_s/C = 0.813$. In the same context, UPC code with the artificial dissipation scheme of Jameson with $k_4 = 1/120$ gives $X_s/C = 0.827$. Useful remarks about the level of dissipation introduced by the upwind schemes can be found in the papers by FFA and VUB, where both contributors have used the same software.

In summary, the less diffusive schemes predict very accurately the separation point, but generally lead to difficult convergence and, quite often, to non-converged solutions. On the other hand, the excessive artificial dissipation may succesfully overcome the convergences problems but moves the separation point downstream of its reference position. In order to overcome convergence difficulties associated with the high-order schemes, NTUA used a blend of first- and second-order accurate schemes which proved to stabilize the solution. Using this hybrid scheme (80 % second-order and 20 % first-order upwinding), the separation point is calculated at about$X_s/C = 0.8$.

It is also interesting to compare the recirculation zone provided by VUB using upwind and central schemes. A same local view of the predicted flowfields, in the vicinity of the trailing edge, is provided for all contributors in figure 1. Figure 2 shows a comparison of the pressure coefficient distribution, while figure 3 presents a comparison of the distribution of the friction coefficient along the airfoil.

3 *TC1*: On the Efficiency of the Parallel Code

The speed-up obtained by the various codes on the various parallel machines are compared in figures 4 and 5 . Some comments are necessary in order to have a better interpretationa of these curves. These are given below:

- For some parallel *FORTRAN* compilers, the optimizer may considerably change the efficiency of the parallel method. Care has to be taken to prevent the optimized parallel run from spoiling either the numerics (convergence) or the physics (consistency with the flow problem we are dealing with). One may see the relevant discussion in the paper by TUS.

- Superlinear speed-up curves may appear, especially in (implicit) parallel algorithms that do not aim at strictly reproducing the sequential convergence. For instance, TUS presents superlinear convergence on the Paragon-Intel, as a consequence of the more efficient preconditioning of the decomposed problem.

- Numerical methods (implicit or Jacobi schemes, i.e. the so-called explicit-like methods), which involve a number of internal iterations, may diminish the communication cost by increasing the number of processors. This is the case of preconditioners which operate separately within each subdomain. They are capable of providing a faster convergence, while requiring a smaller amount of information to be exchanged.

- No emphasis was given to the way partitions have been created, although this may affect the efficiency of the parallel solvers. Patched subdomains, enriched by a series of fake nodes) have been generally used; the patched subdomains seem to be advantageous compared to the overlaid ones (see discussion by URO).

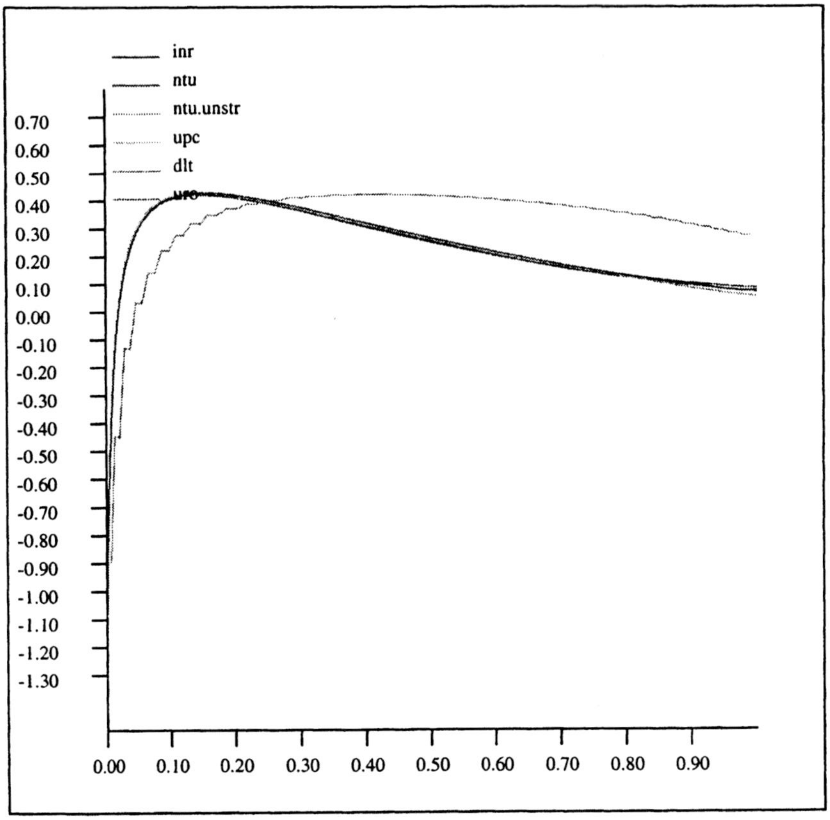

Figure 2 Pressure Coefficient Distribution for TC1

- Artifices to improve the parallel efficiency are in use, especially in solvers involving multi-stage algorithms. For instance, URO's explicit 3-stage Runge-Kutta scheme is combined with the exchange of intermediate data (synchronization points) at the end of the first and the third stages, only. On the contrary, FFA uses communication at every stage and every intermediate grid of the multigrid algorithm.
- The NACA0012 structured grid has been transformed to a pseudo unstructured one, by splitting each quadrilateral into two triangles. The structured grid computation gave a slightly better speed-up. This was attributed to the fact that the computing cost per iteration for the structured grids was greater than that for the unstructured ones (due to the different ways used

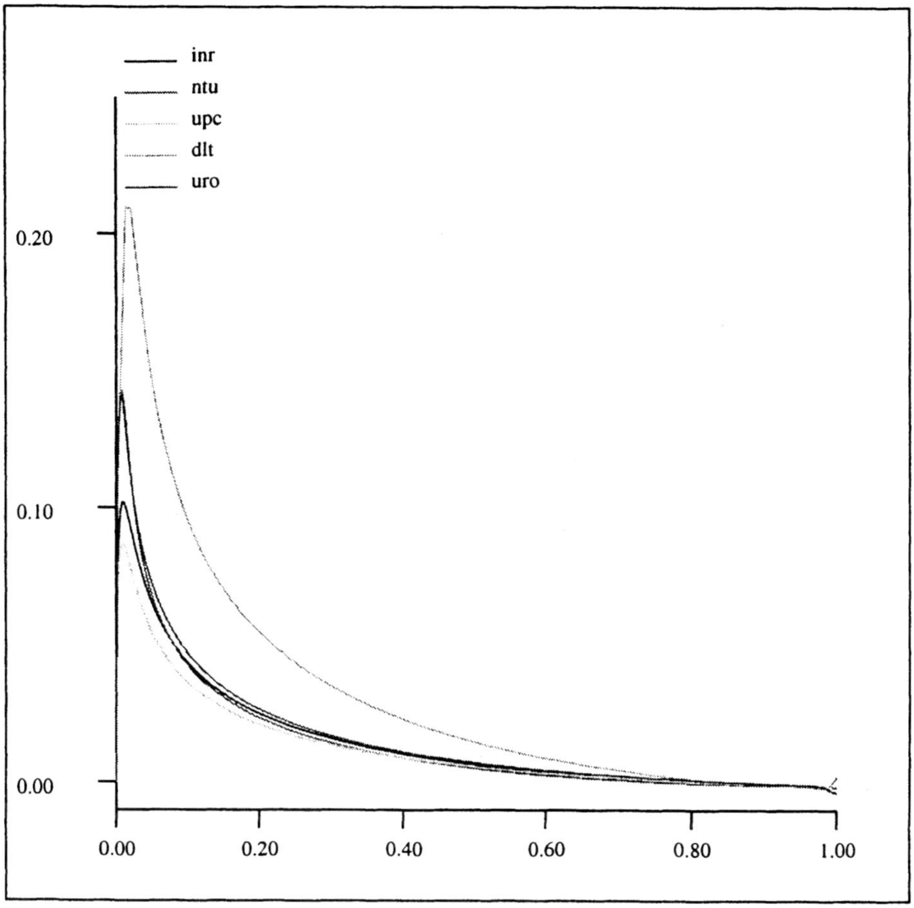

Figure 3 Friction Coefficient Distribution for TC1

to compute the viscous terms), whereas the communication cost increased when dealing with unstructured grids.

- Scalability studies are mainly presented by TUS, UPC and INR. For the scalability study, the same amount of cells or nodes is processed on each processor and successive runs are to be performed by increasing the number of processors. For contributors using unstructured grids, slight deviations are inevitable, due to the grid generation procedure which is not fully controllable (as far as the number of triangles to be generated is concerned). INR reports a degradation of scalability by 13% between 1 and 16 processors in KSR1 (for the Jacobi method, with 36 internal iterations). The degradation is greater (about 17%), if the number of internal iterations increases to 72. UPC reports scalability problems due to the complexity of the multigrid strategy which destroys the current data structure and adds complication through renumeration. The fully implicit BI-CGSTAB method used by TUS makes difficult any attempt to accurately measure the scalability. Approximately only, one may say that the preconditioning and the communication time remain almost constant, for various numbers of equally loaded processors. Differences in the CPU cost are attributed to differences in the implicit scheme operating in different subdomains as well as the cost for calculating global sums which increases the number of processors.

- Some remarks, that can be hardly generalized, concern parallel platforms used by the contributors. For instance, TUS study reveals that, in this particular flow problem with the particular implicit solver, $CRAYT3D$ is proved three times faster than the Intel-Paragon $XP/S - 5$, although its CPU is only two times faster; communications are much faster in the former platform. According to TUS, the comparison between $CRAYT3D$ and $IBM - SP2$ showed that $CRAYT3D$ is faster (speed-up 7.45, for 8 processors) than $IBM - SP2$ (speed-up 6.35, for the same number of processors). It is interesting to note that, for $IBM - SP2$, FFA presents similar speed-up values (6.54, for 8 processors and the standard grid for the NACA0012 case) using its own code. DLT reports a better speed-up (7.37) on $IBM - SP2$, using 8 processors. On the other hand, FFA reports a quite low speed-up value for a cluster of $SGIR4000$ processors (speed-up 3.41, for 8 processors) and this is explained by the fact that the problem is too small to be parallelized on a cluster of workstations. In conformity to the last remark is the conclusion drawn by VUB, measuring a speed-up value equal to 3.6 for a cluster of heterogeneous workstations (PVM, a reasonable extrapolation gives speed-up about 3.2, for 8 processors).

4 Analysis of the Three-Dimensional Flow Cases

The analysis of the 3-D flow around the ONERA M6-wing, at infinite Mach number 0.8, zero incidence and laminar flow conditions (the Reynolds number based on chord was equal to 1000, *TC2*) is provided by DLT. They have used an unstructured grid consisting of 62967 nodes and 349808 tetrahedra which has been decomposed to 32, 64 lnd 128 blocks for parallel calculations on the Intel $IPSC860$. Since it was impossible to use less than 32 processors, the reference speed-up value is that of 32 processors (all of the provided figures should be multiplied by 32). In this respect and with the so-introduced inaccuracy, they have calculated speed-up values equal to about 60.8 and 112 for 64 and 128 processors, respectively.

FFA presents results for the analysis of the same wing, at turbulent flow conditions. The Reynolds number was equal to $11.7 \cdot 10^6$ and the incidence angle

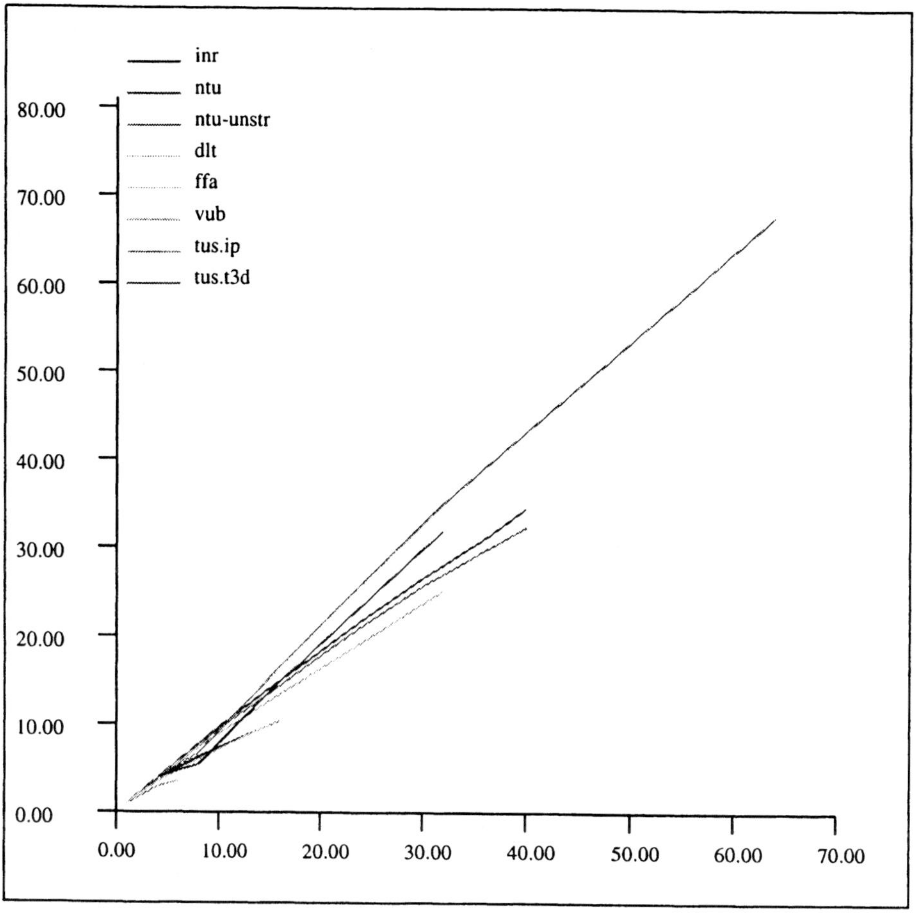

Figure 4 Speed-up curves for TC1

was equal to 3.06 degrees (*TC5*). The calculation was based on the Baldwin-Lomax turbulence model and a structured grid with $1.2 \cdot 10^6$ grid nodes. The reported speed-up is of the order of 7.4 for 8 $IBM - SP2$ processors and about 14.1 for 16 processors. In the relevant paper, one may compare the improvement in the quality of results with respect to a 2^3 times coarser grid.

For the two aforementioned 3-D flow problems, the reader should refer to the corresponding papers included in the same volume.

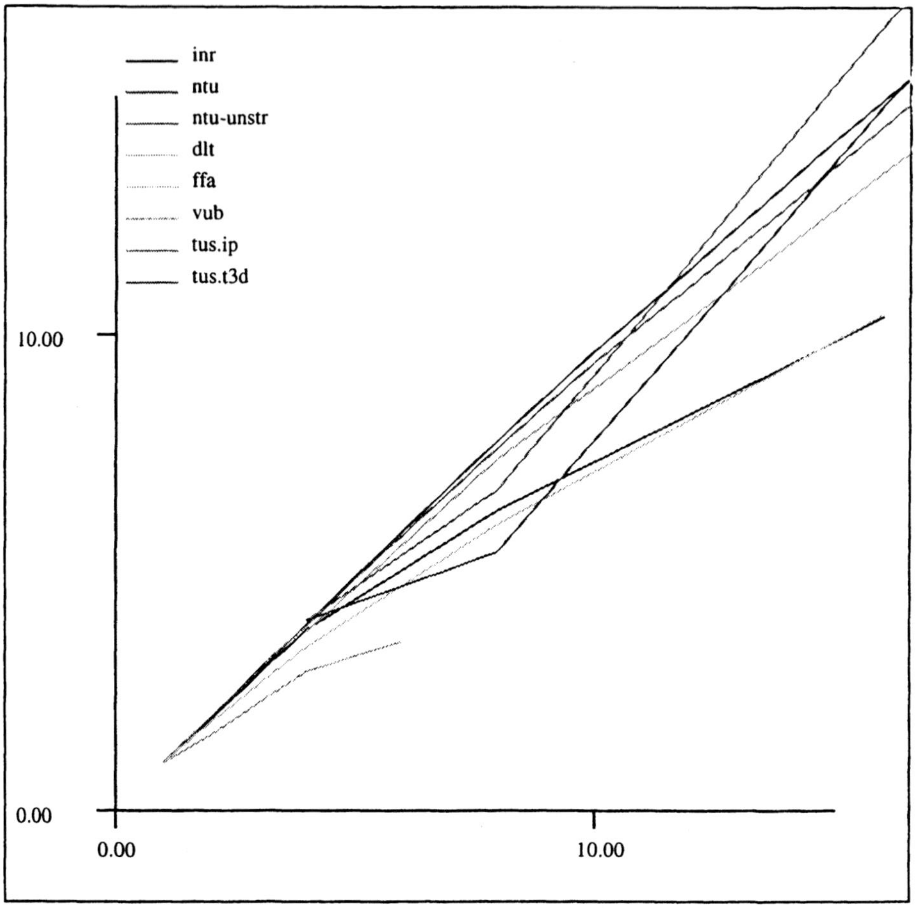

Figure 5 Enlargement of speed-up curves for TC1

Bibliography

[Mavr89] *Mavriplis, P., Jameson, A. and Martinelli, L., (1989) :* , Multigrid Solution of the Navier-Stokes Equations on Triangular Meshes ICASE Rep **89-11**

5 General remarks on the methodologies used for the MPP test cases

From the synthesis achieved with the design data collected in the database different comments on the progress accomplished and future direction of research can be outlined:

- HPCN facilities: availability of definitive advantages provided by HPCN facilities on distributed hardware platforms are for solving large scale problems of relevance to the aerospace industry that cannot be solved otherwise;

- Multidomain techniques: development of multidomain techniques allowing patched viscous and inviscid subdomains for the solution of viscous two and three- dimensional flows on parallel architectures: the local solutions are found to be as accurate as the single domain through appropriate interface treatment;

- Communication: increasing influence of networked communication appears at block interfaces: an intelligent treatment of interfaces taking into account of the physics in the boundary conditions is of major importance in the future

- Distributed multilevel methods: parallelization of multilevel methods in a domain decomposition require mesh decomposer to partition global mesh into non overlapping blocks. Gridless methods are also potential candidates to be investigated in the future in distributed simulation. HPCN facilities will facilitate direct data acquisition block by block with interfaces and favor local and interface solvers operating with non matching grids

- Speed up and robustness: intelligent matching of interfaces with distributed HPCN facilities for optimizers and solvers should increase significantly both the convergence speed up and also the robustness of the solution process and not only the scalability and load balance factors; up to now a crucial ingredient still remains the minimization of the inter communication on cost.

- Accuracy and efficiency remain two cost conflicting targets in modern distributed numerical simulation: optimality of the mesh and approximation schemes remain major concerns within efficiency. Discrepancies of friction coefficients from different solution of the same problem plotted on Fig.4 is a significant example of daily engineering computations. "Better, faster and cheaper" remains a constant target of modern design analysis.

V Conclusion and Perspectives

The aim of this concerted action was to determine through a database workshop event the impact of parallel architectures on the usability of fast laminar and turbulent Navier Stokes flow solvers used in the context of internal and external aerodynamic design.

Previous studies had targeted vector computers but alternative architectures offering high speed at low cost was rapidly proposed by hardware vendors as distributed parallel processor machines.The price to pay by users - scientists and engineers- to benefit of MPP power was not only to simply divide computational domain into parts but also to construct software carefully to ensure scalability, load balancing and low interprocessor communication by intelligent interfaces between sub-domains.

The spirit of this workshop was founded on two simple dual ideas: first consider a wide range of MPP platforms on which to run a particular code; second evaluate the performance of implementation with respect to the size, the data structure and the type of problem.

The focus of the comparative database workshop was to get a clear understanding of the value of parallel architecture machines (levels of scalability) from the parallel solution of the compressible Navier-Stokes equations at the lowest cost for a given accuracy. The synthesis explains how scalability will translate into a reduced elapsed-cpu and communication- time for industrial simulation of complex configurations.

The concerted action covered a sufficiently representative set of different platforms (8) and pre-industrial codes (8) to draw clear understanding conclusions.The task was clearly focused on the Navier Stokes flows, this mathematical model being the key problem which demands a huge amount of computing power for the simulation of real life problems in Industry.

Parallelized Navier Stokes data performed on the above test cases have been analyzed in real time on workstations. They can be improved in the future or serve as a reference with novel Navier Stokes algorithms taking into account the parallel features of the machines not only to perform faster internal loops (scalability) but speed up convergence of the method (less iterations).

1 Perspectives

Aeronautical industry will make in a near future an increasing use of parallel Navier Stokes solvers as starting point for a wide range of complex modelisation problems including turbulence, acoustics, combustion and many coupled multidisciplinary problems adjacent to Aerodynamics -namely aeroelasticity- with unsteady Navier Stokes flows as dominant forthcoming discipline.

In the parallel context a significant improvement of the existing codes is expected in terms of efficiency and robustness due to mathematical treatments of interfaces with domain decomposition methods using efficient preconditionings, new grid strategies like non matching grids at the interfaces, new intelligent treatment of interface boundary conditions with distributed and networked hardware features.

At a more generalized level intelligent parallelized versions of Navier Stokes solvers will be used as flow analysis modules in multi disciplinary design optimisation or active flow control loops.

Irreversible tendencies indicates that European industry will make more and more use of Information Technology strongly due to the economical survival rules of designing high tech products of increasing complexity at the lowest possible cost.

Among information technology outcomes, the parallel Navier Stokes codes on MPP considered in this ECARP concerted action filled a significant gap and will serve in a near future part of the different needs of European airframe manufacturers.

Recently a growing popularity of domain decomposition techniques has been experienced by the scientific and industrial community. This popularity is due to the needs for mathematical tools and numerical algorithms for solving efficiently Partial Differential Equations of various types on distributed computers and networked workstations. Parallel Domain Decomposition methods have become a necessary focused point of high-tech interaction between numerical analysts, applied mathematicians and computer scientists.

Acknowledgments

The Optimum Design contributors are grateful to INRIA systems engineers for providing them with database graphic visualization tools and checking their data to the standards and UPC for the precious help in the preparation of this volume. They acknowledge partners of ECARP for many fruitful and useful discussions for their computational design on mesh adaption and validation issues all along the project.

Annex: List of Partners and Addresses

- A.III.1: Aerospatiale: Aircraft Division, A/DET/EG/Aero, M0142/3, 316 route de Bayonne, F-31060 Tolouse Cedex 03, France
- A.III.2: Alenia Aeronautica D. V. D.: Dept. TEVT-Metodi Fluidodinamici, Corso Marche 41, I-10146 Torino, Italy
- A.III.3: CASA: Dept. Aerodynamics, P$^{\underline{o}}$ John Lenon s/n, E-28906 Getafe, Madrid, Spain
- A.III.4 and B.III.1: Dasa-M: Dept. MT63, Postfach 80 11 60, D-81663 München, Germany
- A.III.6, A.III.7 and B.III.2: Dassault Aviation: 78 Quai Marcel Dassault, F-92214 Saint Cloud, France
- A.III.5: Dasa-Airbus: Dept. Flight Physics EF 10, Hünefeldstr. 1-5. D-28199 Bremen, Germany
- A.III.8: INRIA Rocquencourt: Projet M3N, BP105, F-78153 Le Chesnay Cedex, France
- A.III.9 and B.III.4: INRIA Sophia Antipolis: Unité de Recherche, F-2004 Route des Luciolis, 06560 Valbonne, France
- A.III.10: National Aerospace Laboratory (NLR): Aerodynamics Division / Theoretical Aerodynamics, Anhony Fokkerweg 2, NL-1059 CM Amsterdam, Netherlands
- A.III.11 and B.III.5: National Technical University of Athens (NTUA): Iroon Polytechniou 9, Polytechnioupolis, GR-15773 Athens, Greece
- A.III.12 and B.III.6: Universitat Politècnica de Catalunya (UPC): Strength Materials Dept., Gran Capità s/n, Campus Nord UPC, Mòdul C1, E-08034 Barcelona, Spain
- B.III.3: FFA: The Aeronautical Research Institute of Sweden. S-161 11 Bromma. Sweden.
- B.III.7: Universita di Roma "La Sapienza" (UR): Dipartimento di Meccanica e Aeronautica, Via Eudosiana 18, I-00184 Rome, Italy
- B.III.8: Vrije Universiteit Brussel (VUB): Dept. Fluid Mechanics, Pleinlaan 2, B-1050 Brussels, Belgium

Addresses of the Editors of the Series "Notes on Numerical Fluid Mechanics"

Prof. Dr. Ernst Heinrich Hirschel (General Editor)
Herzog-Heinrich-Weg 6
D-85604 Zorneding
Federal Republic of Germany

Prof. Dr. Kozo Fujii
High-Speed Aerodynamics Div.
The ISAS
Yoshinodai 3-1-1, Sagamihara
Kanagawa 229
Japan

Prof. Dr. Bram van Leer
Department of Aerospace Engineering
The University of Michigan
3025 FXB Building
1320 Beal Avenue
Ann Arbor, Michigan 48109-2118
USA

Prof. Dr. Michael A. Leschziner
UMIST-Department of Mechanical Engineering
P.O. Box 88
Manchester M60 1QD

Prof. Dr. Maurizio Pandolfi
Dipartimento di Ingegneria Aeronautica e Spaziale
Politecnico di Torino
Corso Duca Degli Abruzzi, 24
I-10129 Torino
Italy

Prof. Dr. Arthur Rizzi
Royal Institute of Technology
Aeronautical Engineering
Dept. of Vehicle Engineering
S-10044 Stockholm
Sweden

Dr. Bernard Roux
Institut de Recherche sur les Phénomènes Hors d'Equilibre
(IRPHE)
Technopole de Chateau-Gombert
F-13451 Marseille Cedex 20
France

Brief Instruction for Authors

Manuscripts should have well over 100 pages. As they will be reproduced photomechanically they should be produced with utmost care according to the guidelines, which will be supplied on request.
In print, the size will be reduced linearly to approximately 75 per cent. Figures and diagrams should be lettered accordingly so as to produce letters not smaller than 2 mm in print. The same is valid for handwritten formulae. Manuscripts (in English) or proposals should be sent to the general editor, Prof. Dr. E. H. Hirschel, Herzog-Heinrich-Weg 6, D-85604 Zorneding.

Kolibri 360 GmbH
Ferdinand-Jühlke-Straße 7
99095 Erfurt, Deutschland
produktsicherheit@kolibri360.de